AF785370

Contents

foods MDPI

Article

Ilex latifolia Improves the Anti-Tumor Effectiveness of Rapamycin Against Breast Cancer In Vitro and In Vivo

Zhengnan Ren [1], Yikuan Wu [2], Xiaoying Guo [3], Haizhi Tian [4], Hongjing Ou [1], Zihan Xiong [1], Yu Xiao [1], Longquan Xiao [1], Jing Li [1], Haibo Wu [5] and Xinhui Wang [1,*]

1 College of Food and Biological Engineering, Chengdu University, Chengdu 610106, China; rzhengnan@outlook.com (Z.R.)
2 Institute of Science and Technology, Jiangnan University, Wuxi 214122, China
3 Sichuan Kelun Pharmaceutical Co., Ltd., Chengdu 610599, China
4 School of Food Science and Technology, Jiangnan University, Wuxi 214122, China
5 Turpan Institute of Agricultural Sciences, Xinjiang Academy of Agricultural Sciences, Turpan 838099, China
* Correspondence: wangxinhui@cdu.edu.cn

Abstract: Breast cancer remains one of the leading causes of cancer-related mortality among women worldwide. Although the mTOR inhibitor rapamycin exhibits notable anti-tumor activity, its clinical application is limited by metabolic side effects, particularly dyslipidemia. This study aimed to investigate the potential of *Ilex latifolia* (*I. latifolia*, large-leaf kudingcha), a traditional Chinese tea known for its lipid-lowering properties, to enhance the therapeutic efficacy of rapamycin in breast cancer. The combined effects of *I. latifolia* and low-dose rapamycin on tumor cell proliferation, cell cycle progression, apoptosis, and inflammation were assessed in four breast cancer cell lines and a murine breast cancer model. While low-dose *I. latifolia* alone exhibited limited anti-tumor activity, its combination with low-dose rapamycin synergistically inhibited tumor proliferation, induced cell cycle arrest, promoted apoptosis, and reduced inflammation in vitro. In vivo, dietary supplementation with *I. latifolia* mitigated rapamycin-induced lipid disturbances, reduced tumor growth, enhanced apoptosis, and alleviated inflammation in tumor tissues. These findings highlight *I. latifolia* as a promising dietary adjunct to rapamycin, providing a safer and more effective combinatorial strategy for breast cancer treatment.

Keywords: *Ilex latifolia*; rapamycin; breast cancer; dietary intervention

1. Introduction

Breast cancer ranks among the most prevalent cancers in women worldwide and continues to be a primary contributor to cancer-related deaths [1]. Early detection and advanced treatment strategies, such as surgery, chemotherapy, radiation therapy, and targeted therapies, have significantly improved survival rates [2]. However, recurrence and metastasis continue to present major challenges, particularly in advanced and metastatic cases [2]. The heterogeneity of breast cancer, with various subtypes such as triple-negative breast cancer (TNBC) and human epidermal growth factor receptor 2 (HER2)-positive breast cancer, further complicates therapeutic approaches [3,4]. These variations make applying a one-size-fits-all treatment strategy difficult, necessitating the development of novel therapeutic interventions. In recent years, natural products have garnered increasing interest in cancer research due to their multi-target mechanisms and relatively low toxicity compared to synthetic drugs [5,6]. These compounds often offer complementary therapeutic effects and are being explored as potential adjuvants in cancer treatment.

Rapamycin, a well-known mammalian target of rapamycin (mTOR) inhibitor, has demonstrated considerable anti-cancer potential, particularly in inhibiting tumor cell proliferation, blocking angiogenesis, and inducing autophagy [7]. However, as a single agent, the clinical effectiveness of rapamycin in cancer treatment is limited [8]. Prolonged use of rapamycin can lead to significant side effects, including hyperlipidemia, metabolic disorders, and the emergence of drug resistance [8]. Furthermore, the immunosuppressive effects of rapamycin can increase the risk of infections and compromise overall patient health [9]. These limitations have prompted researchers to explore combination therapies to enhance their therapeutic potential. While several studies have focused on combining rapamycin with synthetic drugs [10–12], there is limited research on using natural food components or herbal products to enhance the anti-tumor effects of rapamycin.

Kudingcha is a traditional herbal tea widely consumed in China and comprises two primary species: *Ilex latifolia* (*I. latifolia*, large-leaf kudingcha) and *Ligustrum robustum* (*L. robustum*, small-leaf kudingcha). Ethnopharmacological and ethnobotanical studies have shown that both *I. latifolia* and *L. robustum* have long been used in traditional medicine for their anti-inflammatory, antioxidant, and lipid-lowering properties [13–18]. Interestingly, previous studies further demonstrated the potential of *I. latifolia* to inhibit tumor cell proliferation, promote apoptosis, and modulate the tumor immune microenvironment [19,20]. However, its pronounced bitterness may limit its acceptability for long-term or high-dose consumption [13]. Given its natural origin, favorable safety profile, and lipid-lowering effects, *I. latifolia* holds promise as a dietary adjunct to rapamycin, potentially enhancing anti-tumor efficacy while mitigating its metabolic side effects.

This study aims to evaluate the adjuvant potential of *I. latifolia* in combination with low-dose rapamycin for breast cancer treatment, using both in vitro and in vivo models. Interestingly, our study has identified a non-toxic dose of *I. latifolia* that does not directly inhibit breast cancer cell growth. However, this dose significantly improves the anti-tumor efficacy of rapamycin. This combination therapy may provide a promising and less-toxic alternative for the treatment of breast cancer, potentially improving patient outcomes without exacerbating side effects.

2. Materials and Methods

2.1. Preparation of *I. latifolia* Extracts

I. latifolia and *L. robustum* were purchased from Efuton Tea Co., Ltd. (Zhejiang, China). The dried tea was steeped in distilled water (1 g tea: 10 mL water) for 30 min, followed by three cycles of extraction. In each cycle, the mixture was heated in a water bath at 100 °C for 1 h, and the supernatant was collected. After each extraction, 10 mL of fresh distilled water was added to the residue, and the process was repeated. The extracts from all three cycles were combined, filtered, and concentrated to one-third of the original volume using a rotary evaporator under reduced pressure. The final extract was sterilized and adjusted to 0.1 g/mL and then stored at −20 °C until further use.

2.2. Characterization of *I. latifolia* Extracts

Five-tenths milliliter of the *I. latifolia* extracts was incorporated into a combination of chromatographic-grade methanol and ultrapure water (2:1, *v*/*v*), thereafter subjected to vortex mixing for 2 min and ultrasonic treatment for 5 min. The solution was subsequently filtered using a 0.22 μm membrane, and 1.5 mL of the filtrate was placed into a chromatography vial for analysis. The liquid chromatography analysis utilized the ACQUITY UPLC

H-Class (Waters, MA, USA), employing mobile phases of 0.1% formic acid in water (Phase A) and 0.1% formic acid in acetonitrile (Phase B). The column was the BEH C18 (Waters, 1.7 μm, 2.1 × 50 mm). Mass spectrometry was conducted utilizing the G2-XS QTof MS instrument (Waters), employing both positive and negative ion modes. All organic reagents were purchased from Sinopharm Chemical Reagent Co., Ltd., (Shanghai, China). The gathered data were analyzed utilizing MassLynx 4.2 software (Waters), and secondary mass spectrometry (MS/MS) fragmentation peaks were first compared with the UNIFI natural product database, followed by a comparison with literature data or standard references for identification.

2.3. Cell Culture

The human breast cancer cell lines HCC1806, MDA-MB-231, MCF-7, and SK-BR-3 (ATCC, VA, USA), representing different molecular subtypes, were cultured under specified conditions. HCC1806 and MDA-MB-231 are TNBC cell lines characterized by the absence of estrogen receptor (ER), progesterone receptor (PR), and human epidermal growth factor receptor 2 (HER2) [21]. In contrast, MCF-7 cells are luminal A subtype, positive for ER and PR, while SK-BR-3 cells belong to the HER2-positive subtype, overexpressing HER2 [22]. HCC1806 cells were maintained in RPMI 1640 medium (Gibco^{TM}, Grand Island, NY, USA) with 10% fetal bovine serum (Gibco^{TM}) and 1% penicillin–streptomycin (Gibco^{TM}). MDA-MB-231, MCF-7, and SK-BR-3 cells were grown in high-glucose Dulbecco's Modified Eagle Medium (DMEM, Gibco^{TM}) with 10% fetal bovine serum and 1% penicillin–streptomycin. MCF-10A cells, a non-tumorigenic, immortalized human mammary epithelial cell line, were used as the negative control (ATCC, VA, USA). These cells were cultured in DMEM supplemented with 5% horse serum (Gibco^{TM}), 20 ng/mL epidermal growth factor (Gibco^{TM}), 0.5 μg/mL hydrocortisone (Merck, Shanghai, China), 100 ng/mL cholera toxin (Merck), and 10 μg/mL insulin (Merck). All cell lines were incubated at 37 °C in a humidified atmosphere containing 5% CO_2. Cell handling and all related procedures were conducted under sterile conditions in a designated biosafety cabinet. Sub-culturing was performed using 0.25% trypsin–EDTA (Gibco^{TM}) when cell confluence reached approximately 80–90%.

2.4. Cell Treatment

After cell adhesion, different concentrations of single and combined treatments were applied. Each treatment group included three replicates. The concentration gradient of tea water extract was 0, 0.125, 0.25, 0.375, and 0.5 mg/mL. The stock solution was diluted with anhydrous ethanol to achieve the final working concentrations before application. The concentration gradient of rapamycin (MedChemExpress LLC, Shanghai, China) was 0, 2, 4, 8, and 16 μmol/L [23]. The combination treatment (combined group) was administered with 0.25 mg/mL tea water extract and 8 μmol/L rapamycin.

2.5. Cell Counting Assay

Cells in the logarithmic growth phase were harvested and resuspended, and the cell suspension was adjusted to a concentration of 2.0 × 10^5 cells/mL. The suspension was then seeded into 6-well plates, with 2 mL of cell suspension per well. After incubating for 12 h in a cell culture incubator, the cells adhered to the surface of the wells. Following the treatment protocol, the cells were exposed to the appropriate drugs at the specified concentrations. After 72 h of continued incubation, the old medium was removed, and the cells were washed three times with 1–2 mL of PBS. Then, 200 μL of trypsin-EDTA solution was added for cell digestion, following standard procedures as previously described. The digestion was terminated by adding 800 μL of complete growth medium. The cell suspension was carefully pipetted to achieve single-cell dispersion. The cell suspension was transferred to a 1.5 mL tube, and 10 μL of the suspension was placed on a hemocytometer. Cell counts were conducted using an inverted microscope.

2.6. CCK-8 Cell Cytotoxicity Assay

Cells in the logarithmic growth phase were harvested and resuspended, and the concentration of the cell suspension was adjusted to 0.5×10^5–1.0×10^5 cells/mL. Using a multi-channel pipette, 100 μL of the suspension (approximately 1.0×10^4 cells) was seeded into each well of a 96-well plate. To minimize errors caused by evaporation, 100 μL of PBS was added to the wells along the edges of the plate. The plate was incubated in a cell culture incubator for 12 h to allow cell adhesion. Following the experimental protocol, the cells were treated with the designated drugs and concentrations. After 72 h of incubation, the old medium was aspirated, and 100 μL of fresh complete growth medium, along with 10 μL of CCK-8 solution, was added to each well. The plate underwent incubation for an additional 2 h. The optical density at 450 nm was measured using a microplate reader. The cell death rate is calculated using the formula provided by the manufacturer of the CCK-8 kit (Dojindo Laboratories Co., Ltd., Kumamoto, Japan).

2.7. Cell Cycle Analysis

During the logarithmic growth phase, cells were harvested and resuspended to a concentration of 2.0×10^5 cells/mL. The cell suspension was then evenly seeded into 6-well plates, with 2 mL per well. After incubating for 12 h in a cell culture incubator at 37 °C and 5% CO_2 to facilitate cell adhesion, the cells were subsequently treated with the designated drugs at specified concentrations, as described in the experimental setup. After 24 h of incubation, cells were collected and resuspended in 300 μL of PBS (pre-cooled at 4 °C overnight). The suspension was transferred into 1.5 mL microcentrifuge tubes, and 700 μL of anhydrous ethanol (pre-cooled at −20 °C overnight) was added dropwise for fixation at 4 °C overnight. After centrifugation, the ethanol was eliminated, and the cells were subsequently washed once with PBS. The cells were then resuspended in 50 μg/mL propidium iodide staining solution containing 0.1% Triton X-100 and 100 μg/mL RNase (Servicebio, Hubei, China). The mixture was incubated in the dark at room temperature for 15 min. Finally, cell cycle distribution was analyzed using a flow cytometer (Attune® NxT, Thermo Fisher Scientific, Waltham, MA, USA), and the data were processed using Flowjo software (v 10.8.1).

2.8. Western Blot

The cells underwent two washes with pre-cooled PBS and were subsequently lysed on ice utilizing the RIPA buffer that contained protease and phosphatase inhibitors. (Beyotime, Shanghai, China). The lysates were centrifuged at 12,000× *g* for 15 min at 4 °C, and the supernatants were collected. Protein concentrations were determined using the BCA protein assay kit (Beyotime), and equal amounts of protein (30–50 μg) were loaded onto SDS-PAGE gels for separation. After electrophoresis, proteins were transferred to PVDF membranes (Merck), which were blocked with 5% non-fat milk in TBST for 1 h at room temperature. The membranes underwent overnight incubation at 4 °C with primary antibodies specific for p21 (27296-1-AP, Proteintech, Chicago, IL, USA), p27 (26714-1-AP, Proteintech), and β-actin (20536-1-AP, Proteintech, used as a loading control). After washing the membranes with TBST, they were incubated with HRP-conjugated secondary antibodies for 2 h at room temperature (Thermo Fisher Scientific). The membranes underwent rewashing with TBST, and protein bands were visualized utilizing an enhanced chemiluminescence (ECL) detection system (ProteinSimple, San Jose, CA, USA). The intensity of the protein bands was quantified using ImageJ software (1.54g), and the relative expression levels of p21 and p27 were normalized to β-actin. Each experiment was performed in triplicate to ensure reproducibility.

2.9. Quantitative PCR Analysis

RNA extraction was performed on treated cells and mouse tumor tissues utilizing TRIzol reagent. (Thermo Fisher Scientific), followed by reverse transcription into cDNA. Gene expression levels were analyzed via qPCR using specific primers (Table 1). *ACTB* and *Actb* were used as internal controls in the cell and animal experiments, respectively. The expression levels of target genes were determined using the $2^{-\Delta\Delta Ct}$ method.

Table 1. Sequences of primers utilized in this research.

Gene	Species	Forward	Reverse
BAX	Human	5′-CCCGAGAGGTCTTTTTCCGAG-3′	5′-CCAGCCCATGATGGTTCTGAT-3′
BCL2	Human	5′-GGTGGGGTCATGTGTGTGG-3′	5′-CGGTTCAGGTACTCAGTCATCC-3′
P53	Human	5′-CAGCACATGACGGAGGTTGT-3′	5′-TCATCCAAATACTCCACACGC-3′
IL6	Human	5′-ACTCACCTCTTCAGAACGAATTG-3′	5′-CCATCTTTGGAAGGTTCAGGTTG-3′
IL1B	Human	5′-ATGATGGCTTATTACAGTGGCAA-3′	5′-GTCGGAGATTCGTAGCTGGA-3′
NFKB1	Human	5′-AACAGAGAGGATTTCGTTTCCG-3′	5′-TTTGACCTGAGGGTAAGACTTCT-3′
ACTB	Human	5′-CATGTACGTTGCTATCCAGGC-3′	5′-CTCCTTAATGTCACGCACGAT-3′
Bax	Mouse	5′-TGAAGACAGGGGCCTTTTTG-3′	5′-AATTCGCCGGAGACACTCG-3′
Bcl2	Mouse	5′-GTCGCTACCGTCGTGACTTC-3′	5′-CAGACATGCACCTACCCAGC-3′
Il6	Mouse	5′-TAGTCCTTCCTACCCCAATTTCC-3′	5′-TTGGTCCTTAGCCACTCCTTC-3′
Il1b	Mouse	5′-GCAACTGTTCCTGAACTCAACT-3′	5′-ATCTTTTGGGGTCCGTCAACT-3′
Actb	Mouse	5′-AATCCCATCACCATCTTCCA-3′	5′-TGGACTCCACGACGTACTCA-3′

2.10. Experimental Animals and Treatment

Six-week-old BALB/c nude mice (GemPharmatech, Jiangsu, China) were housed under controlled conditions (22–25 °C, 12 h light/dark cycle) with free access to food and water (Jiangnan University, Jiangsu, China). Following a one-week adaptation period, mice were randomly assigned to four groups. ($n = 5$ per group). HCC1806 cells were resuspended in serum-free medium to 2.0×10^7 cells/mL and mixed with Matrigel (Corning, NY, USA). Subcutaneous injection of 100 μL of the cell–Matrigel mixture was administered to each mouse. Once tumors reached about 200 mm^3, the groups were fed the respective diet. The *I. latifolia* water extract was filtered and added to the normal maintenance feed at a ratio of 5 g of dried tea per kg of feed powder, which was then pelleted. Rapamycin was added to the feed at a dosage of 20 mg/kg feed. Tumor size was assessed every four days, and after 16 days, serum and tumor tissues were collected for subsequent analysis. All animal procedures adhered to the guidelines established by the Institutional Animal Ethics Committee of Jiangnan University (JN. No20170213-20171221[6]), and the animal studies followed the ARRIVE reporting guidelines [24].

2.11. Immunohistochemistry of Tumor Tissue

Tumor tissues were preserved in 4% paraformaldehyde and subsequently embedded in paraffin. Sections measuring 5 μm were prepared and affixed to slides. Antigen retrieval was conducted by heating the sections in citrate buffer (pH 6.0) at 95 °C for 30 min. Following a 5% BSA blocking step, the sections were incubated overnight at 4 °C with primary antibody anti-Ki67 (28074-1-AP, Proteintech). After PBS washes, slices were incubated with HRP-conjugated secondary antibody for 1 h at room temperature. The signals were visualized with 3,3′-diaminobenzidine, and nuclei were counterstained with hematoxylin. The stained sections were examined under a light microscope, and Ki67-positive cells were quantified.

2.12. Data Analysis

All data are expressed as the mean ± standard deviation (SD). Statistical analysis was conducted utilizing GraphPad Prism (version 9.5.0). A one-way analysis of variance (ANOVA) accompanied by Tukey's post hoc test was employed for comparisons among various groups. Differences with $p < 0.05$ were deemed statistically significant. All experiments were conducted a minimum of three times to guarantee repeatability.

3. Results

3.1. The Main Components of *Ilex latifolia* Extract Are Triterpene Saponins and Phenolic Acid

The chemical composition of *I. latifolia* extract was characterized using Fourier-transform infrared spectroscopy, ultra-performance liquid chromatography coupled with quadrupole time-of-flight mass spectrometry, and the UNIFI natural product database. Phytochemical analysis revealed a complex profile, primarily consisting of polyphenolic and saponin compounds, which together accounted for over 92% of the identified components. Secondary metabolites included alkaloids and terpenoids, while organic acids and lipids were present in smaller amounts. Trace quantities of specialized metabolites, such as aromatic compounds, carbohydrates, steroids, glycosides, amino acids, ketones, and alcohols, were also detected (Table 2). Mass spectrometry data were processed using the UNIFI platform and manually validated. The predominant components identified in the extract were 1,4-dicaffeoylquinic acid, 1-caffeoylquinic acid, esculentoside L, ruberythric acid, and cynanoside O (Table 3).

Table 2. Composition of *Ilex latifolia*.

Constituents Class	Representative Subclass	Relative Content (%)
Saponins	Triterpenoid saponins	46.89349
Polyphenols	Phenolic acids	45.96911
Alkaloids	Pyrrolizidine alkaloids	2.914044
Terpenoids	Sesquiterpenes	1.920293
Organic acids	Citric acid derivatives	1.270804
Lipids	Glyceride	0.630561
Aromatic compounds	Benzene compounds	0.170388
Carbohydrates	Polysaccharides	0.117496
Steroids	Phytosterols	0.04531
Glycosides	Cyanogenic glycosides	0.03653
Amino acids	Essential amino acids	0.016108
Ketones	Quinones	0.009137
Alcohols	Aliphatic Alcohols	0.006725

Table 3. Top 10 predominant components in *Ilex latifolia*.

No	Component Name	t_R (min)	Formula	Molecular Ion (*m/z*)	Peak Area	Ion Mode	Relative Content (%)
1	1,4-dicaffeoylquinic acid	7.41	$C_{25}H_{24}O_{12}$	515.1198	13,226,556	$[M-H]^-$	17.17072603
2	1-caffeoylquinic acid	5.08	$C_{16}H_{18}O_9$	353.0884	8,470,651	$[M-H]^-$	10.99660619
3	esculentoside L	11.70	$C_{48}H_{76}O_{20}$	971.4829	6,761,731	$[M-H]^-$	8.778084822
4	ruberythric acid	6.08	$C_{25}H_{26}O_{13}$	533.1303	4,749,673	$[M-H]^-$	6.166029449
5	cynanoside O	12.82	$C_{48}H_{74}O_{19}$	953.4716	4,566,480	$[M-H]^-$	5.928208144
6	marsdenoside H	13.73	$C_{48}H_{76}O_{19}$	955.4881	3,383,373	$[M-H]^-$	4.39229765
7	clinodiside A	12.04	$C_{48}H_{78}O_{19}$	957.5033	2,835,076	$[M-H]^-$	3.680498027
8	clinodiside B	13.41	$C_{54}H_{88}O_{23}$	1103.5600	2,428,726	$[M-H]^-$	3.152974118
9	nelumboroside A	6.83	$C_{27}H_{30}O_{16}$	609.1455	1,692,296	$[M-H]^-$	2.196940078
10	soyasaponin I	14.33	$C_{48}H_{78}O_{18}$	941.5083	1,652,682	$[M-H]^-$	2.14551315

3.2. *Ilex latifolia, Ligustrum robustum, and Rapamycin Inhibit Breast Cancer Cell Proliferation in a Way Dependent on Dosage*

To evaluate the anti-tumor effects of *I. latifolia, L. robustum*, and rapamycin, we selected four representative breast cancer cell lines: HCC1806 and MDA-MB-231 (triple-negative breast cancer, characterized by high invasiveness and metastatic potential), MCF-7 (estrogen receptor-positive breast cancer, representing hormone-dependent tumors), and SK-BR-3 (HER2-overexpressing breast cancer). These cell lines encompass the major molecular subtypes of breast cancer, reflecting the diversity of breast cancer classifications. The findings indicated that both *I. latifolia* and *L. robustum* displayed substantial inhibitory effects on the proliferation of all four cell lines at concentrations of 0.375 and 0.5 mg/mL, respectively (Figure 1A,B). Rapamycin markedly decreased cell counts across all four cell lines at concentrations of 8 and 16 μM (Figure 1C). Notably, no inhibitory effects were observed on the MCF-10A cells, a normal mammary epithelial cell line, suggesting that these compounds selectively target cancer cells without affecting normal breast cells. These findings highlight the dose-dependent inhibitory effects of *I. latifolia, L. robustum*, and rapamycin on breast cancer cell proliferation.

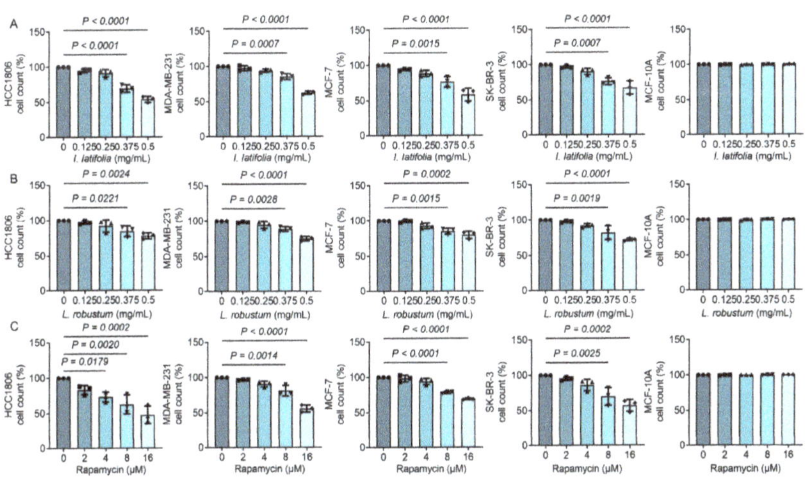

Figure 1. The inhibitory effects of *Ilex latifolia, Ligustrum robustum*, and rapamycin on breast cancer cells. (**A**) Percentage of cell count under different concentrations (0, 0.125, 0.25, 0.375, and 0.5 mg/mL) of *I. latifolia* treatment in HCC1806, MDA-MB-231, MCF-7, SK-BR-3, and MCF-10A cells. (**B**) Percentage of cell count under different concentrations (0, 0.125, 0.25, 0.375, and 0.5 mg/mL) of *L. robustum* treatment in HCC1806, MDA-MB-231, MCF-7, SK-BR-3, and MCF-10A cells. (**C**) Percentage of cell count under different concentrations (0, 2, 4, 8, and 16 μM) of rapamycin treatment in HCC1806, MDA-MB-231, MCF-7, SK-BR-3, and MCF-10A cells. The initial cell count was set to 100%. Data in (**A–C**) were presented as mean ± SD from three independent experiments. p values in (**A–C**) were determined by Tukey's multiple comparisons test following one-way ANOVA. $p < 0.05$ was considered statistically significant. Triangles, diamonds, circles, and squares are used solely to differentiate between experimental groups and do not indicate specific biological meanings. Each dot represents data from an independent experiment.

3.3. *Ilex latifolia Amplifies the Antiproliferative Effect of Rapamycin on Breast Cancer Cells*

To investigate the synergistic effect of *I. latifolia, L. robustum*, and rapamycin, we used a concentration of 0.25 mg/mL for both *I. latifolia* and *L. robustum*, which individually did not influence the growth of breast cancer cells (Figure 1). These were amalgamated with the minimal effective dosage of rapamycin (8 μM, Figure 1). Indeed, neither 0.25 mg/mL

I. latifolia nor *L. robustum* alone could inhibit breast tumor growth (Figure 2A). Interestingly, *I. latifolia* exhibited potent activity in enhancing the antiproliferative effects of rapamycin (Figure 2A). In contrast, combining *L. robustum* with rapamycin did not enhance the inhibitory effect on cell proliferation observed with rapamycin alone (Figure 2A). Additionally, the CCK-8 assay demonstrated that rapamycin induced tumor cell death (Figure 2B). When combined with *I. latifolia*, cell death was further increased in all four breast cancer cell lines (Figure 2B). However, the combination of *L. robustum* with rapamycin did not amplify the anti-tumor effects of rapamycin (Figure 2B). These findings suggest that *I. latifolia* may serve as an effective dietary adjunct to rapamycin for enhancing its anti-tumor activity, while *L. robustum* does not provide the same synergistic effect.

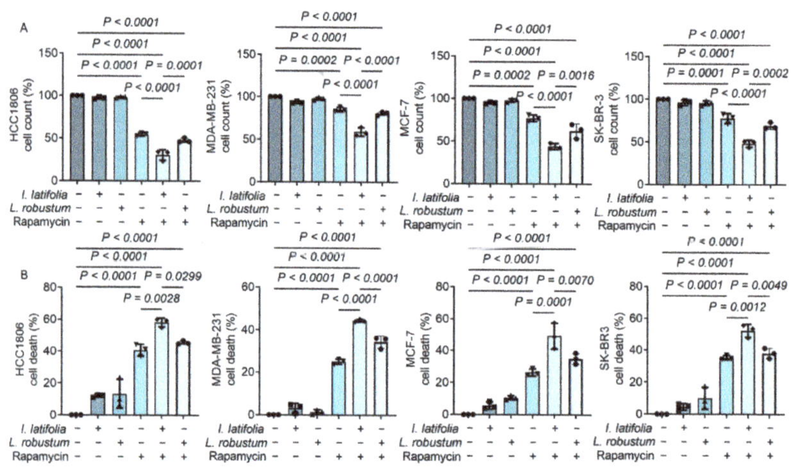

Figure 2. Effect of *Ilex latifolia* or *Ligustrum robustum* combined with rapamycin on breast cancer cells. (**A**) Percentage of cell count after treatment with *I. latifolia* (0.25 mg/mL), *L. robustum* (0.25 mg/mL), rapamycin (8 μM), *I. latifolia* + rapamycin, and *L. robustum* + rapamycin in HCC1806, MDA-MB-231, MCF-7, and SK-BR-3 cells. The initial cell count was set to 100%. (**B**) Percentage of cell death following the same treatments in HCC1806, MDA-MB-231, MCF-7, and SK-BR-3 cells. Data in (**A**,**B**) were presented as mean ± SD from three independent experiments. *p* values in (**A**,**B**) were determined by Tukey's multiple comparisons test following one-way ANOVA. $p < 0.05$ was considered statistically significant. Triangles, diamonds, circles, and squares are used solely to differentiate between experimental groups and do not indicate specific biological meanings. Each dot represents data from an independent experiment.

3.4. *Ilex latifolia* Enhances Rapamycin-Induced Tumor Cell Cycle Arrest

To further investigate the adjuvant role of *I. latifolia* in rapamycin treatment, HCC1806 cells were selected based on their pronounced response in Figure 2. Morphological changes under different treatments were visualized using microscopy. Under normal conditions, HCC1806 cells grew in clusters and adhered tightly to the culture dish. Low concentrations of *I. latifolia* extract or rapamycin alone did not significantly alter cell morphology. However, the combined treatment caused notable morphological alterations at the cell edges, with a significant reduction in cell number (Figure 3A). To evaluate the influence on the cell cycle, flow cytometry analysis was performed. Cells in the G0/G1 phase (2n), S phase (2n~4n), and sub-G1 phase (<2n) were observed, with the G0/G1 phase linked to cell quiescence, the S phase representing DNA replication, and the sub-G1 peak indicating apoptotic cell populations. Rapamycin promoted the G0/G1 phase, inhibited the S phase, and increased the sub-G1 phase (Figure 3B). This indicates that rapamycin therapy activates cellular apoptosis mechanisms and disrupts normal cell cycle progression. Interestingly, the combination of rapamycin and *I. latifolia* further enhances rapamycin-induced tumor cell cycle progression disruption (Figure 3B). Additionally, Western blotting analysis revealed elevated expression levels of P21 and P27 proteins following combination treatment (Figure 3C). These results suggest that *I. latifolia* enhances rapamycin-induced cell cycle arrest, particularly at the G1 phase, thereby inhibiting tumor cell proliferation.

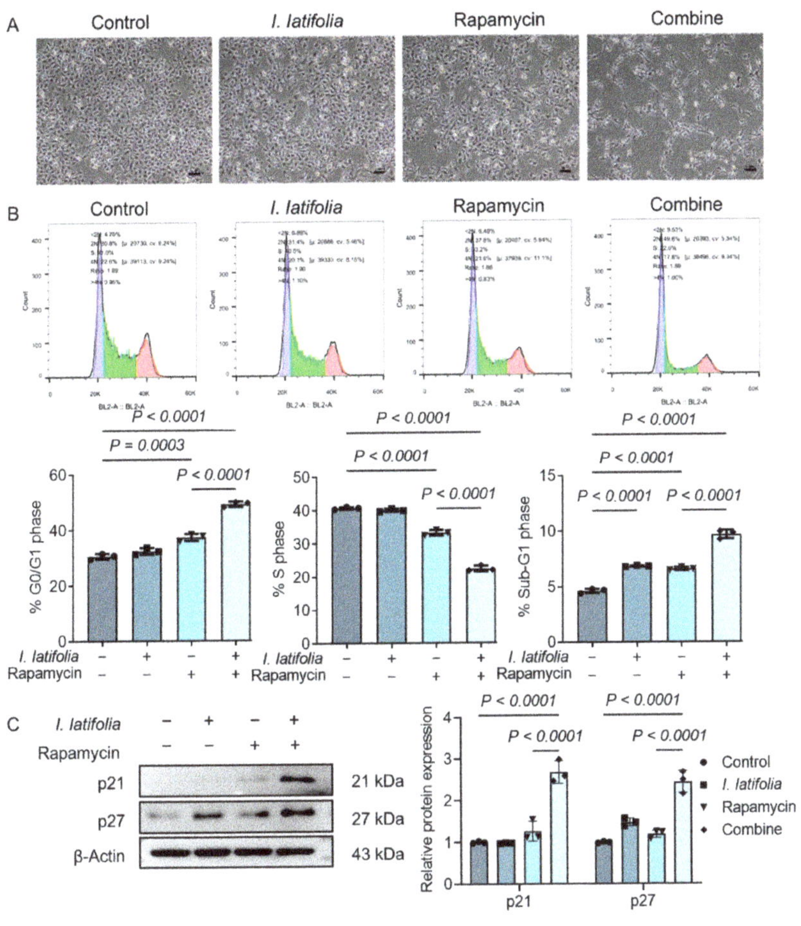

Figure 3. Effects of *Ilex latifolia* combined with rapamycin on the cell cycle. (**A**) Representative images of cell morphology under a microscope (scale bar: 50 μm). (**B**) Flow cytometry analysis of cell cycle distribution. (**C**) Protein expression and quantification of p21 and p27 in HCC1806 cells. Data in (**B**,**C**) were presented as mean ± SD from three independent experiments. *p* values in (**B**,**C**) were determined by Tukey's multiple comparisons test following one-way ANOVA. $p < 0.05$ was considered statistically significant. Triangles, diamonds, circles, and squares are used solely to differentiate between experimental groups and do not indicate specific biological meanings. Each dot represents data from an independent experiment.

3.5. Ilex latifolia Augments Rapamycin-Induced Modulation of Apoptosis and Inflammation in Breast Cancer Cells

To investigate the effects of *I. latifolia* and rapamycin combination treatment, we examined the expression of apoptosis-related genes (*BAX*, *BCL2*, and *P53*) and inflammation-related genes (*IL6, IL1B,* and *NFKB1*) in HCC1806 cells using RT-qPCR. The results revealed that in comparison to the control group, combination treatment with *I. latifolia* and rapamycin significantly upregulated the expression of *BAX* and *P53,* both of which are associated with apoptosis induction (Figure 4A,B). On the other hand, the anti-apoptotic protein *BCL2* was downregulated following the combined treatment (Figure 4C). In terms of inflammatory factors, *IL6, IL1B,* and *NFKB1* expression levels were notably reduced in the combination treatment group relative to the control group, suggesting a potential reduction in tumor-associated inflammation (Figure 4D–F). Notably, the combination treatment group showed more significant effects than rapamycin treatment alone (Figure 4A–F). These findings indicate that the combination of *I. latifolia* and rapamycin enhances apoptosis and modulates inflammatory responses in HCC1806 cells.

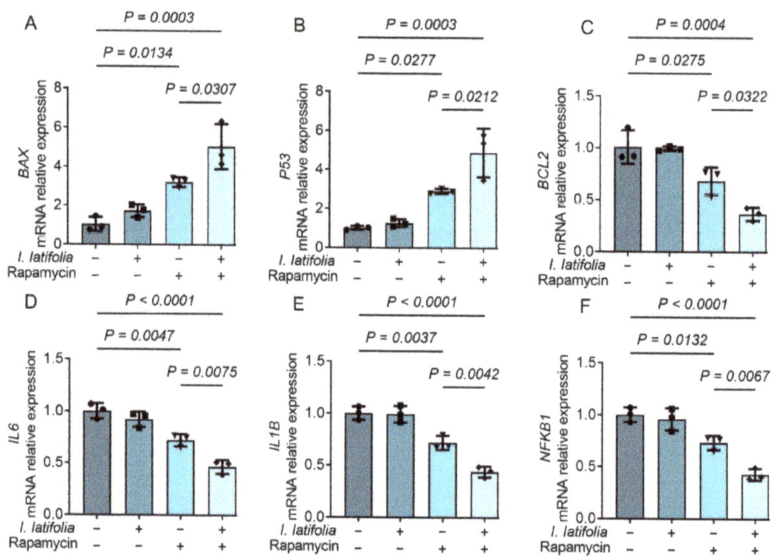

Figure 4. Effects of *Ilex latifolia* combined with rapamycin on apoptosis and inflammation. (**A–C**) Relative expression levels of apoptosis-related genes (**A**) *BAX,* (**B**) *P53,* and (**C**) *BCL2.* (**D–E**) Relative expression levels of inflammation-related genes (**D**) *IL6,* (**E**) *IL1B,* and (**F**) *NFKB1.* Data in (**A–F**) were presented as mean ± SD from three independent experiments. p values in (**A–F**) were determined by Tukey's multiple comparisons test following one-way ANOVA. $p < 0.05$ was considered statistically significant. Triangles, diamonds, circles, and squares are used solely to differentiate between experimental groups and do not indicate specific biological meanings. Each dot represents data from an independent experiment.

3.6. Combined Treatment of Ilex latifolia and Rapamycin Inhibits Breast Cancer Cell Growth in Mice

After confirming the synergistic effect of *I. latifolia* and rapamycin in vitro, we proceeded to further investigate their therapeutic potential in vivo. Mice were randomly allocated to the designated treatment groups, and tumors were surgically removed after the treatment period for analysis (Figure 5A). Visual inspection of the excised tumors revealed notable differences in tumor size across the groups. The combined treatment group showed significantly smaller tumors compared to both the control and single-treatment groups, suggesting a more pronounced anti-tumor effect when *I. latifolia* and rapamycin were used together (Figure 5B). Tumor growth was closely monitored over time, as shown in Figure 5C. Before the initiation of the treatment, no notable disparities in tumor volume were detected among the groups, indicating that the tumors grew at a similar rate. However, as the treatment progressed, the control group had a substantial augmentation in tumor volume, while the tumors in the treatment groups showed slower growth. By day 16, tumors in the rapamycin-treated group were significantly smaller and lighter compared to the control, confirming its effective inhibition of tumor growth (Figure 5C,D). Notably, the combined treatment with *I. latifolia* further reduced tumor volume and weight, augmenting the anti-neoplastic properties of rapamycin (Figure 5C,D). These results indicate that *I. latifolia* enhances the anti-tumor efficacy of rapamycin in vivo.

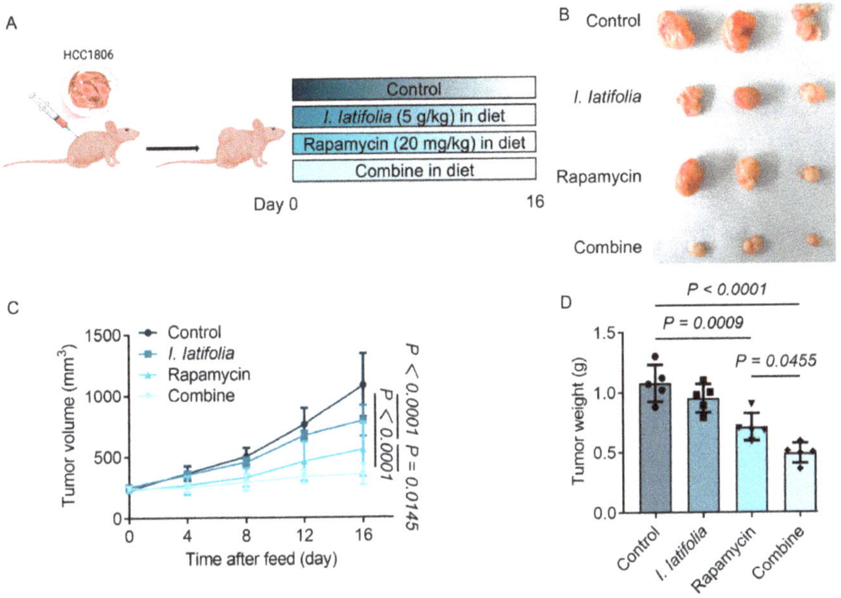

Figure 5. The protective effect of *Ilex latifolia* combined with rapamycin on breast cancer mice. (**A**) Animal protocol (diagram created with BioRender.com). (**B**) Representative photos of tumors. (**C**) Tumor volume changes ($n = 5$). (**D**) Tumor weight ($n = 5$). Data in (**C**) were presented as mean and error ± SD and p values were determined by Tukey's multiple comparisons test following two-way ANOVA. Data in (**D**) were presented as mean ± SD and p values were determined by Tukey's multiple comparisons test following one-way ANOVA. $p < 0.05$ was considered statistically significant. Triangles, diamonds, circles, and squares are used solely to differentiate between experimental groups and do not indicate specific biological meanings. Each dot represents data from an individual mouse.

3.7. *Ilex latifolia* Enhances the Anti-Tumor Effect of Rapamycin in Mice

We further investigated the effects of combining *I. latifolia* with rapamycin on lipid metabolism, tumor cell proliferation, apoptosis, and inflammation. Our results showed that *I. latifolia* significantly improved the lipid abnormalities induced by rapamycin, including reductions in triglycerides (TG), total cholesterol (TC), and low-density lipoprotein cholesterol (LDL-C), and increasing high-density lipoprotein cholesterol (HDL-C) (Figure 6A). These findings suggest that *I. latifolia* can alleviate the lipid-related side effects of rapamycin. Additionally, we evaluated tumor cell proliferation using immunohistochemistry to detect Ki67 expression, a marker of proliferative activity. Ki67-positive cells, stained dark brown to black, were markedly diminished in the combination therapy group relative to the control and single-treatment groups (Figure 6B). Quantitative analysis confirmed that the number of Ki67-positive cells in the combined group was significantly lower than in the other groups (Figure 6B). These results demonstrate that the combination of *I. latifolia* and rapamycin effectively inhibits tumor cell proliferation. Moreover, the expression of genes associated with apoptosis (*Bax* and *Bcl2*) and inflammation (*Il6* and *IL1b*) in tumor tissues was notably modulated by rapamycin (Figure 6C,D). Dietary treatment with *I. latifolia* augmented the pro-apoptotic and anti-inflammatory effects of rapamycin in breast cancer mice. These data highlight the efficacy of this combination medication in improving breast cancer treatment.

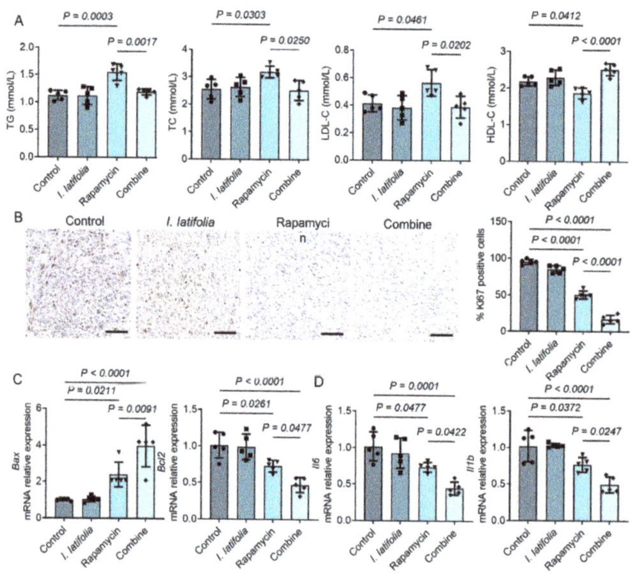

Figure 6. Effect of combined *Ilex latifolia* and rapamycin treatment on breast cancer mice. (**A**) Blood levels of triglycerides (TG), total cholesterol (TC), low-density lipoprotein cholesterol (LDL-C), and high-density lipoprotein cholesterol (HDL-C) in mice ($n = 5$). (**B**) Quantification of Ki-67 staining in tumor tissues ($n = 5$, scale bar: 100 μm). (**C**) Relative expression of apoptosis-related genes *Bax* and *Bcl2* in tumor tissues ($n = 5$). (**D**) Relative mRNA expression of inflammation cytokines *Il6* and *Il1b* in tumor tissues ($n = 5$). Data in (**A**–**D**) were presented as mean ± SD from three independent experiments. p values in (**A**–**D**) were determined by Tukey's multiple comparisons test following one-way ANOVA. $p < 0.05$ was considered statistically significant. Triangles, diamonds, circles, and squares are used solely to differentiate between experimental groups and do not indicate specific biological meanings. Each dot represents data from an individual mouse.

4. Discussion

This study demonstrates that the combination of low-dose rapamycin with *I. latifolia* represents a promising strategy to enhance breast cancer therapy while mitigating rapamycin-induced lipid metabolic disorders. Although low-dose *I. latifolia* alone lacked significant anti-tumor activity, its synergistic effects with low-dose rapamycin were evident both in vitro and in vivo. This combination not only inhibited tumor proliferation, induced apoptosis, and suppressed inflammation but also alleviated rapamycin-associated metabolic side effects. These findings underscore the potential of *I. latifolia* as a dietary supplement to complement rapamycin-based therapies, offering a safer and more effective approach to breast cancer treatment.

Rapamycin, an mTOR inhibitor, has demonstrated promising efficacy in breast cancer therapy by focusing on critical mechanisms implicated in tumor growth and persistence [7,25]. Both preclinical and clinical investigations demonstrate that rapamycin inhibits tumor growth by blocking mTOR-driven cell cycle progression and promoting apoptosis [26]. This is achieved through the downregulation of anti-apoptotic proteins, such as Bcl-2, and the upregulation of pro-apoptotic factors, such as Bax [27]. A meta-analysis of clinical trials demonstrated that rapamycin-based regimens significantly extended progression-free survival in individuals with advanced or treatment-resistant breast carcinoma. However, these benefits were associated with notable side effects, including hyperlipidemia, metabolic disturbances, and immunosuppression [28]. These toxicities limit the long-term use of rapamycin, highlighting the need for strategies to mitigate adverse reactions while improving therapeutic efficacy. Interestingly, natural compounds may serve as effective adjuncts to rapamycin. For example, compounds like resveratrol and berberine have been shown to enhance autophagic flux when combined with rapamycin, potentially overcoming drug resistance in breast cancer models [29,30]. However, clinical validation is necessary to confirm these interactions in human subjects.

I. latifolia, commonly known as large-leaf Kudingcha, is a traditional Chinese tea with over 2000 years of history [13]. Predominantly cultivated in the Hainan and Zhejiang provinces, it is recognized for its unique bioactive components and health-promoting properties [31]. Our analysis identified 1,4-dicaffeoylquinic acid, 1-caffeoylquinic acid, esculentoside L, and ruberythric acid as major bioactive components in the extract of *I. latifolia*. These compounds are known for various biological activities and may contribute to the enhanced anti-tumor effects observed in combination with low-dose rapamycin. Given that our study observed reduced inflammatory markers and increased apoptosis following co-treatment, it is plausible that chlorogenic acid derivatives, especially 1,4-dicaffeoylquinic acid, contribute to these effects via suppression of the NF-κB pathway and modulation of pro-apoptotic signaling. These mechanisms are consistent with previous reports and may underlie the enhanced anti-tumor activity observed in the *I. latifolia* and rapamycin combination [32–34]. Additionally, esculentoside L, a triterpenoid saponin, has been reported to exert anti-inflammatory and pro-apoptotic effects in several diseases, including cancer [35]. Ruberythric acid, an anthraquinone compound, has demonstrated cytotoxic and anti-inflammatory properties [36]. These bioactive constituents likely work together to give *I. latifolia* its anti-breast cancer activity. However, the bitterness of *I. latifolia* limits its widespread acceptance. Incorporating it at minimal effective doses into daily diets as an adjuvant to pharmacological treatments could offer a feasible strategy to enhance therapeutic efficacy while maintaining patient compliance.

Dietary intervention has emerged as a promising strategy in the prevention and management of breast carcinoma, focusing on both dietary patterns and specific functional foods with potential anti-cancer properties [37,38]. Clinical approaches emphasize the role of nutrition in modulating the molecular pathways involved in cancer development, including inflammation and oxidative stress [39]. Dietary interventions incorporating antioxidant-rich foods, such as fruits, vegetables, and whole grains, have demonstrated a reduction in breast carcinoma risk by mitigating oxidative damage and enhancing immune function [40,41]. Additionally, specific bioactive compounds found in foods, such as polyphenols, flavonoids, and omega-3 fatty acids, have demonstrated anti-inflammatory and anti-cancer effects by targeting key signaling pathways like NF-κB [42–44]. In this context, *I. latifolia* represents a promising dietary adjunct to rapamycin, offering both anti-tumor and lipid-regulating benefits. However, its optimal dose range, bioavailability, and long-term safety remain to be fully elucidated. Future food-based interventions should consider these factors, along with personalized nutrition strategies, to maximize therapeutic efficacy.

This study reveals that *I. latifolia* can potentiate the anti-tumor efficacy of low-dose rapamycin, as evidenced by the observed inhibition of tumor proliferation, promotion of apoptosis, attenuation of inflammation, and reduction in tumor burden in both in vitro and in vivo models. While *I. latifolia* alone exhibited limited efficacy, its combination with rapamycin produced synergistic effects and alleviated rapamycin-induced metabolic disturbances. These findings suggest that *I. latifolia* may serve as a potential dietary adjuvant to enhance the therapeutic benefit of rapamycin-based regimens, warranting further investigation into its translational relevance.

Author Contributions: Z.R.: conceptualization, data curation, formal analysis, funding acquisition, methodology, and writing—original draft; Y.W.: conceptualization, resources, and writing—review and editing; X.G., H.T., H.O., Z.X., Y.X., L.X. and J.L.: methodology. H.W.: resources and funding acquisition. X.W.: project administration, funding acquisition, supervision, and writing—reviewing and editing. All authors have read and agreed to the published version of the manuscript.

Funding: This study was supported by funds from the Sichuan Science and Technology Program (Grant No. 2025ZNSFSC1056), the Natural Science Foundation of Jiangsu Province (Grant No. BK20231042), the Joint Fund for Science and Technology Education of Sichuan Province (Grant No. 2024NSFSC2067), the Industry–Education Deep Integration Reform Pilot Project of Sichuan Province, Turpan City Key Research and Development Special Project (Grant No. 2024008), and Undergraduate Innovation Training Program Incubation and Cultivation Project of Chengdu University (Grant No. CDUCX2025299).

Institutional Review Board Statement: All animal procedures adhered to the guidelines established by the Institutional Animal Ethics Committee of Jiangnan University (JN. No20170213-20171221[6], 2017-02-13).

Informed Consent Statement: Not applicable.

Data Availability Statement: The original contributions presented in this study are included in the article. Further inquiries can be directed to the corresponding author.

Conflicts of Interest: Author Xiaoying Guo was employed by the company Sichuan Kelun Pharmaceutical Co., Ltd., where she held the position of Deputy General Manager. She contributed to the Methodology of this study. This study did not receive any commercial funding, and the results will not be used for the future product development of Sichuan Kelun Pharmaceutical Co., Ltd. The remaining authors declare that the research was conducted in the absence of any commercial or financial relationships that could be construed as a potential conflict of interest.

Abbreviations

The following abbreviations are used in this manuscript:

ANOVA	One-way analysis of variance
ER	Estrogen receptor
HDL-C	High-density lipoprotein cholesterol
HER2	Human epidermal growth factor receptor 2
I. latifolia	*Ilex latifolia*
LDL-C	Low-density lipoprotein cholesterol
L. robustum	*Ligustrum robustum*
mTOR	Mammalian target of rapamycin
PR	Progesterone receptor
SD	Standard deviation
TC	Total cholesterol
TG	Triglycerides
TNBC	Triple-negative breast cancer

References

1. Harbeck, N.; Penault-Llorca, F.; Cortes, J.; Gnant, M.; Houssami, N.; Poortmans, P.; Ruddy, K.; Tsang, J.; Cardoso, F. Breast cancer. *Nat. Rev. Dis. Primers* **2019**, *5*, 66. [CrossRef]
2. Waks, A.G.; Winer, E.P. Breast Cancer Treatment: A Review. *JAMA* **2019**, *321*, 288–300. [CrossRef] [PubMed]
3. Zagami, P.; Carey, L.A. Triple negative breast cancer: Pitfalls and progress. *npj Breast Cancer* **2022**, *8*, 95. [CrossRef]
4. Zardavas, D.; Irrthum, A.; Swanton, C.; Piccart, M. Clinical management of breast cancer heterogeneity. *Nat. Rev. Clin. Oncol.* **2015**, *12*, 381–394. [CrossRef] [PubMed]
5. Mitra, S.; Dash, R. Natural Products for the Management and Prevention of Breast Cancer. *Evid. Based Complement. Altern. Med.* **2018**, *2018*, 8324696. [CrossRef] [PubMed]
6. Li, Y.; Li, S.; Meng, X.; Zhang, J.-J.; Li, H.-B. Dietary Natural Products for Prevention and Treatment of Breast Cancer. *Nutrients* **2017**, *9*, 728. [CrossRef]
7. Seto, B. Rapamycin and mTOR: A serendipitous discovery and implications for breast cancer. *Clin. Transl. Med.* **2012**, *1*, 29. [CrossRef]
8. Li, J.; Kim, S.G.; Blenis, J. Rapamycin: One drug, many effects. *Cell Metab.* **2014**, *19*, 373–379. [CrossRef]
9. Yu, F.Y.; Zheng, K.; Wu, Y.F.; Gao, S.-W.; Weng, Q.-Y.; Zhu, C.; Wu, Y.-P.; Li, M.; Qin, Z.-N.; Lou, J.-F.; et al. Rapamycin Exacerbates Staphylococcus aureus Pneumonia by Inhibiting mTOR-RPS6 in Macrophages. *J. Inflamm. Res.* **2023**, *16*, 5715–5728. [CrossRef]
10. Sun, C.Y.; Li, Y.Z.; Cao, D.; Zhou, Y.-F.; Zhang, M.-Y.; Wang, H.-Y. Rapamycin and trametinib: A rational combination for treatment of NSCLC. *Int. J. Biol. Sci.* **2021**, *17*, 3211–3223. [CrossRef]
11. Jobu, Y.; Nishigawa, M.; Furihata, K.; Uchida, K.; Taniuchi, K. Inhibitory effects of the combination of rapamycin with gemcitabine plus paclitaxel on the growth of pancreatic cancer tumors. *Hum. Cell* **2025**, *38*, 44. [CrossRef] [PubMed]
12. Dong, Y.; Gong, W.; Hua, Z.; Chen, B.; Zhao, G.; Liu, Z.; Thiele, C.J.; Li, Z. Combination of Rapamycin and MK-2206 Induced Cell Death via Autophagy and Necroptosis in MYCN-Amplified Neuroblastoma Cell Lines. *Front. Pharmacol.* **2020**, *11*, 31. [CrossRef]
13. Li, L.; Xu, L.J.; Ma, G.Z.; Dong, Y.M.; Peng, Y.; Xiao, P.G. The large-leaved Kudingcha (Ilex latifolia Thunb and Ilex kudingcha C.J. Tseng): A traditional Chinese tea with plentiful secondary metabolites and potential biological activities. *J. Nat. Med.* **2013**, *67*, 425–437. [CrossRef] [PubMed]
14. Wang, C.; Wang, R.; Chen, Y.; Wang, L.; Zhou, S.; Wang, H. Discovery of an EGFR tyrosine kinase inhibitor from Ilex latifolia in breast cancer therapy. *Bioorg. Med. Chem. Lett.* **2019**, *29*, 1282–1290. [CrossRef] [PubMed]
15. Wu, H.; Chen, Y.-L.; Yu, Y.; Zang, J.; Wu, Y.; He, Z. Ilex latifolia Thunb protects mice from HFD-induced body weight gain. *Sci. Rep.* **2017**, *7*, 14660. [CrossRef]
16. Feng, R.-B.; Fan, C.-L.; Liu, Q.; Liu, Z.; Zhang, W.; Li, Y.-L.; Tang, W.; Wang, Y.; Li, M.-M.; Ye, W.-C. Crude triterpenoid saponins from Ilex latifolia (Da Ye Dong Qing) ameliorate lipid accumulation by inhibiting SREBP expression via activation of AMPK in a non-alcoholic fatty liver disease model. *Chin. Med.* **2015**, *10*, 23. [CrossRef]
17. Lau, K.-M.; He, Z.-D.; Dong, H.; Fung, K.-P.; But, P.P.-H. Anti-oxidative, anti-inflammatory and hepato-protective effects of Ligustrum robustum. *J. Ethnopharmacol.* **2002**, *83*, 63–71. [CrossRef]
18. Yi, F.; Zhao, X.-L.; Peng, Y.; Xiao, P.G. Genus llex L.: Phytochemistry, Ethnopharmacology, and Pharmacology. *Chin. Herb. Med.* **2016**, *8*, 209–230. [CrossRef]
19. Hu, T.; He, X.-W.; Jiang, J.-G. Functional Analyses on Antioxidant, Anti-inflammatory, and Antiproliferative Effects of Extracts and Compounds from Ilex latifolia Thunb., a Chinese Bitter Tea. *J. Agric. Food Chem.* **2014**, *62*, 8608–8615. [CrossRef]
20. Chen, J.; Du, Y.; Long, Y.; Tao, D.; Hu, M.; Jiang, Y.; Wan, Y.; Yang, D. Polyphenols in Ilex latifolia Thunb. inhibit human lung cancer cell line A549 by regulation of the PI3K-Akt signaling pathway. *BMC Complement. Med. Ther.* **2022**, *22*, 85. [CrossRef]
21. Chavez, K.J.; Garimella, S.V.; Lipkowitz, S. Triple negative breast cancer cell lines: One tool in the search for better treatment of triple negative breast cancer. *Breast Dis.* **2010**, *32*, 35–48. [CrossRef]
22. Holliday, D.L.; Speirs, V. Choosing the right cell line for breast cancer research. *Breast Cancer Res.* **2011**, *13*, 215. [CrossRef]
23. Yellen, P.; Saqcena, M.; Salloum, D.; Feng, J.; Preda, A.; Xu, L.; Rodrik-Outmezguine, V.; Foster, D.A. High-dose rapamycin induces apoptosis in human cancer cells by dissociating mTOR complex 1 and suppressing phosphorylation of 4E-BP1. *Cell Cycle* **2011**, *10*, 3948–3956. [CrossRef]
24. Percie Du Sert, N.; Hurst, V.; Ahluwalia, A.; Alam, S.; Avey, M.T.; Baker, M.; Browne, W.J.; Clark, A.; Cuthill, I.C.; Dirnagl, U.; et al. The ARRIVE guidelines 2.0: Updated guidelines for reporting animal research. *PLoS Biol.* **2020**, *18*, e3000410.
25. Chang, S.B.; Miron, P.; Miron, A.; Iglehart, J.D. Rapamycin Inhibits Proliferation of Estrogen-Receptor-Positive Breast Cancer Cells. *J. Surg. Res.* **2007**, *138*, 37–44. [CrossRef]
26. Steelman, L.S.; Martelli, A.M.; Cocco, L.; Libra, M.; Nicoletti, F.; Abrams, S.L.; McCubrey, J.A. The therapeutic potential of mTOR inhibitors in breast cancer. *Br. J. Clin. Pharmacol.* **2016**, *82*, 1189–1212. [CrossRef] [PubMed]

27. Al-Astani Tengku Din, T.A.; Shamsuddin, S.H.; Idris, F.M.; Mansor, W.N.A.W.; Jalal, M.I.A.; Jaafar, H. Rapamycin and PF4 induce apoptosis by upregulating Bax and down-regulating survivin in MNU-induced breast cancer. *Asian Pac. J. Cancer Prev.* **2014**, *15*, 3939–3944. [CrossRef] [PubMed]
28. Qiao, L.; Liang, Y.; Mira, R.R.; Lu, Y.; Gu, J.; Zheng, Q. Mammalian target of rapamycin (mTOR) inhibitors and combined chemotherapy in breast cancer: A meta-analysis of randomized controlled trials. *Int. J. Clin. Exp. Med.* **2014**, *7*, 3333–3343.
29. Alayev, A.; Berger, S.M.; Kramer, M.Y.; Schwartz, N.S.; Holz, M.K. The combination of rapamycin and resveratrol blocks autophagy and induces apoptosis in breast cancer cells. *J. Cell. Biochem.* **2015**, *116*, 450–457. [CrossRef]
30. Guo, N.; Yan, A.; Gao, X.; Chen, Y.; He, X.; Hu, Z.; Mi, M.; Tang, X.; Gou, X. Berberine sensitizes rapamycin-mediated human hepatoma cell death in vitro. *Mol Med Rep.* **2014**, *10*, 3132–3138. [CrossRef]
31. Wüpper, S.; Lüersen, K.; Rimbach, G. Chemical Composition, Bioactivity and Safety Aspects of Kuding Tea-From Beverage to Herbal Extract. *Nutrients* **2020**, *12*, 2796. [CrossRef]
32. Liang, N.; Kitts, D.D. Role of Chlorogenic Acids in Controlling Oxidative and Inflammatory Stress Conditions. *Nutrients* **2016**, *8*, 16. [CrossRef]
33. Alcázar Magaña, A.; Kamimura, N.; Soumyanath, A.; Stevens, J.F.; Maier, C.S. Caffeoylquinic acids: Chemistry, biosynthesis, occurrence, analytical challenges, and bioactivity. *Plant J.* **2021**, *107*, 1299–1319. [CrossRef]
34. Tajik, N.; Tajik, M.; Mack, I.; Enck, P. The potential effects of chlorogenic acid, the main phenolic components in coffee, on health: A comprehensive review of the literature. *Eur. J. Nutr.* **2017**, *56*, 2215–2244. [CrossRef]
35. Bailly, C.; Vergoten, G. Esculentosides: Insights into the potential health benefits, mechanisms of action and molecular targets. *Phytomedicine* **2020**, *79*, 153343. [CrossRef]
36. Murthy, H.N.; Joseph, K.S.; Paek, K.Y.; Park, S.Y. Anthraquinone Production from Cell and Organ Cultures of Rubia Species: An Overview. *Metabolites* **2022**, *13*, 39. [CrossRef]
37. Shin, S.; Fu, J.; Shin, W.K.; Huang, D.; Min, S.; Kang, D. Association of food groups and dietary pattern with breast cancer risk: A systematic review and meta-analysis. *Clin. Nutr.* **2023**, *42*, 282–297. [CrossRef]
38. De Cicco, P.; Catani, M.V.; Gasperi, V.; Sibilano, M.; Quaglietta, M.; Savini, I. Nutrition and Breast Cancer: A Literature Review on Prevention, Treatment and Recurrence. *Nutrients* **2019**, *11*, 1514. [CrossRef]
39. Martin, L.J.; Li, Q.; Melnichouk, O.; Greenberg, C.; Minkin, S.; Hislop, G.; Boyd, N.F. A Randomized Trial of Dietary Intervention for Breast Cancer Prevention. *Cancer Res.* **2011**, *71*, 123–133. [CrossRef]
40. Aune, D.; Chan, D.S.; Vieira, A.R.; Rosenblatt, D.A.N.; Vieira, R.; Greenwood, D.C.; Norat, T. Fruits, vegetables and breast cancer risk: A systematic review and meta-analysis of prospective studies. *Breast Cancer Res. Treat.* **2012**, *134*, 479–493. [CrossRef]
41. Xiao, Y.; Ke, Y.; Wu, S.; Huang, S.; Li, S.; Lv, Z.; Yeoh, E.-K.; Lao, X.; Wong, S.; Kim, J.H.; et al. Association between whole grain intake and breast cancer risk: A systematic review and meta-analysis of observational studies. *Nutr. J.* **2018**, *17*, 87. [CrossRef] [PubMed]
42. Losada-Echeberría, M.; Herranz-López, M.; Micol, V.; Barrajón-Catalán, E. Polyphenols as Promising Drugs against Main Breast Cancer Signatures. *Antioxidants* **2017**, *6*, 88. [CrossRef] [PubMed]
43. Park, M.Y.; Kim, Y.; Ha, S.E.; Kim, H.H.; Bhosale, P.B.; Abusaliya, A.; Jeong, S.H.; Kim, G.S. Function and Application of Flavonoids in the Breast Cancer. *Int. J. Mol. Sci.* **2022**, *23*, 7732. [CrossRef] [PubMed]
44. Fabian, C.J.; Kimler, B.F.; Hursting, S.D. Omega-3 fatty acids for breast cancer prevention and survivorship. *Breast Cancer Res.* **2015**, *17*, 62. [CrossRef]

foods

MDPI

Article

Evaluation of Withering Quality of Black Tea Based on Multi-Information Fusion Strategy

Ting An [1], Yongwen Jiang [1], Hanting Zou [1], Xuan Xuan [1], Jian Zhang [2,*] and Haibo Yuan [1,*]

1 National Key Laboratory for Tea Plant Germplasm Innovation and Resource Utilization, Tea Research Institute, Chinese Academy of Agricultural Sciences, Hangzhou 310008, China; anting_mac@163.com (T.A.); jiangyw@tricaas.com (Y.J.); 13500436167@163.com (H.Z.); 2024613022042@stu.zafu.edu.cn (X.X.)

2 School of Intelligent Manufacturing, Huzhou College, Huzhou 313000, China

* Correspondence: zhangjian@zjhzu.edu.cn (J.Z.); 192168092@tricaas.com (H.Y.)

Abstract: The intelligent perception of moisture content (MC) for tea leaves during the black tea withering process is an unsolved task because of the acquisition of limited sample characteristic information. In this study, both the external and internal features of withering samples were simultaneously acquired based on near-infrared spectroscopy (NIRS) and machine vision (MV) technology. Different data fusion strategies, including low-, middle- and high-level strategies, were employed to integrate two types of heterogeneous information. Subsequently, the different fused features were combined with a support vector regression (SVR) algorithm to establish the moisture perception models of withering leaves. The middle-level-variable iterative space shrinkage approach (VISSA) displayed the best performance with 5.7705 for the relative percent deviation (RPD). Therefore, the proposed multi-information fusion strategy could achieve an intelligent perception of tea leaves in the black tea withering process. The integration of NIRS and MV technology overcomes the limitations of single-technology approaches in black tea withering assessment, providing a robust methodology for precision processing and targeted quality control of black tea.

Keywords: MV; NIRS; data fusion strategy; moisture content; black tea withering

1. Introduction

Black tea is one of the world's renowned beverages, widely favored by consumers for its health benefits and flavor [1]. It is worth noting that the health benefits and flavor of black tea are the results of the combined action of various water-soluble substances, such as catechin, caffeine, soluble sugars, amino acids, and tea pigments. There exists a significant correlation between the substance content and the quality of black tea. As we all know, the manufacturing craft determines the quality of black tea. Remarkably, withering is the first process in the preparation of black tea, and its quality lays the foundation for subsequent manufacturing steps. Generally, the MC is the most critical factor in evaluating the withering quality. The withering leaves with insufficient moisture are prone to damage in the rolling process affecting the appearance quality of the finished tea. Conversely, the excessive moisture of tea leaves could result in the loss of some essential components for finished tea. In actual production, experienced tea makers indirectly assess the withering degree based on the visual and tactile feature changes in tea leaves. This method is susceptible to external factors and cannot accurately obtain the MC of withering leaves. Although the moisture analyzer and gravimetric oven could accurately determine the MC of tea samples, they exhibit significant lag time, rendering them unable to guide practical withering production effectively. Hence, an accurate and rapid non-destructive detection method for MC of withering leaves should be developed.

In recent years, non-destructive testing technology has been widely employed in the assessment of agricultural product quality, particularly during the withering process [2–5]. Liang et al. obtained the image information of tea samples under different withering durations using MV technology. The color and texture features of these sample images were extracted for the construction of the MC perception model, which demonstrated a satisfactory result with the correlation coefficient of the prediction set (R_p) for 0.9314 [6]. In a study conducted by our team, the convolutional neural network model based on the confidence levels was proposed for quantitatively predicting the MC values of withering leaves, which demonstrated significant advantages compared with the traditional machine learning strategy [7]. Huang et al. collected tea leaves with different varieties and batches. The MC perception model of tea leaves with different varieties and batches was established based on the Vis-NIR spectroscopy and model transfer strategy [8]. The withering process of black tea is an exceptionally complex procedure involving gradual enzymatic and non-enzymatic reactions. As the samples' MC values diminish, notable changes occur in both their external appearance and internal quality characteristics. From an external characteristics' perspective, the color of withering leaves transitions from bright green to dark green, and their texture curls because of the loss of the MC. MV technology achieves a quantitative perception of moisture in withering leaves based on these appearance phenomena [7]. In terms of internal characteristics perspective, as the withering progresses, some significant changes would occur in certain substances, such as the moisture, chlorophyll, and theaflavin content. These alterations in substances would lead to differences in the stretching vibrations of some functional groups at different spectral bands, which is the reason NIRS can assess the MC in withering leaves [9]. Consequently, both MV and NIRS technology are limited to capturing unilateral digital information of withering leaves. Hyperspectral imaging technology (HSI) technology, integrating spectral and image information, could accurately obtain spectral information at the pixel level of samples and form three-dimensional data cubes. An et al. established cross-validation models of the MC for front and back sizes of single withering leaves and produced the moisture distribution map for the local region of withering leaves [10]. In contrast to An's study, Wei et al. developed the MC prediction model for accumulated withering leaves and created moisture perception maps at different moisture gradients [11]. Remarkably, the MC prediction models constructed by the above methods did not comprise external features of tea leaves. They solely relied on the pixel-level spectral information to produce moisture distribution maps. Hence, obtaining comprehensive samples of digital information and effectively integrating these features are particularly crucial for comprehensively evaluating the withering quality. Wang et al. integrated the visual, olfactory, and gustatory features of tea samples and achieved the qualitative discrimination of withering degree [12]. However, this study did not quantify the sample MC values, which directly affect the withering degree. In a study conducted by Chen et al., the image and spectral features of withering leaves were integrated based on data and feature-level fusion strategy and successfully established the MC perception models of withering leaves [13]. Although these studies have demonstrated the feasibility of assessing the withering quality of black tea based on the low and middle-level fusion strategies, they have not established the MC perception models using decision-level information, nor have they analyzed the contribution of each single information technology to the MC perception model. Hence, developing the MC perception models for withering leaves based on different fusion strategies is a highly meaningful task.

The core principle of data fusion strategies lies in multi-level (low-, mid-, and high-level) integration of heterogeneous information to compensate for the limitations of single techniques. For instance, the NIRS effectively captures the internal chemical compositions of tea leaves yet suffers from limited detection areas and lacks appearance morphological information. Conversely, MV technology monitors macroscopic exterior features across larger areas but cannot obtain internal features. The multi-information fusion strategy enables simultaneous monitoring of both the internal and external features of withering samples, thereby overcoming the limitations of single information technology. Furthermore, this methodology can be extended to acquire and evaluate both internal and external quality parameters of agricultural products, enabling comprehensive quality assessment critical for agricultural products. Hence, the multi-information fusion strategy was employed in this study to evaluate black tea withering quality comprehensively.

In this study, the MV technology was employed to acquire the color and texture features of withering leaves, which were utilized to represent the external visual characteristics of tea leaves. Meanwhile, the NIRS technology was leveraged to capture sensitive bands related to the moisture of withering leaves. Subsequently, the above digital information was integrated by different levels of fusion strategies, leading to the establishment of moisture perception models during the withering process.

2. Materials and Methods

2.1. Samples

In this study, 30 kg 'Jiukeng' fresh leaves were selected for the black tea withering experiment. These fresh tea leaves were harvested on 14 September 2024 from the experimental tea plantation at the Tea Research Institute of the Chinese Academy of Agricultural Sciences. During collection, the freshly harvested leaves were temporarily stored in a breathable bamboo basket with ice bags to prevent thermal accumulation in accumulated leaves that could compromise quality parameters, then immediately transported to the processing laboratory. Subsequently, the surface condensate water was carefully removed from the fresh tea leaves, and these tea leaves were applied for the withering experiment. The withering experiment lasted for 12 h and encompassed three stages: insufficient, moderate, and excessive withering. Generally, fresh leaves containing more than 62% moisture are classified as insufficient withering. Tea leaves with moisture levels ranging from 58% to 62% indicate moderate withering, while those below 58% are considered excessive withering because of excessive moisture loss [13]. In the withering experiment, these fresh tea leaves were put into a withering trough and placed in a special environment maintained at 3 cm sample thickness, 50% humidity, and 30 °C temperature. These accumulated withering leaves were turned every 1 h to ensure uniform dehydration. To characterize the changes in the withering leaves, several withering leaves were collected every 1 h for the collection of the MC values and digital information of each tea sample using the five-point sampling method. For the MC detection, 3 g tea leaves were placed in the moisture analyzer (MA35M-000230V1, Sartorius, Göttingen, Germany), which demonstrated a readability of 0.01% and a sensor precision of 1 mg, to measure the internal MC of withering samples, and this process was repeated 3 times every hour. The moisture values measured from 3 replicate experiments were averaged to represent the MC values at this specific withering time. Subsequently, the image and spectral information were obtained, respectively. Figure 1 illustrates the process of the withering experiment and relevant data analysis.

2.2. Spectra Acquisition by NIRS

The spectral information of withering leaves was collected utilizing a spectrum analyzer (IAS-3100, Intelligent Analysis Service Co., Ltd., Wuxi, China). This spectrum analyzer was constructed based on the dispersion principle of an array MEMS micro-mirror and grating, with a spectral acquisition range covering the near-infrared region of 900–1700 nm. Some parameters, including baseline noise, wavelength repeatability, wavelength accuracy, and absorbance repeatability, were below 0.0001 AU, 0.02 nm, 1 nm, and 0.0002 AU, respectively. Subsequently, several withering leaves (whole leaves) were collected at each withering time using the five-point sampling method. These collected leaves were then evenly divided into 15 samples (approximately 50 g per sample) for the acquisition of the NIR spectrum. Importantly, the spectrums from 5 local regions were acquired for each sample and averaged to represent the spectral features for this specific withering sample. Consequently, a total of 195 spectra of withering samples were obtained, which were subjected to standard normal variate (SNV) for preprocessing. In succession, these withering samples were sent to the MV system for image information collection.

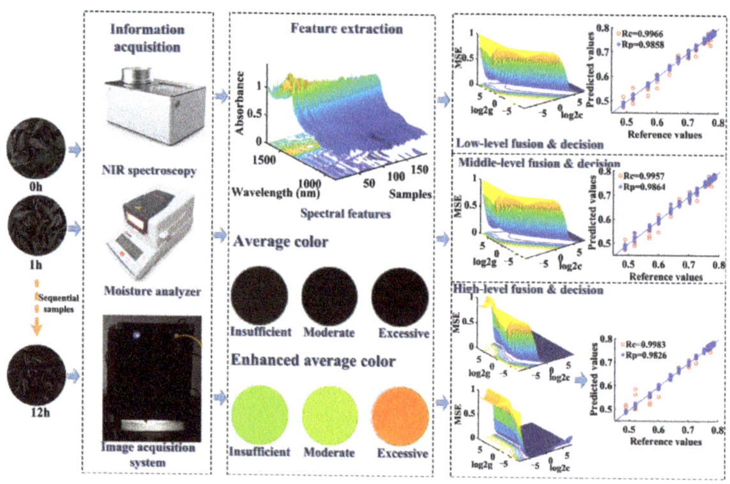

Figure 1. The flow diagram of the withering experiment and data analysis. Note: The closer the color is to yellow, the higher the MSE value; the closer it is to blue, the lower the MSE value.

2.3. Image Acquisition by Self-Built MV System

The MV system was developed by our research team and consists of these components: an industrial-grade camera, an arc-shaped illumination source, a mounting frame, and a software system. The FI-S200C-G industrial camera (Fenghua Imaging Technology Co., Ltd., Ningbo, China), composed by 4 mm low-distortion lens, 1/2.8″ CMOS sensor, and 1080 × 1080 pixels resolution, was operated with an exposure time of 0.09 ms. The monochromatic white DOME-type arc light source provided uniform illumination at 100 lx intensity throughout the imaging process. A total of 195 images of withering samples were acquired based on a self-built MV system. Subsequently, the 12 color and 6 texture features were extracted from withering sample images, respectively. The 12 color features contained the mean value of the red component (R), green component (G), blue component (B), hue (H), saturation (S), visible light (V), brightness component (L), a component (a), b component (b), super green transformation (2G-R-B), the ratio of the red channel to the green channel (R/G), and the color angle (hab*). The 6 texture features included average gray value (m), standard deviation (δ), smoothness (r), third moment (μ), consistency (U) and entropy (e). These color and texture features were subjected to Normalization for processing.

2.4. Data Fusion Strategy

In this study, the spectra and image information were applied to describe the internal and external characteristics of withering leaves, respectively. To comprehensively acquired digital information of withering leaves, a data fusion strategy should be developed. Generally, data fusion strategies are categorized into low (data)-, middle (feature)- and high (decision)-levels [14]. In our research, the spectra and image features of withering leaves were merged into a data matrix, which was regarded as a low-level fusion strategy. For the middle-level fusion strategy, the dimensionality reduction and feature selection were applied to each digital information for the selection of effective variables. The principal component analysis (PCA) is a commonly utilized method for data dimensionality reduction [15]. In contrast to the dimensionality reduction method, the feature selection strategy eliminated irrelevant information from raw data while retaining features that possessed relative independence. In our research, the features selection strategies, including variables combination population analysis and iterative retained information variable algorithm (VCPA-IRIV), the variable iterative space shrinkage approach (VISSA), and competitive adaptive reweighted sampling (CARS) were performed for effective spectra information selection. Meanwhile, the correlation analysis was applied to remove irrelevant color and texture features [10,16–18]. The high-level fusion strategy was used to consolidate the decision results from individual digital information to achieve the "ensemble decision". Multiple linear regression (MLR) was used to integrate the decision information from spectral and image characteristics to explain variations in the moisture of withering leaves in this study. Although the ultimate perception results could be affected by the decision accuracy of each digital information, both NIRS and MV technology can effectively characterize the moisture changes in withering leaves. Hence, the MLR algorithm was applied to integrate the decision results from different digital information:

$$y_{i,ac} = b + k_1 \times x_{1,\,pc} + k_2 \times x_{2,pc} + \cdots + k_n \times x_{n,pc} \tag{1}$$

$$y_{i,pc} = b + k_1 \times x_{1,pc} + k_2 \times x_{2,pc} + \cdots + k_n \times x_{n,pc} \tag{2}$$

$$y_{i,pp} = b + k_1 \times x_{1,pp} + k_2 \times x_{2,pp} + \cdots + k_n \times x_{n,pp} \tag{3}$$

The i and n denote the ith sample and nth digital information, respectively. The y means the MC values of withering leaves. The x represents different digital information. Consequently, the $y_{i,ac}$, $y_{i,pc}$ and $y_{i,pp}$ demonstrate the actual value of the calibration set, the predicted value of the calibration set, and the predicted value of the prediction set, respectively. The $x_{n,pc}$ and $x_{n,pp}$ denote the predicted value of calibration and prediction set based on the nth digital information, respectively. The Equation (1) was used to determine the coefficient k and constant b. The Equation (2) was applied to obtain the $y_{i,pc}$.

2.5. Moisture Perception Model Establishment and Evaluation

In contrast with the support vector machine (SVM) in classification problems, the SVR algorithm attempts to find an "optimal hyperplane" in the feature information [19]. This hyperplane is considered a regression function, aiming to approximate the feature information and predict the targeted outcomes within an acceptable margin of error. Remarkably, the SVR algorithm enhances the robustness of the perception model by maximizing the margin of the regression model. Hence, this algorithm will be applied for the establishment of a withering moisture perception model in this research. In addition, several parameters, including the correlation coefficient of calibration set (R_c), R_p, RPD, the root mean square error of the calibration (RMSEC) and prediction (RMSEP) set, were employed to evaluate the performance of the MC perception model for withering leaves.

2.6. Software

All data analyses, including feature extraction, multi-information fusion, and predictive modeling, were implemented using MATLAB R2017b (The Math Work, Inc., Natick, MA, USA). The statistical visualizations were created using Origin 2021b (OriginLab Corporation, Northampton, MA, USA).

3. Results and Discussion

3.1. Digital Information Representation and Analysis of Different Sensors

Some significant changes occurred in both the internal substances and external characteristics of withering samples. To provide a more comprehensive description of the dynamic changes in withering samples, NIRS and MV technologies were utilized to represent the internal and external characteristics of tea leaves digitally. These features from different sensors were represented and analyzed, respectively.

3.1.1. Internal Features of Withering Leaves

In this study, the average spectra of each sample were utilized to characterize the internal features of withering leaves. Furthermore, the representative spectra of all withering samples were presented in Figure 2a, which indicated that all the sample spectra exhibited a consistent trend in the range of 900–1700 nm. Although the absorbance of the spectra exhibited a consistent changing trend, there still existed some noise in certain bands that were not readily perceptible to the human eye. Consequently, all spectral data for withering leaves were subjected to the SNV algorithm to eliminate the impact of optical scattering. The average spectra of tea samples with different moisture are displayed in Figure 2b. Obviously, the spectral distribution of all samples becomes more concentrated. After the SNV preprocessing, the spectra of withering samples with different moisture still exhibited the same trend, but there were subtle differences in the spectral profiles. The absorbance near 900 nm and 1650 nm increased over the moisture lost. However, the opposite is true near the absorption peak at 1450 nm. Evidently, these spectral absorbance variations directly correlate with the physical state of withering leaves. As moisture loss occurs, some changes in leaf cell membrane permeability alter light scattering properties, accounting for the observed absorbance increases at specific wavelengths, such as 900 nm and 1650 nm. Notably, the band near 1450 nm represents a strong absorption feature for water corresponding to the O–H first overtone, where higher moisture content yields greater absorbance. The moisture loss during withering directly leads to reduced absorption intensity in this spectral region. Additionally, the spectra of withering leaves also displayed several absorption peaks, such as around 960 nm, 1100 nm, 1170 nm, and 1500 nm. The peaks around 960 nm were attributable to the overtone of OH groups [20]. The peaks near 1100 nm and 1500 nm were related to the second and first overtone of OH groups [21]. The peak near 1170 nm corresponded to the overtone of CH groups [22]. These peaks and their corresponding functional groups for water and catechins demonstrated the feasibility that NIRS technology characterizing the internal features of withering leaves.

3.1.2. External Features of Withering Leaves

In actual withering production, the moisture loss progressively concentrates the cell sap, while enzymatic and non-enzymatic reactions drive chlorophyll transformation and theaflavin formation. These biochemical changes induce a visible color transition from bright green to dark green, demonstrating the feasibility of using color-related characteristics, such as basic color, saturation, and brightness components, for assessing withering quality. Furthermore, the moisture loss reduces intercellular space, resulting in increased leaf rigidity, curling morphology, and surface wrinkling. These physical alterations modify light scattering properties, leading to distinct textural changes among samples at different withering times. Consequently, the color and texture information can represent changes in the external features of withering leaves. In this study, the average color and texture features, such as R, G, B, H, S, V, L, a, b, 2G-R-B, R/G, hab* m, δ, r, μ, U, and e, were applied to characterize the external features of withering leaves. Subsequently, a correlation analysis was conducted to characterize the correlations between different variables. As can be seen in Figure 3b, some significant correlations exist among most variables, attributable to the fact that both color and textural features are governed by the same physiological and biochemical processes. The correlations among color features stem from the synergistic effects of the biochemical reactions on pigment composition. For instance, chlorophyll degradation reduces the G value, resulting in coupled variations in derived parameters such as R/G ratio and 2G-R-B. The correlations among texture features are directly associated with physical and structural changes in withering leaves induced by moisture loss. With the loss of moisture, the increasing curling degree of withering samples leads to a synergistic change in texture parameters, evidenced by the significant negative correlation between U and e. The withering of black tea is a complex process, and the correlations among these multidimensional features are a unified manifestation of synergistic interactions between chemical and physical changes during this progression. Therefore, further analysis was required to identify strategies for eliminating redundant information and to assess the contribution of each variable to the moisture perception model.

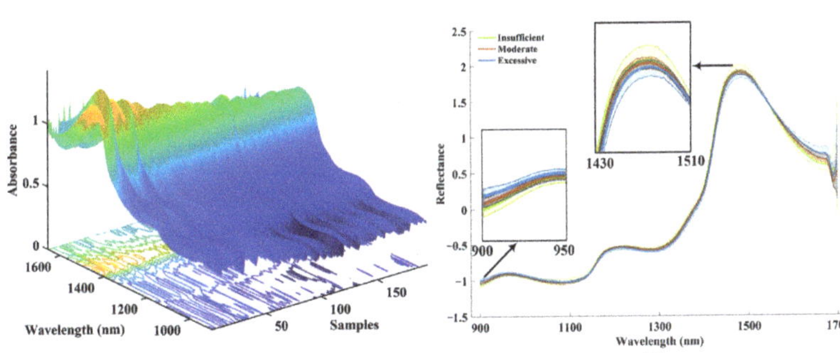

Figure 2. The spectra of withering leaves based on (**a**) raw data and (**b**) the average spectra of different withering degrees based on SNV processing data. Note: The closer the color is to yellow, the higher the absorbance value; the closer it is to blue, the lower the absorbance value.

3.2. Effective Feature Selection of Different Digital Information

According to the above description, not all variable features contributed to the construction of the moisture perception model, and some redundant features could degrade the performance of the perception model. Hence, some effective feature selection strategies would be performed. The CARS, VCPA-IRIV, and VISSA algorithms were performed to select effective variables for spectral information, respectively. The Pearson correlation analysis was conducted for the selection of effective color and texture information.

3.2.1. Feature Selection of Spectral Information

After repeated testing, some notable parameters of CARS, VCPA-IRIV, and VISSA algorithms were determined. The sampling runs and cross-validation folds for the CARS algorithm were 50 and 5, respectively. The group number of cross-validation and maximal PCs for the VCPA-IRIV algorithm were 10 and 5, respectively. The maximal PCs and cross-validation folds for the VISSA algorithm were 15 and 5, respectively. The CARS algorithm selected 41 effective features for the moisture perception model of withering leaves. The compression rate of spectral information was 94.6%, indicating that the CARS algorithm considered most of the variables for raw spectral data as redundant information. Compared with the CARS algorithm, the VCPA-IRIV algorithm selected 86 effective features, which was nearly double the number of features chosen by the CARS algorithm. The compression rate of raw spectra was 88.7%. The VISSA algorithm retained 231 effective features for the moisture perception model of withering leaves, and the compression rate of raw spectra only reached 69.6%. These selected variables are displayed in Figure 3a. Obviously, these selected bands were inattentively distributed. However, some effective variables were selected by these three algorithms. The selected bands at around 970 nm were attributable to second stretching overtones for free OH groups. These selected bands near 1200 nm corresponded to second stretching overtones by the absorption of CH_3 and CH_2 groups. These selected bands at 1400–1420 nm were related to the first stretching overtones and combination bands of free OH groups. These selected bands are located at 1660–1700 nm due to the first stretching overtones by the absorption of CH_3 and CH_2 groups. Although some bands could not represent the overtone of OH, they could represent the overtone of HX, which showed a similar change in the OH groups [23]. Specifically, the selected bands located around 970 nm and 1420 nm, corresponding to the first and second stretching overtones as well as combination bands of OH groups, can directly reflect moisture changes in withering leaves [20,21]. The selected bands near 1200 nm and 1700 nm were related to the catechins content present in withering leaves [22]. Notably, although certain functional groups are closely associated with catechins, both catechins and MC exhibit a significant decreasing trend during withering processing, indicating a strong positive correlation [24]. Consequently, the corresponding spectral bands of these functional groups can directly or indirectly characterize MC changes. This finding provides crucial mechanistic insights for evaluating withering quality using NIRS technology.

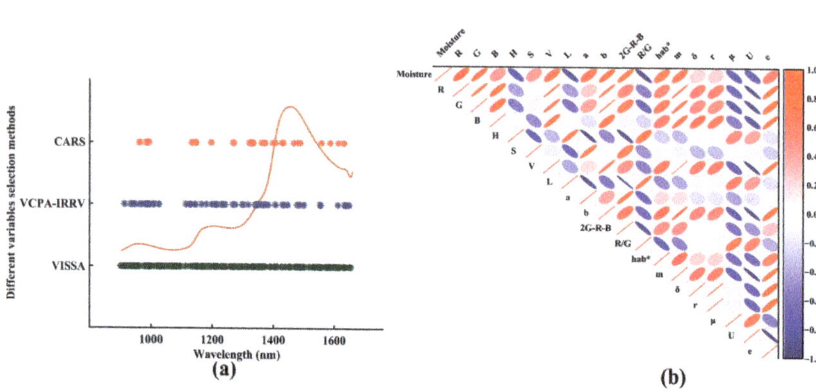

Figure 3. The selected effective features for spectra data (**a**) and image data (**b**).

3.2.2. Feature Selection of Image Information

The Pearson correlation analysis was performed to assess the degree of response for color and texture variables to the changes in the MC in withering leaves, and the results were displayed in Figure 3b. Of this figure, the *p*-values for these color and texture variables were below 0.01, demonstrating a significant correlation between these variables and the changes in the MC in withering leaves. Obviously, the variables including R, G, B, S, V, a, b, 2G-R-B, hab*, m, δ, r, and e exhibited a positive correlation with the changes in the MC in withering leaves and the correlation coefficients of these variables with respect to changes in the MC were 0.79, 0.89, 0.46, 0.45, 0.76, 0.72, 0.87, 0.90, 0.71 0.76, 0.24, 0.25, and 0.6, respectively. Conversely, the variables, such as H, L, R/G, μ, and U, demonstrated a negative correlation with changes in the MC in withering leaves, and the correlation coefficients of these variables with respect to changes in the MC were −0.80, −0.89, −0.92, −0.78, and −0.77, respectively. In addition, the cutoff line of the correlation coefficient was set to 0.6, indicating that the variables with an absolute correlation coefficient greater than 0.6 were considered to be effective features. Hence, the effective color and texture features, including R, G, H, V, L, a, b, 2G-R-B, R/G, hab*, m, μ, U, and e, were retained for the establishment of moisture perception models.

3.3. Moisture Prediction Models with a Single Technology

The moisture perception models for withering leaves were constructed based on the SVR algorithm and each single digital information technology, such as NIRS and MV technologies. Before modeling, the digital information and its corresponding MC values were divided into a calibration set and a prediction set with the radio of 3:1. The performances of established moisture perception models were displayed in Table 1. Remarkably, the R_p values of these established models with single digital information were both higher than 0.95, and the RPD values were both higher than 3.5, indicating that both the NIRS and MV technologies demonstrated strong perception capabilities for the moisture of withering leaves. In addition, the established model using NIRS technology displayed better performance than the established model based on MV technology. The color and texture features indirectly reflected the changes in MC in withering leaves. In contrast to MV technology, the NIRS technology provided changes in spectral reflectance, which was directly related to the OH groups for water. Furthermore, each type of digital information could only provide one-sided sample information. Hence, different data fusion strategies were proposed to integrate digital information from different sensors.

Table 1. SVR models for moisture content of withering leaves using individual data.

Data	No. of Variables	Parameter		Calibration Sets		Prediction Sets		
		c	*g*	R_c	RMSEC	R_p	RMSEP	RPD
NIR	760	0.4353	1.3195	0.9971	0.0078	0.9775	0.0210	4.7258
MV	18	0.4353	0.0206	0.9958	0.0094	0.9657	0.0260	3.6824

3.4. Moisture Perception Models with Multi-Level Data Fusion

In this study, the low-, middle- and high-level data fusion strategies were applied to integrate digital information from different sensors. Subsequently, the fused digital information was combined with the SVR algorithm to construct the moisture perception model for withering leaves.

3.4.1. Low-Level Fusion Strategy

The low-level fusion strategy means all digital information is simply concatenated, with the dimension of the integrated information being a composite of the dimensions of each individual piece of digital information. Table 2 displays the prediction results based on this fusion strategy. The R_c, RMSEC, R_p, RMSEP, and RPD were 0.9966, 0.0082, 0.9858, 0.0180, and 5.6142, respectively. Obviously, the established moisture perception model based on a low-level fusion strategy displayed better prediction performance than the established model using single pieces of digital information, indicating that acquiring and representing comprehensive information of withering leaves contributed to enhancing the performance of the moisture perception model. However, there existed a certain degree of redundancy among different feature variables, which reduced the running speed and performance of the moisture perception model. Consequently, some features that contributed significantly to the moisture perception model need to be further selected to improve the running speed and performance of the prediction model.

Table 2. SVR models for moisture content of withering leaves using multi-level data fusion.

Level	Methods	No. of Variables	Parameter		Calibration Sets		Prediction Sets		
			c	g	R_c	RMSEC	R_p	RMSEP	RPD
Low-level		778	3.0314	256	0.9966	0.0082	0.9858	0.0180	5.6142
Middle-level	PCA	22	0.7579	0.0156	0.9964	0.0086	0.9669	0.0260	3.7149
	CARS	55	0.5743	256	0.9922	0.0125	0.9854	0.0193	5.2756
	VCPA-IRIV	100	1	256	0.9932	0.0116	0.9859	0.0188	5.4320
	VISSA	245	5.2780	256	0.9957	0.0092	0.9864	0.0178	5.7705
High-level	MLR				0.9883	0.0057	0.9826	0.0186	5.3456

3.4.2. Middle-Level-PCA Fusion Strategy

The middle-level-PCA fusion strategy indicates that the NIRS and MV digital information were subject to the PCA algorithm, respectively. Subsequently, the optimal PCs of each information technology were selected based on the model's perceptual performance and these selected optimal PCs were concatenated into a feature matrix for the establishment of a moisture perception model. The cumulative explained and the explained percent for the NIRS and MV digital information using the first 18 PCs are demonstrated in Figure 4a,c, respectively. Obviously, the first 18 PCs of NIRS and MV digital information could represent 99.82% and 100% of the information for raw variables, respectively. This phenomenon indicated that the first 18 PCs of NIRS and MV digital information could represent almost all the content of raw sample information. To obtain the optimal PCs of the two types of digital information, the SVR algorithm was employed to establish the moisture perception model, with the results presented in Figure 4b,d. Consequently, the first 17 PCs of raw spectral digital information and the first 5 PCs of image digital information were directly connected to construct the moisture perception model, with the results shown in Table 2. In contrast to each single piece of digital information, the performance of the middle-level-PCA strategy was not satisfactory, indicating that the appropriate data fusion strategy is critical to the predictive performance of the perception model.

3.4.3. Middle-Level-Variable Selection Fusion Strategy

The middle-level-variable selection fusion strategy indicates that the NIRS and MV digital information are subject to different variable selection algorithms and Pearson correlation analysis, respectively. These selected effective features were concatenated into a feature matrix for the establishment of a moisture perception model. For the middle-level-CARS strategy, a total of 55 variables were selected to establish the moisture model, and the compression rate of raw variables was 92.9%. The R_c, RMSEC, R_p, RMSEP, and RPD were 0.9922, 0.0125, 0.9854, 0.0193, and 5.2756, respectively. For the middle-level-VCPA-IRIV, 100 variables were retained to establish the moisture perception model for withering leaves, and the compression rate of raw variables was 87.1%. The R_c, RMSEC, R_p, RMSEP, and RPD were 0.9932, 0.0116, 0.9859, 0.0188, and 5.4320, respectively. For the middle-level-VISSA strategy, 245 variables were selected for the establishment of the moisture perception model, and the compression rate of raw variables was 68.5%. The R_c, RMSEC, R_p, RMSEP, and RPD were 0.9957, 0.0092, 0.9864, 0.0178, and 5.7705. The results are displayed in Table 2. Obviously, as the number of variables decreases, the performance of the perception model decreases, indicating that some important bands for spectral information were removed.

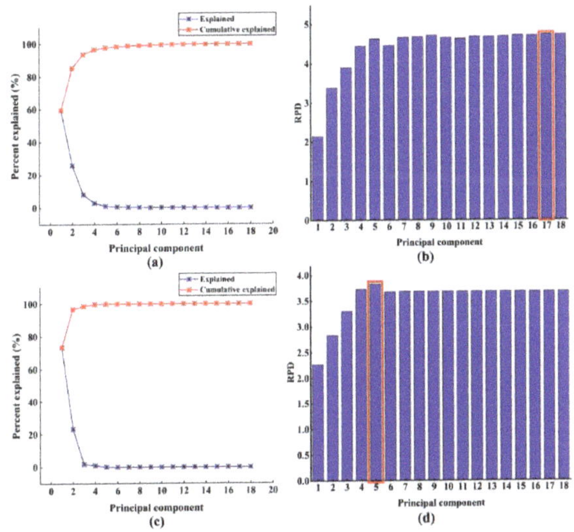

Figure 4. Percent explained and cumulative explained of spectra data (**a**) and image data (**c**) by PCA; SVR prediction models on different numbers of PCs to determine the best RPD value for spectra data (**b**) and image data (**d**).

3.4.4. High-Level Fusion Strategy

The high-level fusion strategy means that the decision results for each single piece of digital information are fused to construct the moisture perception model. The multiple linear fitting equations are displayed using Equation (4). Based on the weight coefficients of the spectra and image information in the linear regression equation of the MC, the order of two sensors to the improvement of model performance was NIRS > MV. The same conclusion was obtained by the modeling results of each single digital information (displayed in Table 1). The values of R_p, RMSEP, and RPD for the moisture perception model using a high-level fusion strategy were 0.9826, 0.0186, and 5.3456, respectively. The high-level fusion strategy demonstrated satisfactory predictive performance, primarily attributed to the reliable decision outcomes provided by each single digital information.

The established MC perception model uses a high-level fusion strategy.

$$y = 0.60339x_1 + 0.42081x_2 - 0.01610 \tag{4}$$

The parameters, including x_1, x_2, and y, mean the predicted value of NIRS, MV, and the actual value of moisture content, respectively.

3.5. Model Comparison and Evaluation

The performance of the moisture perception model based on each single piece of digital information and the fusion digital information were proposed and compared. Obviously, most perception models using data fusion strategies performed better predictive performance than the established models based on single digital information, except for the middle-level fusion-PCA strategy. The results demonstrated that not all data fusion strategies could enhance the performance of the perception model. Although the obtained PCs using the PCA method could represent over 99% of raw data information, not all PCs exhibited specific responses to the research target, resulting in the unsatisfactory predictive performance of the perception model. In contrast to the middle-level-PCA strategy, the middle-level-VISSA strategy was considered the best moisture perception model for withering leaves because it showed the best RPD value. For the color and texture features, 14 variables, including R, G, H, V, L, a, b, 2G-R-B, R/G, hab*, m, μ, U, and e were selected. For these spectral data, 231 effective bands were retained, and they might be related to the second overtone of OH and CH [21,22]. During the withering process of black tea, the moisture loss was very obvious and also accompanied by the enzymatic reaction and reduction of compounds containing hydrogen. Therefore, some other important components, such as catechins, amino acids, and soluble sugars, changed significantly [25,26]. There is a certain correlation between the changes in these substances and the moisture. These effective features selected by the middle-level-VISSA strategy could directly or indirectly represent the changes in moisture in withering leaves. Hence, they could be applied for the evaluation basis for the MC in withering leaves.

Liang et al. and Zhang et al. constructed the moisture perception model of tea leaves based on the digital information for MV and NIRS, respectively [6,27]. Their moisture perception models based on single digital information achieved satisfactory results. However, our proposed moisture perception model exhibited better predictive performance, demonstrating that the fusion of comprehensive feature information of tea samples is indeed beneficial for improving the performance of prediction models. In addition, Chen et al. established the moisture prediction model of withering leaves based on low- and middle-level fusion strategies [13]. They believed that the middle-level-RF (cut of line = 0.8) displayed the best predictive performance, with 5.5596 for RPD. In contrast to Chen's study, our research proposed the high-level fusion strategy to further characterize the contribution of each single piece of digital information to the moisture perception model. In addition, our proposed middle-level-VISSA strategy displayed a higher RPD value, indicating that our constructed moisture perception model has better predictive performance.

In this study, the proposed data fusion strategy effectively overcomes the limitations of individual techniques, including restricted detection range and penetration depth, by integrating complementary information from NIRS (internal features) and MV (external features). Our proposed methods enable comprehensive characterization of withering samples, thereby facilitating the development of more accurate quality assessment models. This strategy significantly enhances the precision of withering quality control in actual production.

4. Conclusions

(1) This study proved the feasibility of the perception of the MC values in withering leaves based on the NIRS combined with MV technology. The multi-information fusion strategy overcomes the limitations of single digital information that can only capture

partial feature information, providing a novel methodology for the comprehensive characterization of tea samples.

(2) NIRS technology demonstrates superior moisture detection capability compared with MV technology. This discovery provides valuable insights into refining fusion information models and developing specialized equipment in subsequent research.

(3) An appropriate data fusion strategy could remove redundant features and improve the accuracy of the moisture prediction model for withering leaves. The middle-level-VISSA strategy achieves the most accurate moisture perception. The selection of multi-information fusion strategies is critically important for accurate perception and precise control of moisture content during withering processing.

Author Contributions: T.A.; formal analysis, writing—original draft, writing—review and editing, methodology and data curation, funding acquisition. Y.J.; data curation, software, and visualization. H.Z.; project administration and visualization. X.X.; supervision and validation. J.Z.; Conceptualization, software, and project administration. H.Y.; funding acquisition validation and supervision. All authors have read and agreed to the published version of the manuscript.

Funding: This study was supported by the Natural Science Foundation of Zhejiang Province (LQN25C130009), the Science and Technology Innovation Project of the Chinese Academy of Agricultural Sciences (CAAS-ASTIP-TRICAAS).

Institutional Review Board Statement: Not applicable.

Informed Consent Statement: Not applicable.

Data Availability Statement: The original contributions presented in this study are included in the article. Further inquiries can be directed to the corresponding authors.

Conflicts of Interest: The authors declare that they have no known competing financial interests or personal relationships that could have appeared to influence the work reported in this paper.

References

1. Richelle, M.; Tavazzi, I.; Offord, E.; Chemistry, F. Comparison of the Antioxidant Activity of Commonly Consumed Polyphenolic Beverages (Coffee, Cocoa, and Tea) Prepared per Cup Serving. *J. Agric. Food Chem.* **2001**, *49*, 3438–3442. [CrossRef] [PubMed]
2. Qiao, M.; Xu, Y.; Xia, G.; Su, Y.; Lu, B.; Gao, X.; Fan, H. Determination of hardness for maize kernels based on hyperspectral imaging. *Food Chem.* **2022**, *366*, 130559. [CrossRef] [PubMed]
3. Tian, X.; Zhang, C.; Li, J.; Fan, S.; Yang, Y.; Huang, W. Detection of early decay on citrus using LW-NIR hyperspectral reflectance imaging coupled with two-band ratio and improved watershed segmentation algorithm. *Food Chem.* **2021**, *360*, 130077. [CrossRef]
4. Li, J.; Zhang, R.; Li, J.; Wang, Z.; Zhang, H.; Zhan, B.; Jiang, Y. Detection of early decayed oranges based on multispectral principal component image combining both bi-dimensional empirical mode decomposition and watershed segmentation method. *Postharvest Biol. Technol.* **2019**, *158*, 110986. [CrossRef]
5. Wei, Y.; Wu, F.; Xu, J.; Sha, J.; Zhao, Z.; He, Y.; Li, X. Visual detection of the moisture content of tea leaves with hyperspectral imaging technology. *J. Food Eng.* **2019**, *248*, 89–96. [CrossRef]
6. Liang, G.; Dong, C.; Hu, B.; Zhu, H.; Yuan, H.; Jiang, Y.; Hao, G. Prediction of Moisture Content for Congou Black Tea Withering Leaves Using Image Features and Nonlinear Method. *Sci. Rep.* **2018**, *8*, 7854. [CrossRef]
7. An, T.; Yu, H.; Yang, C.; Liang, G.; Dong, C. Black tea withering moisture detection method based on convolution neural network confidence. *J. Food Process Eng.* **2020**, *43*, e13428. [CrossRef]
8. Huang, Z.X.; Sanaeifar, A.; Tian, Y.; Liu, L.; Zhang, D.Y.; Wang, H.; Ye, D.P.; Li, X.L. Improved generalization of spectral models associated with Vis-NIR spectroscopy for determining the moisture content of different tea leaves. *J. Food Eng.* **2021**, *293*, 110374. [CrossRef]
9. Zhang, S.; Zuo, Y.; Wu, Q.; Wang, J.; Ban, L.; Yang, H.; Bai, Z. Development and Validation of Near-Infrared Methods for the Quantitation of Caffeine, Epigallocatechin-3-gallate, and Moisture in Green Tea Production. *J. Anal. Methods Chem.* **2021**, *2021*, 9563162. [CrossRef]

10. An, T.; Yu, S.; Huang, W.; Li, G.; Tian, X.; Fan, S.; Dong, C.; Zhao, C. Robustness and accuracy evaluation of moisture prediction model for black tea withering process using hyperspectral imaging. *Spectrochim. Acta A Mol. Biomol. Spectrosc.* **2022**, *269*, 120791. [CrossRef]
11. Wei, Y.Z.; He, Y.; Li, X.L. Tea moisture content detection with multispectral and depth images. *Comput. Electron. Agric.* **2021**, *183*, 106082. [CrossRef]
12. Wang, Y.; Liu, Y.; Cui, Q.; Li, L.; Ning, J.; Zhang, Z. Monitoring the withering condition of leaves during black tea processing via the fusion of electronic eye (E-eye), colorimetric sensing array (CSA), and micro-near-infrared spectroscopy (NIRS). *J. Food Eng.* **2021**, *300*, 110534. [CrossRef]
13. Chen, J.; Yang, C.; Yuan, C.; Li, Y.; An, T.; Dong, C. Moisture content monitoring in withering leaves during black tea processing based on electronic eye and near infrared spectroscopy. *Sci. Rep.* **2022**, *12*, 20721. [CrossRef] [PubMed]
14. Zhu, J.; Fan, X.; Han, L.; Zhang, C.; Wang, J.; Pan, L.; Tu, K.; Peng, J.; Zhang, M. Quantitative analysis of caprolactam in sauce-based food using infrared spectroscopy combined with data fusion strategies. *J. Food Compos. Anal.* **2021**, *104*, 104130. [CrossRef]
15. Gewers, F.L.; Ferreira, G.R.; Arruda, H.F.; Silva, F.N.; Comin, C.H.; Amancio, D.R.; Costa, L.D. Principal Component Analysis: A Natural Approach to Data Exploration. *ACM Comput. Surv.* **2021**, *54*, 1–34. [CrossRef]
16. Deng, B.-C.; Yun, Y.-H.; Liang, Y.-Z.; Yi, L.-Z. A novel variable selection approach that iteratively optimizes variable space using weighted binary matrix sampling. *Analyst* **2014**, *139*, 4836–4845. [CrossRef]
17. Li, J.; Chen, L. Comparative analysis of models for robust and accurate evaluation of soluble solids content in 'Pinggu' peaches by hyperspectral imaging. *Comput. Electron. Agric.* **2017**, *142*, 524–535. [CrossRef]
18. Wang, Y.; Li, L.; Liu, Y.; Cui, Q.; Ning, J.; Zhang, Z. Enhanced quality monitoring during black tea processing by the fusion of NIRS and computer vision. *J. Food Eng.* **2021**, *304*, 110599. [CrossRef]
19. Ye, J.Y.; Yang, Z.X.; Li, Z.L. Quadratic hyper-surface kernel-free least squares support vector regression. *Intell. Data Anal.* **2021**, *25*, 265–281. [CrossRef]
20. Wang, Y.-J.; Li, T.-H.; Li, L.-Q.; Ning, J.-M.; Zhang, Z.-Z. Evaluating taste-related attributes of black tea by micro-NIRS. *J. Food Eng.* **2021**, *290*, 110181. [CrossRef]
21. Lee, M.-S.; Hwang, Y.-S.; Lee, J.; Choung, M.-G. The characterization of caffeine and nine individual catechins in the leaves of green tea (*Camellia sinensis* L.) by near-infrared reflectance spectroscopy. *Food Chem.* **2014**, *158*, 351–357. [CrossRef] [PubMed]
22. Chen, Q.; Chen, M.; Liu, Y.; Wu, J.; Wang, X.; Ouyang, Q.; Chen, X. Application of FT-NIR spectroscopy for simultaneous estimation of taste quality and taste-related compounds content of black tea. *J. Food Sci. Technol. Mysore* **2018**, *55*, 4363–4368. [CrossRef] [PubMed]
23. He, X.; Feng, X.; Sun, D.; Liu, F.; Bao, Y.; He, Y. Rapid and Nondestructive Measurement of Rice Seed Vitality of Different Years Using Near-Infrared Hyperspectral Imaging. *Molecules* **2019**, *24*, 2227. [CrossRef] [PubMed]
24. Bokuchava, M.A.; Skobeleva, N.I. The biochemistry and technology of tea manufacture. *Crit. Rev. Food Sci. Nutr.* **1980**, *12*, 303–370. [CrossRef]
25. Chun, J.U. Rapid Measure of Color and Catechins Contents in Processed Teas Using NIRS. *Korean J. Plant Resour.* **2010**, *23*, 386–392.
26. Jiang, H.; Yu, F.; Qin, L.I.; Zhang, N.A.; Cao, Q.; Schwab, W.; Daxiang, L.; Song, C. Dynamic change in amino acids, catechins, alkaloids, and gallic acid in six types of tea processed from the same batch of fresh tea (*Camellia sinensis* L.) leaves. *J. Food Compos. Anal.* **2019**, *77*, 28–38. [CrossRef]
27. Zhang, M.; Guo, J.; Ma, C.; Qiu, G.; Ren, J.; Zeng, F.; Lü, E. An effective prediction approach for moisture content of tea leaves based on discrete wavelet transforms and bootstraps of shrinkage algorithm. *Appl. Sci.* **2020**, *10*, 4839. [CrossRef]

foods

MDPI

Article

Decoding Light-Spreading Intensity Effects on the Sensory Quality and Volatile Compounds of Green Tea: An Integrated GC-E-Nose and Targeted Metabolomics Analysis

Qiwei Wang [1,†], Jiajing Hu [1,†], Jiahao Tang [1,2], Xianxiu Zhou [1,3], Haibo Yuan [1], Yongwen Jiang [1], Jialing Xie [4,*] and Yanqin Yang [1,*]

1 National Key Laboratory for Tea Plant Germplasm Innovation and Resource Utilization, Tea Research Institute, Chinese Academy of Agricultural Sciences, Hangzhou 310008, China; wangqiwei@tricaas.com (Q.W.); 15926789897@163.com (J.H.); tangjiahao@tricaas.com (J.T.); zhouxianxiu@tricaas.com (X.Z.); 192168092@tricaas.com (H.Y.); jiangyw@tricaas.com (Y.J.)
2 State Key Laboratory of Tea Plant Biology and Utilization, Anhui Agricultural University, Hefei 230036, China
3 College of Tea Science, Yunnan Agricultural University, Kunming 650201, China
4 Hezhou Agriculture and Rural Affairs Bureau, Hezhou 542800, China
* Correspondence: xiejialing1123@163.com (J.X.); yangyq@tricaas.com (Y.Y.)
† These authors contributed equally to this work.

Abstract: Spreading, the preliminary step in the production of green tea, is crucial for achieving superior tea quality. This study investigated the effects of spreading on the sensory quality and volatile compounds in green tea under varying intensities of yellow light, employing GC-E-Nose and targeted metabolomics. A notable improvement in overall sensory quality was noted in tea samples subjected to a higher intensity of 6000 Lux, which was characterized by a delightful floral fragrance. In total, 70 volatile compounds were successfully identified, with 61 volatiles detected across all five light intensities. Moreover, 21 pivotal odorants featuring odor activity value (OAV) levels higher than one were determined, among which β-ionone, β-damascenone, linalool, (*E*, *Z*)-2,6-nonadienal, and phenylethyl alcohol exhibited particularly high OAVs. Correlation analysis indicated that phenylethyl alcohol, linalool, and citral exhibited robust positive correlations with the majority of key odorants, suggesting their vital contribution towards aroma enhancement. These findings offer novel insights into the regulation of tea aroma through the manipulation of light intensity during the processing of green tea.

Keywords: light intensity; green tea; GC-E-Nose; GC-MS/MS; volatile compounds

1. Introduction

Tea is recognized globally as the second most prevalent beverage after water. Depending on the manufacturing techniques and fermentation degrees, tea is primarily classified into six major categories. Among them, green tea stands as the most abundant tea type in terms of production and consumption in China, gaining widespread acceptance among consumers due to its extensive health benefits and distinctive aromas. Representative fragrances, such as floral aroma and chestnut-like aroma, are highly favored by consumers due to their delightful flavor characteristics [1].

Aroma is a crucial criterion in the assessment of tea quality. Essentially, it involves the intricate interplay of diverse concentrations and types of volatile substances. Currently, there are approximately 700 volatile components identified in tea, with more than

100 volatile components primarily found in fresh leaves [2]. The remaining volatile compounds are predominantly produced through diverse processing methods, suggesting that variations in these techniques can significantly affect the aroma of tea [3]. Spreading is an essential process for producing high-quality green tea. During the leaf-spreading process, progressive water loss induces cytoplasmic volume contraction. Endogenous enzyme activity may increase due to the concentration effect—a decrease in water content leads to higher effective concentrations of both enzymes and substrates, consequently accelerating the rate of enzymatic reactions [4]. This subsequently promotes the degradation of large molecules like chlorophyll, proteins, fatty acids, and polysaccharides in fresh leaves and generates secondary metabolites, which significantly affect the aroma, taste, and color of green tea [5]. The formation of a favorable aroma type in premium green tea is intimately associated with the spreading process [6,7].

Light, as a critical environmental modulator, exerts profound regulatory effects on the metabolic pathways of *Camellia sinensis*. In addition, recent advances have revealed the practical applications of light-mediated interventions in postharvest tea processing. For instance, controlled light exposure during black tea withering has been demonstrated to enhance the aroma profiles by modulating enzymatic activities and gene expression patterns [8,9]. Notably, emerging evidence demonstrates that yellow light irradiation during spreading significantly improves tea quality compared to red light and orange light treatments [10,11]. The superior effect of yellow light may result from its regulation of secondary metabolite biosynthesis related to tea flavor and aroma [8,12]. Nevertheless, to our knowledge, the effects of spreading on the sensory attributes and volatile compounds in green tea under yellow light with different intensities (YLDI) have yet to be thoroughly and comprehensively explored, which hinders the optimization of light-mediated regulating strategy in green tea production.

The traditional assessment method for tea quality predominantly depends upon expert sensory analysis, which is susceptible to environmental factors and exhibits individual variability. As an important complement to conventional evaluation, the electronic nose is conceptualized on the principle of bionics and emulates the human olfactory system through an array of sensors, possessing the advantages of no pretreatments and rapid analysis. However, this technique is susceptible to sensor drift and contamination. In contrast to traditional electronic noses, the gas chromatography-based electronic nose (GC-E-Nose) is an odor analysis technology based on gas chromatography principles, facilitating the separation of odor substances through two distinct polarity chromatographic columns [13]. Owing to its distinguishing features of excellent sensitivity and swift analysis speed, GC-E-Nose is routinely utilized in profiling the volatile fingerprints of tea [14].

Metabolomics, as a powerful analytical technique, facilitates the comprehensive exploration of small molecule metabolites through high-throughput detection and data analysis. This approach provides robust data support for the enhancement and refinement of food flavor. Depending on the research objectives and applications, metabolomics is categorized into targeted metabolomics and untargeted metabolomics. The latter is the prevalent strategy for metabolic profiling, providing comprehensive data for unknown metabolites. Nevertheless, this approach is not suitable for the quantitative evaluation of metabolites, necessitating the incorporation of high-resolution mass spectrometry to facilitate the identification and characterization of unidentified metabolites. Targeted metabolomics, also known as quantitative metabolomics, primarily focuses on specific metabolites or metabolic pathways. Gas chromatography-mass spectrometry (GC-MS) serves as the predominant method for targeted analysis of volatile compounds, particularly under the multiple reaction monitoring (MRM) mode. In other words, the triple quadrupole tandem mass spectrometer (MS/MS) provides enhanced sensitivity and superior specificity by detecting precursor ions and characteristic product ions of metabolites. Given these excellent performances, gas chromatography-tandem mass spectrometer (GC-MS/MS) has been extensively utilized in targeted analysis of heterocyclic compounds such as pyrazines, pyrroles, and furans present in green tea [15,16].

The current investigation thoroughly examined the impact of spreading under YLDI on the sensory quality and volatile compounds of green tea, utilizing a combination of GC-E-Nose and targeted GC-MS/MS. Our results are expected to offer a theoretical foundation for the enhancement of processing technologies, thereby facilitating the production of premium-grade green tea more efficiently.

2. Materials and Methods

2.1. Materials and Reagents

The main materials utilized for this study included headspace vials (20 mL; Agilent Technologies Inc., Palo Alto, CA, USA), a manual solid-phase microextraction (SPME) handle, and a DVB/CAR/PDMS fiber (Bellefonte, PA, USA). The reagent used in this study was termed as purified water (Hangzhou Wahaha Group Co., Ltd., Hangzhou, China). The standard reference compounds employed in this study are detailed in Table S1.

2.2. Preparation of Tea Samples

Young shoots of *Camellia sinensis* cultivar 'Longjing Changye', comprising a single bud and a single leaf, were collected from Shengzhou City in Zhejiang Province. The intricate processing involved several stages, including spreading, fixation, rolling, first drying, and second drying, with the specific processing parameters illustrated in Figure S1. During the spreading stage, tea leaves were exposed to yellow light of varying intensities (500 Lux, 1000 Lux, 2000 Lux, 4000 Lux, and 6000 Lux). The equipment for light spreading used in this study was an artificial climatic chamber (PRX-4500, Ningbo Prant Instrument Co., Ltd., Ningbo, China). To ensure reproducibility, three independent processing batches (including experimental and validation groups) were performed following identical protocols in April, June, and October 2023. The completed tea samples were preserved at a temperature as low as −20 °C for subsequent analysis.

2.3. Sensory Evaluation

The sensory evaluation was in accordance with the Chinese standard, namely, GB/T 23776-2018 [17]. Ethical permission was not deemed necessary for this study. A panel of six experienced evaluators, who provided their consent to participate in the sensory assessment and to utilize their information, assessed the tea samples subjected to YLDI. The evaluation protocol was as follows: Initially, the tea samples were positioned on a tea tray, and the appearance attribute, including color, tightness, and uniformity, was assessed. Subsequently, 3 g of the tea sample were steeped in 150 mL of boiling water for 4 min. The color and taste attributes were assessed by filtering the tea infusion into an assessment bowl. The aroma attribute was evaluated using hot, warm, and cold sniffing to confirm the type (such as chestnut-like aroma, floral aroma, fresh scent, or their complex fragrance types), intensity, and persistence of the aroma. Lastly, the infused leaf was evaluated. The comprehensive quality was determined utilizing a 100-point scale, with the following weightings assigned to each attribute: appearance (25%), infusion color (10%), aroma (25%), taste (30%), and infused leaf (10%).

2.4. GC-E-Nose Analysis

The GC-E-Nose, acquired from Alpha M.O.S. (Toulouse, France), was employed to investigate the volatile profiles of tea samples exposed to YLDI. The operational parameters were subtly adjusted by referring to earlier studies [18]. The specific parameters were as follows: 0.5 g of tea samples were placed into a hermetically sealed headspace vial (20 mL). Then, the sample vial was positioned in the designated tray location for extraction. The incubation temperature and duration of the GC-E-Nose were set to 60 °C and 20 min, respectively. The injection port temperature was adjusted to 200 °C, with an injection volume of 5000 μL. An integrated odor concentrator, Tenax TA trap, was employed to enrich the volatiles at 20 °C, with an adsorption time of 27 s. The temperature profiles of the two distinct columns (MXT-5 and MXT-1701 columns) were initially maintained at 50 °C (for a duration of 5 s), followed by a gradual increase at a rate of 0.1 °C/s until reaching 80 °C, then further increased at a rate of 0.4 °C/s until 250 °C was achieved, after which it was held for 10 s. For the flame ionization detector, both were operated at 260 °C.

2.5. HS-SPME Coupled to GC-MS/MS Analysis

The HS-SPME procedures involved placing tea samples (0.5 g) and purified water (5 mL) within a 20-mL headspace vial. The DVB/CAR/PDMS fiber was used for the enrichment of volatile substances. The recommended conditions for the incubation were at an optimal temperature of 60 °C for 1 h. Following this, the fiber underwent desorption for a period of 5 min, with the injector temperature maintained at 230 °C.

The profiling of volatile compounds involved the application of an Agilent 7890A gas chromatograph integrated with an Agilent 7000C series triple quadrupole system (Agilent Technologies Inc., Palo Alto, CA, USA). The operational parameters for GC-MS/MS were established following the methodology outlined by Hu et al. [19]. A DB-5MS capillary column (30 m × 0.25 mm × 0.25 μm) was utilized for the separation of the volatile compounds. The oven temperature was initially set at 40 °C for a duration of 5 min, followed by a gradual increase at a rate of 4 °C/min until reaching 160 °C, where it was maintained for an additional 5 min. The carrier gas of helium was at a rate of 1.0 mL/min. The collision gas, high-purity nitrogen (99.999%), was supplied at a rate of 1.5 mL/min, with the quench gas of helium at a rate of 2.25 mL/min. A splitless injection mode was employed for this analysis. The operational temperature for the ion source remained steady at 230 °C, concurrently ensuring that the transfer line temperature was held at 270 °C. The mass spectrometer operated in MRM mode, ensuring that the mass resolutions for MS1 and MS2 were optimized to their respective widest and unity settings.

The identification of volatile compounds was conducted through a comparative analysis of the retention time and mass spectra against the corresponding reference standards. The accurate quantification of these volatiles was performed using the external standard method [11]. The calibration curves were developed by preparing a series of calibration solutions with different established gradients. Comprehensive details regarding the calibration information are provided in Table S2.

2.6. Odor Activity Value (OAV) Analysis

OAV is an essential tool for assessing the individual contribution of compounds. It is calculated by utilizing the ratio between the specified concentration of each constituent and its corresponding threshold level in water obtained from the literature [19]. Compounds with an OAV of one or higher are recognized as potential contributors to the overall aroma profile.

2.7. Statistical Analysis

All the experiments were conducted in triplicate, and the data were reported as the mean and standard deviation. Graphic representations, including Venn diagrams and bar charts, were performed utilizing Origin 20 software (OriginLab Corporation, Northampton, MA, USA). Partial least squares discrimination analysis (PLS-DA), metabolic trajectory mapping, and orthogonal PLS-DA (OPLS-DA) were conducted using SIMCA-P 14 software (Umetrics, Umea, Sweden). Statistical significance between different samples was determined using one-way analysis of variance (ANOVA), followed by Tukey's multiple comparison test ($p < 0.05$), conducted with SPSS 20 software (SPSS Inc., Chicago, IL, USA). The hierarchical clustering (HCL) heat maps depicting key aroma compounds and their correlation analysis were generated using the Chiplot online tool (https://www.chiplot.online/#Heatmap accessed on 6 May 2024).

3. Results and Discussion

3.1. Sensory Evaluation Analysis on Tea Samples Exposed to YLDI

The sensory quality of tea samples subjected to YLDI are detailed in Table S3 through systematic sensory analysis. With respect to the taste attribute, the tea samples exposed to 6000 Lux, 1000 Lux, and 4000 Lux attained scores exceeding 90. Notably, the samples under 6000 Lux attained the highest score of 93, exhibiting a fresh taste with floral flavor. Regarding the aroma attribute, all five tea samples scored above 90, further confirming that yellow light significantly improved the aroma profile of green tea. Specifically, the tea samples under 500 Lux and 2000 Lux predominantly displayed a tender chestnut-like aroma, whereas the tea samples under 1000 Lux demonstrated a chestnut-like aroma accompanied by floral notes. The samples exposed to 4000 Lux and 6000 Lux were distinguished by a pronounced floral aroma, with the 6000 Lux samples exhibiting a particularly strong and enduring floral fragrance. According to their aroma types, the tea samples under YLDI could be categorized into three groups: tender chestnut-like aroma (500 Lux and 2000 Lux), floral aroma (4000 Lux and 6000 Lux), and chestnut-like aroma with floral aroma (1000 Lux). In terms of overall quality, the tea samples processed under 6000 Lux achieved the highest score, followed by those under 1000 Lux, 4000 Lux, 2000 Lux, and 500 Lux, respectively. These results indicate that higher light intensities are beneficial for enhancing the sensory quality of green tea, particularly in promoting the development of floral notes.

3.2. Analysis of the Effects of YLDI on the Volatile Fingerprints of Green Tea Using GC-E-Nose

The tea samples subjected to YLDI underwent analysis via GC-E-Nose, and the volatile fingerprints were obtained. Distinctive variations were observed in the aroma profiles of tea samples under YLDI (Figure S2). Notably, the tea samples exposed to 1000 Lux exhibited fewer volatile peaks on the MXT-1701 and MXT-5 columns compared to the other four treatments, indicating a unique aroma profile for the 1000 Lux samples. The volatile fingerprints of the 500 Lux and 2000 Lux samples were similar, with only minor differences in peak intensities, indicating comparable odor profiles. Additionally, the 6000 Lux samples exhibited lower peak intensities on the MXT-1701 column than other samples. The above results suggest that different light intensities directly influence the volatile fingerprints of green tea.

To explore the effects of YLDI on the aroma quality of green tea, a supervised PLS-DA was employed for multivariate statistical analysis. This approach aims to construct a correlation model that correlates sample categories with metabolic expression levels [20,21]. As illustrated in Figure 1A, the tea samples subjected to 500 Lux and 2000 Lux were clustered at the first quadrant of the score plot, whereas those exposed to 4000 Lux and

6000 Lux were grouped at the fourth quadrant. The 1000 Lux samples were clustered separately on the third quadrant, aligning with the findings from sensory assessment. The model parameters were $R^2Y = 0.987$ and $Q^2 = 0.882$, respectively, indicating robust explanatory power and predictive ability. In addition, a permutation test was performed with 200 iterations. The parameters of $R^2 = (0.0, 0.717)$ and $Q^2 = (0.0, -0.465)$ validated the model's reliability without overfitting (Figure 1B). In summary, GC-E-Nose serves as a complementary and validation tool for traditional sensory evaluation and can rapidly and accurately distinguish tea samples by integrating with multivariate statistical analysis.

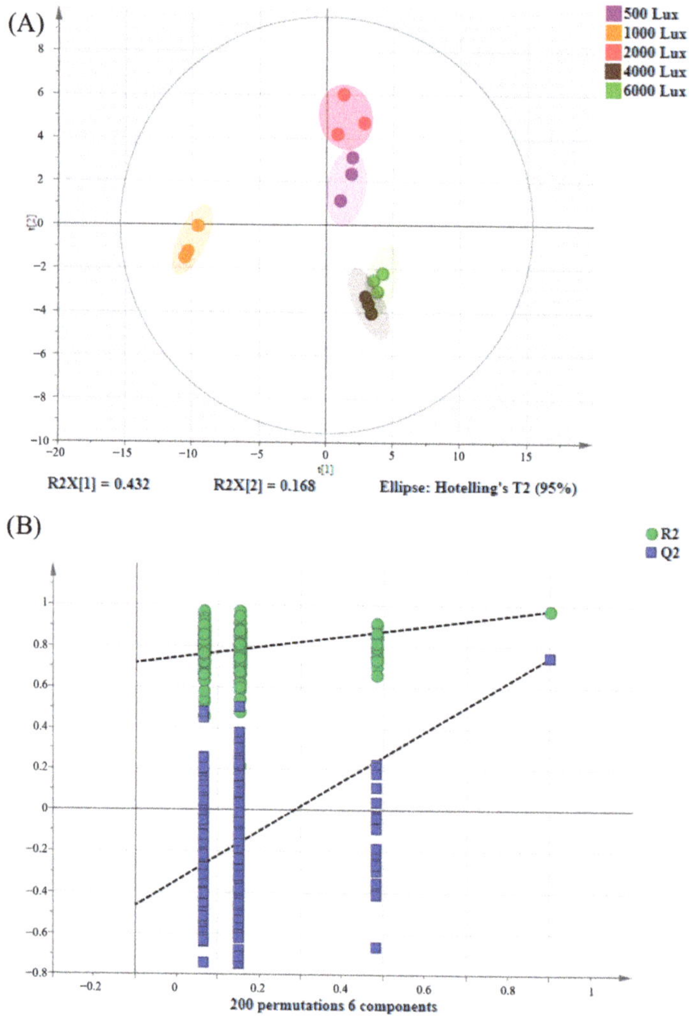

Figure 1. PLS-DA results of tea samples exposed to yellow light with different intensities based on GC-E-Nose. (**A**) Score plots ($R^2Y = 0.987$, $Q^2 = 0.882$); (**B**) Permutation tests with 200 iterations ($R^2 = 0.717$, $Q^2 = -0.465$).

3.3. Analysis of the Effects of YLDI on the Volatile Compounds of Green Tea Using GC-MS/MS

3.3.1. Comparative Analysis of the Volatile Compounds in Tea Samples

In this study, a total of 70 volatile compounds were targeted and analyzed by GC-MS/MS, with detailed quantitative information depicted in Figure 2A and Table S4. Notably, 61 volatile compounds were consistently identified across five different light intensity conditions. In addition, certain volatile substances were detected solely under specific light intensity treatments. For instance, 3-octanone was observed solely in the samples subjected to 4000 Lux and 6000 Lux treatments; (*E*)-2-hexenal was found solely in the samples exposed to 500 Lux and 1000 Lux; *cis*-3-hexenyl salicylate and (3*Z*)-3-hexen-1-yl benzoate were detected solely in samples treated with 1000 Lux.

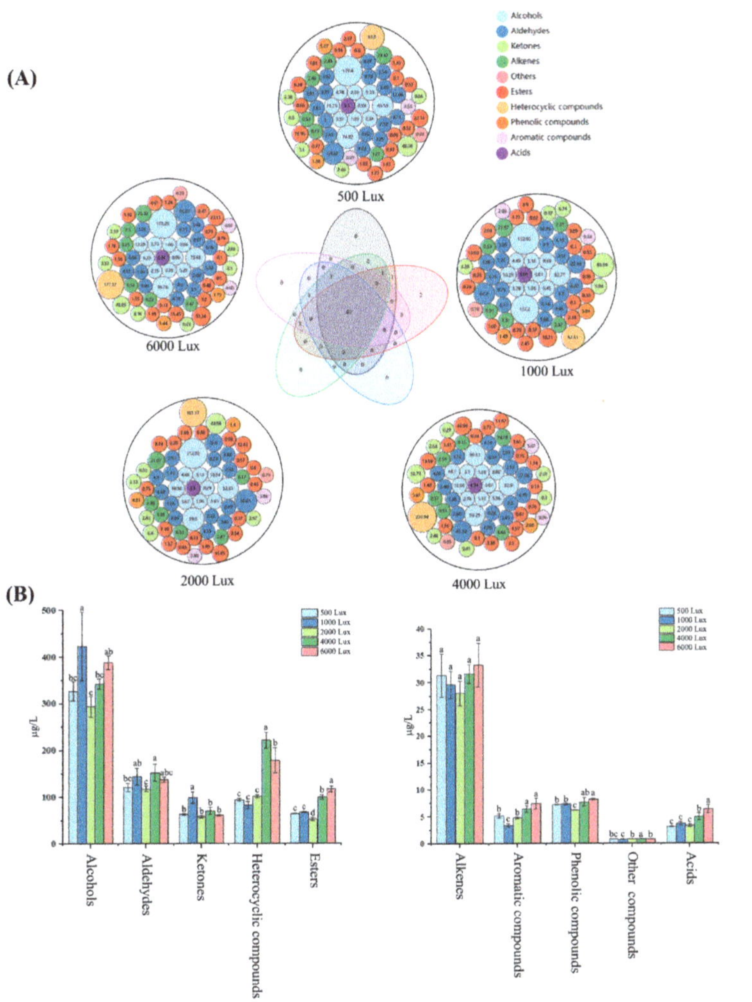

Figure 2. Information of volatile compounds in tea samples exposed to yellow light with different intensities based on GC-MS/MS. (**A**) Volatile compounds in tea samples under different treatments; (**B**) Content comparison of different volatile categories. Different letters indicated significant differences ($p < 0.05$).

The effects of YLDI on different classes of volatile compounds also displayed substantial variations (Figure 2B). Alcohols reached the highest in the samples processed under 1000 Lux, significantly exceeding those under 500 Lux, 2000 Lux, and 4000 Lux ($p < 0.05$). Aldehydes attained the maximum in the tea samples processed under 4000 Lux (152.32 μg/L), followed by the samples processed under 1000 Lux and 6000 Lux, with significantly lower concentrations in the samples processed under 500 Lux and 2000 Lux ($p < 0.05$). Ketones achieved the maximum in the samples subjected to 1000 Lux (99.32 μg/L), which was significantly greater than the levels observed under other light intensities ($p < 0.05$). Heterocyclic compounds were ordered as 4000 Lux > 6000 Lux > 2000 Lux > 500 Lux > 1000 Lux, with the tea samples treated under 4000 Lux significantly higher than other treatments ($p < 0.05$). Esters and Alkenes were the most abundant in tea samples exposed to 6000 Lux. Aromatic hydrocarbons and phenolic compounds were more abundant in the 4000 Lux and 6000 Lux samples, significantly higher than other treatments ($p < 0.05$). Geranic acid exhibited the highest concentration in the 6000 Lux treated sample, significantly higher than other samples ($p < 0.05$). In summary, these results indicate that YLDI significantly influences both the numbers and concentrations of aroma components in green tea, thereby affecting the overall manifestation of the fragrance profile.

3.3.2. Multivariate Statistical Analysis

PLS-DA was employed to further analyze the volatile components obtained from GC-MS. Favorable parameters ($R^2Y = 0.982$ and $Q^2 = 0.896$) were attained, which suggested a robust explanatory capacity and predictive ability. As depicted in Figure 3A, a notable differentiation among the samples was observed, with the distribution patterns of tea samples subjected to five varying light intensities aligning with the findings from the GC-E-Nose analysis. Specifically, the samples exposed to 500 Lux and 2000 Lux were clustered primarily in the upper quadrant of the score plots, while those exposed to 4000 Lux and 6000 Lux were primarily located in the lower left quadrant. Conversely, samples treated with 1000 Lux were situated in the lower right quadrant. Additionally, the reliability of the PLS-DA model was assessed through 200 iterations of permutation testing, which confirmed the absence of overfitting, as depicted in Figure 3B.

To identify the key differential components in tea samples exposed to YLDI, hierarchical cluster analysis was performed. A total of 27 key volatiles were identified based on thresholds of $p < 0.05$ and variable importance in the projection (VIP) > 1. As depicted in Figure 3C, (*E*, *E*)-2,4-nonadienal, butanoic acid, phenylmethyl ester, safranal, and (*E*, *E*)-2,4-decadienal were identified as characteristic volatiles distinguishing the tea samples treated with 500 Lux from those subjected to other light intensities. Citral, nonanal, and hexanol were more abundant in the 1000 Lux samples than other treatments. Nerol exhibited a more pronounced differentiation in the 2000 Lux samples compared to other treatment conditions. Higher levels of theaspirane, (*E*)-2-hexenol, 3-octanone, benzyl alcohol, *D*-limonene, and (*E*)-2-hexenyl butanoate were attained in tea samples exposed to 4000 Lux. The contents of (*E*)-2-nonenal, butanoic acid, hexyl ester, (3*Z*)-3-hexen-1-yl hexanoate, hexyl isovalerate, (*Z*)-3-hexen-1-yl (*Z*)-3-hexenoate, *cis*-3-hexenyl-α-methylbutyrate, and geranic acid displayed elevated levels in the tea samples exposed to 6000 Lux compared to other treatments.

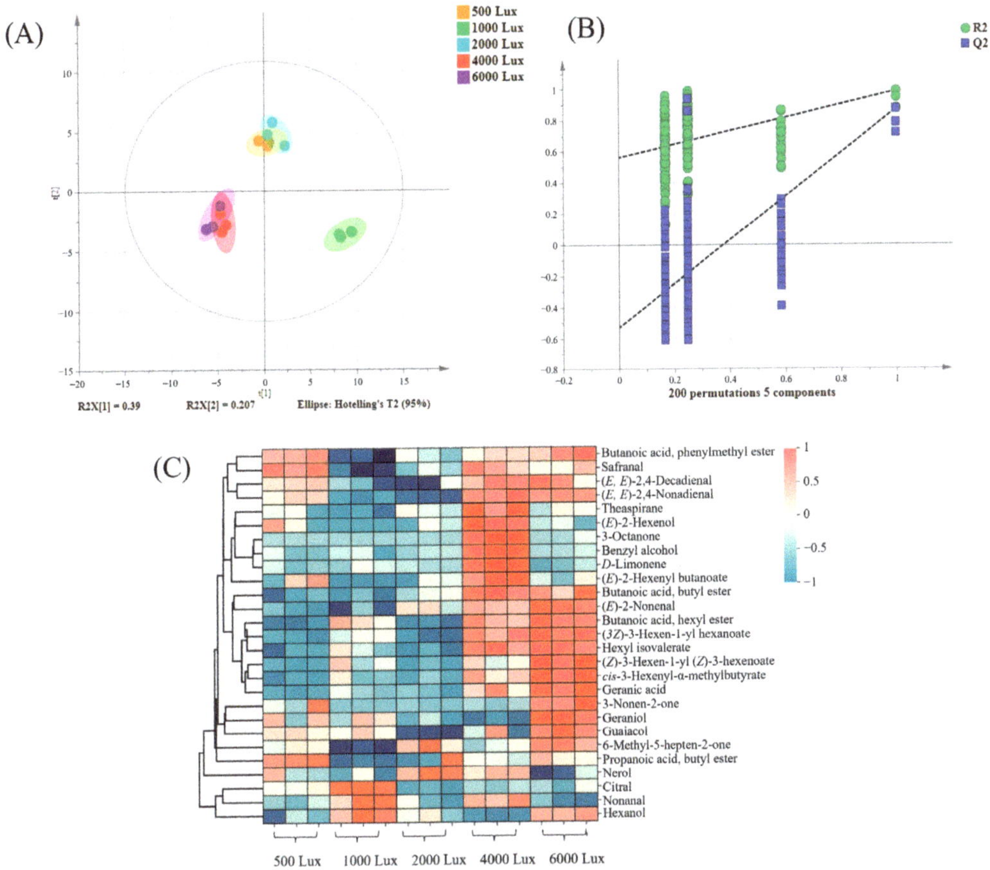

Figure 3. PLS-DA results of volatile compounds in tea samples exposed to yellow light with different intensities based on GC-MS/MS. (**A**) Score plots ($R^2Y = 0.982$, $Q^2 = 0.896$); (**B**) Permutation tests with 200 iterations ($R^2 = 0.562$, $Q^2 = -0.528$); (**C**) Heatmap visualization constructed with VIP > 1.

3.3.3. OAV Analysis

It is universally acknowledged that not all volatile components exert influence on the overall profile of green tea. To screen the pivotal aroma contributors in tea samples exposed to five varying light intensities, OAV analysis was carried out. In general, volatile compounds exhibiting OAV ≥ 1 are deemed as important contributors to the comprehensive aroma, with higher OAV values indicating a stronger contribution [22]. In this investigation, 21 volatile constituents exhibiting OAV ≥ 1 were identified in the tea samples exposed to five distinct light intensities (Figure S3 and Table S5). The top eight volatiles with OAV > 50 in five treatments included (*E*, *Z*)-2,6-nonadienal (OAV: 893.58 ~ 899.69), *β*-damascenone (OAV: 467.90 ~ 840.63), *β*-ionone (OAV: 305.00 ~ 448.11), phenylethyl alcohol (OAV: 170.00 ~ 436.70), linalool (OAV: 88.02 ~ 152.93), (*E*, *E*)-2,4-decadienal (OAV: 92.09 ~ 99.30), (*E*, *E*)-2,4-nonadienal (OAV: 48.81 ~ 66.27), and phenylacetaldehyde (OAV: 50.18 ~ 75.85). Notably, *β*-ionone, (*E*, *E*)-2,4-nonadienal, and (*E*, *E*)-2,4-decadienal exhibited higher OAV values in floral aroma samples exposed to 4000 Lux and 6000 Lux compared to the other three treatments. Additionally, phenylacetaldehyde and linalool attained the highest OAV values (75.85 and 152.93, respectively) under 6000 Lux treatment. These key odorants contributed significantly towards elucidating the floral aroma of green tea, corroborating the findings of Xie et al. [23] and Tian et al. [24]. Interestingly, the OAV of nerolidol was greater than 1 only in tea samples treated with 4000 Lux, while citral exhibited an OAV > 1 exclusively in tea samples treated with 1000 Lux. Additionally, 2-methyl-propanal demonstrated the highest OAV in the tea samples treated with 1000 Lux.

3.3.4. Analysis of the Correlations Between Key Volatile Components and Aroma Quality

To better understand the potential intrinsic relationships between aroma quality and key volatile components, a correlation analysis was performed, with the results illustrated in Figure 4. The aroma quality of tea samples exposed to 1000 Lux showed a significant positive correlation with several volatile compounds, such as 1-octen-3-ol, *β*-damascenone, *cis*-jasmone, phenylethyl alcohol, 2-methyl-propanal, heptanal, decanal, and citral, highlighting their essential influence on the overall aroma quality.

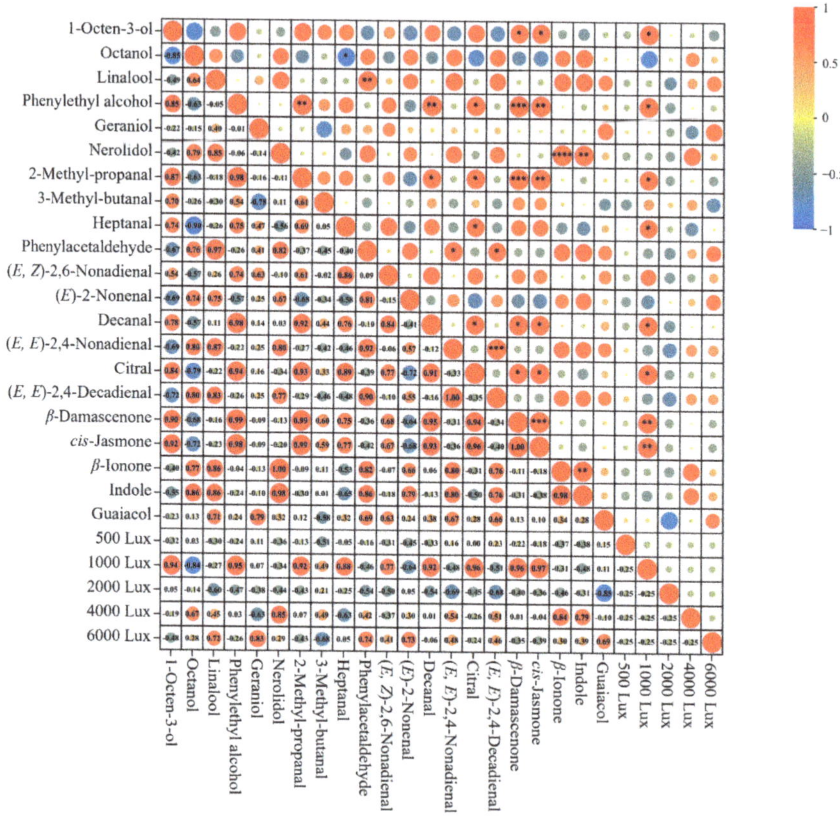

Figure 4. Correlation analysis between 21 key odorants with OAV > 1 and aroma quality of tea samples exposed to yellow light with different intensities. *, $p < 0.05$; **, $p < 0.01$; ***, $p < 0.001$; ****, $p < 0.0001$.

Next, the relationships between these key odorants were further elucidated. Phenylethyl alcohol exhibited strong positive correlations with 2-methyl-propanal (r = 0.98, $p < 0.01$) and decanal (r = 0.98, $p < 0.01$). These compounds generally contribute to the honey, fruity, and floral notes [24,25]. Linalool was significantly correlated with phenylacetaldehyde (r = 0.97,

$p < 0.01$), indicating their combined role in enhancing the sweet, floral, and fruity notes, which were common in high-quality tea aroma profiles [26]. 1-Octen-3-ol was positively correlated with β-damascenone (r = 0.90, $p < 0.05$) and cis-jasmone (r = 0.92, $p < 0.05$), which were pivotal contributors to the floral and fruity aromas [27,28]. The interactions between these ketones and alcohols likely amplify the tea's freshness and complexity [29]. Nerolidol showed a highly significant positive correlation with β-ionone (r = 1.00, $p < 0.001$). Both are derived from carotenoid degradation, and this close relationship points to their synergistic role in contributing to the fruity and violet-like notes of tea [30–32]. Citral was significantly positively correlated with decanal (r = 0.91, $p < 0.05$), 2-methyl-propanal (r = 0.93, $p < 0.05$), and heptanal (r = 0.89, $p < 0.05$). These aldehydes are recognized for their contributions to the citrus and fruity notes, which are commonly found in green tea [24,25,31]. Additionally, a highly significant positive correlation was observed between *(E, E)*-2,4-nonadienal and *(E, E)*-2,4-decadiena ($p < 0.001$), with a correlation coefficient of $r = 1.00$. These two aldehydes, derived from fatty acids, are recognized for imparting fatty and green notes. *Cis*-Jasmone and β-damascenone are pivotal ketones influencing tea aroma [33]. *Cis*-Jasmone exhibited an extremely strong correlation with β-damascenone (r = 1.00, $p < 0.05$). Both *cis*-jasmone and β-damascenone were found to be significantly positively correlated with 2-methyl-propanal, decanal, and citral. It is hypothesized that these ketones and aldehydes function synergistically to enhance the complexity and richness of tea aroma.

In summary, key volatiles such as 2-methyl-propanal, phenylethyl alcohol, heptanal, decanal, citral, *cis*-jasmone, and β-damascenone displayed significant positive correlations with multiple pivotal odorants. These robust interactions indicate their significant contribution to the aroma development of tea samples under YLDI. Further exploration of the molecular interactions involved will enhance our understanding of the mechanisms underlying aroma formation.

3.3.5. Analysis of the Metabolic Pathways of Key Volatile Compounds

By the metabolic pathways involved, the 21 key volatile components discussed above were subcategorized into four classes, as illustrated in Figure S4. According to Xie et al. [23], these comprised three carotenoid-derived volatiles (CDVs), nine fatty acid-derived volatiles (FADVs), six amino acid-derived volatiles (AADVs), and three glycoside-derived volatiles (GDVs).

FADVs are primarily derived from the oxidative degradation of unsaturated fatty acids via the lipoxygenase pathway (Figure 5). The common precursor substances in tea mainly include α-linolenic acid, palmitoleic acid, and linoleic acid [32]. As a crucial odorant in green tea [34], 1-octen-3-ol attained its maximum of 4.45 μg/L under 1000 Lux, then declined as the light intensity increased, mirroring the findings by He et al. [35]. As shown in Figure 6A, the trend for (*E*)-2-nonenal was opposite, with the lowest concentration (4.15 μg/L) under 1000 Lux. As previously highlighted, (*E, E*)-2,4-nonadienal and (*E, E*)-2,4-decadienal demonstrated a high correlation, both exhibiting a trend of initial decrease followed by an increase under different light intensities. *Cis*-jasmone, decanal, heptanal, and (*E, Z*)-2,6-nonadienal attained their highest concentrations at 1000 Lux, whereas octanol peaked at 4000 Lux.

Geraniol and linalool serve as typical GDVs, synthesized via the intermediary geranyl pyrophosphate (geranyl-PP). These enzymes, namely geraniol synthase and linalool synthase, respectively, catalyze this liberation process (Figure 5). As important aroma compounds in black tea, linalool and geraniol impart sweet, floral, fruity, rose-like, and waxy notes [23]. As shown in Figure 6B, linalool and geraniol reached the lowest concentrations under 2000 Lux (52.81 μg/L) and 4000 Lux (99.13 μg/L), respectively, and attained their highest concentrations both under 6000 Lux (91.76 μg/L and 179.25 μg/L, respectively). Citral attained its highest concentration under 1000 Lux (6.02 μg/L), noticeably higher than in other light intensities [36,37]. This divergent pattern may indicate variations in the biosynthetic regulation of these compounds. Specifically, citral is synthesized through the oxidation of geraniol and may be subject to downregulation or degradation in conditions of elevated light intensity. Conversely, linalool and geraniol appear to directly benefit from light-induced metabolic flux within the methylerythritol phosphate (MEP) pathway. Light can induce the up-regulation of the expression of MEP pathway genes, and this evidence conclusively corroborates the light-intensity-dependent metabolic channeling mechanism governing linalool and geraniol biosynthesis [35].

The AADVs, such as 3-methyl-butanal, phenylacetaldehyde, and 2-methyl-propanal, are typical Strecker aldehydes (Figure 5). They are generated via the Strecker degradation pathway, utilizing their respective amino acids of leucine, phenylalanine, and valine as precursors [32]. These AADVs exhibited distinct responses to light intensity. 2-Methyl-propanal attained its maximum (10.96 μg/L) under 1000 Lux treatment; Phenylacetaldehyde exhibited a V-shaped trend, with the highest concentration (91.03 μg/L) under 6000 Lux treatment; 3-Methyl-butanal peaked under 1000 Lux treatment, with a concentration of 27.16 μg/L (Figure 6C). Indole, another AADV, is generated from tryptophan via the action of tryptophan indole-lyase. It was reported as a pivotal floral aroma constituent in jasmine tea [32,38]. Guaiacol was reported to be derived from phenylalanine [39]. Both indole and guaiacol exhibited upward trends with the increasing of light intensity, attaining peak concentrations at 4000 Lux (220.94 μg/L) and 6000 Lux (6.44 μg/L), respectively. It is speculated that high intensity of yellow light promotes the metabolism of phenylalanine-guaiacol and tryptophan-indole [7].

CDVs such as β-ionone, nerolidol, and β-damascenone primarily originate from the oxidative degradation of carotenoids (Figure 5). Specifically, β-damascenone is predominantly produced through the degradation of neoxanthin, whereas β-ionone is generated via the oxidation of β-carotene. Nerolidol is derived from the photo-oxidation of phytofluene and is synthesized via the nerolidol synthase gene [32,40]. In our study, these compounds exhibited divergent responses to light intensity, with variations observed across different treatments. Nerolidol and β-ionone attained their peaks under 4000 Lux treatment (Figure 6D). Ni et al. [40] have found that increasing light intensity can enhance the formation of CDVs, which corroborates our findings. The interconversion of zeaxanthin, antheraxanthin, and violaxanthin, known as the xanthophyll cycle, is regulated by light. Zeaxanthin can be catalyzed into CDVs through the action of carotenoid cleavage dioxygenases. Stronger light intensity promotes the accumulation of zeaxanthin, potentially explaining the highest concentrations of nerolidol and β-ionone under 4000 Lux treatment. In contrast, β-damascenone peaked at 1000 Lux and decreased markedly at higher intensities, which may be attributed to its susceptibility to photo-oxidative degradation. These results imply that although stronger light intensity can enhance the formation of certain CDVs, it may simultaneously accelerate the loss of others, indicating that individual CDVs respond differently to light intensity.

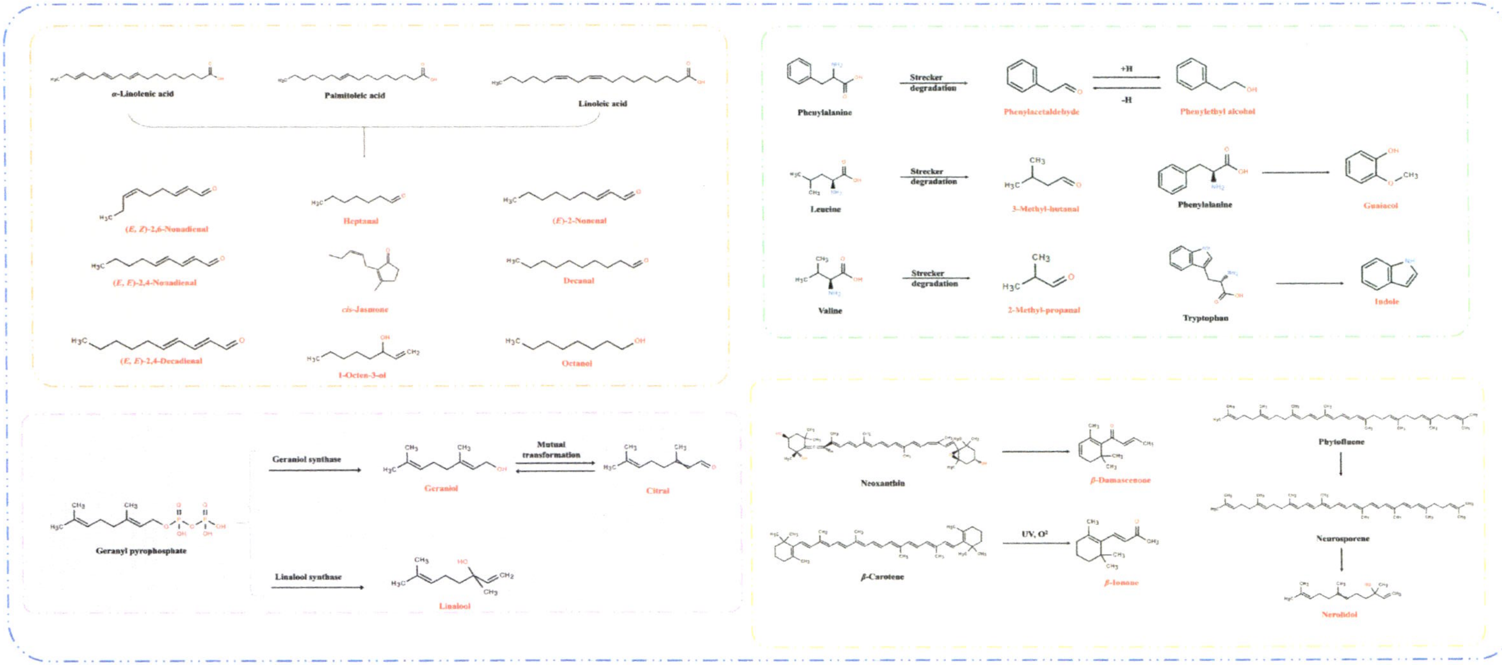

Figure 5. Possible formation pathways of 21 important volatile components with OAV > 1 in tea samples exposed to yellow light with different intensities.

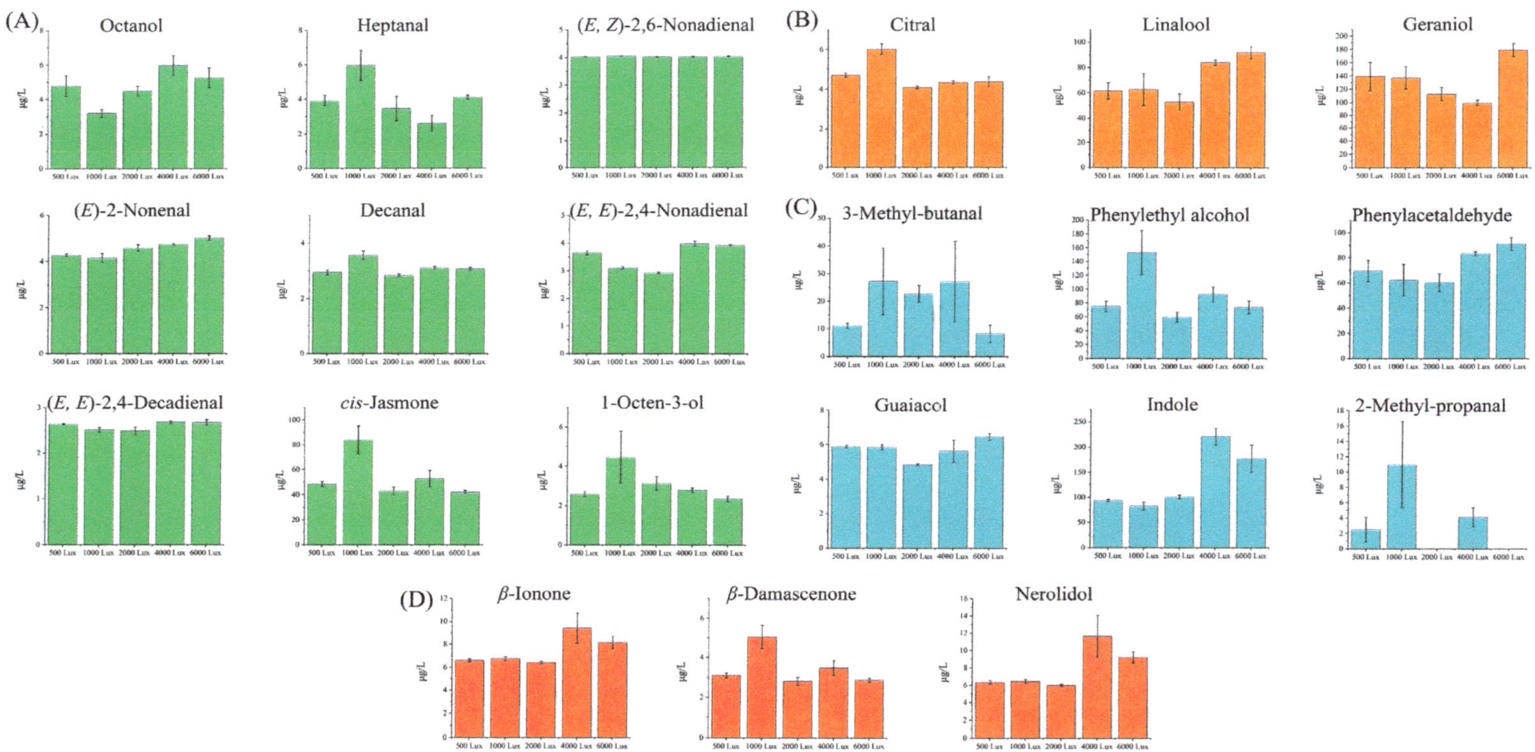

Figure 6. Dynamics changes of key differential volatile compounds in tea samples exposed to yellow light with different intensities. (**A**) Fatty acid-derived volatiles; (**B**) Glycoside-derived volatiles; (**C**) Amino acid-derived volatiles; (**D**) Carotenoid-derived volatiles.

4. Conclusions

This study demonstrated that the aroma quality and volatile compounds of green tea were significantly influenced by spreading under exposure to YLDI. Higher light intensities, particularly 6000 Lux, were found to enhance the overall sensory quality, with a pleasant floral aroma. Substantial variations in volatile profiles were observed in green tea exposed to YLDI via GC-E-Nose and targeted GC-MS/MS analysis. Totally, 70 volatile components were identified, with alcohols higher under 1000 Lux, aldehydes higher under 4000 Lux, and esters and alkenes higher at 6000 Lux. OAV analysis further highlighted the critical role of volatile compounds like citral, β-damascenone, and phenylethyl alcohol in contributing to the floral and fruity aromas, particularly in tea samples under 4000 and 6000 Lux treatments. The metabolic pathways of key aroma volatiles were also investigated, and the results demonstrated that higher light intensities facilitated the accumulation of critical aroma compounds, notably linalool and geraniol. In addition, key volatiles such as 2-methyl-propanal, phenylethyl alcohol, heptanal, decanal, citral, *cis*-jasmone, and β-damascenone exhibited significant positive correlations with multiple pivotal odorants, suggesting their critical role in the aroma development of tea samples under YLDI. The results provide invaluable insights for optimizing conditions to enhance the quality of green tea through light manipulation. In the future, genomics, proteomics, and other advanced techniques will be employed to elucidate the intrinsic formation mechanism of light manipulation.

Supplementary Materials: The following supporting information can be downloaded at: https://www.mdpi.com/article/10.3390/foods14081313/s1, Figure S1: The processing processes of green tea under yellow light with different intensities; Figure S2: Volatile fingerprints on MXT-5 and MXT-1701 columns obtained from GC-E-Nose analysis; Figure S3: OAV comparison of key volatile compounds in tea samples under yellow light with different intensities; Figure S4: Classification of 21 key volatile compounds based on their pathways; Table S1: Relevant standard information of volatile compounds; Table S2: The quantification information of volatile components in this study; Table S3: Results of sensory evaluation of tea samples under yellow light with different intensities; Table S4: The contents of volatile compounds in green tea samples under yellow light with different intensities; Table S5: The OAVs of volatile compounds in green tea samples under yellow light with different intensities. References [2,23,31,41] are cited in the Supplementary Materials.

Author Contributions: Conceptualization, J.X. and Y.Y.; methodology, Q.W.; software, J.T.; validation, Q.W., J.H. and X.Z.; formal analysis, Q.W., J.H. and J.T.; investigation, J.X. and Y.Y.; resources, Y.Y., Y.J. and H.Y.; data curation, J.X.; writing—original draft preparation, Q.W. and J.H.; writing—review and editing, J.X. and Y.Y.; visualization, X.Z.; supervision, Y.J.; project administration, H.Y.; funding acquisition, H.Y. All authors have read and agreed to the published version of the manuscript.

Funding: This research was funded by the Science and Technology Innovation Project of the Chinese Academy of Agricultural Sciences (CAAS-ASTIP-TRICAAS).

Data Availability Statement: The original contributions presented in this study are included in the article/Supplementary Materials. Further inquiries can be directed to the corresponding authors.

Conflicts of Interest: The authors declare no conflicts of interest.

References

1. Yin, P.; Kong, Y.-S.; Liu, P.-P.; Wang, J.-J.; Zhu, Y.; Wang, G.-M.; Sun, M.-F.; Chen, Y.; Guo, G.-Y.; Liu, Z.-H. A Critical Review of Key Odorants in Green Tea: Identification and Biochemical Formation Pathway. *Trends Food Sci. Technol.* **2022**, *129*, 221–232. [CrossRef]
2. Zhai, X.; Zhang, L.; Granvogl, M.; Ho, C.; Wan, X. Flavor of Tea (*Camellia sinensis*): A Review on Odorants and Analytical Techniques. *Compr. Rev. Food Sci. Food Saf.* **2022**, *21*, 3867–3909. [CrossRef]

3. Chen, M.; Guo, L.; Zhou, H.; Guo, Y.; Zhang, Y.; Lin, Z.; Sun, M.; Zeng, W.; Wu, H. Absolute Quantitative Volatile Measurement from Fresh Tea Leaves and the Derived Teas Revealed Contributions of Postharvest Synthesis of Endogenous Volatiles for the Aroma Quality of Made Teas. *Appl. Sci.* **2021**, *11*, 613. [CrossRef]
4. Hou, P.; Wang, F.; Luo, B.; Li, A.; Wang, C.; Shabala, L.; Ahmed, H.A.I.; Deng, S.; Zhang, H.; Song, P.; et al. Antioxidant Enzymatic Activity and Osmotic Adjustment as Components of the Drought Tolerance Mechanism in Carex Duriuscula. *Plants* **2021**, *10*, 436. [CrossRef]
5. Yu, X.; Li, Y.; He, C.; Zhou, J.; Chen, Y.; Yu, Z.; Wang, P.; Ni, D. Nonvolatile Metabolism in Postharvest Tea (*Camellia sinensis* L.) Leaves: Effects of Different Withering Treatments on Nonvolatile Metabolites, Gene Expression Levels, and Enzyme Activity. *Food Chem.* **2020**, *327*, 126992. [CrossRef]
6. Li, Y.; He, C.; Yu, X.; Zhou, J.; Ntezimana, B.; Yu, Z.; Chen, Y.; Ni, D. Study on Improving Aroma Quality of Summer-Autumn Black Tea by Red-Light Irradiation during Withering. *LWT* **2022**, *154*, 112597. [CrossRef]
7. Yu, X.; He, C.; Li, Y.; Zhou, J.; Chen, Y.; Yu, Z.; Wang, Y.; Ni, D. Effects of Different Spreading Treatments on the Formation of Aroma Quality in Green Tea. *Beverage Plant Res.* **2021**, *1*, 14. [CrossRef]
8. Ai, Z.; Zhang, B.; Chen, Y.; Yu, Z.; Chen, H.; Ni, D. Impact of Light Irradiation on Black Tea Quality during Withering. *J. Food Sci. Technol.* **2017**, *54*, 1212–1227. [CrossRef]
9. Xiang, L.; Lin, F.; Sun, W.; Wu, Y.; Chen, M.; Zhang, Q.; Zhou, L.; Weng, R. Effects of LED Yellow Light on the Expression Levels of Aroma Related Genes and the Enzyme Activity in Withering Process of Congou Black Tea. *J. Tea Sci.* **2015**, *35*, 559–566.
10. Fang, Z.-T.; Jin, J.; Ye, Y.; He, W.-Z.; Shu, Z.-F.; Shao, J.-N.; Fu, Z.-S.; Lu, J.-L.; Ye, J.-H. Effects of Different Shading Treatments on the Biomass and Transcriptome Profiles of Tea Leaves (*Camellia sinensis* L.) and the Regulatory Effect on Phytohormone Biosynthesis. *Front. Plant Sci.* **2022**, *13*, 909765. [CrossRef]
11. Xie, J.; Wang, Q.; Hu, J.; Wang, L.; Yu, X.; Yuan, H.; Jiang, Y.; Yang, Y. Uncovering the Effects of Spreading under Different Light Irradiation on the Volatile and Non-Volatile Metabolites of Green Tea by Intelligent Sensory Technologies Integrated with Targeted and Non-Targeted Metabolomics Analyses. *Food Chem.* **2025**, *463*, 141482. [CrossRef]
12. Tian, S.; Zhou, H.; Yao, X.; Lu, L. Finding the Optimal Light Quality and Intensity for the Withering Process of Fuding Dabai Tea and Its Impact on Quality Formation. *LWT* **2024**, *193*, 115713. [CrossRef]
13. Rong, Y.; Xie, J.; Yuan, H.; Wang, L.; Liu, F.; Deng, Y.; Jiang, Y.; Yang, Y. Characterization of Volatile Metabolites in Pu-Erh Teas with Different Storage Years by Combining GC-E-Nose, GC–MS, and GC-IMS. *Food Chem. X* **2023**, *18*, 100693. [CrossRef]
14. Yang, Y.; Qian, M.C.; Deng, Y.; Yuan, H.; Jiang, Y. Insight into Aroma Dynamic Changes during the Whole Manufacturing Process of Chestnut-like Aroma Green Tea by Combining GC-E-Nose, GC-IMS, and GC × GC-TOFMS. *Food Chem.* **2022**, *387*, 132813. [CrossRef]
15. Yang, Y.; Zhang, M.; Hua, J.; Deng, Y.; Jiang, Y.; Li, J.; Wang, J.; Yuan, H.; Dong, C. Quantitation of Pyrazines in Roasted Green Tea by Infrared-Assisted Extraction Coupled to Headspace Solid-Phase Microextraction in Combination with GC-QqQ-MS/MS. *Food Res. Int.* **2020**, *134*, 109167. [CrossRef]
16. Yang, Y.; Xie, J.; Chen, J.; Deng, Y.; Shen, S.; Hua, J.; Wang, J.; Zhu, J.; Yuan, H.; Jiang, Y. Characterization of N,O-Heterocycles in Green Tea during the Drying Process and Unraveling the Formation Mechanism. *Food Control* **2022**, *139*, 109079. [CrossRef]
17. *GB/T 23776-2018*; Methodology for Sensory Evaluation of Tea. Chinese National Standard: Beijing, China, 2018.
18. Wang, L.; Yang, Y.; Xie, J.; Miao, Y.; Wang, Q.; Jiang, Y.; Deng, Y.; Tong, H.; Yuan, H. Aroma Quality Evaluation of High-Quality and Quality-Deficient Black Tea by Electronic Nose Coupled with Gas Chromatography-Mass Spectrometry. *Food Sci.* **2024**, *45*, 274–282. [CrossRef]
19. Hu, J.; Xie, J.; Wang, Q.; Tang, J.; Zhou, X.; Yuan, H.; Jiang, Y.; Yang, Y. Unraveling the Dynamic Variations of Volatile and Non-Volatile Metabolites in Green Tea during the Yellow-Light Irradiation Spreading Process by Targeted and Untargeted Metabolomics. *LWT* **2025**, *215*, 117152. [CrossRef]
20. Wang, Q.; Shi, D.; Hu, J.; Wang, L.; Xie, J.; Jiang, Y.; Yuan, H.; Yang, Y. Unraveling the Dynamic Changes of Volatile Compounds during the Rolling Process of Congou Black Tea via GC-E-Nose and GC-MS. *Front. Sustain. Food Syst.* **2024**, *8*, 1436542. [CrossRef]
21. Li, Q.; Zhang, C.; Wang, H.; Chen, S.; Liu, W.; Li, Y.; Li, J. Machine Learning Technique Combined with Data Fusion Strategies: A Tea Grade Discrimination Platform. *Ind. Crops Prod.* **2023**, *203*, 117127. [CrossRef]
22. Wang, Q.; Xie, J.; Wang, L.; Jiang, Y.; Deng, Y.; Zhu, J.; Yuan, H.; Yang, Y. Comprehensive Investigation on the Dynamic Changes of Volatile Metabolites in Fresh Scent Green Tea during Processing by GC-E-Nose, GC–MS, and GC × GC-TOFMS. *Food Res. Int.* **2024**, *187*, 114330. [CrossRef] [PubMed]
23. Xie, J.; Wang, L.; Deng, Y.; Yuan, H.; Zhu, J.; Jiang, Y.; Yang, Y. Characterization of the Key Odorants in Floral Aroma Green Tea Based on GC-E-Nose, GC-IMS, GC-MS and Aroma Recombination and Investigation of the Dynamic Changes and Aroma Formation during Processing. *Food Chem.* **2023**, *427*, 136641. [CrossRef]

24. Wang, J.; Ouyang, W.; Zhu, X.; Jiang, Y.; Yu, Y.; Chen, M.; Yuan, H.; Hua, J. Effect of Shaking on the Improvement of Aroma Quality and Transformation of Volatile Metabolites in Black Tea. *Food Chem. X* **2023**, *20*, 101007. [CrossRef] [PubMed]
25. Yin, X.; Xiao, Y.; Wang, K.; Wu, W.; Huang, J.; Liu, S.; Zhang, S. Effect of Shaking Manners on Floral Aroma Quality and Identification of Key Floral-Aroma-Active Compounds in Hunan Black Tea. *Food Res. Int.* **2023**, *174*, 113515. [CrossRef]
26. Lu, M.; Sheng, C.; Ke, H.; Li, T.; Liu, Q.; Zhang, J.; Li, L.; Wang, Y.; Ning, J. Revealing the Differences in Aroma of Black Tea under Different Drying Methods Based on GC–MS, GC-O. *Food Chem. X* **2024**, *23*, 101782. [CrossRef]
27. Wu, Y.; Li, T.; Huang, W.; Liu, Q.; Deng, G.; Zhang, J.; Wei, Y.; Wang, Y.; Ning, J. Investigation of the Aroma Profile and Blending Strategy of Lu'an Guapian Teas during Grain Rain Period by Sensory Evaluation Combined with SBSE-GC–MS, GC–O and OAV. *Food Chem.* **2025**, *463*, 141167. [CrossRef]
28. Zhao, R.; Yao, H.; Hou, Z.; Zhou, Q.; Zhao, M.; Wu, C.; Zhang, L.; Xu, C.; Su, H. Sensomics-Assisted Analysis Unravels the Formation of the Fungus Aroma of Fu Brick Tea. *Food Chem.* **2024**, *458*, 140174. [CrossRef] [PubMed]
29. Lin, Y.; Huang, Y.; Zhou, S.; Li, X.; Tao, Y.; Pan, Y.; Feng, X.; Guo, H.; Chen, P.; Chu, Q. A Newly-Discovered Tea Population Variety Processed Bai Mu Dan White Tea: Flavor Characteristics and Chemical Basis. *Food Chem.* **2024**, *446*, 138851. [CrossRef]
30. Guo, X.; Ho, C.-T.; Schwab, W.; Wan, X. Effect of the Roasting Degree on Flavor Quality of Large-Leaf Yellow Tea. *Food Chem.* **2021**, *347*, 129016. [CrossRef]
31. Guo, X.; Ho, C.-T.; Wan, X.; Zhu, H.; Liu, Q.; Wen, Z. Changes of Volatile Compounds and Odor Profiles in Wuyi Rock Tea during Processing. *Food Chem.* **2021**, *341*, 128230. [CrossRef]
32. Ho, C.-T.; Zheng, X.; Li, S. Tea Aroma Formation. *Food Sci. Hum. Wellness* **2015**, *4*, 9–27. [CrossRef]
33. Jiang, Q.; Li, L.; Chen, F.; Rong, B.; Ni, H.; Zheng, F. β-Glucosidase Improve the Aroma of the Tea Infusion Made from a Spray-Dried Oolong Tea Instant. *LWT* **2022**, *159*, 113175. [CrossRef]
34. Yang, Y.; Zhang, M.; Yin, H.; Deng, Y.; Jiang, Y.; Yuan, H.; Dong, C.; Li, J.; Hua, J.; Wang, J. Rapid Profiling of Volatile Compounds in Green Teas Using Micro-Chamber/Thermal Extractor Combined with Thermal Desorption Coupled to Gas Chromatography-Mass Spectrometry Followed by Multivariate Statistical Analysis. *LWT* **2018**, *96*, 42–50. [CrossRef]
35. He, Y.; Li, J.; Mei, H.; Zhuang, J.; Zhao, Z.; Jeyaraj, A.; Wang, Y.; Chen, X.; Li, X.; Liu, Z. Effects of Leaf-Spreading on the Volatile Aroma Components of Green Tea under Red Light of Different Intensities. *Food Res. Int.* **2023**, *168*, 112759. [CrossRef] [PubMed]
36. Chen, J.; Yang, Y.; Deng, Y.; Liu, Z.; Shen, S.; Zhu, J.; Yuan, H.; Jiang, Y. Characterization of the Key Differential Volatile Components in Different Grades of Dianhong Congou Tea Infusions by the Combination of Sensory Evaluation, Comprehensive Two-Dimensional Gas Chromatography-Time-of-Flight Mass Spectrometry, and Odor Activity Value. *LWT* **2022**, *165*, 113755. [CrossRef]
37. Yang, Y.; Zhu, H.; Chen, J.; Xie, J.; Shen, S.; Deng, Y.; Zhu, J.; Yuan, H.; Jiang, Y. Characterization of the Key Aroma Compounds in Black Teas with Different Aroma Types by Using Gas Chromatography Electronic Nose, Gas Chromatography-Ion Mobility Spectrometry, and Odor Activity Value Analysis. *LWT* **2022**, *163*, 113492. [CrossRef]
38. Xie, J.; Wang, Q.; Cui, H.; Wang, L.; Deng, Y.; Yuan, H.; Zhu, J.; Yang, Y.; Jiang, Y. Characterization of Gardenia Tea Based on Aroma Profiles Using GC-E-Nose, GC-O-MS and GC × GC-TOFMS Combined with Chemometrics. *Beverage Plant Res.* **2024**, *4*, e001. [CrossRef]
39. Rambla, J.L.; Tikunov, Y.M.; Monforte, A.J.; Bovy, A.G.; Granell, A. The Expanded Tomato Fruit Volatile Landscape. *J. Exp. Bot.* **2013**, *65*, 4613–4623. [CrossRef]
40. Ni, Z.; Yang, Y.; Zhang, Y.; Hu, Q.; Lin, J.; Lin, H.; Hao, Z.; Wang, Y.; Zhou, J.; Sun, Y. Dynamic Change of the Carotenoid Metabolic Pathway Profile during Oolong Tea Processing with Supplementary LED Light. *Food Res. Int.* **2023**, *169*, 112839. [CrossRef]
41. Van Gemert, L.J. *ODOUR THRESHOLDS—Compilations of Odour Threshold Values in Air, Water and Other Media*; Oliemans Punter& Partners BV: Zeist, The Netherlands, 2011.

foods

MDPI

Article

Flavor Characteristics of Sun-Dried Green Tea in Different Regions of Yunnan: Metabolite Basis and Soil Influencing Factors

Miao Zhou [1,†], Xiujuan Deng [1,†], Qiaomei Wang [1], Zhenzhen Wei [2], Xinhua Wang [1], Wenxia Yuan [1], Limei Li [1], Man Zou [1], Weihao Liu [1], Shijie Lu [1], Yubo Sheng [2,*] and Baijuan Wang [1,*]

[1] College of Tea Science, Yunnan Agricultural University, Kunming 650201, China; 18724979985@163.com (M.Z.); xiujuandeng@foxmail.com (X.D.); wqm19850127@163.com (Q.W.); wangaoyu2011@163.com (X.W.); yuanwenxia2023@163.com (W.Y.); 18214178910@163.com (L.L.); 15770383235@163.com (M.Z.); liuweihao2210@163.com (W.L.); 18087238940@163.com (S.L.)

[2] China Tea (Yunnan) Co., Ltd., Kunming 650217, China; weizhenzhen@cofco.com

* Correspondence: shengyubo@cofco.com (Y.S.); wangbaijuan2023@163.com (B.W.); Tel.: +86-18082708495 (Y.S.); +86-13888976094 (B.W.)

† These authors contributed equally to this work.

Abstract: To elucidate the regional flavor characteristics of sun-dried green tea (SDT) and their underlying influencing factors, a comprehensive analysis was conducted using metabolomics and flavoromics approaches. This study systematically examined SDT samples and their corresponding tea garden soils from 13 distinct regions in Yunnan Province. The results revealed that the SDT samples could be classified into two distinct groups based on their flavor profiles. Compared to the regions of Pa Sha (PS), Bang Dong (BD), Dong Ban Shan (DBS), Dong Guo (DG), Su Hu (SH), Gua Feng Zhai (GFZ), and Wu Liang Shan (WLS), the regions of Xin Nong (XN), Ba Ka Nuan (BKN), Mang Ang (MA), Man Nuan (MN), Bing Dao (BDao), and Bin Shan (BS) exhibited a significant upregulation of the tea polyphenols (TP)/free amino acids (FAA) ratio. The former group was characterized by a sweet mellow taste, while the latter displayed a stronger taste profile. Furthermore, the analysis of volatile compounds demonstrated that geraniol and linalool were significantly upregulated in the PS, BD, DBS, DG, BS, and BDao regions, which were associated with tender and floral aromas. In contrast, isophorone, 2-pentyl furan, 1-octanol, D-limonene, and benzaldehyde were markedly enriched in the XN, BKN, MA, MN, SH, GFZ, and WLS regions, contributing to sweet and honey-like aromatic profiles. Altitude and mineral element phosphorus are potential key factors affecting the regional flavor differences in SDT. Specifically, SDT cultivated at higher altitudes and in soils with elevated available phosphorus content exhibited a greater likelihood of accumulating sweet mellow and floral compounds. This study provides scientific evidence for understanding the characteristic flavor profiles of SDT across different regions, offering valuable insights into the factors contributing to regional flavor differentiation in tea production.

Keywords: sun-dried green tea; flavor characteristic; different regions; tea garden soil

1. Introduction

Yunnan's favorable geography and climate have given rise to Pu-erh tea [1]. In recent years, Pu-erh tea has become increasingly popular with consumers because of its unique health benefits and distinctive flavor [2]. As the foundation of Pu-erh tea's characteristic flavor, the Yunnan large-leaf variety of SDT plays a crucial role in determining the final product quality. Therefore, comprehensive research on the terroir-specific quality attributes of the Yunnan large-leaf variety of SDT is essential for establishing effective quality control measures throughout Pu-erh tea production processes. This investigation is particularly critical for maintaining the unique regional characteristics and ensuring consistent quality in Pu-erh tea manufacturing.

Volatile organic compounds (VOCs) constitute essential components that significantly contribute to the complex flavor profile of SDT. Extensive research has identified alcohols, ketones, and hydrocarbons as the predominant classes of volatile compounds in SDT [3,4]. The samples from each producing area had their own characteristic volatile components [5], such as different townships in Menghai County, where the SDT aroma is dominated by fruity, floral, and woody flavor profiles [6]. Advanced analytical techniques, including headspace solid-phase microextraction-gas chromatography-mass spectrometry (HS-SPME-GC-MS), gas chromatography-ion mobility spectrometry (GC-IMS), and electronic nose analysis, have proven effective in discriminating SDT aroma profiles across different Yunnan tea-growing regions [4,5,7]. However, the specific key flavor compounds underlying these regional variations remain to be fully elucidated.

Research on SDT has revealed significant differences in the chemical composition of tea from different production regions. For instance, SDT from Bulang Mountain has the highest water extracts (WE) and TP content, while the FAA content is lower than that of SDT from Nannuo Mountain [8]. Sensory evaluation coupled with biochemical analysis has demonstrated that SDT from the Yiwu tea region consistently surpasses that from the Linxiang region in overall quality metrics [9]. Furthermore, altitudinal influences on SDT composition have been identified, with low-altitude SDT from Wuliang Mountain containing elevated TP levels, contributing to a more pronounced bitter and astringent taste profile [10]. Conversely, high-altitude SDT exhibits increased FAA content, resulting in enhanced freshness and briskness. Seasonal variations have also been observed, with spring tea containing the highest FAA levels, summer tea being rich in total catechins, and autumn tea being characterized by elevated TP content [11].

The environmental conditions surrounding tea cultivation areas, including soil, light, temperature, and altitude, can create unique microclimates [12]. These microclimates contribute to the unique "local flavor" of Yunnan SDT. Soil, as the basis for the survival and growth of the tea tree, along with its physical and chemical properties, and fertility, are the main factors affecting the growth of tea trees and the quality of tea [13]. FAA in tea are precursors for the production of volatile compounds [14], and the accumulation of FAA is affected by the nitrogen content in the soil. Appropriate application of nitrogen fertilizer can promote the formation of amino acid flavor substances, but excessive application of nitrogen fertilizer may lead to an increase in grassy flavor and reduce the aroma of tea [15,16]. In addition, phosphorus and magnesium can improve tea quality by affecting the synthesis of aromatic substances and promoting the Maillard reaction [16]. Phosphorus fertilization increases the content of TP, while potash application increases the total amount of catechins [17,18]. Application of large amounts of phosphorus and potassium in tea gardens promotes malic acid metabolism in tea buds and induces redistribution of photosynthetic products and carbohydrates in favor of the catechin pathway, which further affects tea quality [19]. Differences in SDT quality are closely linked to the environmental conditions around the tea-producing areas; therefore, the key components responsible for these differences require further investigation.

Given the established correlation between SDT quality variations and the environmental conditions of tea-producing regions, further investigation is warranted to elucidate the key components responsible for these qualitative differences. This research direction would provide valuable insights into the complex interactions between environmental factors and tea quality parameters.

2. Materials and Methods

2.1. Preparation of Tea Samples

The SDT samples in this study came from 13 origins in the Digital Tea Plantation Base established by China Tea (Yunnan) Co. Ltd. (Kunming, China) and Yunnan Agricultural University which were named after the respective townships. Tea samples were collected from five randomly selected sampling points at each tea plantation, with each sampling point having the same depth (rhizosphere, 20 cm). Approximately 1 kg of soil, consisting of a mixture of soil samples, were collected from each sampling point [20]. The specific details of the sample collection sites are provided in Table 1.

Table 1. Information on where the samples were collected.

Number	Origin Number	Region	Altitude	Longitude and Latitude
1	Ba Ka Nuan (BKN)	Banzhang Village, Brown Mountain Township, Menghai County, Xishuangbanna Prefecture, Yunnan Province, China	1650 m	E 100°20′, N 21°40′
2	Man Nuan (MN)	Mananan Village, Brown Mountain Township, Menghai County, Xishuangbanna Prefecture, Yunnan Province, China	1300 m	E 100°16′, N 21°37′
3	Mang Aang (MA)	Menghuang Village, Brown Mountain Township, Menghai County, Xishuangbanna Prefecture, Yunnan Province, China	1600 m	E 100°22′, N 21°35′
4	Xin Nong (XN)	Xinlong Village, Brown Mountain Township, Menghai County, Xishuangbanna Prefecture, Yunnan Province, China	1300 m	E 100°16′, N 21°37′
5	Su Hu (SH)	Suhu Village, Gelanghe Township, Menghai County, Xishuangbanna Prefecture, Yunnan Province, China	1650 m	E 100°45′, N 21°97′
6	Pa Sha (PS)	Pasha Village, Gelanghe Township, Menghai County, Xishuangbanna Prefecture, Yunnan Province, China	1700 m	E 100°45′, N 21°97′
7	Gua Feng Zhai (GFZ)	Mahe Village, Yiwu Township, Mengla County, Xishuangbanna Prefecture, Yunnan Province, China	1200 m	E 101°21′, N 21°51′
8	Bing Dao (BDao)	Bingdao Laozhai, Mengku Town, Shuangjiang County, Lincang City, Yunnan Province, China	1900 m	E 99°90′, N 23°27′
9	Dong Guo (DG)	Dongguo Mengku Township, Shuangjiang County, Lincang City, Yunnan Province, China	1750 m	E 99°90′, N 23°20′
10	Dong Ban Shan (DBS)	Nasjiao Village, Mengku Town, Shuangjiang County, Lincang City, Yunnan Province, China	1800 m	E 100°09′, N 23°90′
11	Bing Shan (BS)	Bingshan Village, Mengku Town, Shuangjiang County, Lincang City, Yunnan Province, China	1700 m	E 99°82′, N 23°65′
12	Bang Dong (BD)	Hexin Village, Bangdong Township, Linxiang District, Lincang City, Yunnan Province, China	2000 m	E 100°21′, N 23°56′
13	Wu Liang Shan (WLS)	Shanchong Village, Jingping Town, Jingdong County, Pu'er City, Yunnan Province, China	1900 m	E 100°62′, N 24°09′

2.2. Sensory Evaluation

The sensory review refers to GB/T23776-2018 ("Tea sensory review method") [21]. Among these, the sensory quantitative descriptive analysis method collects sensory primitives related to the aroma and taste of SDT, combines the flavor and quality characteristics of Yunnan large-leaf variety of SDT, and comprehensively formulates the SDT sensory quantitative evaluation table. The aroma evaluation record table has 11 evaluation dimensions: pure and normal, strong and rich, high and upward, lasting, clean and refreshing, flowery aroma, honey aroma, sweet aroma, tender aroma, and grass odor. The taste evaluation record table has 10 evaluation dimensions–strong, thick, mellow, pure, sweet, smooth, sweet aftertaste, fresh and brisk, bitter, and astringent. The quantitative evaluation of sensory characteristics is based on a scale of 0–5 to rate their intensity (0 for none, 5 for the strongest intensity) [22].

2.3. Analysis of Nonvolatile Compounds

Referring to GB/T8305-2013 ("Determination of water extracts of tea") [23], GB/T8314-2013 ("Determination of total free amino acids of tea") [24], and GB/T8313-2018 ("Detection methods for tea polyphenols and catechins in tea") [25], the water extracts (WE), total free amino acids (FAA), and tea polyphenols (TP) were determined, respectively. The anthrone sulfuric acid method was used to determine soluble sugars (SS) [26].

The determination of catechin, flavonoid, and purine alkaloid fractions was carried out by high-performance liquid chromatography (HPLC) following the method described by Yang [27]. The instrument used in this study is a 1200 high-speed liquid chromatograph equipped with a C18 column (4.6 mm × 100 mm, 2.7 μm, Agilent (Santa Clara, CA, USA)). This method employed a mobile phase A consisting of 0.261% phosphoric acid and 5% acetonitrile and a mobile phase B of 80% methanol. The elution gradient proceeded as follows: mobile phase B increased linearly from 10% to 45% between 0 and 16 min; from 16 to 22 min, mobile phase B increased linearly from 45% to 65%; mobile phase B was held constant at 65% from 22 to 25.9 min; from 25.9 to 29 min, mobile phase B increased linearly to 100%; and mobile phase B was maintained at 100% from 29 to 30 min. The column temperature was maintained at 35 °C. Each sample was extracted and analyzed in triplicate.

2.4. Determination of Volatile Components

Electronic nose determination was performed according to the test method of Wu Shanshan et al. [7]. Each tea sample was repeated 3 times.

Headspace solid-phase microextraction (HS-SPME) combined with gas chromatography-mass spectrometry (GC-MS) was utilized to separate and identify the volatile compounds of sun-dried green tea in different regions of Yunnan. The instruments used in this study include a 7890A-5975C headspace solid-phase microextraction GC-MS (Agilent, USA), a DB-WAX column (30 m × 0.25 mm × 0.25 μm, Agilent, USA), and a 65 μm solid-phase microextraction head (PDMS/DVB, Supelco (Bellefonte, PA, USA)). The headspace solid-phase microextraction process was as follows: 1 g of tea sample was placed in a 20 mL headspace vial. A total of 1 μg of ethyl acetate (decanoic acid ethyl ester) was added as the internal standard, followed by the addition of 6 mL of boiling water. The mouth of the vial was then sealed. The CTC autosampler was set with the following conditions: 60 °C, stabilized for 10 min; 65 μm polydimethylsiloxane/divinylbenzene (PDMS/DVB) extraction head; extraction at 60 °C for 30 min with a rotational speed of 250 rpm. For gas chromatography-mass spectrometry (GC-MS) coupled detection, the inlet port temperature was set to 230 °C and the desorption time was 5 min.

GC conditions were as follows: column, DB-WAX (30 m × 0.25 mm × 0.25 μm); carrier gas, He; oven, 50 °C (5 min) to 230 °C (7 min) at 6 °C/min; column temperature, 50 °C (5 min) to 230 °C (7 min); temperature rising rate, 6 °C/min; split ration, no split; MS conditions, mass spectrometry conditions; ion source, EI; gas interface temperature, 280 °C; ion source temperature, 230 °C; quadrupole temperature, 150 °C. Qualitative and quantitative techniques were employed to identify and measure volatile compounds following the approach outlined by Deng et al. [28,29].

For the calculation of relative odor activity value (ROAV), the component with the highest contribution to the flavor of the sample has a $ROAV_{max}$ of 100.00, while the ROAV values of other components are calculated using Formula (1) [30]:

$$ROAV_i = \frac{C_i}{C_{max}} \times \frac{T_{max}}{T_i} \times 100 \quad (1)$$

where C_i is the relative content of each volatile component, %; T_i is the sensory threshold of each volatile component, mg/kg; C_{max} is the relative content corresponding to the volatile component that contributes the most to the aroma of the tea sample; and T_{max} is the sensory threshold corresponding to the volatile component that contributes the most to the aroma of the tea sample.

2.5. Determination of Soil Nutrient Content

For the soil agrochemical analysis, the soil pH was determined using the glass composite electrode method with a soil-to-water ratio of 1:2.5 (m/m). Soil organic matter (SOM) was measured by the potassium dichromate oxidation-external heating method. Total nitrogen (TN) was analyzed using the Kjeldahl method, and alkaline nitrogen (AN) was determined by the alkaline dissolution diffusion method. Total phosphorus (TP) was quantified by the alkali fusion-molybdenum antimony (AMA) colorimetric method, while available phosphorus (AP) was measured using the molybdenum antimony colorimetric method with HCl-NH4F leachate. Total potassium (TK) was analyzed by the alkali fusion flame photometric method, and available potassium (AK) was determined by the flame photometric method with NH4Ac leachate. Additionally, cation exchange capacity (CEC) was measured using the ammonium acetate exchange method.

According to the national environmental technical conditions for tea origin (NY/T853-2004) [31], the fertility grading standards for tea garden soil and the soil nutrition diagnostic indexes for high-quality, high-efficiency, and high-yield tea gardens [32], the soil nutrients of tea gardens were graded.

2.6. Statistical Analysis

Excel 2021 was used to process the raw data, MetaboAnalyst (https://www.metaboanalyst.ca/) was used for orthogonal partial least squares discriminant analysis, and Origin2022 and Chiplot (https://www.chiplot.online/) were used for visual representation of the data.

3. Results and Discussion

3.1. Sensory Flavor Differences in SDT Across Different Regions

Based on the comprehensive analysis of the review results, a flavor wheel (Figure 1A) and corresponding aroma and taste radar charts (Figure 1B,C) for the Yunnan large-leaf variety of SDT were developed. The analytical results revealed significant regional variations in the color, aroma, and taste profiles of SDT across the 13 investigated regions in Yunnan Province. Regarding chromatic characteristics, the dry tea leaves exhibited a predominant coloration spectrum ranging from dark green to brownish-green, with tightly rolled and substantial leaf morphology. The tea infusion demonstrated a characteristic bright yellow to orange hue, while the brewed leaves displayed a yellowish-green to brownish-green coloration with a soft and lustrous appearance. In the context of tea quality assessment, flavor and aroma profiles emerged as critical evaluation parameters, constituting primary determinants in the comprehensive quality evaluation system of SDT. These sensory attributes were found to play a predominant role in the overall quality assessment framework, underscoring their significance in the organoleptic evaluation of SDT.

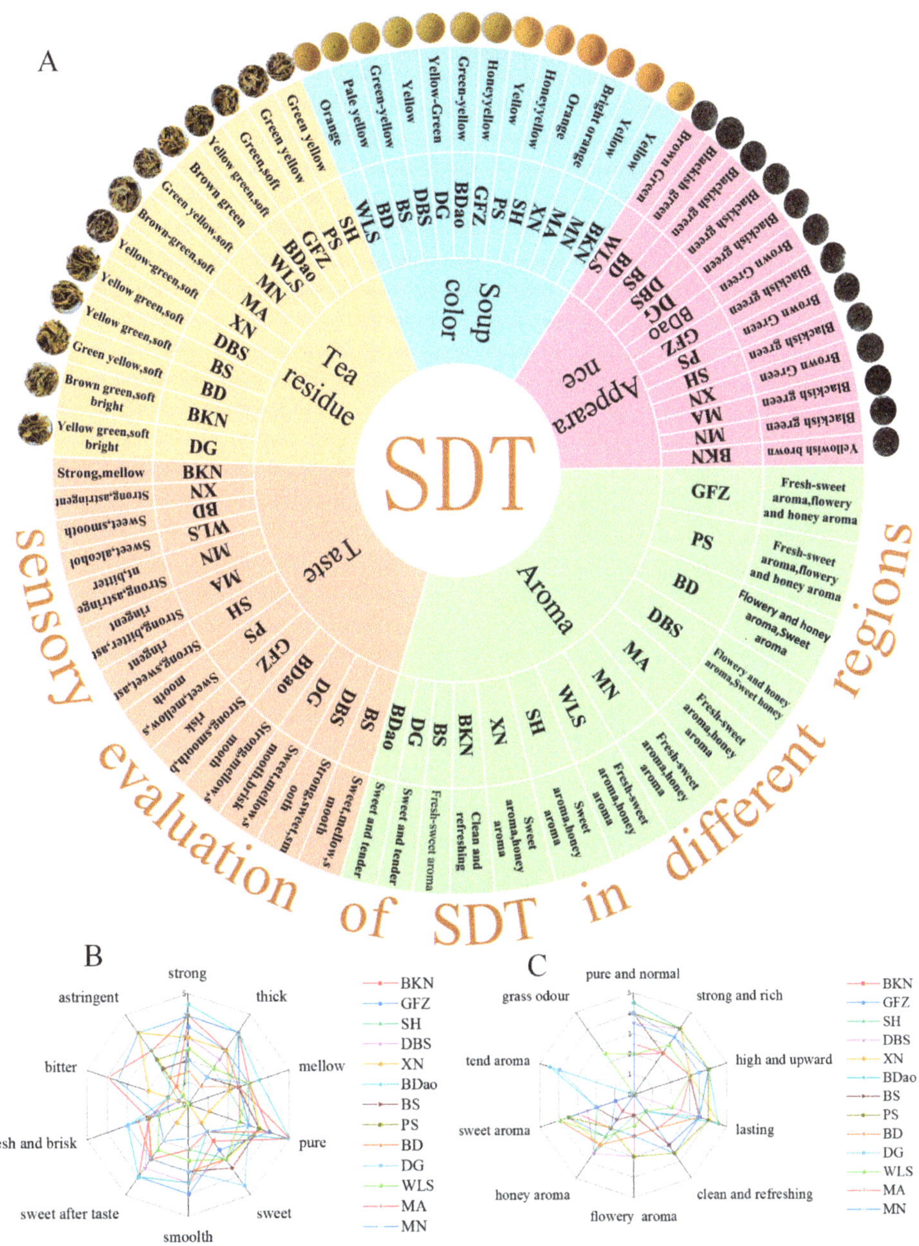

Figure 1. Sensory flavor characteristics of SDT: flavor wheel (**A**); taste radar graph (**B**); aroma radar graph (**C**).

In terms of the taste of the tea, the BDao, GFZ, BKN, SH, MA, MN, and XN samples exhibited heavy and thick attributes, of which the BDao, GFZ, and BKN samples had a strong sweet aftertaste, the MA, MN, and XN samples had an obvious or stronger bitterness and astringency, and the SH sample had a more astringent taste. The BS, PS, WLS, BD, DBS, and DG samples exhibited sweet and mellow attributes, with a strong sweet aftertaste. Among these, BS and PS had a sweet, mellow, and slippery taste with some astringency; BD, DBS, and DG had a sweet and smooth taste, with BD and DBS being more mellow and brisk.

In terms of the aroma of the tea, the WLS, MA, and MN samples predominantly exhibited clean sweetness and a honey aroma. The GFZ and PS samples predominantly exhibited clean sweetness, along with floral and honey aromas. The BKN sample predominantly exhibited a clean and refreshing aroma, while SH and XN were mainly characterized by sweet and honey aromas. The BD and DBS samples predominantly exhibited sweet, floral, and honey aromas, while BS had a mainly clean sweetness. The DG and BDao samples predominantly exhibited sweet and tender aromas.

3.2. Regional Differences in Non-Volatile Compounds of SDT

The taste properties of tea result from the comprehensive effect of water-soluble substances in the tea infusion on human taste receptors. These are mainly affected by the content and ratio of taste substances, such as polyphenols, alkaloids, amino acids, sugars, and organic acids [33]. Quantitative analysis revealed substantial variation in the concentration ranges of water extract (WE), total polyphenols (TP), free amino acids (FAA), soluble sugars (SS), caffeine (CA), gallic acid (GA), and epigallocatechin gallate (EGCG) across SDT samples from 13 distinct regions. Specifically, the measured ranges were as follows: WE (49.79–52.86%), TP (20.45–26.49%), FAA (1.52–2.87%), SS (4.11–6.10%), CA (36.94–77.12 mg/g), GA (0.48–7.98 mg/g), and EGCG (65.54–85.26 mg/g).

The total polyphenol (TP) content in SDT exhibited significant regional variation across the 13 studied areas, with the MN group demonstrating the highest concentration (26.49 ± 0.27%), followed by the MA group (26.02 ± 0.98%). Conversely, the free amino acid (FAA) content was lowest in the MN group (1.52 ± 0.33%), with the MA group showing the second-lowest concentration (1.67 ± 0.14%). Consequently, the TP/FAA ratio reached its maximum in the MN group (18.09 ± 4.35), followed by the MA group (15.84 ± 1.29).

Samples from GFZ and SH regions were characterized by pronounced bitterness and astringency, accompanied by a distinct heavy and thick mouthfeel. Notably, the TP/FAA ratios in the WLS, PS, DG, DBS, SH, GFZ, and BD groups were consistently below 10, with the GFZ group exhibiting the lowest ratio (7.51 ± 0.90), followed by the SH group (7.82 ± 0.75). The SH group simultaneously displayed the highest FAA content (2.87 ± 0.28%) among all the samples. Interestingly, despite their initial bitterness, the GFZ and SH samples developed a noticeable sweet aftertaste.

The WLS group exhibited unique chemical characteristics, including the lowest flavonoid (FN) content (3.69 ± 0.31%), which correlated with reduced bitterness perception. This group also demonstrated the highest concentrations of total catechins (235.94 ± 0.85 mg/g), gallocatechin gallate (GCG) (18.92 ± 0.48 mg/g), and catechin (CA) (77.12 ± 0.98 mg/g), resulting in a predominantly sweet and mellow flavor profile with a persistent sweet aftertaste. This sensory characteristic may be attributed to the formation of hydrogen bonds between alkaloids and catechins in the tea infusion, with the resulting hydrogen bond complexes potentially enhancing the mellow and fresh characteristics of the tea infusion.

Cluster analysis revealed that the SDT samples from 13 distinct regions could be categorized into two primary groups based on their chemical composition profiles. The first group (designated as L1) comprised tea samples from the PS, BD, DBS, DG, SH, GFZ, and WLS regions, while the second group (L2) included samples from the XN, BKN,

MA, MN, BDao, and BS regions (Figure 2A). Quantitative analysis demonstrated that the TP/FAA (total polyphenols to free amino acids) ratio in L1 group samples was consistently below 10, correlating with a predominantly sweet and mellow flavor profile. In contrast, L2 group samples exhibited TP/FAA ratios exceeding 10, corresponding to a heavier and thicker flavor characteristic.

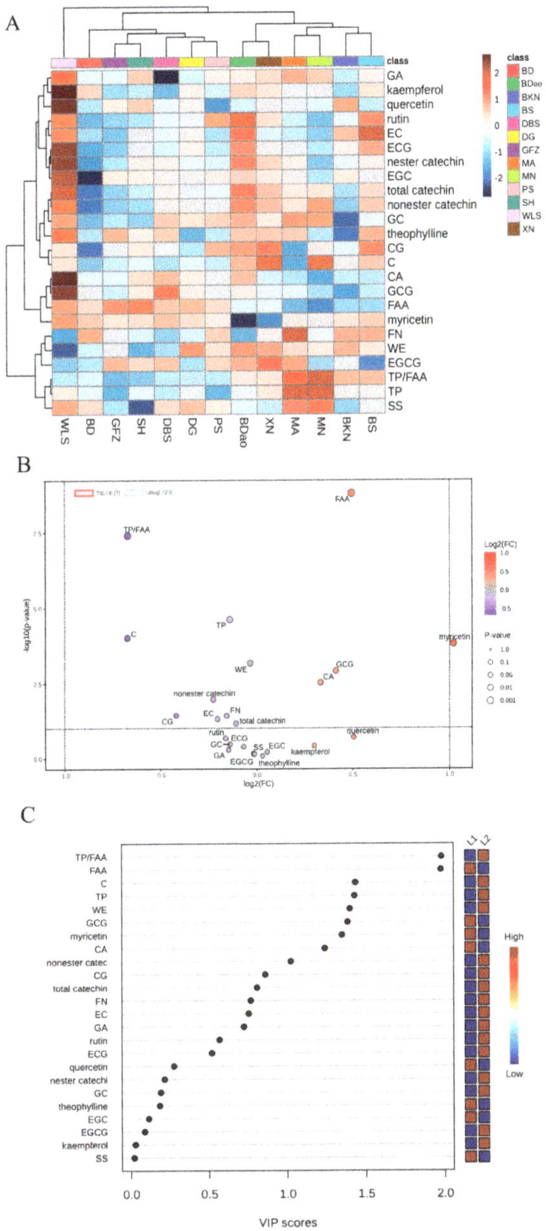

Figure 2. Analysis of regional differences in non-volatile compounds in SDT: heat map (**A**); volcano map (**B**); and VIP value map (**C**).

Orthogonal partial least squares discriminant analysis (OPLS-DA) was subsequently conducted to identify key differential metabolites between the L1 and L2 groups. The model demonstrated satisfactory goodness of fit, with parameter values of R2X:0.745, R2Y:0.931, and Q2:0.892. According to established validation criteria in multivariate analysis, both R2 (coefficient of determination) and Q2 (cross-validated predictive ability) values exceeding the threshold of 0.5 indicate acceptable model performance in both explanatory and predictive capacities. This analysis identified six significant non-volatile metabolites meeting the criteria of fold change (FC) > 1.5 or <0.8, variable importance in projection (VIP) > 1, and p-value < 0.05 (Figure 2B,C). Comparative analysis revealed significant upregulation of TP/FAA and catechin (C) content in the L2 group samples relative to the L1 group. Conversely, FAA, myricetin, GCG, and CA were significantly down regulated in L2 samples. These differential metabolites are potentially responsible for the observed regional variations in SDT taste profiles, providing valuable insights into the chemical basis of regional flavor differentiation in sun-dried tea products.

3.3. Regional Differences in Volatile Compounds of SDT

The aromatic profiles of tea leaves exhibit significant regional variation, playing a pivotal role in determining the flavor quality of SDT [12]. To characterize the volatile organic compounds (VOCs) in SDT, this study employed headspace solid-phase microextraction-gas chromatography-mass spectrometry (HS-SPME-GC-MS). This analytical approach identified 463 distinct volatile compounds, which were systematically classified into 12 chemical categories: 78 alkenes, 66 alcohols, 67 esters, 70 alkanes, 47 ketones, 42 aldehydes, 22 nitrogen-containing compounds, 21 heterocyclic oxygen compounds, 18 carboxylic acids, 10 phenols, 18 aromatic compounds, and 4 miscellaneous compounds (Figure 3). Quantitative analysis revealed substantial variation in volatile composition across different regions. The PS group exhibited the highest diversity of volatile compounds (133 compounds) and the greatest total mass concentration (14.13 mg/g). In contrast, the DG and BS groups demonstrated the lowest compound diversity, each containing 97 volatile compounds. Notably, linalool emerged as the predominant volatile compound across all 13 regional samples, with its relative mass concentration ranging from 0.67 mg/g (XN group) to 26.69 mg/g (DG group).

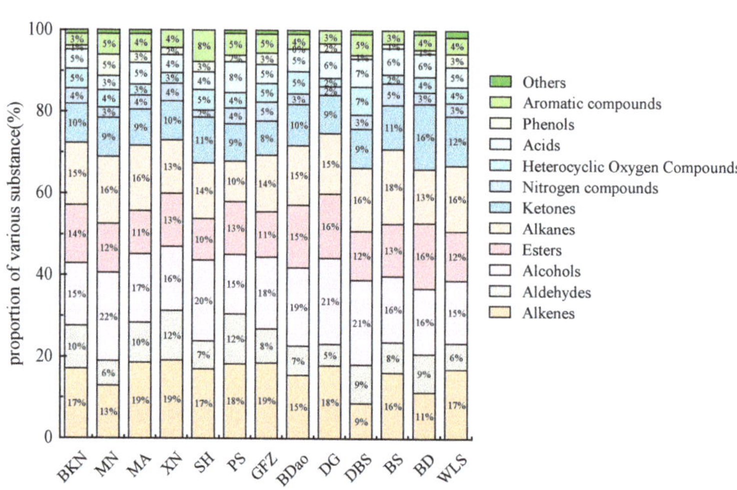

Figure 3. Classification of volatile compounds in SDT.

Although numerous volatile compounds have been identified in tea, only a limited subset significantly contributes to its sensory profile [34]. These influential volatile compounds, which substantially impact the aromatic characteristics of tea, are scientifically defined as "key aroma substances" [35]. Through comprehensive analysis incorporating aroma threshold values and sensory characterization, 118 aroma-active compounds were identified. Subsequent calculation of their relative odor activity values (ROAV) enabled the selection of 36 key aroma-active compounds with ROAV $\geq$ 1.00 (Table 2). Cluster analysis of these 36 key aroma-active compounds (Figure 4A) demonstrated distinct regional differentiation, classifying the 13 studied regions into two primary groups. Group C1 comprised PS, BD, DBS, DG, BS, and BDao regions, characterized by predominant tender and floral aroma profiles. Group C2 included XN, BKN, MA, MN, SH, GFZ, and WLS regions, exhibiting distinctive clean-sweet and honey-like aromatic characteristics. This classification pattern exhibited remarkable consistency with the cluster analysis results obtained from electronic nose detection data (Figure 4B).

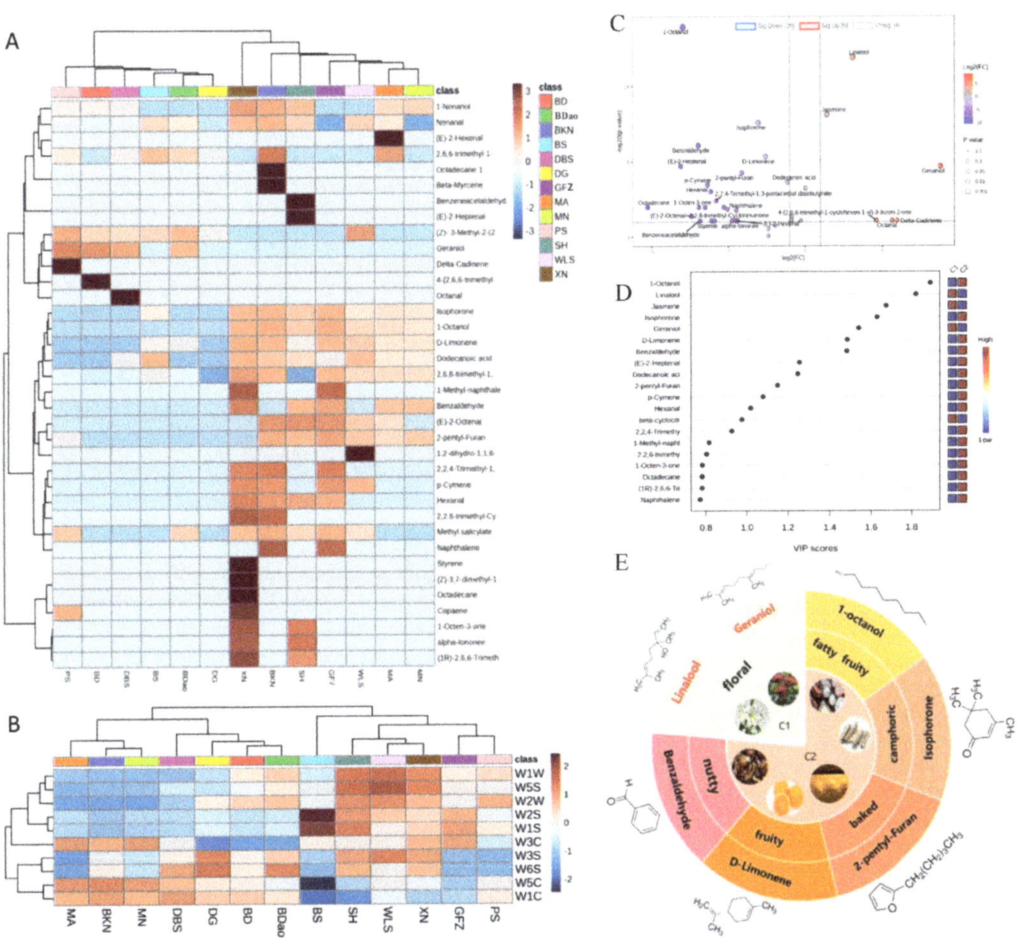

Figure 4. Regional difference analysis of SDT volatile compounds: cluster heat map of 36 aroma active substances (**A**); electronic nose 10 sensor clustering heat map (**B**); volcano chart (**C**); VIP value diagram (**D**); key differential substances (**E**).

Table 2. Key active aroma for taxa C1 and C2.

No.	VM	RI	CAS	Class	Odor Description	OT (ug/g)	RMC (ug/g)		ROAV		FC	p	VIP
							C1	C2	C1	C2			
1	1-Octanol	1280	111-87-5	alcohols	pungent	0.0220	-	2.27	-	2.79	0.00	0.00	1.89
2	Linalool	1269	78-70-6	alcohols	floral, lily of the valley or magnolia scent	0.0060	11.90	1.42	100	100	8.35	0.01	1.82
3	Jasmone	1623	488-10-8	ketones	floral, jasmine, and celery seed aroma	0.0003	15.85	5.91	15.85	6.80	2.68	0.21	1.67
4	Isophorone	1139	78-59-1	ketones	Saussurea costus	0.0110	0.18	1.53	-	1.89	0.12	0.01	1.63
5	Geraniol	1537	106-24-1	alcohols	rose, geranium	0.0066	3.95	-	4.29	-	438.53	0.01	1.54
6	D-Limonene	936	5989-27-5	alkenes	Fruity, sweet fruity (lemon, citrus peel)	0.0340	0.47	2.79	1.16	3.44	0.17	0.01	1.48
7	Benzaldehyde	1245	100-52-7	aldehydes	Fruity, sweet	0.0300	-	1.14	-	1.43	0.01	0.01	1.48
8	(E)-2-Heptenal	1066	2548-87-0	aldehydes	methylphenidate (stimulant drug used in treating asthma)	0.0030	-	2.53	-	3.73	0.00	0.06	1.26
9	Dodecanoic acid	2171	143-07-7	acids	fruity	0.0072	1.44	3.19	2.13	3.85	0.45	0.23	1.25
10	2-pentyl furan	978	3777-69-3	heterozoa	caramel	0.0060	0.19	3.39	1.05	4.37	0.06	0.01	1.15
11	p-Cymene	1007	99-87-6	alkanes	solvent, gasoline, citrus	0.0114	-	0.75	-	1.47	0.01	0.08	1.08
12	Hexanal	953	66-25-1	aldehydes	Fresh and fruity	0.0050	-	0.65	-	1.31	0.01	0.08	1.02
13	β-cyclocitral	1106	432-25-7	aldehydes	Fruity (almond, mango), light, sweet, floral (rose)	0.0050	4.17	4.05	4.34	15.00	1.03	0.98	0.98
14	2,2,4-Trimethyl-1,3-pentanediol diisobutyrate	1560	6846-50-0	esters	-	0.0140	-	0.51	-	1.36	0.02	0.18	0.93
15	1-Methyl-naphthalene	1567	90-12-0	aromatic compounds	Camphoraceous notes, aged; stimulating	0.0080	-	0.36	-	1.10	0.02	0.18	0.82
16	2,2,6-trimethyl-Cyclohexanone	1052	2408-37-9	ketones	Honey scented, lemon scented	0.1000	-	0.31	-	1.24	0.03	0.33	0.81
17	1-Octen-3-one	1045	4312-99-6	ketones	Metallic Aroma	0.0050	-	1.12	-	1.85	0.01	0.20	0.78
18	Octadecane	1754	127-41-3	ketones	floral (violet), woody	0.0016	-	11.42	-	12.16	0.00	0.19	0.78
19	(1R)-2,6,6-Trimethylbicyclo[3.1.1]hept-2-ene	1325	7785-70-8	alkenes	sandalwood	0.0053	-	0.84	-	1.58	0.01	0.24	0.78
20	Naphthalene	1434	91-20-3	aromatic compounds	Tar, camphor breath	0.0500	-	0.21	-	1.15	0.04	0.34	0.77
21	Octanal	1022	124-13-0	aldehydes	fat, soap, lemon, green	0.0007	0.44	-	1.314	-	49.27	0.30	0.68
22	(E)-2-Octenal	1162	593-45-3	alkanes	Chemical odors	0.0200	-	1.54	-	2.46	0.01	0.38	0.66
23	4-(2,6,6-trimethyl-1-cyclohexen-1-yl)-3-buten-2-one	1618	14901-07-6	ketones	floral	0.0070	0.22	-	1.08	-	24.02	0.30	0.64
24	Delta-Cadinene	1446	483-76-1	alkenes	Medicinal and woody	0.0015	0.56	-	1.42	-	62.04	0.30	0.63
25	Beta-Myrcene	918	123-35-3	alkenes	fruity	0.0150	-	0.011	-	1.07	0.83	0.38	0.59
26	Methyl salicylate	1466	119-36-8	esters	floral, Fruity	0.0400	0.56	1.02	1.25	1.65	0.55	0.40	0.56
27	(E)-2-Hexenal	975	6728-26-3	aldehydes	green, leaf	0.0031	-	0.21	-	1.09	0.04	0.38	0.56
28	Nonanal	1115	124-19-6	aldehydes	Sweet rose-like floral scent	0.0035	6.83	12.75	6.83	18.68	0.54	0.34	0.56
29	Styrene	1001	100-42-5	alkenes	balsamic, gasoline	0.0036	-	0.55	-	1.44	0.02	0.38	0.56
30	(Z)-3,7-dimethyl-1,3,6-Octatriene	977	3338-55-4	alkenes	floral, Herbal	0.0340	-	0.19	-	1.08	0.05	0.38	0.56
31	Benzeneacetaldehyde	1350	122-78-1	aldehydes	Fresh and sweet	0.0063	-	1.06	-	1.94	0.01	0.38	0.50
32	alpha-Ionone	1537	18829-55-5	aldehydes	pungent	0.0130	-	0.254	-	1.14	0.04	0.38	0.50
33	1,2-dihydro-1,1,6-trimethyl-Naphthalene	1441	30364-38-6	aromatic compounds	Fruity (citrus, lemon)	0.0025	-	0.63	-	1.52	0.01	0.38	0.45
34	1-Nonanol	1370	143-08-8	alcohols	Fresh, floral, fatty.	0.0053	0.66	2.58	1.18	3.76	0.26	0.15	0.42
35	Safranal	1356	116-26-7	aldehydes	Saffron, Herbal Woody Fragrance	0.0030	2.46	13.23	2.64	16.86	0.19	0.07	0.36
36	Copaene	1206	3856-25-5	alkenes	Lemon-scented woody, earthy, and piney aroma	0.0060	0.19	0.98	1.05	1.87	0.19	0.47	0.09

- Not identified. VM, volatile metabolites; RI, retention index; class, chemical classification originates from (https://www.chemsrc.com/casindex/, accessed on 7 June 2024); odor description, odor description found in the literature with the database (https://www.flavornet.org/flavornet.html, accessed on 6 January 2025; https://www.thegoodscentscompany.com/search2.html, accessed on 7 January 2025); OT, odor thresholds in water; threshold query using https://www.vcf-online.nl/VcfHome.cfm (accessed on 8 June 2024), References: [6,12,22,28,29,34,35]; RMC, relative mass concentration; ROAV, relative odor activity value; FC, fold change; VIP, variable importance in projection.

Orthogonal partial least squares discriminant analysis (OPLS-DA) provides a robust statistical approach for the systematic classification of distinct sample groups. To elucidate the critical volatile compounds contributing to regional differentiation in SDT aroma profiles, OPLS-DA was employed to analyze the volatile organic compounds (VOCs) between C1 and C2 groups. The model demonstrated satisfactory goodness of fit, with parameter values of R2X = 0.726, R2Y = 0.996, and Q2 = 0.993. According to established validation criteria in multivariate analysis, both R2 (coefficient of determination) and Q2 (cross-validated predictive ability) values exceeding the threshold of 0.5 indicate acceptable model performance in both explanatory and predictive capacities. Through this analysis, seven key differential aroma-active compounds were identified based on stringent selection criteria (FC > 1.5 or < 0.8, VIP > 1, and p-value < 0.05) (Figure 4C–E).

Comparative analysis revealed significant upregulation of five compounds in the C2 group relative to C1: isophorone (woody), 2-pentyl furan (caramel), 1-octanol (fatty), D-limonene (fruity), and benzaldehyde (fruity). Conversely, two floral compounds, geraniol and linalool, were significantly downregulated in the C2 group. These identified volatile compounds represent the primary chemical determinants underlying the regional differentiation in SDT aroma characteristics across production areas. The aroma and flavor profiles of green teas are largely determined by their processing methods. Sun-dried green tea, characterized by its unique "sunlight" aroma, retains a higher concentration of volatile alcohols such as linalool and geraniol, which contribute to its balanced and complex aroma profile. By employing geranyl pyrophosphate as the precursor substrate, geraniol and linalool were synthesized via the catalysis of geraniol synthase and linalool synthase, respectively [36]. These aroma compounds are particularly advantageous for Pu-erh tea fermentation, as the oxidation of linalool enhances the formation of woody aroma compounds, enriching the sensory qualities of Pu-erh tea [37].

3.4. Regional Differences in Soil Nutrients

The soil nutrient analysis results (Figure 5) revealed distinct characteristics across the 13 tea garden regions. The soil pH values ranged from 4.45 to 5.16, with 92.31% of the tea gardens meeting the criteria for high-quality fertility tea gardens, indicating optimal conditions for tea tree growth. Notably, the XN tea garden exhibited a pH of 4.45, which is below the threshold of 4.50, suggesting potential soil acidification issues. Total nitrogen (TN) content varied between 0.78 and 6.10 g/kg, with 84.61% of the tea gardens conforming to high-quality fertility standards. However, the MN and MA tea gardens exhibited TN levels classified as grade III. Available nitrogen (AN) content ranged from 54.67 to 214.67 mg/kg, with 61.54% of the tea gardens meeting high-quality fertility standards. The remaining gardens were distributed as follows: BKN, XN, and BS at grade II, and MN and MA at grade III, indicating generally sufficient soil nitrogen content across all sites. Total phosphorus (TP) and available phosphorus (AP) contents were measured at 0.13–0.77 g/kg and 0.83–6.75 mg/kg, respectively, with both parameters falling within grade II to III levels. Total potassium (TK) content ranged from 0.61 to 2.58 g/kg, with 69.23% of the tea gardens meeting high-quality fertility standards. The remaining gardens BKN, XN, PS, and WLS exhibited TK levels at grade II, accounting for 30.77% of the total. Available potassium (AK) content varied between 33.33 and 403.67 mg/kg, with 61.54% of the tea gardens conforming to high-quality fertility standards. The remaining gardens were distributed as follows: BKN, GFZ, and DBS at grade II, and BDao and BD at grade III. Soil organic matter (SOM) content ranged from 18.4 to 94.9 g/kg, with 84.62% of the tea gardens meeting high-quality fertility standards. The MN and MA tea gardens exhibited SOM levels at grade II, accounting for 15.38% of the total sample.

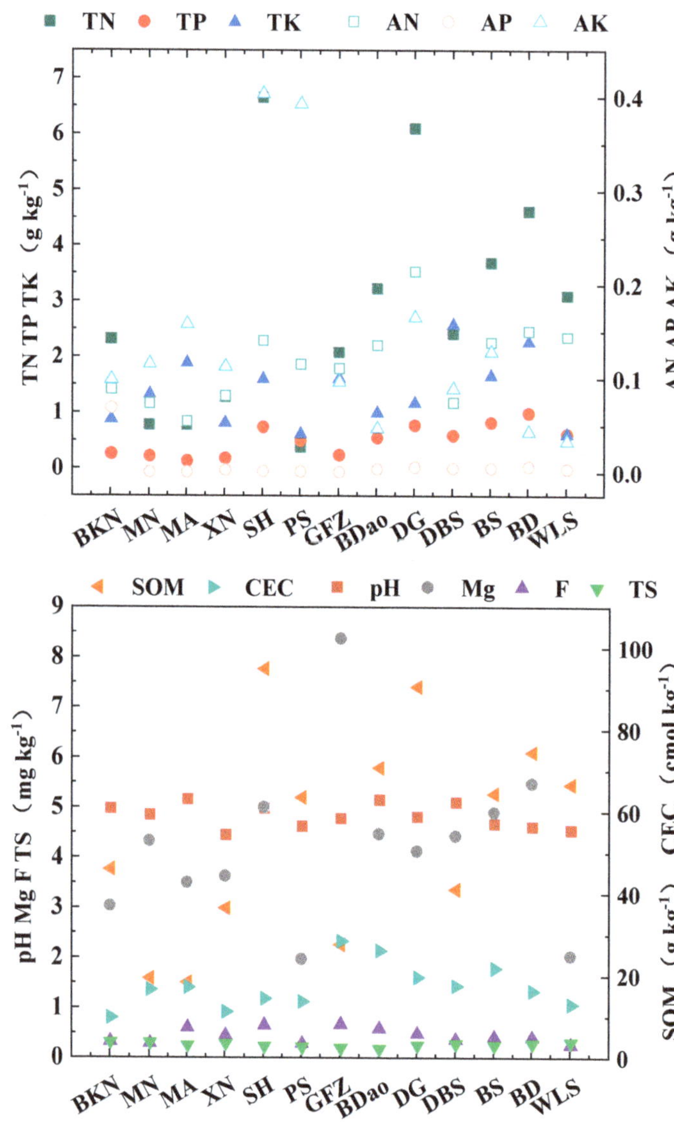

Figure 5. Map of soil nutrient differences.

Cluster analysis of soil nutrient profiles revealed distinct groupings among the sampling sites. Specifically, SH, BD, DBS, DG, BS, and BDao were classified into the T1 group, while XN, BKN, MA, MN, PS, GFZ, and WLS formed the T2 group (Figure 6A). Statistical analysis demonstrated significantly higher soil AP content in the T1 group (3.94–6.75 mg/kg) compared to the T2 group (0.69–3.00 mg/kg). OPLS-DA was subsequently conducted to identify key differentiating factors between the two groups. The model demonstrated satisfactory goodness of fit, with parameter values of R2X:0.621, R2Y:0.878, and Q2:0.827. According to established validation criteria in multivariate analysis, both R^2 (coefficient of determination) and Q2 (cross-validated predictive ability) values exceeding the threshold of 0.5 indicate acceptable model performance in both explanatory and predictive capacities. This analysis identified five mineral elements meeting the following criteria: FC > 1.5 or <0.8, VIP > 1, $p < 0.05$. These elements were determined to be the principal mineral components contributing to the intergroup differentiation (Figure 6B,C). Comparative analysis revealed significant upregulation of multiple soil parameters in the T1 group relative to the T2 group, including TP, AP, TN, SOM, and TK. These findings suggest that these soil characteristics may serve as key indicators for the observed grouping patterns and potentially influence the regional differentiation in tea quality parameters.

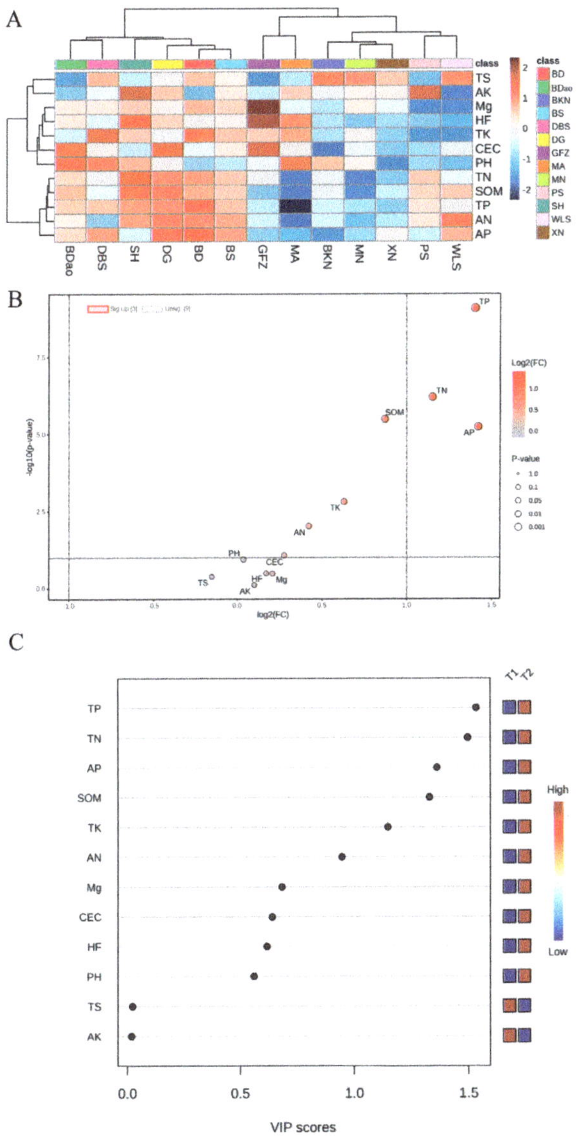

Figure 6. Soil nutrient clustering heat map (**A**), volcano map (**B**), and VIP value map of OPLS-DA (**C**).

3.5. Factors Affecting SDT Quality Differences Across Regions

The biosynthesis of aroma and flavor compounds in SDT is significantly influenced by regional altitudinal variations and soil nutrient composition. Empirical evidence from previous studies, corroborated by the current investigation, has demonstrated substantial heterogeneity in soil nutrient profiles across different regions of Yunnan Province [13,38]. Furthermore, the enhancement of tea quality has been established as a critical factor in improving the economic value of tea products [39]. To systematically investigate the environmental determinants underlying regional quality variations in SDT, this study employed Pearson correlation analysis and interactive Mantel tests to examine the relationships between soil nutrient content, altitudinal factors, and tea quality components across 13 distinct regions.

The analytical results demonstrated significant positive correlations between soil pH and several biochemical components in SDT, including total polyphenols (TP), myricetin, quercetin, protein content, kaempferol, and GA. Soil parameters, particularly SOM, TN, and AN, exhibited positive correlations with FAA, SS, rutin, EC, and CA in SDT, while showing negative correlations with total polyphenols (TP). Furthermore, SOM, total phosphorus (TP), and AP were positively correlated with rutin, EC, CA, and ECG, but negatively correlated with EGCG (Figure 7A). These findings partially diverge from previous research by Yang [38], which reported positive correlations between SOM, TN, AN and FAA, positive correlations with total polyphenols (TP), and negative correlations with pH and total phosphorus (TP) in Pu'er tea gardens.

Altitude demonstrated significant positive correlations with rutin, myricetin, kaempferol, CA, and EC in SDT, while exhibiting a negative correlation with bitter-tasting EGCG. Analysis of non-volatile compounds revealed distinct taste profiles between sample groups: L1 group samples (TP/FAA < 10) were characterized by a sweet and mellow taste, whereas L2 group samples (TP/FAA > 10) exhibited a heavier and thicker taste profile. Notably, with the exception of GFZ and BDao, L1 group samples were predominantly from higher altitudes compared to L2 group samples, suggesting altitude as a critical environmental factor influencing SDT taste characteristics. Specifically, high-altitude samples tended to demonstrate sweeter and mellower profiles, while low-altitude samples exhibited heavier and thicker taste characteristics.

These variations are potentially attributable to the synergistic effects of light intensity and temperature on tea plant metabolism at different altitudes [12]. The photodegradation of theanine into ethylamine and glutamic acid, which serve as precursors for catechin synthesis [40], combined with reduced light intensity and duration in high-altitude regions due to increased cloud cover and precipitation, may contribute to lower TP/FAA ratios in tea leaves. Given the established role of TP/FAA ratio in determining the balance between heavy/thick and sweet/aftertaste characteristics of tea infusion [41,42], the observed lower TP content and higher theanine levels in L1 group samples likely account for their sweeter and mellower taste profile compared to L2 group samples.

The mineral element phosphorus (P) and altitude emerged as the most significant factors influencing the differentiation in aroma-active substances in SDT. Statistical analysis revealed significant positive correlations between phosphorus content/altitude and floral compounds, specifically geraniol and linalool, while demonstrating negative correlations with isophorone (woody), 2-pentyl furan (caramel), 1-octanol (fatty), D-limonene (fruity), and benzaldehyde (fruity) (Figure 7B). Comparative analysis between groups showed significant upregulation of geraniol and jasmone in C1 relative to C2, whereas isophorone, 2-pentyl-furan, 1-octanol, D-limonene, and benzaldehyde were significantly downregulated.

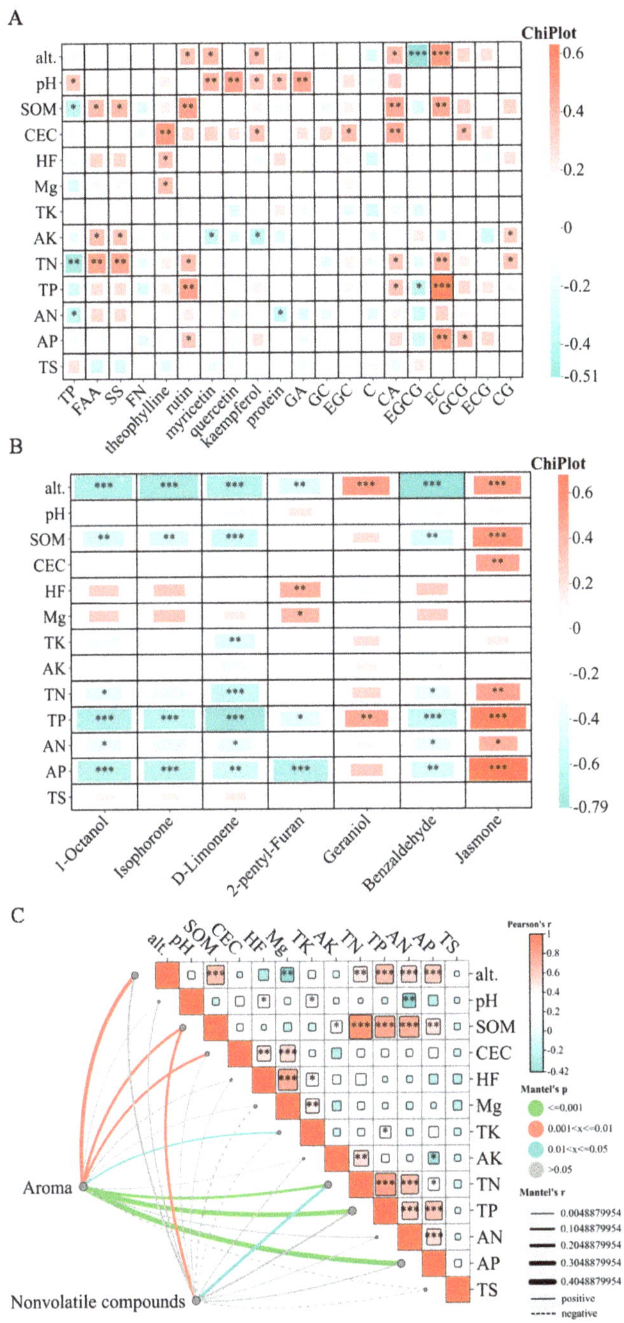

Figure 7. Correlation heat map between environmental factors (**A**,**B**) and interactive Mantel test correlation heat map (**C**). *: $p < 0.05$, **: $p < 0.01$, ***: $p < 0.001$.

The clustering analysis of soil nutrients and volatile compounds, excluding the PS and SH regions, demonstrated consistent patterns. The T1 group exhibited higher soil available phosphorus (AP) content compared to T2, while the T1 and C1 groups generally showed higher altitude levels than their T2 and C2 counterparts, with the exception of the WLS region. These findings suggest that soil phosphorus content and altitude serve as potential key markers influencing the floral, woody, caramel, fatty, and fruity characteristics of SDT. Specifically, regions characterized by elevated AP content and higher altitudes demonstrated a greater propensity for the accumulation of floral compounds.

These results align with previous research demonstrating that high-altitude tea samples contain elevated levels of alcohols with floral and fruity characteristics, while low-altitude regions, such as those producing Xinyangmaojian, exhibit a higher proportion of woody compounds [10,43]. The consistency of these findings across different studies reinforces the significance of edaphic and altitudinal factors in shaping the aromatic profile of tea products.

The results of interaction analysis also verified the above conclusions (Figure 7C). This may be due to the long-term growth of tea trees in acidic soil environments. The available phosphorus content in acidic soils is usually at a low level [44]. In order to adapt to this low phosphorus condition, tea trees will promote the synthesis of secondary metabolites, and these secondary metabolites play a positive role in assisting the synthesis of tea flavor substances in the subsequent process. Tea aroma serves as a critical determinant of sensory evaluation and quality attributes. Its formation mechanisms involve stress responses during both pre-harvest natural growth and post-harvest processing of tea plants (*Camellia sinensis*). Stressors enhance the biosynthesis, accumulation, and emission of aroma compounds by activating metabolic pathways, exerting dual functionality: improving tea aroma quality and strengthening stress adaptation through chemical defense mechanisms [45]. Previous studies have shown that the enrichment of phosphorus contributes to the formation of terpenes, aliphatic compounds, and aromatic alcohols such as linalool. Phosphorus can effectively promote the formation of flavonoids in tea and the increase of TP, FAA, and caffeine content, thereby improving the nutritional value and aroma intensity of tea [46,47]; however, the phosphorus content needs to be within a certain range, and high concentrations of phosphorus content are negatively correlated with tea quality [48]. In a related study conducted by Liu et al. [49], in Wuyi Mountain tea plantations, tea plants were subjected to two fertilization treatments: high phosphorus (HP, 112.5 kg P_2O_5/ha) and low phosphorus (LP, 60 kg P_2O_5/ha). The findings indicated that excessive phosphorus application (112.5 kg P_2O_5/ha) significantly reduced the accumulation of key aroma metabolites, such as 3-octen-2-ol, 4,8-dimethyl-1,7-nonadien-4-ol, and 3-ethyl-4-methylpyrrole-2,5-dione, in the high-mountain Shuixian tea variety, highlighting the importance of balanced phosphorus management for optimizing tea quality. The aroma characteristics of SDT 'local flavor' can be regulated to a certain extent by regulating the application amount of phosphate fertilizer.

4. Conclusions

This study revealed significant regional variations in the quality of SDT. Cluster analysis of non-volatile compounds categorized the samples from 13 regions into two distinct groups—L1 (SH, GFZ, WLS, BD, DBS, DG, and PS) and L2 (BDao, XN, MA, MN, BKN, and BS)—characterized by sweet mellow and strong taste profiles, respectively. Notably, the TP/FAA ratio was significantly upregulated in the L2 group compared to the L1 group. Similarly, cluster analysis of key volatile aroma-active compounds classified the samples into C1 (PS, BD, DBS, BS, BDao, and DG) and C2 (XN, BKN, SH, GFZ, WLS, MA, and MN) groups. The C1 group exhibited dominant tender and floral aromas, whereas the C2 group was characterized by clean-sweet and honey-like aromas. Comparative analysis demonstrated significant upregulation of floral compounds, specifically geraniol and jasmone, in the C1 group, while the C2 group showed significant downregulation of isophorone (woody), 2-pentyl furan (caramel), 1-octanol (fatty), D-limonene (fruity), and benzaldehyde (fruity).

Correlation analysis identified altitude and soil phosphorus content as key factors influencing regional differences in SDT flavor quality. Altitude and available phosphorus (AP) content exhibited positive correlations with floral compounds, as well as with rutin, myricetin, kaempferol, CA, and EC in SDT. Conversely, negative correlations were observed with woody, caramel, fatty, and fruity characteristics, as well as with the bitter and astringent compound EGCG. SDT cultivated in high-altitude regions with elevated AP content demonstrated enhanced sweet mellow characteristics and greater enrichment of floral aroma compounds.

These findings provide scientific data supporting the study of regional flavor characteristics in Yunnan SDT. Phosphorus content plays a critical role in regulating SDT aroma compounds, although its underlying mechanisms require further investigation. Future research should focus on achieving targeted regulation of SDT's regional flavor profiles through controlled variable experiments and multi-omics approaches. This study contributes to a deeper understanding of the factors influencing SDT quality and offers a foundation for optimizing cultivation practices enhancing flavor characteristics.

Author Contributions: Conceptualization, investigation, methodology, and writing—original draft preparation, M.Z. (Miao Zhou) and X.D.; methodology, Q.W. and Z.W.; formal analysis, X.W. and W.Y.; resources, L.L. and M.Z. (Man Zou); data curation, W.L. and S.L.; visualization, Q.W. and X.W.; writing—review and editing, Y.S.; funding acquisition, B.W. All authors have read and agreed to the published version of the manuscript.

Funding: This research was funded by the Yunnan Tea Industry Artificial Intelligence and Big Data Application Innovation Team (202405AS350025); the research and development and demonstration project of data sensing technology and equipment for high-altitude mountain smart agriculture (202302AEO9002001); the study on the screening mechanism of phenotypic plasticity characteristics of large-leaf tea plants in Yunnan driven by Abased on data fusion (202301AS070083); the Yunnan Menghai County Smart Tea Industry Science and Technology Mission (202304BI090013); and the Yunnan Provincial Department of Science and Technology, Joint Agricultural Project (202401BD070001-043).

Institutional Review Board Statement: In compliance with the "Ethics Review Measures for Life Sciences and Medical Research Involving Humans" issued by the National Health Commission of China in 2023 (Document No. 4), specifically Article 32, our study qualifies for exemption from ethical review as it does not present any potential risk to the participants.

Informed Consent Statement: This study involves a sensory evaluation of food taste using healthy food products without additives. It does not collect sensitive personal data or involve commercial interests. All participants were adults, and their data were anonymized. Participants were fully informed and provided written consent prior to participation, with the option to withdraw at any time.

Data Availability Statement: The original contributions presented in the study are included in the article, further inquiries can be directed to the corresponding authors.

Conflicts of Interest: Authors Zhenzhen Wei and Yubo Sheng were employed by the company China Tea (Yunnan) Co., Ltd. The remaining authors declare that the research was conducted in the absence of any commercial or financial relationships that could be construed as a potential conflict of interest.

References

1. Wang, Z.; Zhang, S.; Chen, Y.; Xia, Y.; Wang, H.; Jin, R.; Wang, C.; Fan, Z.; Wang, Y.; Wang, B. Detection of small foreign objects in Pu-erh sun-dried green tea: An enhanced yolov8 neural network model based on deep learning. *Food Control* **2025**, *168*, 110890.
2. Ma, C.; Gao, C.; Li, Y.; Zhou, X.; Fan, G.; Tian, D.; Huang, Y.; Li, Y.; Zhou, H. The characteristic aroma compounds of gaba sun-dried green tea and raw Pu-erh tea determined by headspace solid-phase microextraction gas chromatography-mass spectrometry and relative odor activity value. *Foods* **2023**, *12*, 4512. [CrossRef] [PubMed]
3. Zhang, W.J.; Liu, C.; Yang, R.J.; Zheng, T.T.; Zhao, M.M.; Ma, L.; Yan, L. Comparison of volatile profiles and bioactive components of sun-dried Pu-erh tea leaves from ancient tea plants on Bulang mountain measured by gc-ms and hplc. *J. Zhejiang Univ. Sci. B* **2019**, *20*, 563–575. [CrossRef] [PubMed]
4. Zhao, M.M.; Yang, R.B.; Lv, C.Y. Aroma components of raw materials tea in lincang: A comparative study based on electronic nose and GC-MS technology. *Chin. Agric. Sci. Bull.* **2018**, *34*, 113–122.
5. Liu, X.Y.; Wang, J.; Peng, Y.; Lu, C.Y.; Li, R.Y. Volatile component analysis of sun-dried green tea in Menghai county based on GCIMS. *Sci. Technol. Food Ind.* **2021**, *42*, 233–240.
6. Xiao, Q.M.; Zhang, X.Y.; Wang, J.; Lu, C.Y. Comparative analysis on aromatic components of sun-dried green tea from different townships in Menghai county. *Sci. Technol. Food Ind.* **2023**, *44*, 332–339.
7. Wu, S.S.; Li, F.; Xiong, C.Y.; Ren, H.T.; Ding, Q.H.; Zhao, J.R.; Yang, X.M. Analysis of aroma of sun-dried green tea in different areas of Yunnan province based on electronic nose technology. *J. Tea Commun.* **2022**, *49*, 88–95.
8. Wang, J.; Nian, B.; Liu, X.Y.; Xie, J.; Lu, C.Y. Comparative analysis on quality of autumn sun-dried green tea from main tea producing areas in Menghai of Yunnan. *J. Tea Commun.* **2021**, *48*, 98–104.
9. Luo, H.; Zong, J.F.; Liu, C.Y.; Wang, Q.H.; Yang, L.; Wu, H.Q.; Guan, X.Y.; Zhou, L. Quality differences of ancient tree sun-dried green tea between Yiwu and Linxiang tea areas in Yunnan province. *J. Tea Commun.* **2023**, *50*, 500–508.
10. Chen, H.Y.; Chen, J.H.; Xiong, L.G.; Li, Q.; Jiang, H.J.; Zhang, Q.G.; Huang, J.N.; Liu, Z.H. Analysis of the difference in sensory quality and main biochemical components of Yunnan large leaf sun-dried green tea at different altitudes. *J. Chin. Inst. Food Sci. Technol.* **2023**, *23*, 276–287.
11. Li, F.; Chen, C.; Tian, Y.; Yang, X.; Jin, N.; Li, M. Seasonal variation of biochemical components of different cultivars of *camellia sinensis var. Assamica* in Yunnan. *J. Food Biotechnol.* **2022**, *41*, 88–95.
12. Huang, S.; Tao, L.; Xu, L.; Shu, M.; Qiao, D.; Wen, H.; Xie, H.; Chen, H.; Liu, S.; Xie, D.; et al. Discrepancy on the flavor compound affect the quality of Taiping houkui tea from different production regions. *Food Chem. X* **2024**, *23*, 101547. [PubMed]
13. Lan, M.Q.; Ma, Z.X.; Xiong, C.Y.; Zhou, Z.J.; Yang, D.L.; Yang, X.M. Study on the correlation between physical and chemical properties of soils and tea contents in different regions of pu'er tea garden. *Chin. J. Trop. Agric.* **2023**, *43*, 7–12.
14. Chia, M.A.; Lombardi, A.T.; Melão, M.D.G.G.; Parrish, C.C. Effects of cadmium and nitrogen on lipid composition of chlorella vulgaris (*trebouxiophyceae, chlorophyta*). *Eur. J. Phycol.* **2013**, *48*, 1–11.
15. Zhang, W.; Ni, K.; Long, L.; Ruan, J. Nitrogen transport and assimilation in tea plant (*Camellia sinensis*): A review. *Front. Plant Sci.* **2023**, *14*, 1249202.
16. Liu, M.-Y.; Burgos, A.; Ma, L.; Zhang, Q.; Tang, D.; Ruan, J. Lipidomics analysis unravels the effect of nitrogen fertilization on lipid metabolism in tea plant (*Lamellia sinensis* L.). *BMC Plant Biol.* **2017**, *17*, 1–10.
17. Lei, Q.; Yuan, L. Effect of potassium fertilizer on the absorption of mineral nutrients and yield and quality of young tea. *Chin. J. Soil Fertil.* **2008**, *5*, 41–44.
18. Zhang, W.L.; Xie, H.; Li, T.; Wang, D.D. Effects of different soil texture and management pattern on soil properties and tea quality of tea garden. *Hubei Agric. Sci.* **2017**, *56*, 1824–1827.
19. Wei, K.; Liu, M.; Shi, Y.; Zhang, H.; Ruan, J.; Zhang, Q.; Cao, M. Metabolomics reveal that the high application of phosphorus and potassium in tea plantation inhibited amino-acid accumulation but promoted metabolism of flavonoid. *Agronomy* **2022**, *12*, 1086. [CrossRef]
20. Yan, M.J.; Cheng, Y.Z.; Lin, C.; Wu, Y.Q.; Zheng, X.L.; Wang, F. Effects of continuous chemical fertilizer combined with organic manure on soil bacterial community structure and functional groups in tea gardens. *J. Agro-Environ. Sci.* **2023**, *43*, 2959–2973.
21. *GB/T23776-2018*; Methodology for Sensory Evaluation of Tea. General Administration of Quality Supervision, Inspection and Quarantine of the People's Republic of China: Beijing, China, 2018.
22. Wei, H.; Lan, T.M.; Miao, Y.W.; Meng, Q.; Kun, J.R.; Zhang, Y.R.; Tong, H.R. Analysis of the effect of different full firing methods on the aroma of Jinmudan congou black tea based on sensomics characterization. *J. Tea Sci.* **2023**, *43*, 109–123.
23. *GB/T 8305-2013*; Tea–Determination of Water Extracts Content. Administration of Quality Supervision, Inspection and Quarantine of the People's Republic of China: Beijing, China, 2013.

24. *GB/T8314-2013*; Tea–Determination of Free Amino Acids Content. Administration of Quality Supervision, Inspection and Quarantine of the People's Republic of China: Beijing, China, 2013.
25. *GB/T8313-2018*; Determination of Total Polyphenols and Catechins Content in Tea. National Standard of the People's Republic of China: Beijing, China, 2018.
26. Liu, K.Y.; Wang, L.Y.; An, J.S.; Wang, X.H.; Luo, H.; Fan, J.K.; Chen, L.J.; Ma, Y.; Zhao, M. Analysis of quality and microbial communities of Pu-erh tea through inoculation fermentation with bacillus licheniformis. *J. Food Sci. Technol.* **2022**, *40*, 108–118.
27. Yang, C.H.; Yuan, W.X.; Liu, X.H.; Zhang, S.H.; He, J.J.; Li, L.; Huang, W.; Wang, B.J. Quality characteristics of ancient tree tea in seven villages of Jingmai mountain. *J. Yunnan Agric. Univ. Nat. Sci.* **2024**, *39*, 111–119.
28. Deng, X.; Huang, G.; Tu, Q.; Zhou, H.; Li, Y.; Shi, H.; Wu, X.; Ren, H.; Huang, K.; He, X.; et al. Evolution analysis of flavor-active compounds during artificial fermentation of Pu-erh tea. *Food Chem.* **2021**, *357*, 129783.
29. Xu, J.Y.; Deng, X.J.; Wu, Y.M.; Zhou, M.; Du, C.; Wang, Q.M.; Xia, Y.X.; He, J.J.; Yuan, W.X.; Wu, W.D.; et al. Characteristic changes and potential markers of flavour in raw Pu-erh tea with different ageing cycles analysed by HPLC, HS-SPME-GC-MS, and OAV. *Foods* **2025**, *14*, 829. [CrossRef]
30. Xu, M.T.; Gu, M.Y.; Chen, J.H.; Wei, M.X.; Chen, Q.L.; Wu, W.X.; Zheng, Y.C.; Ye, N.X. Flavor quality analysis of 'jinmudan' and 'jinguanyin' high aroma black tea using metabolomics. *Food Sci.* **2024**, *45*, 150–161.
31. *NY/T 853-2004*; Agriculture Industry Standard. Agricultural Environmental Quality Supervision, Yunnan Provincial Agricultural Environmental Protection Monitoring Station. Environmental and Technical Conditions for Tea Production. Ministry of Agriculture: Beijing, China, 2005; p. 8.
32. Han, W.Y.; Ruan, J.Y.; Lin, Z.; Wu, X.; Xu, Y.W.; Shi, Y.Z.; Ma, L.F. The major nutritional limiting factors in tea soils and development of tea speciality fertilizer series. *J. Tea Sci.* **2002**, *22*, 70–74.
33. Zhang, L.; Cao, Q.Q.; Granato, D.; Xu, Y.Q.; Ho, C.T. Association between chemistry and taste of tea: A review. *Trends Food Sci. Tech.* **2020**, *101*, 139–149.
34. Zheng, Y.; Hu, Q.; Wu, Z.; Bi, W.; Chen, B.; Hao, Z.; Wu, L.; Ye, N.; Sun, Y. Volatile metabolomics and coexpression network analyses provide insight into the formation of the characteristic cultivar aroma of oolong tea (*Camellia sinensis*). *LWT* **2022**, *164*, 113666. [CrossRef]
35. Hao, Z.; Feng, J.; Chen, Q.; Lin, H.; Zhou, X.; Zhuang, J.; Wang, J.; Tan, Y.; Sun, Z.; Wang, Y.; et al. Comparative volatiles profiling in milk-flavored white tea and traditional white tea Shoumei via HS-SPME-GC-TOFMS and OAV analyses. *Food Chem. X* **2023**, *18*, 100710.
36. Tang, J.H.; Hu, J.J.; Zhou, X.X.; Wang, Q.W.; Jiang, Y.W.; Yuan, H.B.; Wang, Y.J.; Yang, Y.Q. Influence of different shaping techniques on the aroma quality and volatile metabolites of green tea revealed by gas chromatography electronic nose and gas chromatography-tandem mass spectrometry. *Foods* **2025**, *14*, 816. [CrossRef] [PubMed]
37. You, Q.S.; Shi, Y.L.; Zhu, Y.; Yang, G.Z.; Yan, H.; Lin, Z.; Lu, H. Effect of different processing technologies on the key aroma-active compounds of green tea. *Food Sci.* **2023**, *44*, 194–200.
38. Yang, G.R.; Wang, X.Q.; Xie, J.; Lu, C.Y.; Li, Y.M. Analysis of the relationship between soil nutrients and tea main quality components of ancient tea arboretum and modern tea garden in Yunnan province. *J. Tea Sci.* **2015**, *35*, 574–582.
39. Ye, J.; Wang, Y.; Wang, Y.; Hong, L.; Jia, X.; Kang, J.; Lin, S.; Wu, Z.; Wang, H. Improvement of soil acidification in tea plantations by long-term use of organic fertilizers and its effect on tea yield and quality. *Front. Plant Sci.* **2022**, *13*, 1055900.
40. Gong, A.D.; Lian, S.B.; Wu, N.N.; Zhou, Y.J.; Zhao, S.Q.; Zhang, L.M.; Cheng, L.; Yuan, H.Y. Integrated transcriptomics and metabolomics analysis of catechins, caffeine and theanine biosynthesis in tea plant (*Camellia sinensis*) over the course of seasons. *BMC Plant Biol.* **2020**, *20*, 1–14.
41. Pu, S.; Xia, L.; Deng, S.; Zhou, Y. Research on the chemical composition change of Yunnan big leaf species sun-dry tea in different months. *Chin. Agric. Sci. Bull.* **2015**, *31*, 247–250.
42. Wang, M.; Yang, J.; Li, J.; Zhou, X.; Xiao, Y.; Liao, Y.; Tang, J.; Dong, F.; Zeng, L. Effects of temperature and light on quality-related metabolites in tea [*Camellia sinensis* (L.) *Kuntze*] leaves. *Food Res. Int.* **2022**, *161*, 111882.
43. Sun, M.F.; Guo, G.Y.; Zang, L.; Qi, H.Z. Analysis of Xinyangmaojian tea aroma components from different altitude by gC-MS. *J. Henan Agric. Sci.* **2014**, *43*, 181–185.
44. Sun, L.L.; Zhang, M.S.; Liu, X.M.; Mao, Q.Z.; Shi, C.; Kochian, L.V.; Liao, H. Aluminium is essential for root growth and development of tea plants (*Camellia sinensis*). *J. Integr. Plant Biol.* **2020**, *62*, 984–997.
45. Zhao, M.Y.; Wang, L.; Wang, J.M.; Jin, J.Y.; Zhang, N.; Lei, L.; Gao, T.; Jing, T.T.; Zhang, S.R.; Wu, Y.; et al. Induction of priming by cold stress via inducible volatile cues in neighboring tea plants. *J. Integr. Plant Biol.* **2020**, *62*, 1461–1468.
46. Hua, L.Y. Exploration of the effect of phosphorus on the quality of tea leaves. *J. Anhui Agric. Sci.* **2014**, *42*, 10052–10053.
47. Liu, Y.; Zhang, M.; Li, Y.; Zhang, Y.; Huang, X.; Yang, Y.; Zhu, H.; Xiong, H.; Jiang, T. Influence of nitrogen fertilizer application on soil acidification characteristics of tea plantations in karst areas of southwest China. *Agriculture* **2023**, *13*, 849. [CrossRef]

48. Zhou, Z.; Liu, Y.; Zhang, L.M.; Xu, R.N.; Sun, L.L.; Liao, H. Soil nutrient status in Wuyi tea region and its effects on tea quality-related constituents. *Sci. Agric. Sin.* **2019**, *52*, 1425–1434.
49. Liu, Y.; Zhuo, Z.; Shi, C.; Liu, B.; Xue, J.; Liang, L.; Wang, Y.; Xu, R.; Liu, B.; Xu, M.; et al. Scent of tea: Metabonomic and ionomic basis of regional aroma of Wuyi rock tea. *Sci. China Life Sci.* **2021**, *52*, 273–284.

foods

MDPI

Article

Effect of Drying Methods on Aroma Profiling of Large-Leaf Green Tea (*Camellia sinensis* var. Assamica) Determined by HS-SPME-GC-MS

Zhengfei Luo [1,†], Linlong Ma [2,†], Yangtao Zhang [3], Yanhong Liu [1], Rui Yang [1], Xuean Dai [1], Tiantian Wang [1], Changmi Lv [1], Lifeng Zuo [1], Yanli Liu [2], Dan Cao [2], Haibo Yuan [4], Longfeng Yu [1,*] and Xiaofang Jin [2,*]

1 Yunnan Key Laboratory of Tea Germplasm Conservation and Utilization in the Lancang River Basin, College of Biotechnology and Engineering, West Yunnan University, Lincang 677000, China; luophy@126.com (Z.L.); 15125951384@163.com (Y.L.); 15887891889@163.com (R.Y.); lzf13759358903@163.com (X.D.); 19527071516@163.com (T.W.); 17874128563@163.com (C.L.); m19869498400@163.com (L.Z.)
2 Key Laboratory of Tea Resources Comprehensive Utilization of Ministry of Agriculture and Rural Affairs, Fruit and Tea Research Institute, Hubei Academy of Agricultural Sciences, Wuhan 430064, China; malinlong@hbaas.com (L.M.); zkslyl@163.com (Y.L.); skyiswide@163.com (D.C.)
3 Lincang Inspection Testing and Certification Institute, Lincang 677000, China; 13988383009@163.com
4 Tea Research Institute, Chinese Academy of Agricultural Sciences, Hangzhou 310008, China; 192168092@tricaas.com
* Correspondence: longfeng_yu@126.com (L.Y.); jxf1130@126.com (X.J.)
† These authors contributed equally to this work.

Abstract: Drying methods play a crucial role in the formation of green tea aromas. This study investigated the aroma characteristics and volatile component profiles of large-leaf green tea under hot-air drying, pan-fired drying, and sun drying. The results revealed significant differences in the sensory aroma characteristics and volatile components of the large-leaf green tea among the three drying methods. The pan-fire-dried green tea (PDGT) exhibited a distinct roasted aroma, while the hot-air-dried green tea (HDGT) and sun-dried green tea (SDGT) displayed a faint scent and lasting aroma characteristics, with the SDGT additionally featuring a noticeable sun-dried odor. A total of 48 differential volatile components were identified, among which β-Ionone, (E)-β-Ionone, 2,2,6-Trimethylcyclohexanone, Dihydroactinidiolide, BenzeneacetAldehyde, 2-Pentylfuran, 1,1,6-Trimethyl-1,2-dihydronaphthalene, δ-Cadinene, β-Myrcene, Geranylacetone, o-Cymene, 6-Methyl-5-hepten-2-one, (E)-β-Ocimene, and BenzAldehyde were identified as the primary contributors to the aroma differences among the three large-leaf green teas. Additionally, 43 differential volatile compounds were found to be significantly correlated with at least one of the aroma types (floral, sweet, green, faint scent, nutty, or roasted). The findings of this study provide a theoretical foundation for understanding the formation of aroma qualities in large-leaf green tea and offer valuable insights for improving its aromatic characteristics.

Keywords: large-leaf tea variety; green tea; drying methods; volatile components

1. Introduction

Green tea is the most produced and consumed type of tea in China, and is highly favored by consumers for its unique flavor qualities and significant health benefits [1,2]. Aroma as a key factor for evaluating the quality of green tea, greatly influences the selection and acceptance of tea by tea drinkers [2,3]. Green tea is made from fresh tea leaves through five key processes: spreading, fixation, rolling, and drying [4]. Among them, drying, as the final processing step, plays a crucial role in the formation and fixation of the tea's aroma quality [5]. Under high temperatures, low-boiling-point, grassy-smelling alcohols and aldehydes volatilize significantly, while higher-boiling-point alcohols, ketones, and acids are retained better, thereby altering the content and relative proportions of aroma components [6,7]. Simultaneously, high temperatures enhance the oxidative degradation, isomerization, and Maillard reactions in the substances within the tea, ultimately forming an extremely complex yet harmonious green tea aroma [4,8].

Different tea plant varieties exhibit significant differences in the content and composition of aroma precursors, as well as the characteristics of endogenous glycosidase enzymes. Even when processed using the same techniques, the resulting tea aromas can vary markedly [9]. The large-leaf tea variety (*C. sinensis* var. assamica) is widely cultivated in the southwestern and southern regions of China. Known for its rich content of intrinsic compounds, it produces teas with distinct qualities compared to those made from *C. sinensis* var. sinensis [10,11]. Processing green tea from large-leaf tea varieties primarily employs three traditional drying methods: hot-air drying, pan-fired drying, and sun drying [12]. Among them, sun drying is more commonly applied to large-leaf tea varieties than to *C. sinensis* var. sinensis [13]. For example, sun-dried green tea produced from the Yunnan large-leaf tea variety is particularly renowned in Yunnan. It can be drunk directly or used as a raw material for processing Pu-erh tea [14,15].

Different drying methods significantly alter the aroma of green tea. For example, pan-fired drying produces a richer array of volatile compounds compared to hot-air drying and is conducive to the formation of chestnut-like and roasted aromas in green tea [16]. During sun drying, the extensive degradation of fatty acids and carotenoids into aldehydes and terpenoid ketones contributes to the faint scent characteristic of sun-dried green tea [17]. However, previous research has primarily focused on *C. sinensis* var. sinensis, with most attention given to pan-fired drying, hot-air drying, and novel drying techniques [16,18]. There has been no systematic study on the aromatic qualities of large-leaf green tea under different drying methods. Therefore, this study utilized fresh leaves of the Mengku large-leaf tea variety (one bud and two leaves) as material. After spreading, fixation, and rolling, the leaves were processed using three drying methods: hot-air drying, pan-fired drying, and sun drying, resulting in hot-air drying green tea (HDGT), pan-fired drying green tea (PDGT), and sun-drying green tea (SDGT). The volatile components of these three large-leaf green teas were systematically analyzed using headspace solid-phase microextraction combined with gas chromatography–mass spectrometry (HS-SPME-GC-MS) and sensory evaluation methods, and key aroma-active components were identified. The aim of the research was to understand the aroma qualities of large-leaf green tea under different drying methods, and provide a theoretical basis for enhancing its aroma quality.

2. Materials and Methods

2.1. Experimental Materials

In April 2023, fresh leaves (FL) of the Mengku large-leaf variety (one bud and two leaves) were harvested at the experimental base of West Yunnan University (located in Mengku Town, Shuangjiang County). The FL underwent spreading, fixation, and rolling, followed by processing into green tea using three drying methods: hot-air drying, pan-fired drying, and sun drying. Specifically, the spreading process involved placing fresh leaves at a density of 0.75 kg/m^2 under conditions of room temperature 18 °C~20 °C, and approximately 75% relative humidity until the moisture content in the leaves decreased to around 65%. For fixation, the spread leaves were processed in a drum fixation machine (6CST-80, Hangzhou Chunjiang Tea Machinery Co., Ltd., Zhejiang, China) at 280 °C for 2 min. Rolling was carried out using a rolling machine (6CR-65, Zhejiang Chunjiang Tea Machinery Co., Ltd., Hangzhou, China) for 30 min. For drying, HDGT was dried in a hot-air dryer (6CHM-901, Zhejiang Fuyang Tea Machinery Co., Ltd., Hangzhou, China) at 110 °C~120 °C for 10~15 min, followed by 80 °C for 15~20 min. PDGT was processed in a drum fixation machine (6CST-80, Zhejiang Chunjiang Tea Machinery Co., Ltd., Hangzhou, China) at 180 °C~250 °C for 25 min, followed by 90 °C~95 °C for 30 min. SDGT was dried in bamboo trays under sunlight at 25 °C~30 °C, with the leaves turned every hour. The moisture content of all tea samples was controlled to below 6%, and the moisture content of tea samples using a rapid moisture analyzer (HS153, Mettler Toledo, Zurich, Switzerland). Finally, 500 g samples of FL, HDGT, PDGT, and SDGT were collected, each in triplicate, and stored in a −80 °C freezer for further use.

2.2. Experiment Methods

2.2.1. Sensory Evaluation and Quantitative Descriptive Analysis (QDA)

The aroma characteristics of the tea samples were subjected to sensory evaluation according to GB/T 23776-2018 [19]. Meanwhile, six experienced evaluators (three females and three males) conducted QDA of the tea samples' aromas, using the method developed by Yue et al. (2023) [20]. In brief, precisely 3.0 g of each tea sample was weighed and brewed with 150 mL of boiling water for 5 min. The tea infusion was then transferred into a black-stoppered conical flask and randomly coded. After assessing the aroma characteristics of all tea samples, six primary aroma types (green, floral, sweet, nutty, roasted, and faint scent) were selected for QDA. A 10-point scale was used to rate the intensity of these six aroma attributes in the tea infusion, where 0 = absent, 3 = weak, 5 = moderate, 7 = strong, and 10 = very strong.

2.2.2. Determination of Volatile Components Content by HS-SPME-GC-MS

The tea samples were retrieved from the −80 °C freezer and subjected to freeze-drying. They were then ground using liquid nitrogen and vortex mixed to ensure homogeneity. Precisely 0.5 g of each sample was weighed into a headspace vial, followed by the addition of 5 mL of saturated NaCl solution and 20 µL of internal standard solution (10 $\mu g \cdot mL^{-1}$ 3-Hexanone-2,2,4,4-d4). The mixture was shaken at a constant temperature of 60 °C for 5 min. A 120 µm DVB/CWR/PDMS fiber was inserted into the headspace vial for HS-SPME for 15 min, followed by desorption at 250 °C for 5 min. The extracted compounds were then separated and identified using GC-MS. GC–MS analysis was carried out using a GC system (8890, Agilent Technologies, Santa Clara, CA, USA) coupled with an MS detector (7000D, Agilent Technologies, Santa Clara, CA, USA). Chromatographic conditions: A DB-5MS capillary column (30 m × 0.25 mm × 0.25 µm, Agilent J&W Scientific, Folsom, CA, USA) was used, with high-purity helium (purity ≥ 99.99%) as the carrier gas at a flow rate of 1.2 mL/min. The injector temperature was set to 250 °C, and the injection was performed in splitless mode with a solvent delay of 3.5 min. The temperature program was as follows: hold at 40 °C for 3.5 min, increase to 100 °C at a rate of 10 °C/min, then to 180 °C at 7 °C/min, and finally to 280 °C at 25 °C/min, holding for 5 min. Mass spectrometry conditions: An electron ionization source was used, with the ion source temperature set to 230 °C, the quadrupole temperature to 150 °C, and the MS interface temperature to 280 °C. The electron energy was 70 eV, and the scanning mode was selective ion monitoring, with precise scanning of qualitative and quantitative ions.

2.2.3. Quantification of Volatile Components and Odor Activity Values (OAVs)

For each component, one quantitative ion and two or three qualitative ions were selected. In the mass spectra of the samples after background subtraction, the presence of both the selected qualitative and quantitative ions served as the identification criterion. The quantitative ion was used for integration and calibration to enhance the accuracy of quantification. The relative content of volatile component i (X_i) was calculated using the following formula:

$$X_i = \frac{V_s \times C_s \times I_i}{1000 \times M \times I_s} \quad (1)$$

In this equation, vs. is the volume of the internal standard (μL), C_s is the concentration of the internal standard (μg·mL^{-1}), M is the mass of the sample (g), I_s is the peak area of the internal standard, and I_i is the peak area of component i in the sample.

The OAV of each volatile component i (OAV_i) in the sample was calculated using the following formula:

$$OAV_i = \frac{X_i}{T_i} \quad (2)$$

In this equation, X_i is the concentration of component i (μg·g^{-1}), and T_i is the odor threshold of compound component i (μg·g^{-1}).

2.2.4. Data Analysis

All experiments were conducted with three biological replicates. Bar graphs were generated using GraphPad Prism 5 software. Principal component analysis (PCA) was performed using SIMCA 14.1 software. Differential analysis was conducted using SPSS 19.0 software. Radar charts, heatmaps, Venn diagrams, and correlation analyses were created using an online platform available at https://www.chiplot.online (accessed on 12 February 2025).

3. Results and Discussion

3.1. Sensory Evaluation of Aroma in Large-Leaf Green Tea Under Different Drying Methods

As shown in Figure 1A, significant differences were observed in the sensory aroma characteristics of large-leaf green tea processed by the three different drying methods. PDGT exhibited a pronounced roasted aroma, while HDGT and SDGT displayed faint scents and lasting aroma characteristics, with SDGT additionally featuring a distinct sun-dried odor. The aroma profiles of these large-leaf green teas prepared under different drying methods are consistent with previous findings in *C. sinensis* var. Sinensis [16,21]. To provide a more detailed and accurate description of the aroma characteristics of the three types of large-leaf green tea, six aroma attributes were evaluated by QDA. As shown in Figure 1B, significant differences were observed in all six aroma attributes among the three tea samples, with nutty, green, roasted, and faint scents showing highly significant differences ($p < 0.01$). PDGT exhibited the most intense roasted and nutty aromas, while the other four attributes were the weakest. HDGT scored the highest in faint scent, with the other five attributes falling between the values of PDGT and SDGT. SDGT demonstrated the most intense green, floral, and sweet aromas, along with the weakest roasted and nutty aromas. These results indicate that different drying methods uniquely influence the sensory characteristics and intensity of aromas in large-leaf green tea. These effects may stem from changes in the content of certain volatile compounds under different drying conditions, leading to distinct aromatic profiles [22].

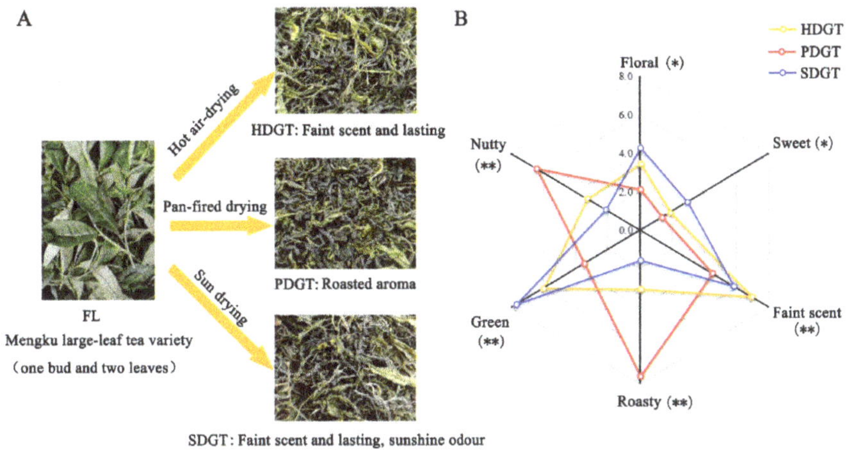

Figure 1. Sensory evaluation of aroma in large-leaf green tea under different drying methods. (**A**) Sensory description of aroma. (**B**) QDA radar chart. "*" indicates significant differences ($p < 0.05$), "**" indicates highly significant differences ($p < 0.01$).

3.2. Analysis of Volatile Components Profile in Large-Leaf Green Tea Under Different Drying Methods

The volatile components of PDGT, HDGT, SDGT, and FL were analyzed and identified using HS-SPME-GC-MS (Figure 2A). A total of 568 volatile components were detected, including 145 terpenoids, 90 esters, 74 heterocyclics, 53 ketones, 50 hydrocarbons, 42 aldehydes, 38 alcohols, 27 aromatics, and 49 others. The content of volatile components varied significantly among the four tea samples, with FL and SDGT exhibiting notably higher levels compared to PDGT and HDGT. PCA showed that FL was clearly distinguished from PDGT, HDGT, and SDGT, indicating that the drying process significantly altered the aroma components of FL when processed into green tea (Figure 2B). Meanwhile, PDGT and HDGT were distinctly separated from SDGT but clustered closer together, suggesting a certain similarity in their volatile components, while both differed significantly from SDGT. These results showed that different drying methods significantly influence the volatile component content of large-leaf green tea. The effect of sun drying on the volatile components of the green tea meant that it could be clearly distinguished from the samples that underwent hot-air drying and pan-fired drying.

As shown in Figure 2C, after FL is processed into PDGT, HDGT, and SDGT, the total content of volatile components significantly decreases by 54.25%, 56.52%, and 46.45%, respectively. The total volatile content of SDGT was 46.92 $\mu g \cdot g^{-1}$, significantly higher than that of PDGT and HDGT, while no significant difference was observed between PDGT and HDGT. The original volatile components in fresh tea leaves undergo substantial degradation and loss under high temperatures, leading to an overall decline in volatile content [6,7]. Meanwhile, similar results have also been found in other dried materials. For example, Zhang et al. [23] found that the operation of drying makes most of the volatile components in fresh *Malus asiatica* nakai decrease or vanish. Compared to SDGT, the higher drying temperatures of PDGT and HDGT resulted in significantly increased degradation and loss of volatile components, whereas these volatiles were better preserved in SDGT [6,7].

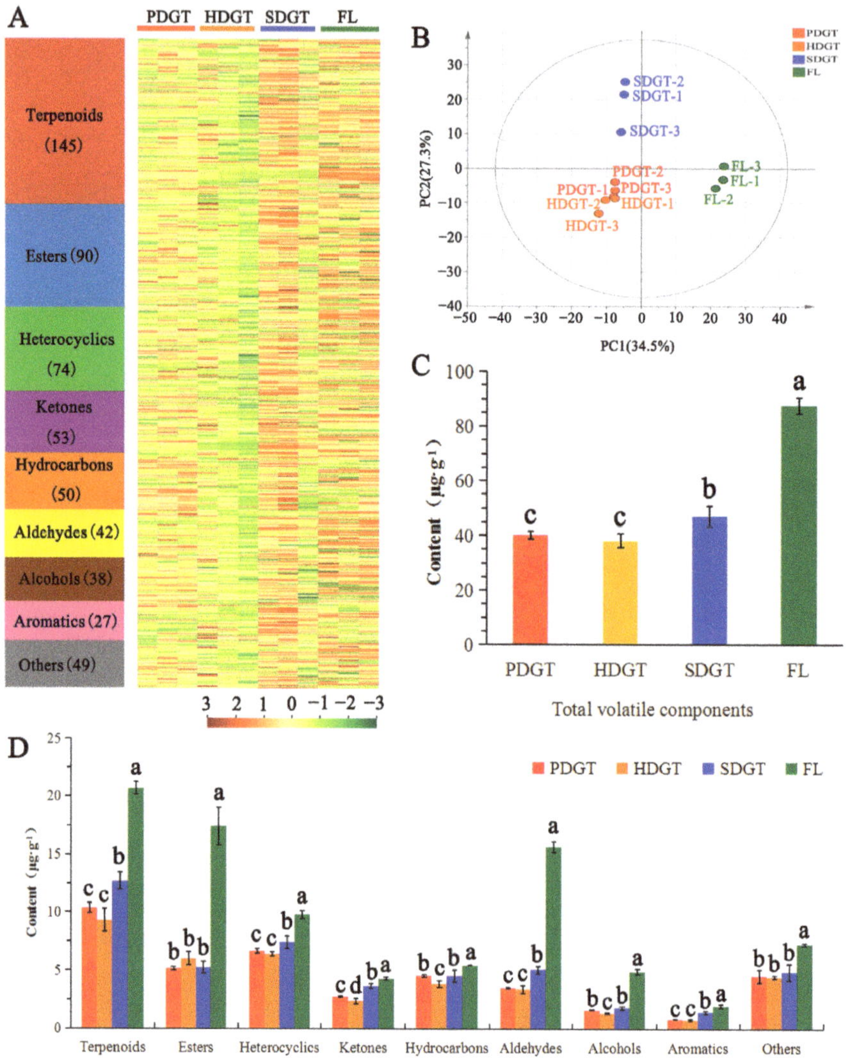

Figure 2. Analysis of volatile components in large-leaf green tea under different drying methods. (**A**) Heat map of the content of volatile components; (**B**) PCA score plot derived from the concentrations of volatile components; (**C**) the total content of volatile components; (**D**) the content of volatile component categories. Different lowercase letters indicate significant differences ($p < 0.05$).

As indicated by the contents of different volatile component categories (Figure 2D), the significant degradation and loss of terpenoids, esters, and aldehydes in FL were key factors contributing to the lower total volatile content in PDGT, HDGT, and SDGT. Compared to FL, the contents of terpenoids in PDGT, HDGT, and SDGT decreases by 49.98%, 54.96%, and 38.58%, the content of esters decreases by 70.39%, 65.43%, and 69.76%, the content of aldehydes decreases by 77.52%, 78.24%, and 67.55%. These results exhibit substantial discrepancies when compared to the changes in aroma during the drying process of chili peppers. In chili peppers, the hot-air drying process led to an increase in acids, furans, and sulfide contents, while the levels of alcohols, esters, and olefins decreased [24].

Among the three drying methods, the contents of terpenoids, heterocyclics, aldehydes, and aromatics in the SDGT were significantly higher than those in the PDGT and HDGT, while no significant differences were observed between PDGT and HDGT. The contents of hydrocarbons and alcohols in PDGT and SDGT were significantly higher than those in HDGT, with no significant difference between PDGT and SDGT. Ketones showed significant differences among the three drying methods, with SDGT having the highest content at 3.68 $\mu g \cdot g^{-1}$, followed by PDGT at 2.76 $\mu g \cdot g^{-1}$, and HDGT with the lowest at 2.39 $\mu g \cdot g^{-1}$. No significant differences were observed in the contents of esters and others among the three drying methods.

3.3. Analysis of Differential Volatile Components in Large-Leaf Green Tea Under Different Drying Methods

The OPLS-DA method was applied to differentiate and analyze the volatile components among PDGT, HDGT, SDGT, and FL. The screening of differential volatile components was based on a variable importance in the project (VIP) ≥ 1, a fold change ≥ 2 or fold change ≤ 0.5, and $p < 0.05$, and excluding volatile components with low contents ($\leq$0.05 $\mu g \cdot g^{-1}$) across all four samples. As shown in Figure 3A, a total of 114 key differential volatile components were identified, including 32 terpenoids, 17 ketones, 16 esters, 14 heterocyclics, 11 aldehydes, 6 hydrocarbons, 6 alcohols, 3 aromatics, and 9 others. Most of these differential volatile components were significantly higher in FL compared to the three different drying methods of large-leaf green tea. The Venn diagram analysis revealed that the number of differential volatile components when comparing PDGT vs. FL, HDGT vs. FL, and SDGT vs. FL was 77, 81, and 79, respectively, while there were 4, 12, and 9, unique differential components, respectively (Figure 3B). Figure 3C shows that there were 48 differential volatile components among the three large-leaf green teas under the different drying methods. The number of differential volatile components when comparing PDGT vs. HDGT, PDGT vs. SDGT, and HDGT vs. SDGT was 7, 31, and 37, respectively, while the number of unique differential components was 3, 8, and 10, respectively. These findings indicate that the processing of fresh leaves into green tea results in significant changes in aroma components, and different drying methods influence the formation of the aroma in large-leaf green tea.

To further screen the markers in large-leaf green tea under different drying methods, the top six components with the largest fold changes in the comparisons between PDGT and HDGT, PDGT and SDGT, and HDGT and SDGT were selected for analysis based on their fold changes. In the PDGT vs. HDGT comparison, the six components with the largest fold changes were α-Ocimene, (E)-β-Ocimene, diisobutyl phthalate, ethotoin, 2,4-Diamino-6-nitrotoluene, and dihydroactinidiolide. Among these, α-Ocimene, (E)-β-Ocimene, and dihydroactinidiolide were present in higher amounts in PDGT compared to HDGT, while the others were lower in PDGT than in HDGT (Figure 3D). In the PDGT vs. SDGT comparison, the six components with the largest fold changes were 2,5-Octanedione, cyclooctanemethanol, 2,2,6-Trimethylcyclohexanone, 2-Methoxyethyl cyanoacetate, benzeneacetaldehyde, and 2,2,4,6,6-Pentamethylheptane. Among these, 2,5-Octanedione, 2-Methoxyethyl cyanoacetate, and 2,2,4,6,6-Pentamethylheptane were present in higher amounts in PDGT compared to SDGT, while the others were lower in PDGT than in SDGT (Figure 3E). In the HDGT vs. SDGT comparison, the six components were 2,5-Octanedione, (E)-β-Ocimene, α-Ocimene, cyclooctanemethanol, 2,2,6-Trimethylcyclohexanone, and 6-Methyl-5-hepten-2-one. Among these, only 2,5-Octanedione was present in higher amounts in HDGT compared to SDGT, while the others were lower in HDGT than in SDGT (Figure 3F).

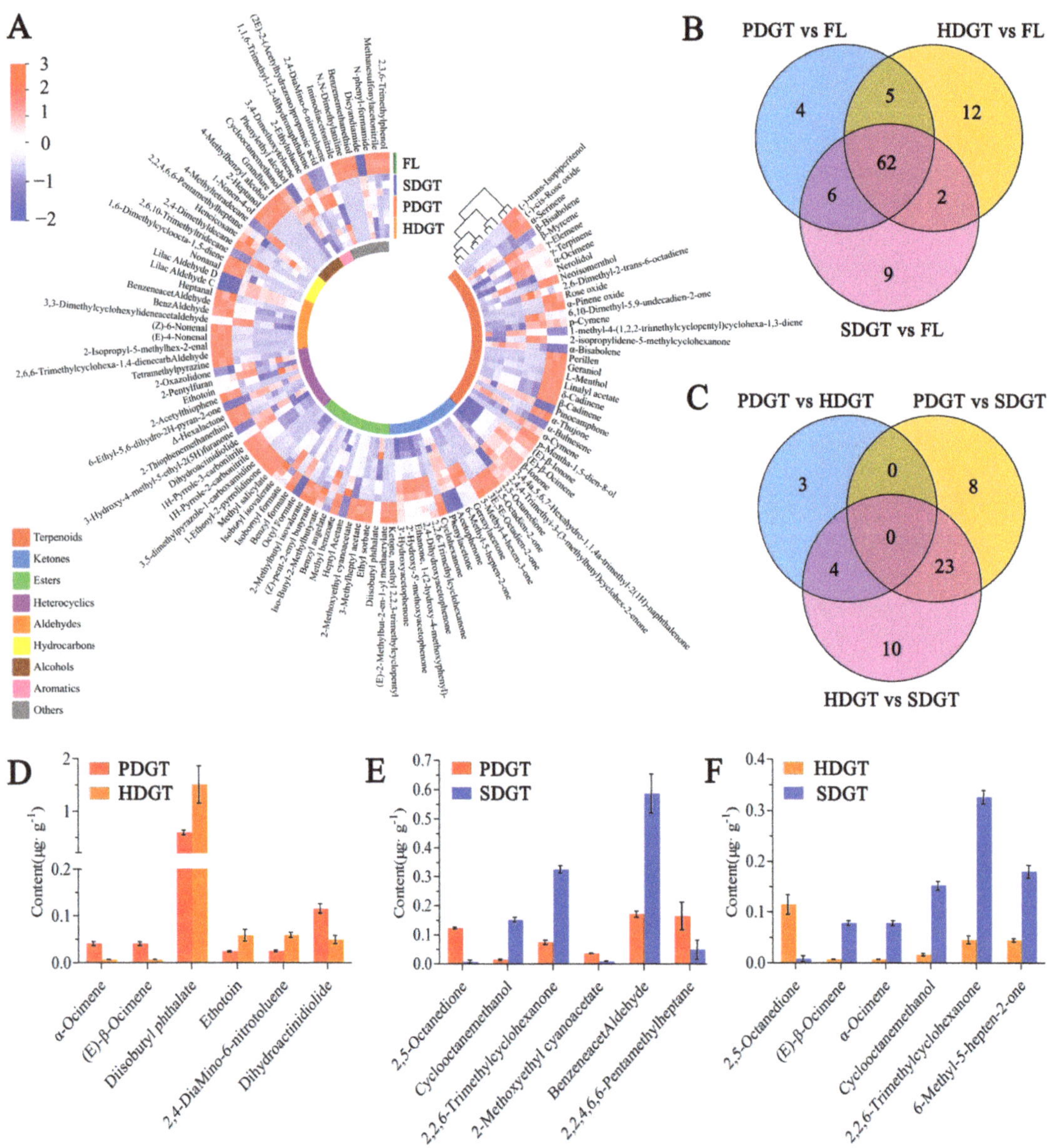

Figure 3. Analysis of differential volatile components in large-leaf green tea under different drying methods. (**A**) Heat map of the content of differential volatile components; (**B**) Venn diagram of the number of differential volatile components in the comparison groups PDGT vs. FL, HDGT vs. FL, and SDGT vs. FL; (**C**) Venn diagram of the number of differential volatile components in the comparisons between PDGT and HDGT, PDGT and SDGT, and HDGT and SDGT; (**D**) the content of the top six components with the largest fold changes in the comparison groups PDGT vs. HDGT; (**E**) The content of the top six components with the largest fold changes in the comparison groups PDGT vs. SDGT; (**F**) the content of the top six components with the largest fold changes in the comparison between HDGT and SDGT.

3.4. Analysis of OAVs of Differential Volatile Components in Large-Leaf Green Tea Under Different Drying Methods

The OAV is a comprehensive indicator that considers both the concentration of volatile components and their odor thresholds, avoiding the misconception of solely relying on concentration to reflect their contribution to aroma. It is a method used to evaluate the importance of volatile components to the aroma quality of samples. In this study, the odor thresholds of the volatile components were primarily obtained from the existing literature [25–27]. From the 48 differential volatile components identified from the three different drying methods used for the large-leaf green tea, 14 volatile components with relatively high OAVs were selected, i.e., those with an OAV > 1 in at least one tea sample. As shown in Table 1, the mean OAVs of the 14 volatile components, in descending order, were β-Ionone, (E)-β-Ionone, 2,2,6-Trimethylcyclohexanone, Dihydroactinidiolide, Benzeneacetaldehyde, 2-Pentylfuran, 1,1,6-Trimethyl-1,2-dihydronaphthalene, δ-Cadinene, β-Myrcene, Geranylacetone, o-Cymene, 6-Methyl-5-hepten-2-one, (E)-β-Ocimene, and Benzaldehyde. Most of these components have been previously proven to be key contributors to tea aroma formation. For example, β-ionone is a key aroma component in Xinyang Maojian tea [28], and (E)-β-ionone is a key aroma component in Xiaokeng green tea [29]. Generally, substances with OAV > 1 are considered to contribute to aroma, while those with OAV > 10 are considered to make a significant contribution to aroma [25]. In this study, β-Ionone, (E)-β-Ionone, 2,2,6-Trimethylcyclohexanone, dihydroactinidiolide, benzeneacetaldehyde, 2-Pentylfuran, and 1,1,6-Trimethyl-1,2-dihydronaphthalene had OAVs greater than 10 in all three of the drying methods used for the large-leaf green tea. δ-Cadinene had OAVs greater than 10 in PDGT and HDGT, while β-Myrcene and Geranylacetone had OAVs greater than 10 in SDGT.

Table 1. Analysis of OAV of differential volatile components in Large-leaf green tea under different drying methods.

No.	Compounds	Threshold ($ug \cdot g^{-1}$) [21–23]	PDGT	HDGT	SDGT	Mean
1	β-Ionone	0.000007	133,873.74	77,360.89	162,112.84	124,449.16
2	(E)-β-Ionone	0.0002	4685.58	2707.63	5673.95	4355.72
3	2,2,6-Trimethylcyclohexanone	0.0001	745.71	454.17	3270.06	1489.98
4	Dihydroactinidiolide	0.0021	54.75	23.70	88.61	55.69
5	BenzeneacetAldehyde	0.0063	27.23	38.26	93.28	52.93
6	2-Pentylfuran	0.006	15.10	10.30	38.19	21.20
7	1,1,6-Trimethyl-1,2-dihydronaphthalene	0.0025	11.07	17.78	29.55	19.47
8	δ-Cadinene	0.0015	16.08	10.90	6.76	11.25
9	β-Myrcene	0.015	5.76	7.31	14.82	9.30
10	Geranylacetone	0.01	4.39	3.34	12.98	6.90
11	o-Cymene	0.01144	3.74	4.20	8.58	5.51
12	6-Methyl-5-hepten-2-one	0.05	1.11	0.89	3.60	1.87
13	(E)-β-Ocimene	0.034	1.20	0.21	2.30	1.24
14	BenzAldehyde	0.35	0.42	0.36	1.01	0.59

Note: PDGT represents pan-fire-dried green tea, HDGT represents hot-air-dried green tea, and SDGT represents sun-dried green tea.

To further understand the differences in aroma formation in the large-leaf green tea prepared under different drying methods, the OAVs of 14 volatile components in PDGT, HDGT, and SDGT were compared and analyzed. In SDGT, the OAVs of all 14 volatile components, except for δ-Cadinene, were higher than those in PDGT and HDGT. Most of these substances have been confirmed to contribute to the formation of floral, sweet, and green aromas in tea. For example, (E)-β-Ionone is an important substance for the formation of floral aromas in the microwave second drying of green tea [30], β-Myrcene is a substance that helps enhance the sweet aroma of black tea [31], and Dihydroactinidiolide is a major contributor to the grassy and woody aromas in sun-dried green tea [21]. This may be the main reason why SDGT scored higher in sensory evaluations for floral, sweet, and green aroma types compared to PDGT and HDGT. Meanwhile, there is a significant antagonistic effect among volatile substances. The lower OAVs of these components in PDGT and HDGT may weaken the masking effects of faint scent, nutty, and roasted aroma types, ultimately making these aromas more perceptible [32,33]. Additionally, the OAVs of β-Ionone, (E)-β-Ionone, 2,2,6-Trimethylcyclohexanone, Dihydroactinidiolide, 2-Pentylfuran, δ-Cadinene, Geranylacetone, 6-Methyl-5-hepten-2-one, (E)-β-Ocimene, and BenzAldehyde in PDGT were higher than those in HDGT. Among these, 2-Pentylfuran, Geranylacetone, benzeneacetaldehyde, and (E)-β-Ionone have been confirmed to contribute to the formation of chestnut-like aromas in green tea [34,35]. This may be an important reason why PDGT scored higher in sensory evaluations for nutty and roasted aroma types compared to HDGT.

3.5. Correlation Analysis of Differential Volatile Components and Main Aroma Types

To further understand the key volatile components contributing to the formation of the main aroma types in large-leaf green tea, a correlation analysis was conducted between the content of 48 differential volatile components and the sensory QDA scores of six aroma types in large-leaf green tea processed using three drying methods. As shown in Figure 4, a total of 43 differential volatile components were significantly positively or negatively correlated with at least one of the key six aroma types: floral, sweet, green, faint scent, nutty, and roasted. Among these, the majority of differential volatile components showed significant positive correlations with floral, sweet, and green aroma types, and their odor descriptions were consistent with the corresponding substances, further confirming the reliability of the study results. The low content of floral, sweet, and green aroma components reduced their masking effect on nutty and roasted aroma components (which showed no significant differences), resulting in a pronounced negative correlation between most differential components and nutty/roasted aroma characteristics [32,33].

As shown in Figure 4, the δ-Cadinene, α-Serinene, nerolidol, 2,5-Octanedione, 2-Methoxyethyl cyanoacetate, α-Bisabolene, 2,2,4,6,6-Pentamethylheptane, and β-Cadinene 8 differential volatile components showed significant negative correlations with floral, sweet, and green aroma types but positive correlations with nutty and roasted aroma types. Additionally, the correlation between differential volatile components and the faint scent aroma type of large-leaf green tea was not strong, showing significant positive correlations only with diisobutyl phthalate, 2,4-Dihydroxyacetophenone, and ethotoin, with correlation coefficients of 0.785, 0.874, and 0.800, respectively. Among the 14 volatile components with relatively high OAVs, β-Myrcene, 6-Methyl-5-hepten-2-one, benzAldehyde, benzeneacetAldehyde, 2,2,6-Trimethylcyclohexanone, 1,1,6-Trimethyl-1,2-dihydronaphthalene, and o-Cymene showed significant positive correlations with floral, sweet, and green aroma types. Among these, β-Myrcene, benzeneacetAldehyde, 1,1,6-Trimethyl-1,2-dihydronaphthalene, and o-Cymene showed significant negative correlations with nutty and roasty aroma types, with correlation coefficients all reaching 0.700. Furthermore, geranylacetone and 2-Pentylfuran, and (E)-β-Ocimene showed significant positive correlations with sweet aroma types, with correlation coefficients of 0.850, 0.857, and 0.674. Geranylacetone and 2-Pentylfuran showed a significant positive correlation with the green aroma type, with correlation coefficients of 0.688 and 0.669. Dihydroactinidiolide, (E)-β-Ionone, and β-Ionone showed no significant relationship with any of the six aroma types.

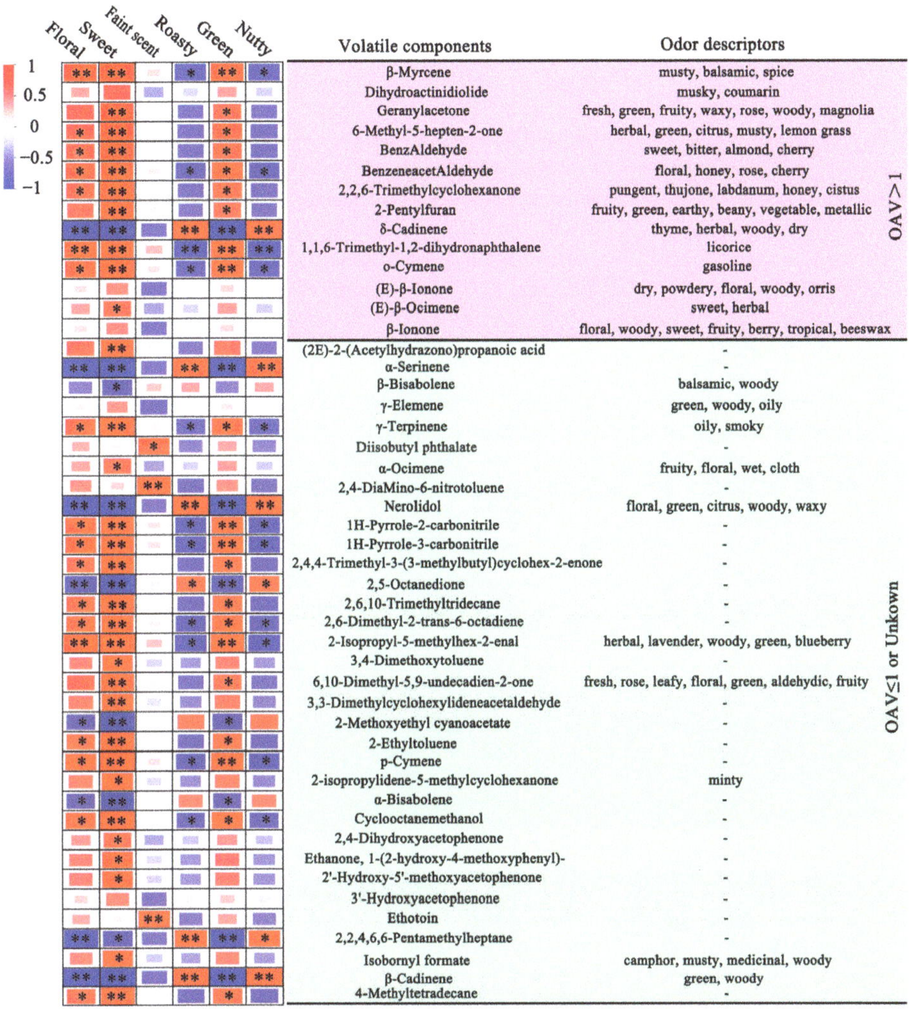

Figure 4. Correlation analysis of differential volatile components and main aroma types. Odor descriptors were obtained from reputable databases: The Good Scents Company (http://www.thegoodscentscompany.com), Perflavory Information System (http://perflavory.com (accessed on 10 December 2024)), The LRI & Odor Database—Odor Data (http://www.odour.org.uk/odour/index.html (accessed on 10 December 2024)), and Food Flavor Lab (http://foodflavorlab.cn/#/home (accessed on 10 December 2024)). "-" signifies that the odor descriptor was not found in the search. "*" indicates significant differences ($p < 0.05$), "**" indicates highly significant differences ($p < 0.01$).

4. Conclusions

In this study, the differences in volatile components of large-leaf green tea prepared under three drying methods were systematically analyzed using HS-SPME-GC-MS combined with sensory evaluation. The results showed that the three drying methods significantly influenced the sensory aroma characteristics of large-leaf green tea. Pan-fire-dried green tea exhibited a distinct roasted aroma, while hot-air-dried and sun-dried green tea displayed faint and lasting aroma characteristics, with the sun-dried green tea additionally featuring a noticeable sun-dried odor. A total of 568 volatile components were detected in the large-leaf green tea prepared under the three drying methods, and 48 differential volatile components were identified. Among these, the effect of sun drying on the volatile components of green tea was clearly distinguished from that of hot-air drying and pan-fire drying. Fourteen volatile components with relatively high OAV were identified as the primary contributors to the aroma differences among the three types of tea. Forty-three differential volatile components were found to be significantly positively or negatively correlated with at least one of the aroma types (floral, sweet, green, faint scent, nutty, or roasted). Most of these differential volatile components showed positive correlations with floral, sweet, and green aromas, while exhibiting negative correlations with nutty and roasted aromas. This study provides a scientific analysis of the effects of three drying methods on the aroma of large-leaf green tea, offering a theoretical basis for improving its aroma quality.

Author Contributions: Conceptualization, Y.L. (Yanhong Liu) and L.Y.; data curation, Z.L. and Y.Z.; formal analysis, L.M., R.Y., H.Y. and T.W.; funding acquisition, L.Y. and X.J.; investigation, X.D. and C.L.; methodology, L.Z. and Y.L. (Yanli Liu); project administration, Y.L. (Yanhong Liu) and L.Y.; resources, Y.Z., L.Z. and L.Y.; software, D.C. and Y.L. (Yanli Liu); supervision, Y.L. (Yanhong Liu) and X.J.; validation, Z.L., R.Y. and C.L.; visualization, L.M. and Y.L.; writing—original draft, Z.L. and L.M.; writing—review and editing, Z.L. and X.J. All authors have read and agreed to the published version of the manuscript.

Funding: This work was supported by the Scientific Research Fund Project of Yunnan Provincial Department of Education (2024J1110), "Xinglin Rencai" Program of Lincang City-Special Project for Young Scientific and Technological Talents, the Open Fund of Yunnan Key Laboratory of Tea Germplasm Conservation and Utilization in the Lancang River Basin, the Yunnan Provincial Science and Technology Talent and Platform Plan—Academician (Expert) Workstation (202305AF150058), the Lincang City Science and Technology Innovation Team (202204AC100001-TD01).

Institutional Review Board Statement: In compliance with the "Ethics Review Measures for Life Sciences and Medical Research Involving Humans" issued by the National Health Commission of China in 2023 (Document No. 4), specifically Article 32, our study qualifies for exemption fromethical review as it does not present any potential risk to the participants.

Informed Consent Statement: Informed consent was obtained from all subjects involved in the study.

Data Availability Statement: The original contributions presented in the study are included in the article. Further inquiries can be directed to the corresponding authors.

Conflicts of Interest: The authors declare no conflicts of interest.

References

1. Liu, Z.; Chen, Q.; Zhang, C.; Ni, L. Comparative study of the anti-obesity and gut microbiota modulation effects of green tea phenolics and their oxidation products in high-fat-induced obese mice. *Food Chem.* **2021**, *367*, 130735. [PubMed]
2. Liu, N.; Shen, S.; Huang, L.; Deng, G.; Wei, Y.; Ning, J.; Wang, Y. Revelation of volatile contributions in green teas with different aroma types by GC–MS and GC–IMS. *Food Res. Int.* **2023**, *169*, 112845.
3. Yang, Z.; Baldermann, S.; Watanabe, N. Recent studies of the volatile compounds in tea. *Food Res. Int.* **2013**, *53*, 585–599.
4. Wan, X. *Tea Biochemistry*, 3rd ed.; China Agriculture Press: Beijing, China, 2003. (In Chinese)
5. Kumar, K.R.; Dashora, K.; Kumar, S.; Dharmaraja, S.; Sanyal, S.; Aditya, K.; Kumar, R. A review of drying technology in tea sector of industrial, non-conventional and renewable energy based drying systems. *Appl. Therm. Eng.* **2023**, *224*, 120118.

6. Teshome, K.; Debela, A.; Garedew, W. Effect of drying temperature and duration on biochemical composition and quality of black tea (*Camellia sinensis* L.) O. Kuntze at Wush Wush, South Western Ethiopia. *Asian J. Plant Sci.* **2014**, *12*, 235–240.
7. Wang, B.; Qu, F.; Wang, P.; Zhao, L.; Wang, Z.; Han, Y.; Zhang, X. Characterization analysis of flavor compounds in green teas at different drying temperature. *LWT* **2022**, *161*, 113394.
8. Chen, Z.; Li, Z.; Zhao, Y.; Zhu, M.; Li, J.; Wang, K. A meta-analysis of dynamic changes of key aroma compounds during black tea processing. *Food Biosci.* **2024**, *58*, 103784.
9. Han, Z.; Rana, M.; Liu, G.; Gao, M.; Li, D.; Wu, F.; Li, X.; Wan, X.; Wei, S. Green tea flavour determinants and their changes over manufacturing processes. *Food Chem.* **2016**, *212*, 739–748.
10. Yang, R.; Ye, C.; Xu, J.; Jiang, Y. Simultaneous analysis of purine alkaloids and catechins in *Camellia sinensis*, *Camellia ptilophylla* and *Camellia assamica var. kucha* by HPLC. *Food Chem.* **2007**, *100*, 1132–1136.
11. Liu, B.; Wang, L.; Zhou, J.; Tang, Y.; Cheng, H.; Wang, P.; Jiang, Y. Fingerprinting construction and genetic diversity analysis of Yunnan Dayezhong tea germplasm resources by ISSR markers. *J. Plant Genet. Res.* **2008**, *9*, 458–464.
12. Liu, F.; Wang, Y.; Li, C.; Wang, X.; Tang, X.; Li, T. Research status and prospect of tea drying technology. *Chin. Agric. Sci. Bull.* **2015**, *31*, 210–215.
13. Lv, H.; Zhang, Y.; Lin, Z.; Liang, Y. Processing and chemical constituents of Pu-erh tea: A review. *Food Res. Int.* **2013**, *53*, 608–618. [CrossRef]
14. Lu, M.; Jiang, Y.; Zhao, W.; Zhang, J.; Chen, Z.; Ning, J. Effects of different drying methods on color difference, taste and chemical components of Yunnan Congou black tea. *LWT* **2025**, *223*, 117696. [CrossRef]
15. Liu, B.; Chen, X.; Wu, X.; Zhang, W.; Wang, Z. Study of Pu'er raw materials grade classification by PCA and PLS-DA. *J. Tea Sci.* **2015**, *35*, 179–184.
16. Tu, Z.; Liu, Y.; Lin, J.; Lv, H.; Zhou, W.; Zhou, X.; Qian, Y.; Zeng, X.; He, W.; Ye, Y. Comparison of volatile and nonvolatile metabolites in green tea under hot-air drying and four heat-conduction drying patterns using widely targeted metabolomics. *Food Chem. X* **2023**, *19*, 100767. [CrossRef]
17. Yang, Y.; Kun, J.; Dai, H.; Tong, H. Changes in aroma compounds and selected precursors of sun-dried green tea during processing. *Food Sci.* **2024**, *45*, 133–142.
18. Li, N.; Yao, Z.; Ning, J.; Sun, L.; Lin, Q.; Zhu, X.; Li, C.; Zheng, X.; Jin, J. Comparison of different drying technologies for green tea: Changes in color, non-volatile and volatile compounds. *Food Chem. X* **2024**, *24*, 101935. [CrossRef]
19. *GB/T 23776-2018*; Methodology for Sensory Evaluation of Tea. Chinese National Standard: Beijing, China, 2018.
20. Yue, C.; Cao, H.; Zhang, S.; Hao, Z.; Wu, Z.; Luo, L.; Zeng, L. Aroma characteristics of Wuyi rock tea prepared from 16 different tea plant varieties. *Food Chem. X* **2023**, *17*, 100586.
21. You, Q.; Shi, Y.; Zhu, Y.; Yang, G.; Yan, H.; Lin, Z.; Lv, H. Effect of different processing technologies on the key aroma-active compounds of green tea. *Food Sci.* **2023**, *44*, 194–200.
22. Lu, M.; Sheng, C.; Ke, H.; Li, T.; Liu, Q.; Zhang, J.; Li, L.; Wang, Y.; Ning, J. Revealing the differences in aroma of black tea under different drying methods based on GC–MS, GC-O. *Food Chem. X* **2024**, *24*, 101782. [CrossRef]
23. Zhang, L.; Cao, S.; Ren, G.; Duan, X. The volatile compounds changes of *Malus asiatica* nakai under different drying methods. *J. Future Foods* **2025**, *5*, 276–282. [CrossRef]
24. Liu, M.; Hu, L.; Deng, N.; Cai, Y.; Li, H.; Zhang, B.; Wang, J. Effects of different hot-air drying methods on the dynamic changes in color, nutrient and aroma quality of three chili pepper (*Capsicum annuum* L.) varieties. *Food Chem. X* **2024**, *22*, 101262.
25. Zhu, J.; Niu, Y.; Xiao, Z. Characterization of the key aroma compounds in Laoshan green teas by application of odour activity value (OAV), gas chromatography-mass spectrometry-olfactometry (GC-MS-O) and comprehensive two-dimensional gas chromatography mass spectrometry (GC×GC-qMS). *Food Chem.* **2021**, *339*, 128136.
26. Yang, Y.; Ai, L.; Mu, Z.; Liu, H.; Yan, X.; Ni, L.; Zhang, H.; Xia, Y. Flavor compounds with high odor activity values (OAV > 1) dominate the aroma of aged Chinese rice wine (Huangjiu) by molecular association. *Food Chem.* **2022**, *383*, 132370. [CrossRef] [PubMed]
27. Gemert, L. *Odour Thresholds: Compilations of Odour Threshold Values in Air, Water and Other Media*; Oliemans Punter & Partners BV: Zeist, The Netherlands, 2011.
28. Cui, J.; Zhai, X.; Guo, D.; Du, W.; Gao, T.; Zhou, J.; Schwab, W.; Song, C. Characterization of key odorants in Xinyang Maojian green tea and their changes during the manufacturing process. *J. Agric. Food Chem.* **2022**, *70*, 279–288.
29. Gan, S.; Chen, Y.; Zhao, L.; Zhao, X.; Qisu, T.; Zhai, X.; Dai, Q. Characterization of the aroma-active compounds in Xiaokeng green tea by three pretreatment methods combined with gas chromatography-olfactometry (GC-O). *Food Res. Int.* **2024**, *187*, 114359. [CrossRef] [PubMed]
30. Wang, H.; Ouyang, W.; Yu, Y.; Wang, J.; Yuan, H.; Hua, J.; Jiang, Y. Analysis of non-volatile and volatile metabolites reveals the influence of second-drying heat transfer methods on green tea quality. *Food Chem. X* **2022**, *14*, 100354. [PubMed]

31. Yang, Y.; Xie, J.; Wang, Q.; Deng, Y.; Zhu, L.; Zhu, J.; Yuan, H.; Jiang, Y. Understanding the dynamic changes of volatile and non-volatile metabolites in black tea during processing by integrated volatolomics and UHPLC-HRMS analysis. *Food Chem.* **2024**, *432*, 137124. [CrossRef]
32. Zhu, J.; Chen, F.; Wang, L.; Niu, Y.; Xiao, Z. Evaluation of the synergism among volatile compounds in oolong tea infusion by odour threshold with sensory analysis and E-nose. *Food Chem.* **2017**, *221*, 1484–1490.
33. Hattori, S.; Akagaki, H.; Ujimori, T. Identification of volatile compounds which enhance odor notes in Japanese green tea using the OASIS (original aroma simultaneously input to the sniffing port) method. *Food Sci. Technol. Res.* **2005**, *11*, 171–174.
34. Zhu, Y.; Lv, H.; Shao, C.; Kang, S.; Zhang, Y.; Guo, L.; Dai, W.; Tan, J.; Peng, Q.; Lin, Z. Identification of key odorants responsible for chestnut-like aroma quality of green teas. *Food Res. Int.* **2018**, *108*, 74–82. [CrossRef] [PubMed]
35. Yin, P.; Xiao, Z.; Yu, F.; Guo, G. Research progress of chestnut-like aroma quality of green tea. *Food Res. Dev.* **2020**, *3*, 219–224.

foods

MDPI

Article

Predicting the Degree of Fresh Tea Leaves Withering Using Image Classification Confidence

Mengjie Wang [1,2,†], Yali Shi [1,†], Yaping Li [2], Hewei Meng [2], Zezhong Ding [1,2], Zhengrui Tian [1,2], Chunwang Dong [1] and Zhiwei Chen [1,*]

[1] Tea Research Institute of Shandong Academy of Agricultural Sciences, Jinan 250100, China; chuanyi2023@163.com (M.W.)

[2] College of Mechanical and Electrical Engineering, Shihezi University, Shihezi 832003, China

* Correspondence: zv.chen@foxmail.com

† These authors contributed equally to this work.

Abstract: Rapid and non-destructive detection methods for the withering degree of fresh tea leaves are crucial for ensuring high-quality tea production. Therefore, this study proposes a fresh tea withering degree detection model based on image classification confidence. The moisture percentage of fresh tea leaves is calculated by developing a weighted method that combines confidence levels and moisture labels, and the degree of withering is ultimately determined by incorporating the standard for wilted moisture content. To enhance the feature extraction ability and classification accuracy of the model, we introduce the Receptive-Field Attention Convolution (RFAConv) and Cross-Stage Feature Fusion Coordinate Attention (C2f_CA) modules. The experimental results demonstrate that the proposed model achieves a classification accuracy of 92.7%. Compared with the initial model, the detection accuracy was improved by 0.156. In evaluating the predictive performance of the model for moisture content, the correlation coefficients (Rp), root mean square error (RMSEP), and relative standard deviation (RPD) of category 1 in the test set were 0.9983, 0.006278, and 39.2513, respectively, and all performance were significantly better than PLS and CNN methods. This method enables accurate and rapid detection of tea leaf withering, providing crucial technical support for online determination during processing.

Keywords: fresh tea leaves; moisture content; confidence weighting; degree of withering; YOLOv8

1. Introduction

In the process of black tea production, withering is the first vital step and the key to making good quality black tea [1,2]. Wilting refers to a series of characteristic changes in the physical parameters of tea leaves, such as moisture content and color, under specific temperature and humidity conditions [3]. As withering progresses, the water content of fresh tea leaves gradually decreases, the leaves become soft, they shrink, their color darkens, and the grassy flavor gradually disappears, resulting in a distinctive fragrance [4–6]. Moisture content is an important indicator of the degree of withering of fresh tea leaves. The industry standard for wilting moisture content in fresh tea leaves is 58–66%. In actual production engineering, the degree of withering mainly relies on a combination of manual judgment and leaf moisture content detection. However, manual judgment is easily influenced by factors such as light, personal experience, and subjective judgment, hence it cannot accurately and objectively determine the degree of withering of fresh tea leaves. In addition, such judgments about color and texture are qualitative rather than standardized [7], making it difficult to distinguish subtle differences between colors and textures, which can easily lead to significant errors and inaccurate judgments. As such, traditional detection methods have the weaknesses of low accuracy, lack of objectivity, and slow speed [8].

In recent years, machine vision [9] combined with machine learning has been used effectively to extract image colors and textures for tasks such as image classification, detection and segmentation, achieving efficient image processing and analysis in various fields of agriculture. Wijaya et al. (2023) [10] used electronic noses and optimized machine learning algorithms to effectively evaluate and predict the quality of non-infused tea aroma, achieving the identification of the quality of non-infused Gambung Green Tea. Li et al. (2023) [11] used the least squares support vector machine qualitative model analysis to quickly detect the fermentation degree of Pu'er tea online, achieving a prediction accuracy of 99.3% on the fermentation prediction set. However, in terms of moisture content detection, Zhang et al. (2024) [12] collected near-infrared spectral data in seven stages of green tea processing and established a water content prediction model for green tea using partial least squares regression and backpropagation neural network, achieving high detection performance. Wang et al. (2023) [13] extracted nine color and five texture features from tea images and constructed two preprocessing methods and two optimization algorithms to predict the moisture content of fresh tea leaves through cross combination. However, machine learning detection methods can only input fixed fresh tea features (such as color and texture), which limits the model's ability to learn features.

Deep learning [14], as a branch of machine learning, is based on the core idea of extracting advanced features of images through multi-level neural networks, thereby achieving tasks such as target classification, recognition, and segmentation. In recent years, deep learning has developed rapidly and applied in various aspects of the tea fields. Chen et al. (2024) [15] proposed an improved YOLOV7 tea bud detection network with multiple sizes and targets in the field of view. By introducing attention mechanism and multi-feature fusion module, an average accuracy of 94.43% was achieved. Based on the regional convolutional neural network Mask-RCNN, Wang et al. (2023) [16] proposed a harvesting point localization method to classify, regress, and segment buds by training regression classification and preselected frames and eliminating the quantization error using regional feature aggregation. An et al. (2020) [17] proposed a withered water detection method based on convolutional neural network confidence to achieve effective detection of moisture content in tea leaves.

The YOLOv8 image classification model, known for its balanced trade-off between speed and accuracy, is utilized as the core algorithm in this study to predict the moisture content of withered fresh tea leaves. A novel weighted method is developed, which combines image classification confidence scores with moisture labels to calculate the percentage of moisture in the tea leaves. This is then correlated with the established standard for wilting moisture content in tea leaves, enabling the determination of the degree of wilting. This approach facilitates rapid, non-destructive detection of the withering level of fresh tea leaves. Furthermore, to enhance both the efficiency and accuracy of the detection process, the model was optimized by sacrificing a small portion of inference speed, significantly improving detection precision. This advancement provides a reliable tool for assessing the quality of tea in subsequent stages of processing.

2. Materials and Methods

2.1. Image Acquisition and Preprocessing

The withered fresh tea leaf images used in this study come from the Tea Research Institute of the Chinese Academy of Agricultural Sciences. The tea variety is Jin Guan Yin, collected in March 2024. In this study, we developed a data collection methodology that involved the synchronized acquisition of both high-resolution images and corresponding moisture measurements across 13 consecutive temporal intervals. Continuous time interval sample data can reflect the impact of changes in moisture content on image color, texture, and other features, helping the model better learn subtle changes in tea phenotype. The data collection and experimental process is shown in Figure 1.

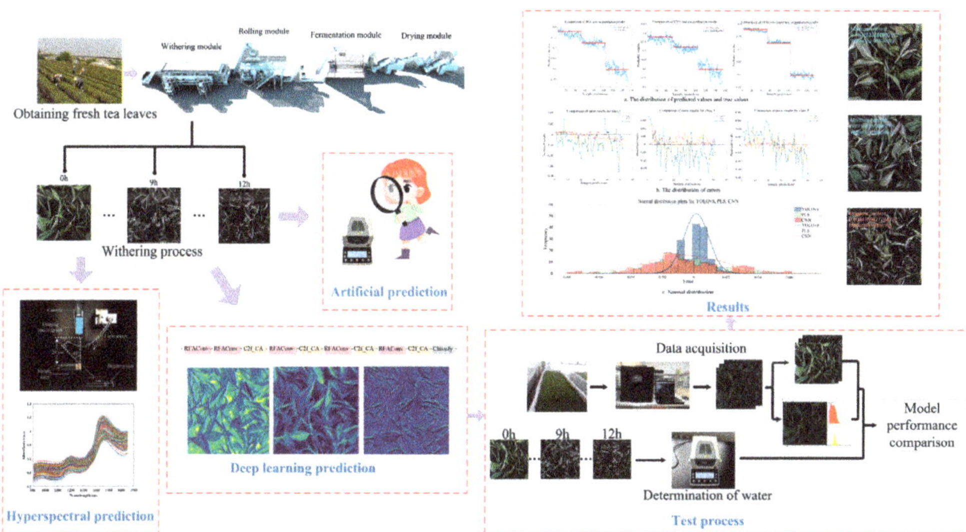

Figure 1. Data collection and algorithm process.

During image acquisition, the acquisition system consists of an artificial climate chamber, an industrial camera (Intel RealSense D405, Intel Corporation, Santa Clara, CA, USA), a uniform light source, and a sample container. During the collection process, the camera is placed 28 cm away from the sample. Samples are collected once every hour, and images of the samples are taken at 13 consecutive time points. A moisture analyzer (XY-110MW) measures the moisture value at each collection time point. Three measurements are recorded at each time point, and the average value is taken as the withering moisture label for that time point. This study obtained a total of 709 wilted fresh tea images and 13 moisture category labels. The data with category labels 1, 5, and 9 were used as an external test set (a total of 153 images). The remaining 10 categories (a total of 556 images) were divided into training sets (a total of 501 images) and validation sets (a total of 55 images) in a ratio of 9:1 to ensure the generalization effect of the model. To prevent the model from overfitting during the training process, a total of 2255 image samples were obtained in this study by cropping, rotating, flipping, and changing the brightness as data enhancement methods of images in the training set.

The experimental data of PLS [18] and CNN [19] were extracted from 709 images using a tea visual image acquisition and analysis software (Software Copyright No. 2014 SR 149549) developed based on Matlab GUI module extracted nine color metrics and six texture features. The nine color metrics include average red component, average green component, average blue component, color angle, average saturation, average luminance component, super-green transformation, ratio of the average red component to average green component, and average hue. The six texture features are average grayscale value, standard deviation, grayscale histogram smoothness, third-order moment bias, grayscale histogram consistency, and grayscale histogram entropy.

Figure 2 shows the moisture percentage and the criteria for determining the degree of withering under continuous time. In Figure 2, the blue shadow indicates moderate withering (moisture content between 58% and 66%), the area above the blue shadow indicates insufficient withering (moisture content greater than 66%), and the area below the blue shadow indicates excessive withering (moisture content less than 58%). The coordinates of the intersection point between the fold line and the blue shadow indicate that, at time t, the fresh tea leaves have reached a moderate withering value.

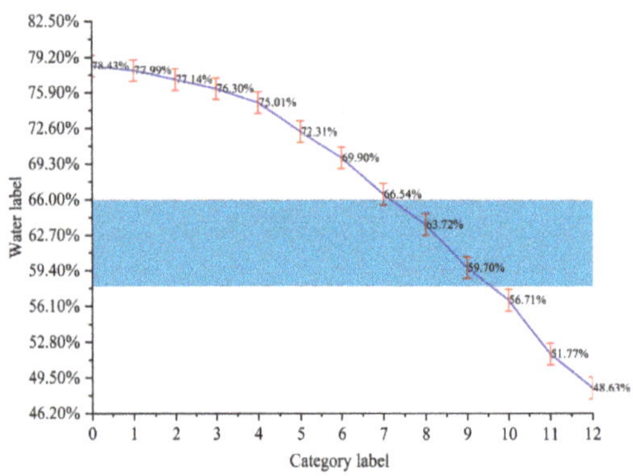

Figure 2. Moisture values of fresh tea leaves under continuous time and the standard for determining the degree of withering.

2.2. *Model Selection and Optimization for Judging the Degree of Withering*

2.2.1. Model Selection

YOLOv8 is a major update of YOLOv5 in 2023, which has greatly improved the detection performance of the model [20]. The core of YOLOv8 image classification is to extract surface features from different feature layers of the input image through convolutional layers, then gradually extract abstract features of the image through multiple convolutions, achieving accurate image classification. However, selecting different variants in different application scenarios is the key to maximizing the advantages of YOLOv8. In image classification models, based on increasing network depth and width, the model is divided into five versions: YOLOv8n, YOLOv8s, YOLOv8m, YOLOv8l, and YOLOv8x. The greater the depth, the higher the accuracy and the lower the timeliness. In this study, we selected YOLOv8s as the prediction model for moisture content and improved its basic structure to enhance the prediction accuracy. The results of the models trained with different versions are presented in Section 3.1.

The structure of YOLOv8s is different from that of mainstream image classification models, as it consists of a backbone and head module. The model structure is shown on the left side of Table 1. To enhance the model's prediction accuracy for the withering degree of tea leaves, this study replaces the standard convolution with the Receptive-Field Attention Convolution (RFAConv) to address the issue of parameter sharing within the model. Additionally, the CA mechanism is incorporated into the C2f module to mitigate the problems of information loss and inaccurate feature fusion during the model's feature integration process. The improved model results are shown on the right side of Table 1.

Table 1. YOLOv8s and improved YOLOv8s network layers.

Network Structure		Network Layer Before Optimization			Improved Network Layer		
	Layer number	Network Layer	Repeats	Kernels size	Network Layer	Repeats	Kernels size
Backbone	(1)	Conv	1	64	RFAConv	1	64
	(2)	Conv	1	128	RFAConv	1	128
	(3)	C2f	3	128	C2f_CA	3	128
	(4)	Conv	1	256	RFAConv	1	256
	(5)	C2f	6	256	C2f_CA	6	256
	(6)	Conv	1	512	RFAConv	1	512
	(7)	C2f	6	512	C2f_CA	6	512
	(8)	Conv	1	1024	RFAConv	1	1024
	(9)	C2f	3	1024	C2f_CA	3	1024
Head	Layer number	Network Layer	Repeats	Output	Network Layer	Repeats	Output
	(1)	Classify	1	10	Classify	1	10

Perceived Field Attention Convolutional RFAConv

RFAConv is a combination of the RFA attention mechanism and convolution, designed to address the issue of parameter sharing in convolutional kernels [21]. Unlike traditional convolutions, RFAConvs dynamically adjust the receptive field for each input image, enabling the model to more accurately learn local features. The RFA mechanism uses average pooling to capture global information within each receptive field. It then applies a 1×1 convolution to facilitate feature interaction. Finally, it uses the Softmax function to weigh the importance of each receptive field. This approach effectively captures spatial feature differences and overcomes the limitations of parameter sharing in convolutional neural networks. The calculation method is detailed in Equation (1).

$$F = Softmax\left(g^{1\times1}(AvgPool(X))\right) \times ReLU\left(Norm\left(g^{k\times k}(X)\right)\right) = A_{rf} \times F_{rf} \quad (1)$$

Here, $g^{1\times1}$ is 1×1 grouped convolution, X is the input feature map, $Norm$ is normalization, K is the size of the convolution kernel, F represents the product of the attention feature map A_{rf} and the receptive field spatial feature F_{rf}.

In RFAConv implementation, the RFA-generated feature maps are reshaped to eliminate spatial overlap, aggregating all receptive field block features. This allows the model to assign specific convolutional kernels to each receptive field, enhancing its ability to adapt to varying input complexities. The method's structure is detailed in Figure 3.

Cross-Stage Feature Fusion Coordinate Attention

The feature fusion process in the C2f module of the YOLOv8s classification model suffers from minor information loss and inaccuracies. To address this issue, the Coordinate Attention (CA) [22] is introduced into the C2f module, resulting in the development of the Cross-Stage Feature Fusion Coordinate Attention (C2f_CA) module. By utilizing CA-block's attention to width and height dimensions, the model's sensitivity to subtle variations in the image is enhanced, leading to more accurate recognition and classification. The C2f_CA module enables the model to extract higher-level information from the feature map, retain more useful feature details, and reduce information loss during the feature transmission process, thereby improving the model's detection accuracy. The structural diagrams of the module before and after optimization are shown in Figure 4.

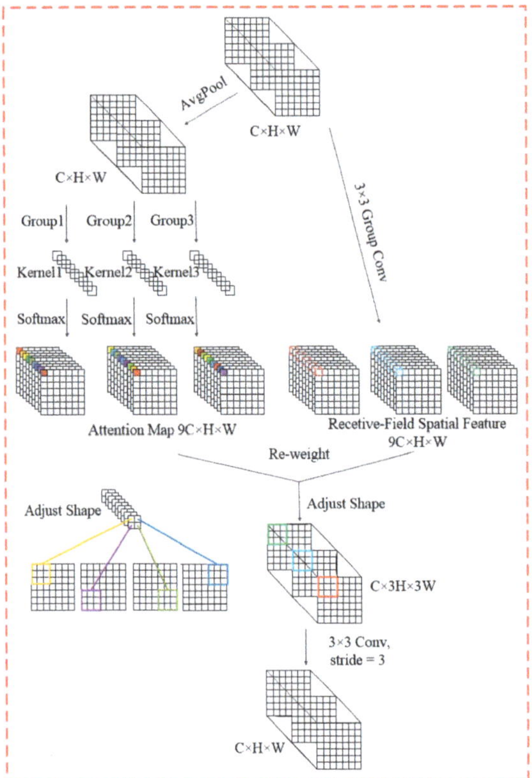

Figure 3. Structure diagram of RFAConv.

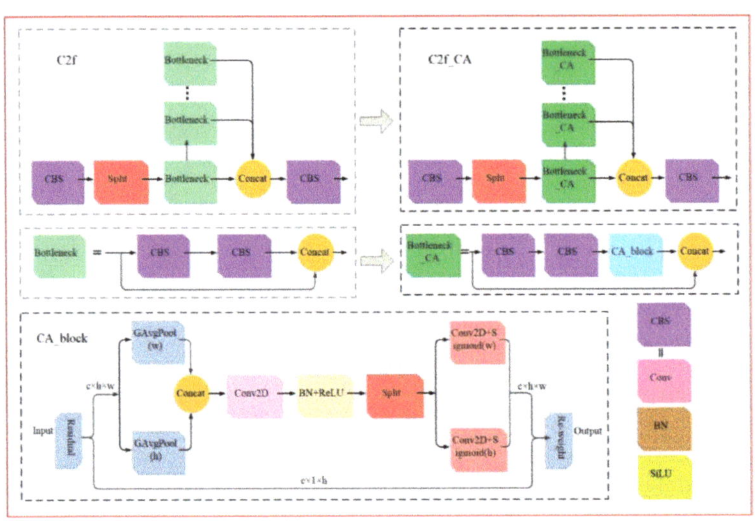

Figure 4. Structural diagram of C2f before and after optimization.

2.3. Quantitative Prediction of Moisture Content and Determination of Wilting Degree Based on Classification Confidence

In this study, a model is developed by using tea leaf images from the training set as input and moisture content and withering degree as output, as shown in Equation (2). The model outputs predictions for the moisture content and withering degree of tea leaves by performing weighted calculations of image classification confidence and moisture content labels, in conjunction with the standard for tea leaf withering moisture content. The confidence is calculated by measuring the similarity between the input image features and the predicted values. The higher the confidence, the more reliable the image classification results, which enhances the accuracy of moisture prediction based on classification confidence.

$$Y = \sum_{i=0}^{n} Y_i * N_i \tag{2}$$

Here, Y represents the moisture content of the predicted sample, Y_i represents the true i-sample moisture category label, N_i represents the predicted i-sample category confidence, and n is the total number of sample categories entered for training; in this study, $n = 9$.

2.4. Evaluation

To better evaluate the predictive performance of the model on the moisture content of fresh tea leaves, this study used correlation coefficient (Rp), relative standard deviation (RPD), and root-mean-square error (RMSEP) as evaluation indicators for the model. The larger the Rp and RPD value, the smaller the RMSEP value, the better the predictive performance of the model [23]. Among these, RPD is the ratio of standard deviation to root mean square error of prediction, which can better reflect the performance of the model. When its value is greater than 2, it indicates that the model can well judge the degree of withering; when it is between 1 and 2, it indicates that the model has an average predictive ability for this parameter; when the RPD is less than 1, it indicates that the model cannot be used for this task. The calculation formulas for Rp, RMSEP, and RPD are shown in Equations (3)–(6).

$$Rp = \sqrt{1 - \frac{\sum_{i=1}^{n} (\hat{y}_i - y_i)^2}{\sum_{i=1}^{n} (\hat{y}_i - \overline{y})^2}} \tag{3}$$

$$RMSEP = \sqrt{\frac{\sum_{i=1}^{n} (y_i - \hat{y}_i)^2}{n}} \tag{4}$$

$$RPD = \frac{SD}{RMSEP} \tag{5}$$

$$SD = \sqrt{\frac{1}{N-1} \sum_{i=1}^{N} (y_i - \overline{y})^2} \tag{6}$$

Among them, $\hat{y}_i$ represents the predicted value at time i, y_i represents the true value at time i, and $\overline{y}$ represents the average value of 10 types of label data, with SD as the standard deviation.

2.5. Implementation Details

During the model training process, this study adopted the Pytorch learning framework and conducted an experiment in Windows 11 using the improved YOLOv8s as the benchmark model. The CPU was AMD Ryzen 5 7500F 6-Core Processor (3.70 GHz, 6 cores/12 threads, 32 MB), and the running memory was 32 GB; the GPU was NVIDIA GeForce RTX 4060 Ti 16G. During the training process, the hyperparameters for model training include a learning rate of 0.01, batch size of 16, epochs of 200, input image resolution of 224 × 224, momentum of 0.937, and a weight of 0.0005, using Adam optimizer.

3. Results and Discussion

3.1. Comparison Between Different Versions of YOLOv8

In this study, we conducted experiments on all five versions of the model, and the results are shown in Table 2. Among these, YOLOv8s exhibited a superior balance of performance characteristics. Specifically, YOLOv8s achieved an accuracy of 0.771, a value surpassed only by YOLOv8m, YOLOv8l, and YOLOv8x. It is crucial that YOLOv8 achieves this performance with a model size of only 4.856 MB, which is significantly smaller than the aforementioned versions. This reduced parameter count facilitates deployment in resource-constrained environments, such as mobile and embedded systems, where computational resources are limited.

Table 2. Detection results of the model.

Model	Acc	Para (MB)	GFLOPs (G)	Speed (ms)	
				Preprocessing	Inference
YOLOv8n	0.729	1.384	3.3	0.2	1.3
YOLOv8s	0.771	4.856	12.5	0.1	1.4
YOLOv8m	0.771	15.048	41.6	0.2	1.7
YOLOv8l	0.786	34.524	98.7	0.2	1.9
YOLOv8x	0.786	53.539	153.8	0.3	2.5

Furthermore, YOLOv8s demonstrated impressive computational efficiency. While its inference time of 1.4 ms was slightly longer than the lightweight YOLOv8n (1.3 ms), YOLOv8s offered substantially improved accuracy. In addition, YOLOv8s achieved the fastest preprocessing time of 0.1 ms, further enhancing its real-time applicability. Based on these findings, YOLOv8s was selected as the foundational model for moisture content prediction, and its architecture was subsequently modified to improve predictive accuracy.

3.2. Comparative Experiments Before and After Model Optimization

To evaluate the superiority of the improved model, this study compared YOLOv8s with the improved model, and the results are shown in Table 3. The improved model reached a Top-1 accuracy of 0.927. This is a 0.156 accuracy increase over YOLOv8s. However, model parameters (Params) increased by 0.224 M, and floating-point operations (GFLOPs) increased by 1.2 G. Image processing speed rose by 0.2ms, and inference speed rose by 3.6ms. These results indicate that the model significantly improves accuracy while slightly increasing computation and detection time. This has great benefits for predicting moisture content based on confidence. Therefore, it is the best choice to increase calculation and detection time a little, greatly improving accuracy. Figure 5 illustrates the differences between the performance of the improved and initial models in identifying the degree of tea leaf wilting. This figure further highlights the superior performance of the improved model in real-time monitoring of tea leaf wilting.

Table 3. Comparative experiments before and after model optimization.

Model	top1_acc	top5_acc:	Params (M)	GFLOPs (G)	Speed (ms)	
					Preprocessing	Inference
YOLOv8s	0.771	1	4.856	12.5	0.1	1.4
Ours	0.927	1	5.08	13.7	0.3	5

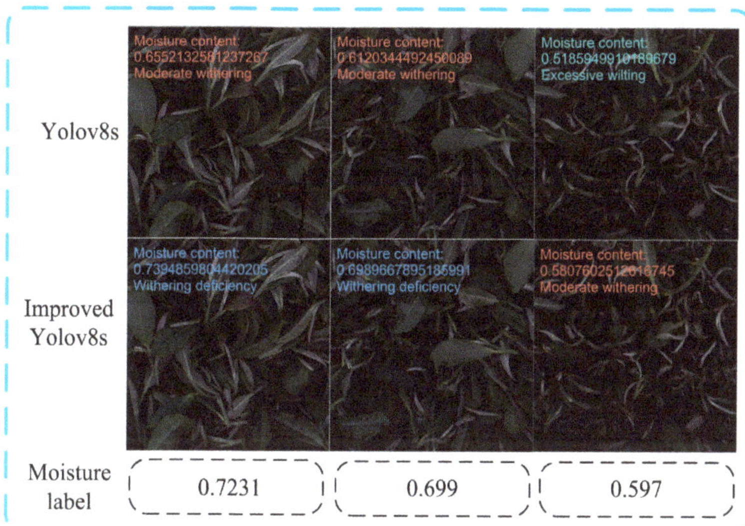

Figure 5. Comparison of model detection results before and after improvement.

3.3. Improve the Model to Predict Moisture Content and Wilting Degree

Table 4 shows an example of the external test set. It shows the image category labels and moisture values in the training set, the prediction confidence of the three category samples in the external test set, and the predicted moisture content of fresh tea leaves in the external test set according to Equation (2). From Table 4, it can be seen that the errors between the predicted and true moisture values of the three types of external test sets are all less than 0.01, indicating that the model also achieves good detection results for untrained categories. In addition, the sum of the confidence levels of the three categories in the external test set in Table 4 is not completely equal to 1. This is because in this study, when predicting the external test set, the confidence level of a single image and the moisture value were only retained until eight decimal places, resulting in rounding errors. Figure 6 shows the entire process of the improved model prediction.

Table 4. Test set confidence and predicted value labels.

Training Set Category	Moisture Label	Test Set Confidence		
		1	5	9
0	0.7843	0.91888142	0	0
1	0.7714	0.08109235	0.00000007	0
2	0.763	0.00002619	0.00961427	0
3	0.7501	0.00000001	0.41952419	0
4	0.699	0	0.57085168	0.00000004
5	0.6654	0	0.00000980	0.00098720
6	0.6372	0	0	0.53591472
7	0.5671	0	0	0.45765832
8	0.5177	0	0	0.00543968
9	0.4863	0	0	0.00000003
Sum of Confidence		0.99999997	1.00000001	0.99999999
Predictive value		0.78325333	0.72105269	0.60449594
True value		0.7799	0.7231	0.597

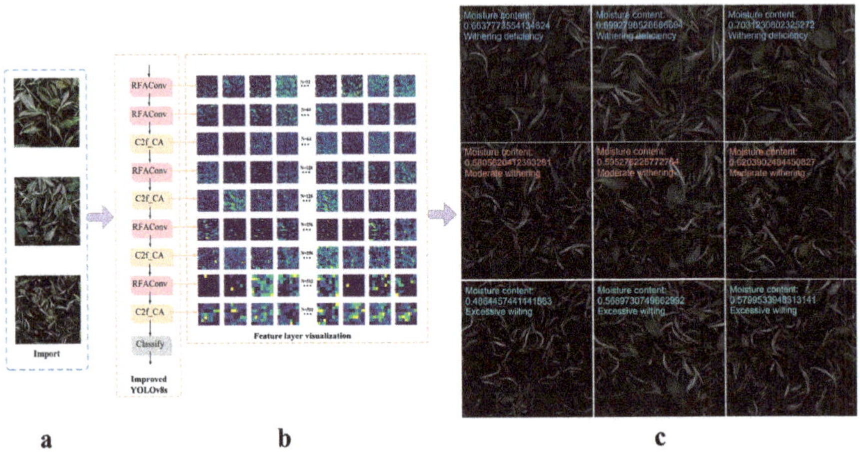

Figure 6. The entire process of improved model prediction. Model input (a). The network structure of the improved model and the visualization of feature maps at each layer (b). Identification results of the model (c).

Table 5 shows the evaluation metrics for the relationship between the predicted results and the true values for the external validation set of the model. These data reveal that the RPD predicted by the model in this study is greater than 2, which well reflects the predictive performance of our model. However, compared with the other two categories, there is an obvious gap between the evaluation index base of category 9 in the external test set of the model. There may be two reasons for this: one is that the color and texture of fresh tea leaves change very little in continuous time, resulting in a larger prediction error of the model; the other is that a large error exists between the moisture labeled values in continuous time before and after category 9, leading to a bias toward the side with a high level of confidence in the computation process, which results in a larger prediction error.

Table 5. Evaluation indicators of different models for the test set prediction results.

	Ours			PLS			CNN		
	1	5	9	1	5	9	1	5	9
Rp	0.9983	0.993	0.9474	0.982	0.966	0.9488	0.9726	0.9277	0.905
RMSEP	0.006278	0.00694	0.018411	0.018615	0.014287	0.024974	0.022124	0.018567	0.03413
RPD	39.2513	20.1693	7.363	12.0667	7.974	6.4651	9.8045	5.5216	4.7987

3.4. Comparison of Moisture Content Prediction Using Different Models

Figure 7 shows the prediction results of PLS, CNN, and YOLOv8s on the external test set. Figure 7a presents the distribution of predicted values versus true values. It is evident that the predictions from our model are closest to the true values, demonstrating that our model has better generalization and robustness compared to PLS and CNN models. Figure 7b shows the prediction errors of the three models. From the error distribution, we can see that the error fluctuations between predicted and true values for the PLS and CNN models are much larger than those of our model. Although our model has notable outliers in Class 5 and Class 9, overall it achieves the best accuracy and is more suitable for online collection of wilted moisture content data from fresh tea leaf images. Figure 7c displays the normal distribution of prediction errors for the three models, helping us further explore the error intervals. Here, the horizontal axis represents the error range and the vertical axis represents the frequency of results within that range. From the normal distribution, it is clear that, compared to PLS and CNN, our model's errors are more concentrated, mainly within the range of [−0.02, 0.02]. Additionally, the comparison of the three models' Rp, RMSEP, and RPD values in Table 5 further confirms that our model performs well in predicting untrained categories.

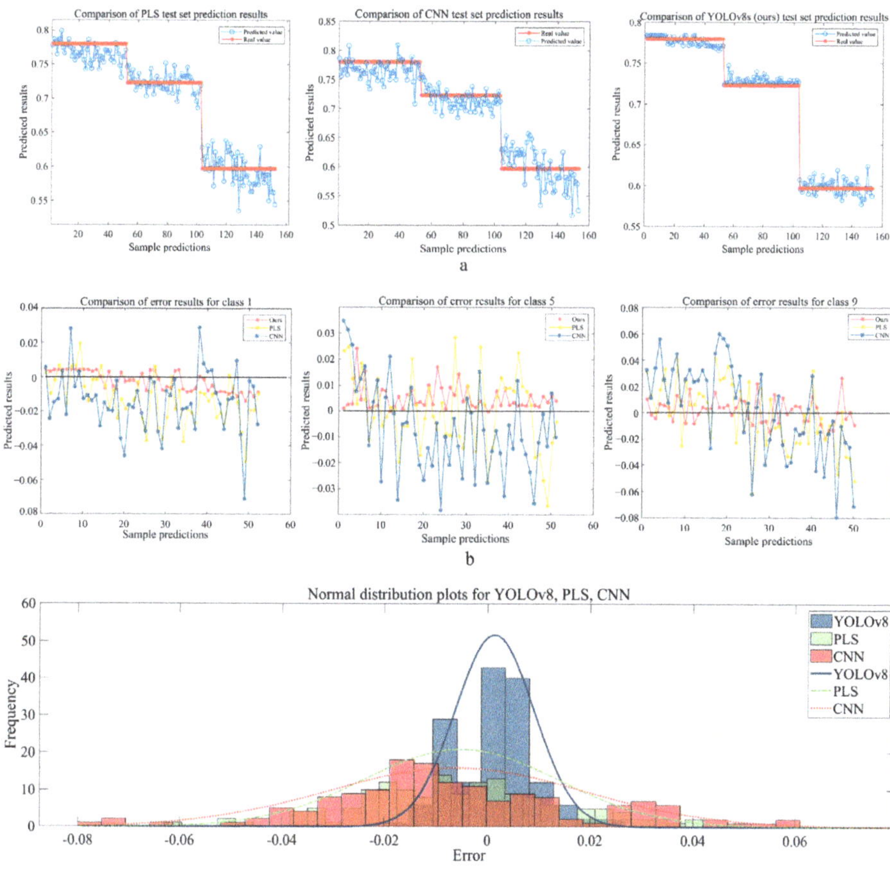

Figure 7. Prediction results of PLS, CNN, and YOLOv8 external test sets. (**a**) The distribution of predicted values and true values. (**b**) The distribution of errors. (**c**) Normal distribution.

4. Conclusions

This study addresses the need for detecting moisture content and determining wilting levels in fresh tea leaves. To improve the model's prediction accuracy and reduce moisture content prediction errors, we introduced the RFAConv and C2f_CA modules. These additions significantly enhanced the model's ability to extract and process features, resulting in a 0.156 improvement in detection accuracy. While this method increased the model's parameters (Params) by 0.224 M and floating-point operations (GFLOPs) by 1.2 G, the average prediction time remained efficient at 5.3 ms. This balance ensures high detection accuracy while maintaining fast prediction speed.

In terms of experimental evaluation, this study achieved the best performance on the external test set 1, with Rp, RMSEP, and RPD values of 0.9983, 0.006278, and 39.2513, respectively. Further analysis revealed that the prediction errors were primarily within the range of [−0.02, 0.02], demonstrating high precision and stability. The results indicate that this model offers significant advantages in real-time detection of moisture content in fresh tea leaves and in determining their wilting levels. It effectively improves the accuracy of wilting degree judgment, providing reliable technical support for tea leaf quality assessment.

Although the proposed tea leaf withering degree detection model, though satisfactory, has limitations in accuracy, single-variety focus, and single-modal analysis. In future research, we will expand the diversity of research subjects and construct a multimodal perception framework. By deeply integrating and analyzing multidimensional data, the prediction accuracy and stability of the model can be further optimized.

Author Contributions: M.W.: validation, writing—original draft. Y.S.: data curation, methodology. Y.L.: software, investigation. H.M.: formal analysis. Z.D.: formal analysis. Z.T.: software. C.D.: software, investigation. Z.C.: writing—review and editing, conceptualization, funding acquisition, supervision. All authors have read and agreed to the published version of the manuscript.

Funding: This research was funded by the Agricultural Improved Variety Project of Shandong Province (2023LZGCQY015), the Key R&D Projects in Zhejiang Province (2023C02043), the Major Project of Shandong Academy of Agricultural Sciences Innovation Project (R&D of specialty crop picking robots), the Agricultural Science and Technology Research Project of Jinan City (GG202401), and the Technology System of Modern Agricultural Industry in Shandong Province (SDAIT19).

Data Availability Statement: The original contributions presented in the study are included in the article, further inquiries can be directed to the corresponding author.

Acknowledgments: I would like to thank Chunwang Dong and Zhiwei Chen for providing valuable revision suggestions during the discussion stage of revising the paper.

Conflicts of Interest: The authors declare that they have no known competing financial interests or personal relationships that could have appeared to influence the work reported in this paper.

References

1. Huang, W.; Fang, S.; Su, Y.; Xia, D.; Wu, Y.; Liu, Q.; Zou, L.; Lu, C.; Deng, W.-W.; Ning, J. Insights into the mechanism of different withering methods on flavor formation of black tea based on target metabolomics and transcriptomics. *LWT* **2023**, *189*, 115537. [CrossRef]
2. Deb, S.; Jolvis Pou, K.R. A review of withering in the processing of black tea. *J. Biosyst. Eng.* **2016**, *41*, 365–372. [CrossRef]
3. Wu, H.; Chen, Y.; Feng, W.; Shen, S.; Wei, Y.; Jia, H.; Wang, Y.; Deng, W.; Ning, J. Effects of three different withering treatments on the aroma of white tea. *Foods* **2022**, *11*, 2502. [CrossRef] [PubMed]
4. Huang, W.; Fang, S.; Wang, J.; Zhuo, C.; Luo, Y.; Yu, Y.; Li, L.; Wang, Y.; Deng, W.-W.; Ning, J. Sensomics analysis of the effect of the withering method on the aroma components of Keemun black tea. *Food Chem.* **2022**, *395*, 133549. [CrossRef] [PubMed]
5. Liang, G.; Dong, C.; Hu, B.; Zhu, H.; Yuan, H.; Jiang, Y.; Hao, G. Prediction of moisture content for congou black tea withering leaves using image features and nonlinear method. *Sci. Rep.* **2018**, *8*, 7854. [CrossRef]
6. Jabeen, S.; Alam, S.; Saleem, M.; Ahmad, W.; Bibi, R.; Hamid, F.S.; Shah, H.U. Withering timings affect the total free amino acids and mineral contents of tea leaves during black tea manufacturing. *Arab. J. Chem.* **2019**, *12*, 2411–2417. [CrossRef]
7. Madhavi, K.V.; Tamilkodi, R.; Dinakar, R.B.; JayaSudha, K. An innovative technique for content based image retrieval using color and texture features. *Int. J. Innov. Res. Comput. Commun. Eng.* **2013**, *1*, 1257–1263.
8. Chen, A.; Chen, H.Y.; Chen, C. Use of temperature and humidity sensors to determine moisture content of oolong tea. *Sensors* **2014**, *14*, 15593–15609. [CrossRef]
9. Steger, C.; Ulrich, M.; Wiedemann, C. *Machine Vision Algorithms and Applications*; John Wiley & Sons: Hoboken, NJ, USA, 2018.
10. Wijaya, D.R.; Handayani, R.; Fahrudin, T.; Kusuma, G.P.; Afianti, F. Electronic Nose and Optimized Machine Learning Algorithms for Non-infused Aroma-based Quality Identification of Gambung Green Tea. *IEEE Sens. J.* **2023**, *24*, 1880–1893. [CrossRef]

11. Li, T.; Lu, C.; Huang, J.; Chen, Y.; Zhang, J.; Wei, Y.; Wang, Y.; Ning, J. Qualitative and quantitative analysis of the pile fermentation degree of Pu-erh tea. *LWT* **2023**, *173*, 114327. [CrossRef]
12. Zhang, Z.; Deng, L. An Adaptive Genetic Algorithm Optimizes Double-Hidden Layer BPNN for Rapid Detection of Moisture Content of Green Tea in Processing. *J. Food Process. Preserv.* **2024**, *2024*, 5876051.
13. Wang, F.; Xie, B.; Lü, E.; Zeng, Z.; Mei, S.; Ma, C.; Guo, J. Design of a Moisture content detection system for yinghong no. 9 tea leaves based on machine vision. *Appl. Sci.* **2023**, *13*, 1806. [CrossRef]
14. LeCun, Y.; Bengio, Y.; Hinton, G. Deep learning. *Nature* **2015**, *521*, 436–444.
15. Chen, T.; Li, H.; Chen, J.; Zeng, Z.; Han, C.; Wu, W. Detection network for multi-size and multi-target tea bud leaves in the field of view via improved YOLOv7. *Comput. Electron. Agric.* **2024**, *218*, 108700.
16. Wang, T.; Zhang, K.; Zhang, W.; Wang, R.; Wan, S.; Rao, Y.; Jiang, Z.; Gu, L. Tea picking point detection and location based on Mask-RCNN. *Inf. Process. Agric.* **2023**, *10*, 267–275.
17. An, T.; Yu, H.; Yang, C.; Liang, G.; Chen, J.; Hu, Z.; Hu, B.; Dong, C. Black tea withering moisture detection method based on convolution neural network confidence. *J. Food Process Eng.* **2020**, *43*, e13428.
18. Bastien, P.; Vinzi, V.E.; Tenenhaus, M. PLS generalised linear regression. *Comput. Stat. Data Anal.* **2005**, *48*, 17–46.
19. Chen, L.; Li, S.; Bai, Q.; Yang, J.; Jiang, S.; Miao, Y. Review of image classification algorithms based on convolutional neural networks. *Remote Sens.* **2021**, *13*, 4712. [CrossRef]
20. Terven, J.; Córdova-Esparza, D.M.; Romero-González, J.A. A comprehensive review of yolo architectures in computer vision: From yolov1 to yolov8 and yolo-nas. *Mach. Learn. Knowl. Extr.* **2023**, *5*, 1680–1716. [CrossRef]
21. Zhang, X.; Liu, C.; Yang, D.; Song, T.; Ye, Y.; Li, K.; Song, Y. Rfaconv: Innovating spatital attention and standard convolutional operation. *arXiv* **2023**, arXiv:2304.03198.
22. Hou, Q.; Zhou, D.; Feng, J. Coordinate attention for efficient mobile network design. In Proceedings of the IEEE/CVF Conference on Computer Vision and Pattern Recognition, Nashville, TN, USA, 20–25 June 2021; pp. 13713–13722.
23. Chen, Q.; Zhang, D.; Pan, W.; Ouyang, Q.; Li, H.; Urmila, K.; Zhao, J. Recent developments of green analytical techniques in analysis of tea's quality and nutrition. *Trends Food Sci. Technol.* **2015**, *43*, 63–82.

foods

MDPI

Article

Magnesium Regulation Increases the Content of Characteristic Volatile Compounds and Enhances the Intensity of Odor Characteristics in Tea Tree Leaves

Jianghua Ye [1,†], Qiqi Weng [2,†], Yulin Wang [2], Weiting Cheng [1,3], Junbin Gu [2], Qi Zhang [1], Bitong Zhu [1], Qiyan Liu [1], Xiaoli Jia [1], Juanying Wang [4] and Haibin Wang [1,2,*]

[1] College of Tea and Food, Wuyi University, Wuyishan 354300, China; jhye1998@wuyiu.edu.cn (J.Y.)
[2] College of Life Science, Longyan University, Longyan 364012, China
[3] College of Horticulture, Fujian Agriculture and Forestry University, Fuzhou 350002, China
[4] Institute of Agro-Bioengineering, Guizhou University, Guiyang 550025, China
* Correspondence: 82015018@lyun.edu.cn
† These authors contributed equally to this work.

Abstract: Magnesium plays an important role in regulating the growth and quality of tea trees. However, the effect of magnesium regulation on changes in the aroma quality of tea tree leaves remains unknown. In this study, the volatile compounds of tea leaves under different magnesium concentrations were determined to obtain the characteristic volatile compounds that changed significantly and their odor characteristics and to explore the effect of magnesium regulation on the aroma quality of tea leaves. The results showed that magnesium significantly enhanced the content of 15 characteristic volatile compounds in tea tree leaves, especially heptyl formate and t-geraniol. The 15 characteristic volatile compounds mainly showed five kinds of odor characteristics, such as fruity, green, floral, pungent, woody, and burnt; magnesium regulation critically increased floral, fruity, and green odor characteristics. This study lays an important foundation for the application of exogenous magnesium ions to regulate tea aroma quality.

Keywords: tea tree; magnesium regulation; characteristic volatile compounds; odor characteristics

1. Introduction

In China, tea trees play a pivotal role as a key economic plant to promote the development of Chinese agricultural economy [1]. The growth of tea trees depends on the uptake of several elements to ensure their normal physiology and development [2]. Among them, magnesium plays a crucial role as one of the essential elements for tea tree growth [3]. According to a previous study, maintaining magnesium ion content in plants in the range of 1.5~3.5 mg/g is most favorable for plant growth [4]. And proper application of magnesium fertilizer has been shown to be effective in enhancing plant yield [5]. It has been reported that increased application of magnesium improves the efficiency of nutrient uptake and utilization by the root system of tea trees, which in turn increases the annual yield of tea by more than 15.8% or more [6]. The economic benefits of tea are closely related to its yield and quality, so improving tea quality while ensuring yield is essential to improving its economic benefits [7]. Magnesium plays an important role in the synthesis and distribution of secondary metabolites in tea tree leaves. However, while increasing magnesium supply enhances tea yield, it may lead to decreases in catechins, total phenols, and total free amino acids in tea leaves [8]. Chen et al. [9] found that adequate magnesium supply favored the growth of the tea tree root system and increased the leaf starch content but inhibited the accumulation of secondary metabolites, thus affecting the quality of tea tree leaves. Similarly, Xu et al. [10] found that magnesium fertilization could promote the accumulation of theanine and reduce the content of tea polyphenols in tea tree leaves, which made the tea taste more refreshing and reduced bitterness and astringency. Zhang et al. [11] explored the effect of magnesium on tea quality with the help of transcriptomics technology and concluded that increasing the magnesium supply would inhibit the accumulation of theanine and caffeine, which in turn was detrimental to the overall quality of tea. Taken together, several studies have shown that although magnesium helps to increase tea yield, it adversely affects the synthesis of major biochemical quality components of tea, such as tea polyphenols, catechins, and caffeine, which may indirectly hinder the enhancement of tea quality.

Aroma, as an important characteristic of tea quality, plays a crucial role in the formation and grading of tea quality [12]. In research on the effect of magnesium on tea quality, many scholars have focused on changes in tea biochemical quality indexes after magnesium regulation, which is the basis for evaluating tea quality [8,9,11]. However, little research has been reported on the effect of magnesium regulation on tea aroma quality. The formation of tea aroma characteristics and its intensity are closely related to the type and content of volatile compounds in tea tree leaves [13]. By determining the content of volatile compounds in tea tree leaves, the aroma quality of tea can be effectively evaluated [14]. Gas chromatography-mass spectrometry (GC-MS) has shown significant advantages in the determination of volatile compounds due to its excellent performance, and it has been widely used in the analysis of volatile compounds in tea [15]. For example, Fu et al. [16] used GC-MS to analyze the volatile compounds of 22 types of fresh leaves of oolong tea and revealed the main substances responsible for the formation of the aroma differences between different varieties of oolong tea. Zhang et al. [17] analyzed the volatile compounds of different aroma types of Longjing tea by GC-MS and identified the key volatile compounds that differentiated between the different aromas. Accordingly, this study used HS-SPME and GC-MS to analyze the type and content of volatile compounds in tea tree leaves after magnesium regulation at different concentrations and further identified the characteristic volatile compounds and their odor characteristics that changed significantly after magnesium regulation, aiming to explore the effects of magnesium regulation on the aroma quality of tea and provide a solid theoretical foundation for the enhancement of the aroma quality of tea through the regulation of exogenous magnesium ions.

2. Materials and Methods

2.1. Chemical Reagents and Instruments

The chemical reagents used in this study, such as KNO_3 (Sinopharm Chemical Reagent Co., Ltd., Beijing, China), $(NH_4)_2SO_4$ (Sinopharm Chemical Reagent Co., Ltd., Beijing, China), K_2SO_4 (Sinopharm Chemical Reagent Co., Ltd., Beijing, China), $FeSO_4$ (Sinopharm Chemical Reagent Co., Ltd., Beijing, China), $Al_2(SO_4)_3$ (Sinopharm Chemical Reagent Co., Ltd., Beijing, China), KH_2PO_4 (Sinopharm Chemical Reagent Co., Ltd., Beijing, China), $CaCl_2$ (Sinopharm Chemical Reagent Co., Ltd., Beijing, China), and $MgSO_4$ (Sinopharm Chemical Reagent Co., Ltd., Beijing, China) were analytically pure. The instruments used were a fully automated solid-phase microextractor (HS-SPME, SPME Arrow, CTC Analytics AG, Zwingen, Switzerland), headspace vials (Agilent, Palo Alto, CA, USA), DVB/CWR/PDMS extraction heads (Agilent, Palo Alto, CA, USA), an analyzing tube aging device (Fiber Conditioning Station, CTC Analytics AG, Zwingen, Switzerland), gas chromatography (GC, 8890B, Agilent, CA, USA), mass spectrometry (MS, 7000D, Agilent, CA, USA), and DB-5MS capillary column (30 m × 0.25 mm × 0.25 μm, Agilent, CA, USA).

2.2. Experimental Design and Sample Collection

In this study, one-year-old tea tree seedlings (Rougui, about 30 cm in height and 0.3 cm in diameter) with uniform growth were selected. Seedlings were transplanted to complete nutrient solution for 45 d and then used in the magnesium regulation experiment. The complete nutrient solution was slightly modified from the formulation of Sun et al. [18], and the concentrations of KNO_3, $(NH_4)_2SO_4$, K_2SO_4, $FeSO_4$, $Al_2(SO_4)_3$, KH_2PO_4, $CaCl_2$, and $MgSO_4$ were 125, 187.5, 25, 16, 200, 100, 100, and 100 μmol/L, respectively. The pH of the nutrient solution was adjusted to 4.5.

The root systems of the recovered tea tree seedlings were rinsed with deionized water and then transplanted to three different magnesium concentrations in complete nutrient solution for culture. Three independent replicates were established for each magnesium concentration, and magnesium concentrations in the three different treatments were 0 mmol/L (ML), 0.4 mmol/L (MM), and 0.8 mmol/L (MH), while the concentrations of other ions were consistent with the above complete nutrient solution. After transplanting, the tea trees were placed in an artificial climatic chamber for cultivation, with continuous aeration of the culture solution for 24 h per day, a light period of 12 h (8:00 a.m.~20: 00 p.m.), an intensity of 1500 lux, a temperature of 25 °C, and a humidity of 75%. After magnesium regulation, tea tree seedlings were treated for 21 d. During this period, the same culture solution was changed every 7 days. At the end of magnesium regulation, one bud and two leaves of tea tree seedlings were collected and immediately stored in liquid nitrogen for extraction and determination of volatile compounds, with three independent replicates for each treatment.

2.3. HS-SPME Extraction of Volatile Compounds

Samples of tea trees treated with different magnesium concentrations were ground and mixed using liquid nitrogen, respectively. For each sample, 500 mg was weighed and placed in a headspace bottle, and 10 mL of saturated NaCl solution was added to collect tea volatile compounds using fully automated HS-SPME. Three independent replicates were performed for each sample. The extraction method was described as follows: the extraction head was a 120 μm DVB/CWR/PDMS solid-phase microextraction head, the extraction head was placed in a resolving tube aging device for 2 h before sampling, and the temperature was set at 250 °C. The volatile compounds were collected by placing the headspace vials containing the tea tree leaf samples in a thermostatic water bath at 60 °C, oscillating for 5 min. The extraction head was immediately inserted into the headspace bottle, and the extraction time was 15 min, and the sample was then used for gas chromatography-mass spectrometry (GC-MS) determination and analysis [19].

2.4. Determination of Volatile Compounds by GC-MS

An Agilent 8890B gas chromatograph (GC) coupled with a 7000D mass spectrometer (MS) was used to determine volatile compounds in tea leaves. The capillary column used for GC was an Agilent DB-5MS and the carrier gas was helium. First, the solid-phase microextraction head was placed in the inlet port of the GC and desorbed for 5 min, and the temperature was set at 250 °C. The flow rate of the carrier gas for the GC was 1.2 mL/min, and the sample was injected without splitting, with a solvent delay of 3.5 min. The programmed heating mode of the GC was set to first hold the sample at 40 °C for 3.5 min, and then linearly warm it up to 100 °C at 10 °C/min, and then linearly warm it up to 180 °C at 7 °C/min, and then linearly warm it up to 280 °C at 25 °C/min for 5 min. MS was performed by electron bombardment of the ion source, in which the temperature of the ion source was set at 230 °C, the temperature of the mass spectrometry connector was set at 280 °C, the temperature of the quadrupole was set at 150 °C, and the electron energy was 70 eV. The mass range of the mass spectrometry scan was in the range of 10–1000 m/z and the scan time was 30 min. MS was performed in selective ion detection mode, with qualitative and quantitative ions scanned precisely.

After GC-MS determination, the sample data were collected and processed using MassHunter software (B 08.00). This was done by selecting 2~3 qualitative ions for each compound and comparing them with the NIST20 mass spectrometry database. A compound was characterized if all qualifying ions appeared in the mass spectra of the compound after subtracting the background and the retention time was consistent with the database standard [20]. After characterization of the compound, a quantitative ion was selected, integrated, and corrected, and the compound was quantified based on the obtained peak area [21].

2.5. Statistical Analysis

Raw data on volatile compounds of tea leaves were initially organized and counted using Excel 2020 and subsequently unified for in-depth analysis and graphing with the help of Rstudio software (v 4.2.3) [22]. Box plots for volatile compound content analysis were plotted using gghalves 0.1.4; multi-group volcano plots between ML, MM, and MH were generated through the combined application of ggplot2 and ggnewscale; K-means clustering analysis of differential volatile compounds was executed using stats 4.2.0; OPLS-DA models were constructed for ML, MM, and MH with the help of ropls and mixOmics and the corresponding graphs were produced; bubble feature plots of volatile compounds were produced using ggplot2 3.4.0; Mulberry diagram analysis for volatile compound classification was implemented using networkD3 0.4.1; TOPSIS analysis of volatile the compound contribution rates was performed by dplyr 1.1.4; odor wheels for characteristic volatile compounds were jointly produced using circlize version 0.4.15 and vegan version 2.6.4; and conventional radar plots and principal component analysis plots were produced relying on fmsb 0.7.6 and ggbiplot 0.55, respectively. To assess variability between samples or indicators, variance analysis and paired student's t-tests were used, and differences were considered to reach a significant level when $p < 0.05$.

3. Results and Discussion

3.1. Effect of Magnesium Regulation on Volatile Compounds of Tea Tree Leaves

Volatile compounds, which are mainly derived from secondary metabolites of tea trees, are the main compounds that form the aroma characteristics of tea [23]. Elemental supply plays an important role in the synthesis of volatile compounds in tea trees; however, there are significant differences between different elements in regulating the synthesis of volatile compounds in tea trees [24]. For example, Al and Pb are toxic and detrimental to the growth of tea trees, reducing their ability to synthesize volatile compounds [25,26], while Se and Mo are beneficial in increasing tea yield, volatile compound content, and tea aroma [27,28]. In this study, the volatile compounds of tea tree leaves were measured after treatment with different magnesium concentrations (Figure 1A) and it was found that a total of 772 volatile compounds was detected (Table S1), which could be categorized into 15 groups, namely, terpenoids (19.04%), esters (17.23%), ketones (12.44%), heterocyclic compounds (11.27%), alcohols (10.88%), aldehydes (5.83%), acids (5.18%), amines (3.63%), aromatics (3.37%), hydrocarbons (3.37%), phenols (2.85%), ethers (2.20%), nitrogen compounds (1.55%), halogenated hydrocarbons (0.65%), and sulfur compounds (0.52%), with the highest number of terpenoid compounds. Analysis of total volatile compounds after treatment with different magnesium concentrations revealed (Figure 1B) that ML, MM, and MH did not differ significantly ($p > 0.05$) in the total content of volatile compounds. The principal component analysis (Figure 1C) showed that principal component 1 and principal component 2 could effectively distinguish ML, MM, and MH in different regions, with an overall contribution of 68.35%. It can be seen that although there was no significant difference in the total content of volatile compounds in tea leaves after treatment with different magnesium concentrations, different groups of volatile compounds underwent significant changes in their content. Further analysis revealed (Figure 1D) that the volatile compounds most relevant to ML were mainly ketones, acids, halogenated hydrocarbons, alcohols, and aldehydes, while phenols were relevant to MM, and terpenoids, sulfur compounds, aromatics, and nitrogen compounds were relevant to MH. Qiu et al. [29] analyzed the effect of different amounts of applied nitrogen on the volatile compound content in tea leaves and found that an increase in nitrogen dosage had no significant effect on the total content of volatile compounds in tea leaves, but the terpenoids content was significantly increased. Xu et al. [10] analyzed the effect of magnesium fertilizer on the quality of tea leaves in the field and also found that an increase in the amount of magnesium had a lesser effect on the total content of volatile compounds in tea leaves. A similar phenomenon was also found in this study, which showed that magnesium regulation changed the content of different types of volatile compounds in tea tree leaves, although the effect on the total amount of volatile compounds in tea tree leaves was small.

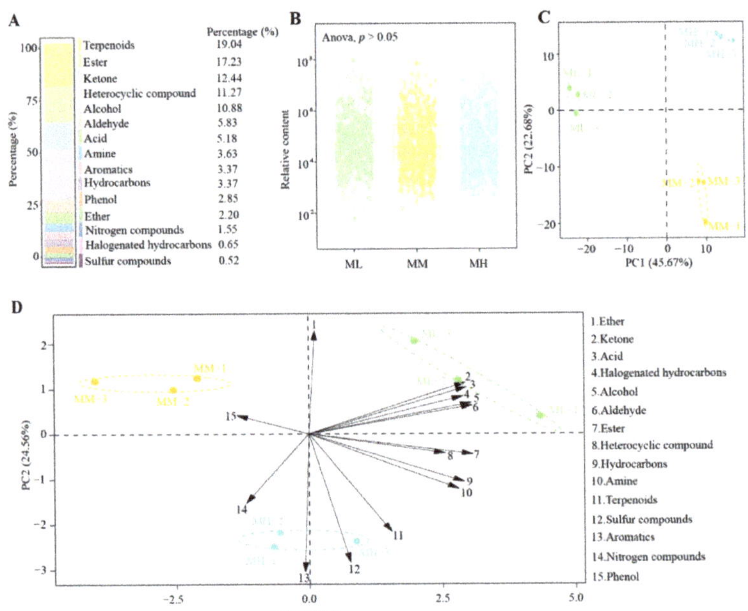

Figure 1. Effect of magnesium regulation on the type and content of volatile compounds in tea tree leaves. Note: ML: Mg^{2+} concentration 0 mmol/L; MM: Mg^{2+} concentration 0.4 mmol/L; MH: Mg^{2+} concentration 0.8 mmol/L; (**A**) Types of volatile compounds of tea tree leaves and their percentage of quantity; (**B**) Effect of magnesium regulation on the volatile metabolite content of tea tree leaves; (**C**) Principal component analysis of different samples under magnesium regulation; (**D**) Principal component analysis of volatile compounds of different samples after categorization under magnesium regulation.

3.2. Screening of Characteristic Volatile Compounds of Tea Tree Leaves Under Magnesium Regulation

There are many types of volatile compounds in tea; however, the formation of its special odor characteristics depends on a small number of characteristic volatile compounds [30]. Shen et al. [31] found that the volatile compounds in An tea mainly consisted of esters, acids, and hydrocarbons, which accounted for more than 50% of the total volatile compounds, of which seven volatile compounds played a key role in the odor characteristics. Liang et al. [32] found that the key aroma active compounds in Wuyi rock tea were mainly 11 aromatic volatile compounds, especially linalool and geraniol. It can be seen that characteristic volatile compounds play an important role in the formation of special odor characteristics of tea. Accordingly, in this study, based on the determination of volatile compounds, firstly, multi-group volcano plots and K-means clustering analysis were used to screen for significant changes in differential volatile compounds in tea tree leaves at different magnesium concentrations. The results showed (Figure 2A) that a total of 357 volatile compounds significantly changed in tea tree leaves with increasing magnesium concentration, and subsequent K-means cluster analysis of these 357 volatile compounds revealed that 197 volatile compounds showed a significant upward trend and 160 showed a significant downward trend with increasing magnesium concentration. Based on the above significantly changed volatile compounds, OPLS-DA models of ML, MM, and MH were constructed to screen for key differential volatile compounds, and the results showed (Figure 2B) that the constructed OPLS-DA models effectively distinguished the three different treatments in different regions. The fit ($R^2Y = 0.990$) and predictability ($Q^2 = 0.799$) of the models reached a significant level ($p < 0.005$) and could be used to screen for key differential volatile compounds. From the S-Plot of the model, 130 key differential volatile compounds (VIP > 1) were derived that significantly differentiated ML, MM, and MH. Further analysis of the classification and content of key volatile compounds revealed (Figure 2C) that the 130 key volatile compounds could be classified into 15 groups, namely, acids, alcohols, aldehydes, amines, aromatics, esters, ethers, halogenated hydrocarbons, heterocyclic compounds, hydrocarbons, ketones, nitrogen compounds, phenols, sulfur compounds, and terpenoids, of which only the content of nitrogen compounds showed a decreasing trend as the magnesium concentration increased, while the content of the other 14 groups of volatile compounds showed an increasing trend. Key volatile compounds were analyzed using a bubble feature map, and 35 characteristic volatile compounds were screened and obtained (Figure 3A). The 35 characteristic volatile compounds could be mainly classified into 10 groups (Figure 3B), which were terpenoids (28.57%), ester (14.29%), heterocyclic compounds (14.29%), acids (8.57%), alcohols (8.57%), aldehydes (8.57%), hydrocarbons (5.71%), ketones (5.71%), amines (2.86%), and aromatics (2.86%), and in short, terpenoids accounted for the largest proportion, followed by esters and heterocyclic compounds.

The transformation relationship between compounds is extremely complex. For example, ketones, hydrocarbons, and aldehydes can be converted into alcohols, while alcohols and acids can form esters after esterification and may also be dehydrogenated to form isoprenoid structures, which in turn leads to the synthesis of terpenoids [33]. The conduct of these reactions involves the participation of a multitude of enzymes in a complex process [34]. As reported, terpenoids, esters, and heterocyclic compounds are the main components of volatile compounds in tea tree leaves, and they are also important volatile compounds for the formation of the special odor characteristics of tea, and their content directly affects the intensity of odor characteristics of tea [35,36]. It was evident that the key to the effect of magnesium regulation on volatile compounds in tea tree leaves was also focused on terpenoids, esters, and heterocyclic compounds, and changes in the content of these three groups of volatile compounds may affect the formation of odor characteristics of tea.

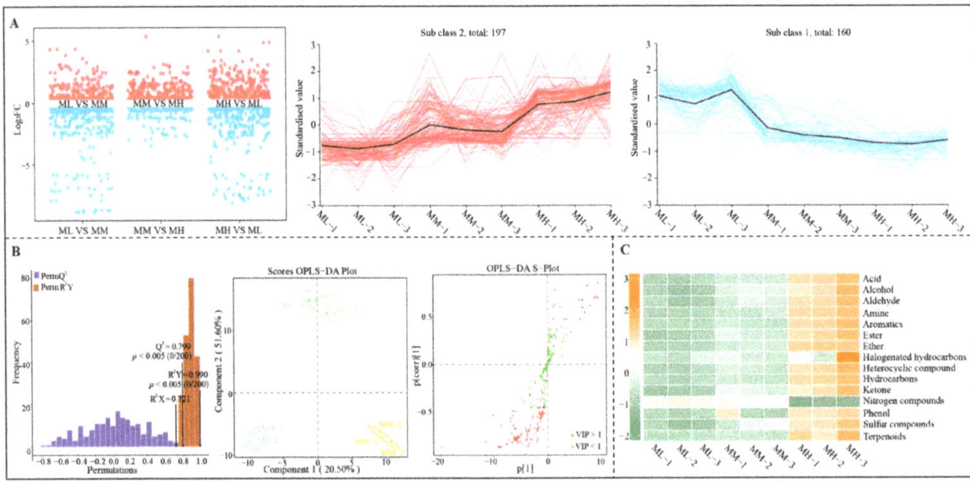

Figure 2. Screening and content analysis of key differential volatile compounds in tea tree leaves under magnesium regulation. Note: ML: Mg^{2+} concentration 0 mmol/L; MM: Mg^{2+} concentration 0.4 mmol/L; MH: Mg^{2+} concentration 0.8 mmol/L; (**A**) Volcano plot screening and K-Means clustering analysis of differential volatile compounds under magnesium regulation; (**B**) OPLS-DA model of ML, MM, and MH for screening of key differential volatile compounds; (**C**) Classification and content analysis of key differential volatile compounds.

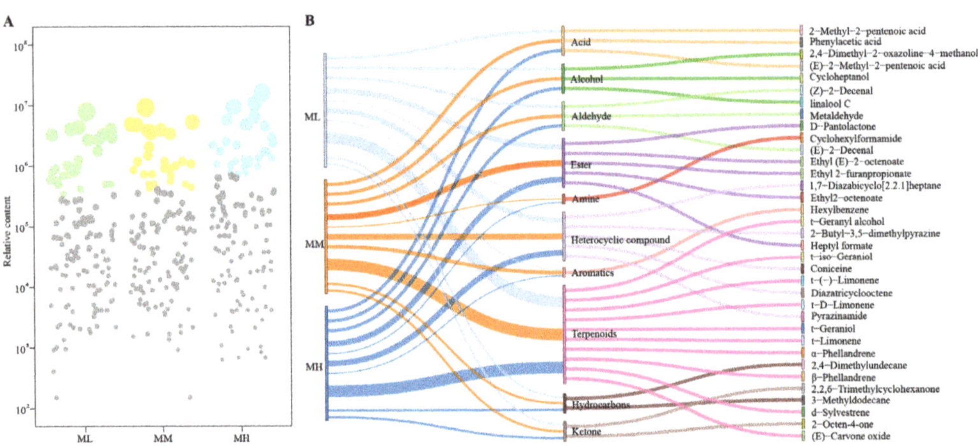

Figure 3. Screening and content analysis of characteristic volatile compounds of tea tree leaves under magnesium regulation. Note: ML: Mg^{2+} concentration 0 mmol/L; MM: Mg^{2+} concentration 0.4 mmol/L; MH: Mg^{2+} concentration 0.8 mmol/L; (**A**) Screening of characteristic volatile compounds under magnesium regulation by bubble feature map; (**B**) Classification of characteristic volatile compounds and their content analysis.

3.3. Contribution Weighting Analysis of Characteristic Volatile Compounds in Differentiating Between Different Magnesium Concentrations

On the basis of the previous analysis, this study further used TOPSIS weighting analysis to determine the contribution weights of the characteristic volatile compounds in distinguishing between different magnesium concentrations, and the results showed (Figure 4A) that for the 35 characteristic volatile compounds, only 15 of them had contribution weights greater than 10%, which were heptyl formate (83.20%), t-geraniol (66.50%), metaldehyde (46.78%), 2-octen-4-one (45.78%), t-geranyl alcohol (43.93%), t-iso-geraniol (43.93%), 2,4-dimethyl-2-oxazoline-4-methanol (37.31%), 1,7-diazabicyclo [2.2.1]heptane (36.30%), cycloheptanol (35.23%), 2,2,6-trimethylcyclohexanone (18.97%), D-pantolactone (13.13%), (E)-carvone oxide (11.34%), t-(-)-limonene (10.46%), t-D-limonene (10.46%), and t-limonene (10.46%). Moreover, the content of all the above 15 characteristic volatile compounds showed an increasing trend with the increase in magnesium concentration (Figure 4B). It was evident that after magnesium regulation, these 15 characteristic volatile compounds played a key role in distinguishing between different magnesium treatments, especially heptyl formate and t-geraniol, both of which were higher in contribution and content.

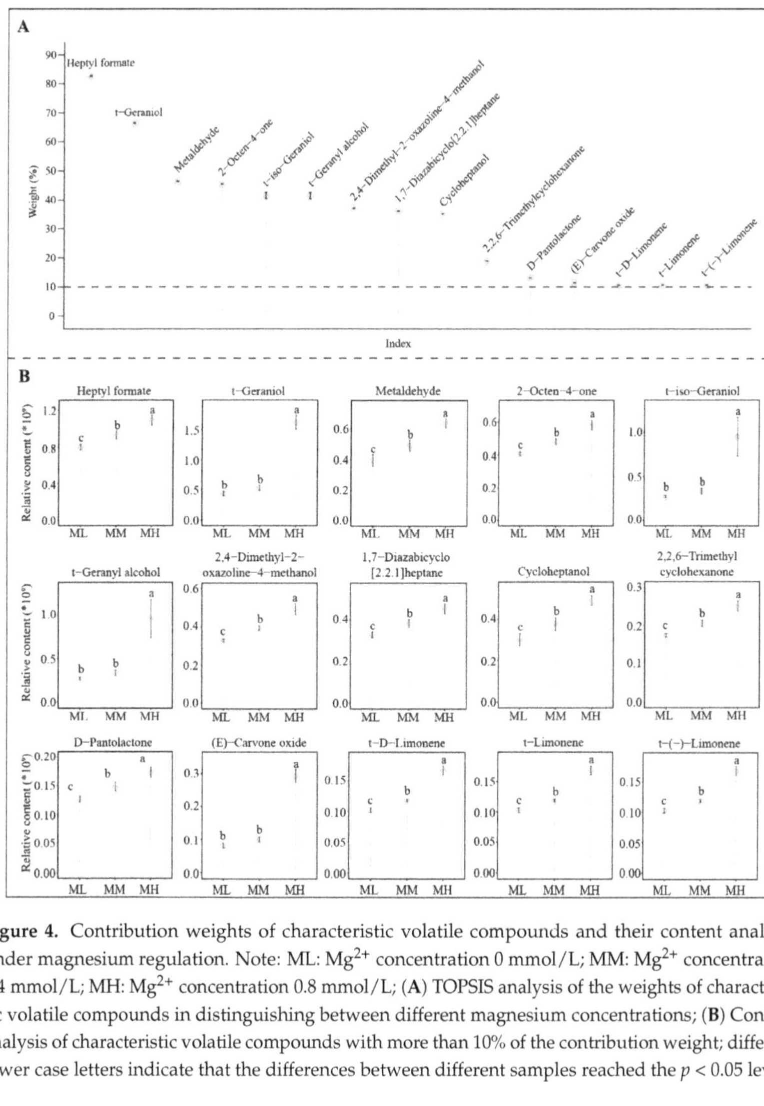

Figure 4. Contribution weights of characteristic volatile compounds and their content analysis under magnesium regulation. Note: ML: Mg^{2+} concentration 0 mmol/L; MM: Mg^{2+} concentration 0.4 mmol/L; MH: Mg^{2+} concentration 0.8 mmol/L; (**A**) TOPSIS analysis of the weights of characteristic volatile compounds in distinguishing between different magnesium concentrations; (**B**) Content analysis of characteristic volatile compounds with more than 10% of the contribution weight; different lower case letters indicate that the differences between different samples reached the $p < 0.05$ level.

Further transformation analysis of the 15 characteristic compounds with high contribution rates revealed (Figure 5) that cycloheptanol could be converted to heptyl formate by oxidation, while heptyl formate could be converted to 2-octen-4-one by oxidation and synthesis. Secondly, acetic acid could synthesize three characteristic volatile compounds, such as 2-octen-4-one, metaldehyde, and 2,4-dimethyl-2-oxazoline-4-methanol, under suitable conditions, and 2,4-dimethyl-2-oxazoline-4-methanol was converted to 1,7-diazabicyclo [2.2.1]heptane by synthesis. Moreover, the transformation analysis of the characteristic compounds also revealed that t-geraniol, t-geranyl alcohol, and t-iso-geraniol were isomers of each other, which were converted to heptyl formate, 2,2,6-trimethylcyclohexanone, or isomers of limonene (t-limonene, t-(-)-limonene and t-D-limonene), whereas 2,2,6-trimethylcyclohexanone was converted by synthesis to D-pantolactone, and isomers of t-limonene were converted to (E)-carvone oxide by oxidation. It was evident that t-geraniol and its isomers played an important role in the formation of characteristic volatile compounds, which were precursors for the transformation of a large number of characteristic volatile compounds.

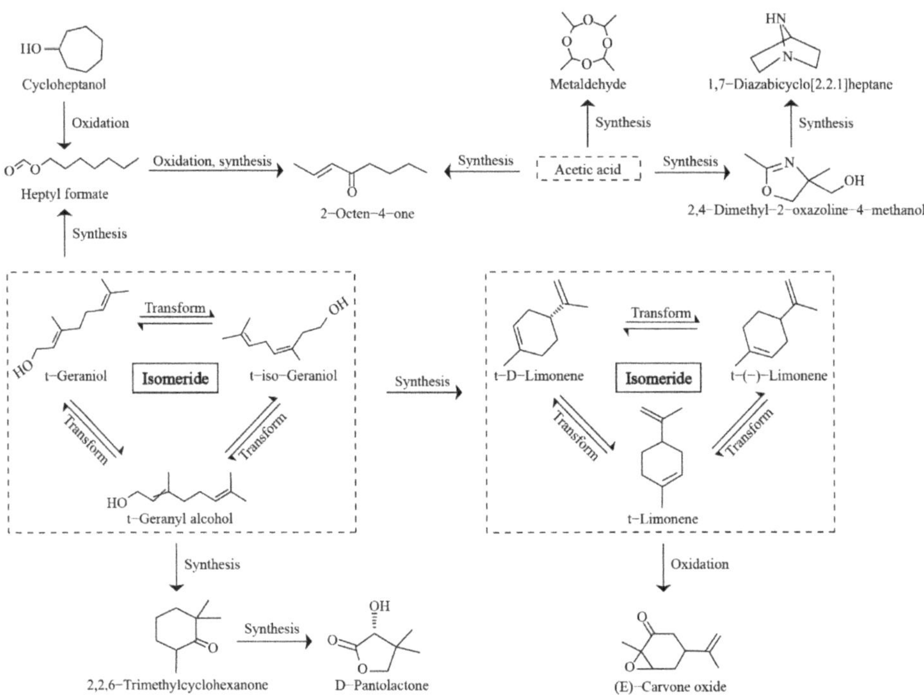

Figure 5. Transformation relationship of characteristic volatile compounds.

3.4. *Odor Analysis of Characteristic Volatile Compounds of Tea Tree Leaves Under Magnesium Regulation*

The formation of odor characteristics of tea depends on the volatile compounds contained in tea, and different volatile compounds present different types of odor characteristics, and the intensity of odor characteristics is closely related to the content of volatile compounds [37,38]. For example, 2-octen-4-one has fruity odor characteristics [39], heptyl formate has green odor characteristics [40], geraniol has floral odor characteristics [41], 2,2,6-trimethylcyclohexanone has pungent odor characteristics [42], (-)-limonene has woody odor characteristics [43], and D-pantolactone has burnt odor characteristics [44]. Further analysis of the odor characteristics of the 15 characteristic volatiles in this study revealed (Figure 6A) that these 15 characteristic volatile compounds under magnesium regulation exhibited six main odor characteristics, namely fruity, green, floral, pungent, woody, and burnt. Among them, fruity was mainly contributed by 2-octen-4-one, cycloheptanol, t-D-limonene, and t-limonene; green was mainly contributed by heptyl formate, metaldehyde, and (E)-carvone oxide; floral was mainly contributed by t-geraniol, t-geranyl alcohol, and t-iso-geraniol; pungent was mainly contributed by 2,4-dimethyl-2-oxazoline-4-methanol and 2,2,6-trimethylcyclohexanone; and woody and burnt were mainly contributed by t-(-)-limonene and D-pantolactone(-), respectively. Further analysis of the intensity of the odor characteristics of tea leaves after magnesium regulation showed (Figure 6B) that, with the increase in magnesium concentration, the odor characteristics of fruity, green, floral, pungent, woody, and burnt in tea were all significantly enhanced, especially floral, fruity, and green. It was evident that magnesium was beneficial to improving the intensity of odor characteristics of tea.

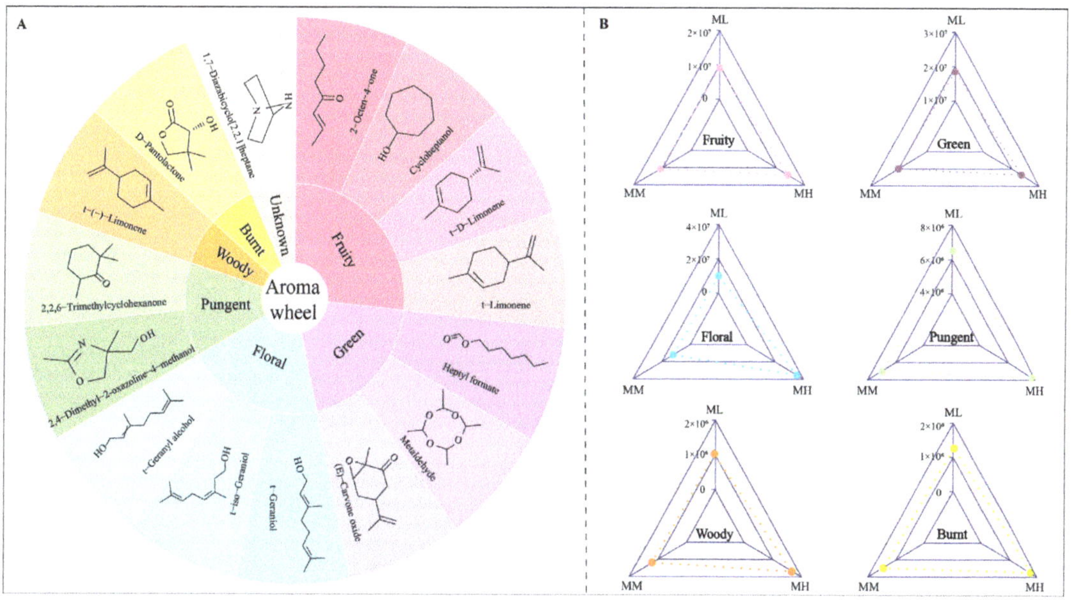

Figure 6. Effect of magnesium regulation on odor characteristics and their intensities in tea tree leaves. Note: ML: Mg^{2+} concentration 0 mmol/L; MM: Mg^{2+} concentration 0.4 mmol/L; MH: Mg^{2+} concentration 0.8 mmol/L; (**A**) Odor analysis of characteristic volatile compounds of tea tree leaves under magnesium regulation; (**B**) Intensity analysis of odor characteristics of tea under magnesium regulation.

4. Conclusions

In this study, the effect of magnesium regulation on volatile compounds and odor characteristics of tea tree leaves was analyzed, and the results showed (Figure 7) that magnesium regulation significantly changed the content of volatile compounds in tea tree leaves, especially terpenoids, esters, and heterocyclic compounds. After magnesium regulation, the characteristic volatile compounds that changed significantly in tea tree leaves were mainly 15 kinds, and their contents showed an increasing trend with the increase of magnesium ion concentration. Among them, heptyl formate and t-geraniol played a key role in distinguishing between different concentrations of magnesium ions. After magnesium regulation, the characteristic volatile compounds of tea tree leaves mainly exhibited five types of odor characteristics, such as fruity, green, floral, pungent, woody, and burnt, the intensities of which all showed a significant upward trend with the increase of magnesium concentration, especially floral, fruity, and green. In conclusion, magnesium is beneficial to enhancing the synthesis of volatile compounds in tea leaves and the intensity of odor characteristics such as floral, fruity, and green. This study provides an important theoretical basis for the application of exogenous magnesium ions to regulate the growth of tea trees and improve the aroma quality of tea leaves.

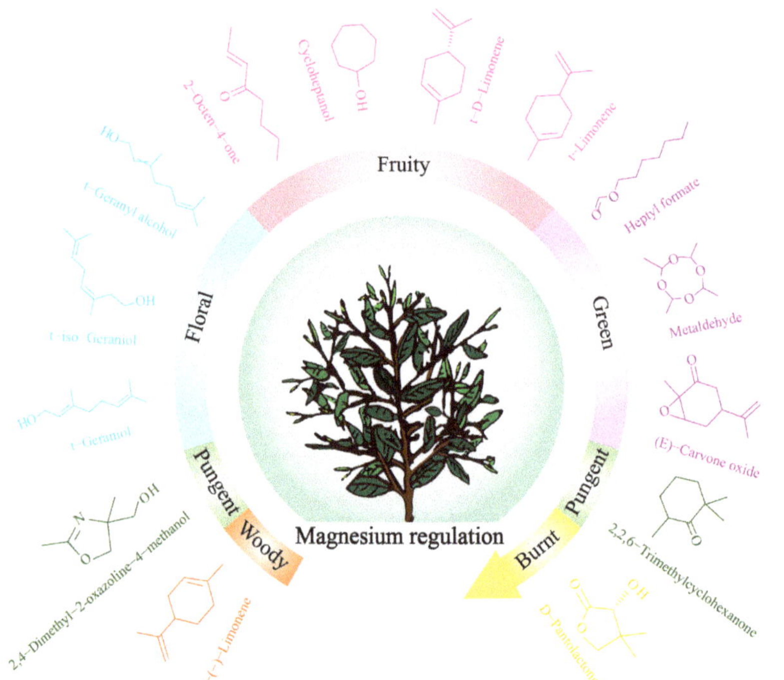

Figure 7. Effect of magnesium regulation on characteristic volatile compounds and odor characteristics of tea tree leaves.

Supplementary Materials: The following supporting information can be downloaded at: https://www.mdpi.com/article/10.3390/foods14061043/s1, Table S1. Raw data of volatile compounds in tea leaves.

Author Contributions: J.Y.: Writing—original draft, Conceptualization, Funding acquisition, Visualization, Methodology. Q.W.: Writing—original draft, Methodology, Data curation, Formal analysis. Y.W., W.C. and J.G.: Methodology, Validation, Writing—original draft. Q.Z., B.Z. and Q.L.: Data curation, Methodology, Validation, Writing—review & editing. X.J.: Writing—review & editing, Funding acquisition, Resources. J.W.: Methodology, Writing—review & editing. H.W.: Writing—review & editing, Project administration, Supervision, Resources, Funding acquisition, Conceptualization. All authors have read and agreed to the published version of the manuscript.

Funding: This work was supported by Natural Science Foundation of Fujian Province (2024J01861, 2024N0009, 2024J01866); Youth Science and Technology Talent Project of Education Department of Guizhou Province (KY 2022-153); Construction of First-Class Undergraduate Specialty (Tea Science) in Fujian Province (SJZY2019004); Nanping City Science and Technology Plan Project (NP2021KTS06, NP2021KTS05, N2021Z012); Innovation and Entrepreneurship Training Program for College Students (202411312002); and Wuyi University Horizontal Research Program (2023-WHFW-017).

Institutional Review Board Statement: Not applicable.

Informed Consent Statement: Not applicable.

Data Availability Statement: The original contributions presented in the study are included in the article/Supplementary Material, further inquiries can be directed to the corresponding authors.

Conflicts of Interest: The authors declare no conflicts of interest.

References

1. Zhao, L.; Ruan, J.; Shi, X. Local industrial policies and development of agricultural clusters: A case study based on a tea cluster in China. *Int. Food Agribus. Manag. Rev.* **2021**, *24*, 267–288. [CrossRef]
2. Wen, B.; Zhang, X.; Ren, S.; Duan, Y.; Zhang, Y.; Zhu, X.; Wang, Y.; Ma, Y.; Fang, W. Characteristics of soil nutrients, heavy metals and tea quality in different intercropping patterns. *Agrofor. Syst.* **2020**, *94*, 963–974. [CrossRef]
3. Sun, L.; Zhang, Z.; Li, Y.; Ruan, J.; Karak, T.; Yang, T. Mineral nutrients on tea yield and quality formation. *Front. Plant Sci.* **2023**, *14*, 1192432. [CrossRef] [PubMed]
4. Shaul, O. Magnesium transport and function in plants: The tip of the iceberg. *Biometals* **2002**, *15*, 307–321. [CrossRef]
5. Wang, Z.; Hassan, M.U.; Nadeem, F.; Wu, L.; Zhang, F.; Li, X. Magnesium fertilization improves crop yield in most production systems: A meta-analysis. *Front. Plant Sci.* **2020**, *10*, 495191. [CrossRef] [PubMed]
6. He, D.; Chen, X.; Zhang, Y.; Huang, Z.; Yin, J.; Weng, X.; Yang, W.; Wu, H.; Zhang, F.; Wu, L. Magnesium is a nutritional tool for the yield and quality of oolong tea (*Camellia sinensis* L. *) and reduces reactive nitrogen loss. Sci. Hortic.* **2023**, *308*, 111590. [CrossRef]
7. Tang, S.; Fu, H.; Pan, W.; Zhou, J.; Xu, M.; Han, K.; Chen, K.; Ma, Q.; Wu, L. Improving tea (*Camellia sinensis*) quality, economic income, and environmental benefits by optimizing agronomic nitrogen efficiency: A synergistic strategy. *Eur. J. Agron.* **2023**, *142*, 126673. [CrossRef]
8. Li, J.; Li, Q.H.; Zhang, X.Y.; Zhang, L.Y.; Zhao, P.L.; Wen, T.; Zhang, J.; Xu, W.; Guo, F.; Zhao, H.; et al. Exploring the effects of magnesium deficiency on the quality constituents of hydroponic-cultivated tea (*Camellia sinensis* L.) Leaves. *J. Agric. Food Chem.* **2021**, *69*, 14278–14286. [CrossRef]
9. Chen, L.; Liu, M.; Cai, Y.; Wu, L.; Zhang, Q. Magnesium deficiency differentiated effects on the roots growth and shoot metabolism by regulating the distribution of photosynthetic products in tea plants. *Sci. Hortic.* **2024**, *338*, 113748. [CrossRef]
10. Xu, J.; Wu, L.; Tong, B.; Yin, J.; Huang, Z.; Li, W.; Li, X. Magnesium supplementation alters leaf metabolic pathways for higher flavor quality of oolong tea. *Agriculture* **2021**, *11*, 120. [CrossRef]
11. Zhang, Y.; Zhang, Q.; Wang, Y.; Lin, S.; Chen, M.; Cheng, P.; Wang, Y.; Du, M.; Jia, X.; Wang, H.; et al. Effects of magnesium on transcriptome and physicochemical index of tea leaves. *Plants* **2023**, *12*, 1810. [CrossRef] [PubMed]
12. Zhai, X.; Zhang, L.; Granvogl, M.; Ho, C.T.; Wan, X. Flavor of tea (*Camellia sinensis*): A review on odorants and analytical techniques. *Compr. Rev. Food Sci. F* **2022**, *21*, 3867–3909. [CrossRef]
13. Zeng, L.; Jin, S.; Xu, Y.Q.; Granato, D.; Fu, Y.Q.; Sun, W.J.; Yin, J.; Xu, Y.Q. Exogenous stimulation-induced biosynthesis of volatile compounds: Aroma formation of oolong tea at postharvest stage. *Crit. Rev. Food Sci.* **2024**, *64*, 76–86. [CrossRef]
14. Hou, Z.; Chen, Z.; Li, L.; Chen, H.; Zhang, H.; Liu, S.; Zhang, R.; Song, Q.; Chen, Y.; Su, Z.; et al. Comparison of volatile compounds in Jingshan green tea scented with different flowers using GC-IMS and GC-MS analyses. *Foods* **2024**, *13*, 2653. [CrossRef]
15. Chen, Z.; Li, Z.; Zhao, Y.; Zhu, M.; Li, J.; Wang, K. A meta-analysis of dynamic changes of key aroma compounds during black tea processing. *Food Biosci.* **2024**, *58*, 103784. [CrossRef]
16. Fu, Z.; Hao, S.; Zhou, J.; Feng, W.; Zhu, M.; Wu, J.; Zhang, Y.; Yu, Y. Profiling volatile compounds in fresh leaves of 22 major oolong tea germplasm cultivated in Fujian of China. *Sci. Hortic.* **2024**, *327*, 112849. [CrossRef]
17. Zhang, H.; Zhang, J.; Liu, S.; Li, T.; Wei, Y.; Gu, Z.; Su, Z.; Ning, J.; Wang, Y.; Hou, Z. Characterization of the key volatile compounds in longjing tea (*Camellia sinensis*) with different aroma types at different steeping temperatures by GC-MS and GC-IMS. *LWT-Food Sci. Technol.* **2024**, *200*, 116183. [CrossRef]
18. Sun, L.; Zhang, M.; Liu, X.; Mao, Q.; Shi, C.; Kochian, L.V.; Liao, H. Aluminium is essential for root growth and development of tea plants (*Camellia sinensis*). *J. Integr. Plant Biol.* **2020**, *62*, 984–997. [CrossRef]
19. Ye, J.; Zhang, Q.; Cheng, P.; Wang, Y.; Zou, J.; Lin, S.; Li, M.; Jia, M.; Chen, Y.; Jia, X.; et al. Aviation mutagenesis alters the content of volatile compounds in Dahongpao (*Camellia sinensis*) leaves and improves tea quality. *Foods* **2024**, *13*, 946. [CrossRef]

20. Yuan, H.; Cao, G.; Hou, X.; Huang, M.; Du, P.; Tan, T.; Zhang, Y.; Zhou, H.; Liu, X.; Liu, L.; et al. Development of a widely targeted volatilomics method for profiling volatilomes in plants. *Mol. Plant* **2022**, *15*, 189–202. [CrossRef]
21. Ye, J.; Zhang, Q.; Li, M.; Wang, Y.; Jia, M.; Hong, L.; Chen, Y.; Pang, X.; Jia, X.; Wang, H. Tea quality of the mysterious "Dahongpao mother tree" (*Camellia sinensis*). *Foods* **2024**, *13*, 1548. [CrossRef] [PubMed]
22. Wickham, H. *Advanced R*, 2nd ed.; Chapman & Hall's R Series; Taylor and Francis: Abingdon, UK, 2019.
23. Zhao, S.; Cheng, H.; Xu, P.; Wang, Y. Regulation of biosynthesis of the main flavor-contributing metabolites in tea plant (*Camellia sinensis*): A review. *Crit. Rev. Food Sci.* **2023**, *63*, 10520–10535. [CrossRef] [PubMed]
24. Ran, W.; Li, Q.; Hu, X.; Zhang, D.; Yu, Z.; Chen, Y.; Wang, M.; Ni, D. Comprehensive analysis of environmental factors on the quality of tea (*Camellia sinensis* var. *sinensis) fresh leaves. Sci. Hortic.* **2023**, *319*, 112177. [CrossRef]
25. Reyrolle, M.; Bareille, G.; Epova, E.N.; Barre, J.; Bérail, S.; Pigot, T.; Desauziers, V.; Gautier, L.; Le Bechec, M. Authenticating teas using multielement signatures, strontium isotope ratios, and volatile compound profiling. *Food Chem.* **2023**, *423*, 136271. [CrossRef]
26. Samynathan, R.; Shanmugam, K.; Nagarajan, C.; Murugasamy, H.; Ilango, R.V.J.; Shanmugam, A.; Venkidasamy, B.; Thiruvengadam, M. The effect of abiotic and biotic stresses on the production of bioactive compounds in tea (*Camellia sinensis* (L.) *O. Kuntze). Plant Gene* **2021**, *27*, 100316. [CrossRef]
27. Xu, J.; Zhang, Y.; Zhang, M.; Wei, X.; Zhou, Y. Effects of foliar selenium application on Se accumulation, elements uptake, nutrition quality, sensory quality and antioxidant response in summer-autumn tea. *Food Res. Int.* **2024**, *175*, 113618. [CrossRef]
28. Zohora, K.F.T.; Arefin, M.R. Tea and tea product diversification: A review. *Turkish J. Agric. Food Sci. Technol.* **2022**, *10*, 2334–2353. [CrossRef]
29. Qiu, Z.; Liao, J.; Chen, J.; Li, A.; Lin, M.; Liu, H.; Huang, W.; Sun, B.; Liu, J.; Liu, S.; et al. Comprehensive analysis of fresh tea (*Camellia sinensis* cv. Lingtou Dancong) leaf quality under different nitrogen fertilization regimes. *Food Chem.* **2024**, *439*, 138127. [CrossRef]
30. Yin, P.; Kong, Y.S.; Liu, P.P.; Wang, J.J.; Zhu, Y.; Wang, G.M.; Sun, M.F.; Chen, Y.; Guo, G.Y.; Liu, Z.H. A critical review of key odorants in green tea: Identification and biochemical formation pathway. *Trends Food Sci. Technol.* **2022**, *129*, 221–232. [CrossRef]
31. Shen, S.; Wu, H.; Li, T.; Sun, H.; Wang, Y.; Ning, J. Formation of aroma characteristics driven by volatile components during long-term storage of An tea. *Food Chem.* **2023**, *411*, 135487. [CrossRef]
32. Liang, Y.; Wang, Z.; Zhang, L.; Dai, H.; Wu, W.; Zheng, Z.; Lin, F.; Xu, J.; Huang, Y.; Sun, W. Characterization of volatile compounds and identification of key aroma compounds in different aroma types of Rougui Wuyi rock tea. *Food Chem.* **2024**, *455*, 139931. [CrossRef] [PubMed]
33. Hadacek, F. Secondary metabolites as plant traits: Current assessment and future perspectives. *Crit. Rev. Plant Sci.* **2002**, *21*, 273–322. [CrossRef]
34. Kessler, A.; Kalske, A. Plant secondary metabolite diversity and species interactions. *Annu. Rev. Ecol. Evol. Syst.* **2018**, *49*, 115–138. [CrossRef]
35. An, H.; Ou, X.; Zhang, Y.; Li, S.; Xiong, Y.; Li, Q.; Huang, J.; Liu, Z. Study on the key volatile compounds and aroma quality of jasmine tea with different scenting technology. *Food Chem.* **2022**, *385*, 132718. [CrossRef]
36. Yang, Y.; Zhu, H.; Chen, J.; Xie, J.; Shen, S.; Deng, Y.; Zhu, J.; Yuan, H.; Jiang, Y. haracterization of the key aroma compounds in black teas with different aroma types by using gas chromatography electronic nose, gas chromatography-ion mobility spectrometry, and odor activity value analysis. *LWT-Food Sci. Technol.* **2022**, *163*, 113492. [CrossRef]
37. Qin, D.; Wang, Q.; Jiang, X.; Ni, E.; Fang, K.; Li, H.; Wang, Q.; Pan, C.; Li, B.; Wu, H. Identification of key volatile and odor-active compounds in 10 main fragrance types of Fenghuang Dancong tea using HS-SPME/GC-MS combined with multivariate analysis. *Food Res. Int.* **2023**, *173*, 113356. [CrossRef]
38. Ma, L.; Sun, Y.; Wang, X.; Zhang, H.; Zhang, L.; Yin, Y.; Wu, Y.; Du, L.; Du, Z. The characteristic of the key aroma-active components in white tea using GC-TOF-MS and GC-olfactometry combined with sensory-directed flavor analysis. *J. Sci. Food Agric.* **2023**, *103*, 7136–7152. [CrossRef]
39. Zeng, Y.; Nong, B.; Xia, X.; Zhang, Z.; Wang, Y.; Xu, Y.; Feng, R.; Guo, H.; Liang, Y.; Chen, C.; et al. Metabolome and transcriptome unveil the correlated metabolites and transcripts with 2-acetyl-1-pyrroline in fragrant rice. *Int. J. Mol. Sci.* **2024**, *25*, 8207. [CrossRef]
40. Chen, Q.; Yu, P.; Li, Z.; Wang, Y.; Liu, Y.; Zhu, Y.; Fu, H. Re-rolling treatment in the fermentation process improves the aroma quality of black tea. *Foods* **2023**, *12*, 3702. [CrossRef]
41. Zhang, Y.; Zhi, H.; Qu, L.; Su, D.; Luo, J. Analysis of flower volatile compounds and odor classification of 17 tree peony cultivars. *Sci. Hortic.* **2024**, *338*, 113665. [CrossRef]

42. Sun, Z.; Lin, Y.; Yang, H.; Zhao, R.; Zhu, J.; Wang, F. Characterization of honey-like characteristic aroma compounds in Zunyi black tea and their molecular mechanisms of interaction with olfactory receptors using molecular docking. *LWT-Food Sci. Technol.* **2024**, *191*, 115640. [CrossRef]
43. Vieira, A.J.; Beserra, F.P.; Souza, M.C.; Totti, B.M.; Rozza, A.L. Limonene: Aroma of innovation in health and disease. *Chem. Biol. Interact.* **2018**, *283*, 97–106. [CrossRef] [PubMed]
44. Arsa, S.; Theerakulkait, C. Preparation, aroma characteristics and volatile compounds of flavorings from enzymatic hydrolyzed rice bran protein concentrate. *J. Sci. Food Agric.* **2018**, *98*, 4479–4487. [CrossRef] [PubMed]

foods

MDPI

Article

Effects of Liupao Tea with Different Years of Aging on Glycolipid Metabolism, Body Composition, and Gut Microbiota in Adults with Obesity or Overweight: A Randomized, Double-Blind Study

Yuyang Wang, Qiang Hu, Botian Chen and Defu Ma *

Department of Social Medicine and Health Education, School of Public Health, Peking University Health Science Center, Beijing 100191, China; qianghu@stu.pku.edu.cn

* Correspondence: madefu@bjmu.edu.cn; Tel.: +86-10-82801743; Fax: +86-10-82801519

Abstract: Background: Liupao tea (LPT) is a traditionally fermented dark tea from Guangxi, China and the effects of different aging periods of LPT on metabolic health remain inadequately explored. Methods: This randomized, double-blind, longitudinal study enrolled 106 adults with obesity or overweight who were assigned to consume LPT of different ages over a 90-day period. Participants were randomly divided into four groups, each consuming LPT that had been aged for 1 year, 4 years, 7 years, or 10 years. The metabolic parameters, body composition, and gut microbiota were assessed at baseline and after the 90-day intervention. Results: All LPT groups experienced significant reductions in systolic blood pressure (SBP) and diastolic blood pressure (DBP), with the 10-year-aged group showing the most notable SBP decrease ($p < 0.001$). Total cholesterol (TC) and low-density lipoprotein cholesterol (LDL-C) levels decreased significantly in the 1-, 4-, and 10-year-aged groups ($p < 0.05$), while high-density lipoprotein cholesterol (HDL-C) increased in the 7-year-aged group ($p < 0.05$). Body weight, body fat mass (BFM), body mass index (BMI), waist circumference (WC), body fat percentage (BFP), and visceral fat area (VFA) significantly declined across all groups ($p < 0.05$). Gut microbiota analysis showed changes in specific genera, though overall diversity remained stable. No significant differences were found in metabolic or microbiota outcomes between the different aged groups. Conclusions: LPT consumption effectively improves blood pressure, lipid profiles, and body composition in adults with obesity without adverse liver effects. The aging duration of LPT does not significantly alter these health benefits, challenging the belief that longer-aged LPT is superior.

Keywords: Liupao tea; weight management; lipid profile; gut health; randomized trial

1. Introduction

Metabolic diseases, including obesity, diabetes, cardiovascular diseases, and fatty liver, have emerged as major global public health concerns [1,2]. According to the latest data from the Non-Communicable Disease Risk Factor Collaboration (NCD-RisC) and the World Health Organization (WHO), the global prevalence of obesity has shown a steady upward trend, affecting approximately 880 million adults (504 million women and 374 million men) affected by obesity in 2022 [3]. Obesity and its associated metabolic disorders significantly increase the incidence and mortality of cardiovascular diseases, diabetes, stroke, and other conditions, while also imposing a substantial socioeconomic burden [4]. Recent reports indicate that in 2022, the global economic loss caused by overweight and obesity, including healthcare costs and productivity losses, accounted for 0.05% to 2.42% of global GDP across various countries [5]. Therefore, identifying effective interventions to prevent and manage obesity and its related metabolic diseases has become a critical issue in global health.

Lifestyle improvements, especially dietary interventions, have been shown to play a crucial role in the prevention and management of metabolic diseases [6–8]. Fermented teas, through microbial fermentation, yield unique bioactive compounds that potentially modulate metabolic health. As a specific type of black tea, Pu-erh tea has demonstrated the ability to alter body composition and energy efficiency in mice fed a high-fat diet. It has been shown to alleviate metabolic endotoxemia, reduce systemic and multi-organ inflammation, and improve glucose and lipid metabolic disorders [9]. The theabrownins present in Pu-erh tea have been found to significantly reduce serum total cholesterol (TC) and low-density lipoprotein (LDL) levels by inhibiting hepatic cholesterol synthesis and promoting lipid catabolism [10,11]. Ya'an Tibetan tea, a traditionally fermented tea, was demonstrated by Ye et al. to significantly improve weight management and fat accumulation in HFD-induced obese rats. It enhances antioxidant enzyme activity, reduces inflammation, and reverses HFD-induced hepatic damage [12]. Kombucha, a non-alcoholic fermented tea beverage produced through symbiotic bacterial and yeast fermentation, has been reported to exert beneficial effects on metabolic health. According to a literature review by Costa, kombucha consumption can significantly mitigate oxidative stress and inflammation, enhance hepatic detoxification processes, and provide potential benefits for the management and treatment of obesity and its related comorbidities [13].

Liupao tea (LPT), a type of dark tea produced in Guangxi, China, is known for its unique fermentation process and aging procedure, which endow it with distinctive chemical components and bioactivities [14,15]. Compared to green tea, LPT has higher levels of alkaloids and fatty acids, and contains relatively higher amounts of ellagic acid, hypoxanthine, and theobromine [16]. Ding et al. found that LPT extract significantly alleviated hyperglycemia, insulin resistance, and dyslipidemia in streptozotocin-induced diabetic rats, with its blood glucose-lowering effect comparable to that of metformin when administered at high doses [17].

Studies have shown that bioactive compounds in tea, such as polyphenols, theaflavins, and tea polysaccharides, possess multiple health effects, including antioxidant, anti-inflammatory, blood glucose-lowering, lipid-lowering, blood pressure-regulating, and gut microbiota-modulating activities [18–21]. In traditional Chinese medicine, LPT is believed to have "lipid-lowering, spleen-strengthening, and digestive-promoting" effects [22]. The microbial metabolites and chemical components (such as tea polyphenols, free amino acids, and water-soluble sugars) produced during the fermentation process of LPT change over time, and these changes play an important role in metabolic regulation [23,24]. However, there is still limited research on the effects of different aging years of LPT on body composition, glycolipid metabolism, and gut microbiota, particularly the potential effects of different aging years of LPT in populations with metabolic abnormalities such as obesity, hyperglycemia, and dyslipidemia.

This study investigates the effects of LPT, aged for different amounts of time (1, 4, 7, and 10 years), on glycolipid metabolism, body composition, and gut microbiota in adults with obesity or overweight through a longitudinal analysis. It is hypothesized that LPT consumption, regardless of length of aging, will demonstrate significant improvements in metabolic outcomes, including blood pressure, lipid profiles, and glucose regulation, while simultaneously reducing body fat accumulation and reshaping gut microbiota composition. Consequently, this research seeks to elucidate the mechanisms by which LPT modulates metabolic health through gut microbiota changes, thereby providing novel dietary intervention strategies for the management of metabolic disorders.

2. Materials and Methods

2.1. Study Design

This is a randomized, double-blind, longitudinal study with a 90-day intervention period, exploring the effects of LPT aged for different numbers of years on glycolipid metabolism, body composition, and gut microbiota in adults with obesity or overweight. This study was approved by the Biomedical Ethics Committee of Peking University (NO.IRB00001052-23203) and registered with the Chinese Clinical Trial Registry (Trial Registration number: ChiCTR2400089348). Figure 1 illustrates the study flow, from recruitment to data analysis.

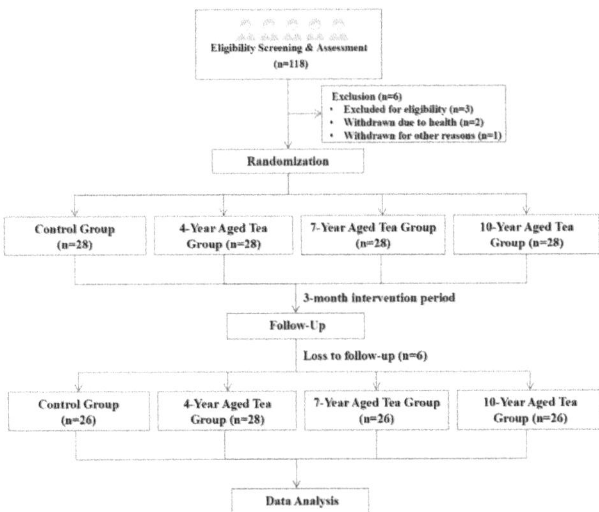

Figure 1. Study flowchart from participant recruitment to data analysis.

2.2. Participant Recruitment and Eligibility Criteria

The recruitment process began on 1 April 2024, with the posting of project promotional posters in the Xueyuan Road community in Haidian District, Beijing, and concluded on 30 April 2024. The inclusion criteria were as follows: (1) age between 18 and 70 years; (2) BMI $\geq$ 24.0 kg/m^2; (3) no severe liver, kidney, gastrointestinal diseases, or diabetic complications; (4) no use of medications or dietary supplements to lower blood lipids, blood glucose, or body weight in the past three months, and agreement not to use such substances during the study; (5) voluntary participation after receiving a detailed explanation of the project and signing the informed consent form.

Exclusion criteria included: (1) pregnant, breastfeeding, or pre-conception women; (2) serious heart, liver, kidney, hematologic diseases, or psychiatric disorders; (3) use of lipid-lowering, glucose-lowering, or weight-loss medications or supplements in the last three months that may affect the study outcomes; (4) excessive smoking or alcohol consumption; (5) special dietary habits such as vegetarianism or ketogenic diets (high-fat, low-carb); (6) inability to follow study requirements; (7) allergies to tea components. Informed consent was obtained from all participants under the guidance of the researchers to ensure that each participant fully understood the main content of the study, its associated risks, and potential benefits.

2.3. Sample Size Estimation

The sample size for this study was estimated based on previous research on dark tea, where BMI decreased from 28.67 ± 2.91 kg/m^2 to 26.86 ± 3.17 kg/m^2. Using a longitudinal design, with a significance level (α) of 0.05 and a power ($1-\beta$) of 80%, the required sample size per group was calculated to be 24 participants. Accounting for a 10% dropout rate, the final sample size for each group was 26, resulting in a total of 104 participants [25].

2.4. Intervention Materials

The LPT samples used in the experiment all originated from Zhaoping County, Guangxi Zhuang Autonomous Region. During the aging process, the samples were stored in a constant temperature and humidity warehouse with a temperature range of 20–25 °C, relative humidity of 60–65%, ensuring proper air circulation and no direct sunlight exposure. The LPT used in the experiment was provided free of charge by Guangxi Zhaoping County General Mountain Agricultural Technology Co., Ltd. (Hezhou, China). The main water-soluble components identified in these LPT samples for the different years of aging are summarized in Table S1.

2.5. Randomization and Blinding

Participants were randomly assigned into four groups based on the aged years of LPT: 1-year, 4-year, 7-year, and 10-year-aged tea groups. Randomization was performed by a statistician not involved in the trial using computer-generated random numbers to ensure impartiality and randomness in group assignment. The randomization list was then handed over to the sample packager, who packaged the tea samples (6 g per bag) according to the assigned groups while ensuring uniformity in the external packaging of each tea bag. The study followed a double-blind design, where both participants and researchers were unaware of the specific tea aging groups to avoid bias. Participants were notified to collect the pre-packaged tea from the study office on the 2nd day before the intervention, as well as on the 28th and 58th days of the intervention, and their names and collection times were recorded.

2.6. Intervention and Adherence Monitoring

During the intervention, participants consumed 6 g of LPT (one tea bag) daily. The tea was brewed using a standardized 300 mL teapot, with 250 mL of boiling water for a 30–45 s infusion, followed by immediate consumption once the tea had cooled to a warm temperature. Participants were allowed to re-brew the tea until the color of the tea became pale. To optimize the intervention's effectiveness, participants were instructed to drink the tea 1–2 h after each meal, and to complete their daily consumption before 3 p.m. to minimize potential disruption to nighttime sleep.

To improve adherence, a dedicated WeChat group was created where participants uploaded daily photos of their tea consumption as proof of compliance and kept a personal tea diary. Research staff monitored the group daily, providing reminders to participants who missed their tea consumption. If participants experienced any intolerance or adverse effects, they were instructed to immediately stop drinking the tea and contact the researchers. Participants who failed to comply with the intervention protocol or who could not be reached for three consecutive days were withdrawn from the study, and the reason for discontinuation or dropout was documented. Throughout the study, participants were advised to maintain their usual diet, exercise, and sleep routines without making any significant lifestyle changes. Adherence to the intervention was closely monitored, with daily follow-up and reminders provided to participants. Any non-compliance or adverse events were recorded and addressed promptly. If participants were lost to follow-up or dropped out of the study, the reasons for discontinuation were documented.

2.7. Questionnaire Survey and Clinical Parameter Measurement

At baseline, demographic data, including gender, age, marital status, education level, smoking habits, alcohol consumption, and family history of chronic diseases, were collected via questionnaires. The following measurements were taken at baseline and after 90 days of intervention. The clinical parameters collected in this study include blood pressure, lipid metabolism indicators, glucose metabolism indicators, and liver function markers. Before blood pressure measurement, participants were required to sit quietly in a calm environment for 5 min. The upper arm was exposed and kept level during measurement. Blood pressure was measured three times, and the average of the three measurements was taken as the final value. On the morning of the test, approximately at 8:00 A.M., fasting venous blood (after fasting for more than 8 h) was drawn from participants into centrifuge tubes. After centrifugation and aliquoting, the samples were stored in a −80 °C ultralow freezer until testing. Fasting blood glucose (FBG) was measured using the hexokinase method, hemoglobin A1c (HbA1c) was measured by ion-exchange chromatography, and fasting insulin levels (INS) were measured using chemiluminescent immunoassay. TC, triglycerides (TG), high-density lipoprotein cholesterol (HDL-C), low-density lipoprotein cholesterol (LDL-C), apolipoprotein A1 (APOA1), and apolipoprotein B (APOB) were assessed by enzyme methods. To evaluate the safety of LPT intervention, key biomarkers of liver health, including total bilirubin (T-BIL), alanine aminotransferase (ALT), aspartate aminotransferase (AST), serum alkaline phosphatase (ALP), serum creatine kinase (CK), and serum lactate dehydrogenase (LDH), were also measured. All participant samples were collected by professional medical staff at Beijing Meinian Health Examination Center and sent to a specialized laboratory for analysis using an automated biochemical analyzer.

2.8. Body Composition Measurement

Height was measured with participants barefoot and standing upright, with their back against a wall. A right-angle board was placed on the participant's head, and the measurement was repeated twice, with the average of the two used as the final height. Waist circumference (WC) was measured with a standard soft tape at 1 cm above the navel. The measurement was repeated twice, and if the difference between the two was less than 0.5 cm, the average value was taken as the final waist circumference. Body composition was assessed using the IOI353 body composition analyzer (Jawon Medical Co., Ltd., Seoul, Republic of Korea). After verifying participant details (enrollment number, gender, and age) on the device, participants were instructed to remove their outerwear, stand barefoot on the analyzer's electrodes, and grip the hand electrodes at a 30-degree angle to their body. Participants were instructed to remain still during the measurement. Body mass index (BMI) is calculated by dividing the subject's weight (kg) by the square of their height (m). The analyzer measured body weight, body fat mass (BFM), muscle mass (MM), body fat percentage (BFP), lean body mass (LBM), and visceral fat area (VFA).

2.9. DNA Extraction and Amplification Library Construction

Total genomic DNA was extracted from the samples using the SDS method to ensure that the quality of the extracted DNA was suitable for subsequent amplification and sequencing. Specific primer pairs (515F and 806R) were used to amplify the V3-V4 regions of the 16S rRNA gene. The PCR reaction mixture contained 15 μL of Phusion® High-Fidelity PCR Master Mix (New England Biolabs, Inc., Ipswich, MA, USA), 0.2 μM of each primer, and 10 ng of genomic DNA. The PCR conditions were as follows: an initial denaturation at 98 °C for 1 min, followed by 30 cycles at 98 °C for 10 s, 50 °C for 30 s, and 72 °C for 30 s, with a final extension at 72 °C for 5 min. After electrophoresis on a 2% agarose gel to confirm the PCR products, the samples were purified using magnetic beads and mixed in equal proportions. The constructed libraries were evaluated for quality using a Qubit® 2.0 Fluorometer (Thermo Fisher Scientific, Inc., Waltham, MA, USA) and the Agilent Bioanalyzer 2100 system (Agilent Technologies, Inc., Santa Clara, CA, USA). The qualified libraries were sequenced on an Illumina NovaSeq platform (Illumina, Inc., San Diego, CA, USA), generating 250 bp paired-end reads.

2.10. Bioinformatics Analysis and Diversity Analysis

After splitting the sequencing data based on barcode and primer sequences, paired-end reads were assembled using FLASH (Version 1.2.11, https://ccb.jhu.edu/software/FLASH/, accessed on 15 October 2024), generating the raw tag sequences. The raw data were then quality-filtered using fastp (Version 0.23.1) to obtain clean tags. The effective tags were obtained by removing chimeric sequences through alignment with the Silva database (Version 2.15.0) using Vsearch. Denoising of the effective tags was performed using the DADA2 algorithm or the deblur module in QIIME2, resulting in amplicon sequence variants (ASVs). Taxonomic annotation of the ASVs was performed using the Silva138.1 database for 16S rRNA gene sequences, with supplementary taxonomic information retrieved from the NCBI database when necessary. Based on the normalized data, both α-diversity and β-diversity analyses were performed to evaluate the diversity within and between the microbial communities across the samples. Based on the species annotation and abundance information at the genus level for all samples, the top 35 genera were selected. Clustering was performed at the species level using their abundance data in each sample to identify the relative abundance of species across the groups.

2.11. Statistical Analysis

For the basic demographic information, clinical parameters, and body composition indicators of the participants, continuous variables that followed a normal distribution are expressed as mean ± standard deviation (SD), with group differences assessed using *t*-tests or one-way analysis of variance (ANOVA). Categorical variables are presented as frequencies (n) and percentages (%), with group differences compared using the chi-square test. If the ANOVA results show significant differences ($p < 0.05$), Bonferroni correction or the least significant difference (LSD) method will be used to conduct post hoc pairwise comparisons. The missing data of dropouts were handled using multiple imputation by chained equations (MICE) and the robustness of the results was assessed using intention-to-treat (ITT) analysis.

Alpha diversity was calculated using indices such as Shannon, Simpson, and Chao1, and was evaluated using QIIME2 software to assess the richness and diversity of microbial communities within the samples. Beta diversity was used to compare microbial community compositions between different samples. Based on species annotation results and feature sequence abundance, sequences within the same taxonomic group were merged to generate a species abundance table (profiling table). Furthermore, UniFrac distances were calculated based on the phylogenetic relationships between feature sequences, including both unweighted and weighted UniFrac distances. These distances, which are calculated based on the evolutionary information of microbial sequences between samples, formed a distance matrix. This UniFrac distance matrix, whether weighted or unweighted, was then transformed into a new set of orthogonal axes, with the first principal coordinate representing the factor of maximum variation, the second principal coordinate representing the second largest variation factor, and so on. Beta diversity differences were visualized using principal coordinates analysis (PCoA), revealing differences between samples (or groups). Two-dimensional PCoA results were displayed using the ade4 and ggplot2 packages (version 2.15.3) in R software (version 4.1.3).

To evaluate the significance of differences in community structure between groups, the adonis and anosim functions in QIIME2 were used for analysis. The Mann–Whitney U test was used to compare significantly different species at each taxonomic level (phylum, class, order, family, genus, species) before and after the intervention across the LPT of different aging years. To investigate the functional roles of microbial communities and identify differences between groups, PICRUSt2 software (Version 2.1.2-b) was used for functional annotation. KO (KEGG Orthology) functions were ranked by relative abundance at baseline and follow-up, and the top 10 KO IDs were selected. The Wilcoxon signed-rank test was used to compare changes within the same samples over time.

All statistical analyses were performed using R statistical software version 4.1.3, with statistical significance defined as $p < 0.05$.

3. Results

3.1. Basic Characteristics of Participants

Table 1 showed the comparison of baseline characteristics between the total population and different aged intervention groups. The total sample size was 106 participants, including 26 in the 1-year-aged group, 28 in the 4-year-aged group, 26 in the 7-year-aged group, and 26 in the 10-year-aged group. The gender distribution in the total population was 26.4% male and 73.6% female. There were no significant differences between the groups in terms of age, gender, marital status, education level, monthly income, alcohol consumption, and tea-drinking habits. Among the participants, 79.2% had T2DM, 67% had hypertension, and 61.3% had dyslipidemia. No significant differences were observed across the groups in the distribution of these medical history ($p = 0.100$, $p = 0.756$, and $p = 0.526$).

Table 1. Baseline characteristics of the total population and different aged LPT intervention groups [a].

Characteristic	Total Population Group (*n* = 106)	1-Year-Aged Group (*n* = 26)	4-Year-Aged Group (*n* = 28)	7-Year-Aged Group (*n* = 26)	10-Year-Aged Group (*n* = 26)	*p*
Age (years), mean ± SD	50.9 ± 17.4	51.3 ± 17.9	49.5 ± 17.7	51.2 ± 17.2	51.8 ± 17.5	0.968
Gender, *n* (%)						0.974
Male	28 (26.4)	7 (26.9)	8 (28.6)	7 (26.9)	6 (23.1)	
Female	78 (73.6)	19 (73.1)	20 (71.4)	19 (73.1)	20 (76.9)	
Marital Status, *n* (%)						0.790
Single	17 (16.0)	4 (15.4)	6 (21.4)	4 (15.4)	3 (11.5)	
Married/cohabiting	79 (74.5)	21 (80.8)	20 (71.4)	19 (73.1)	19 (73.1)	
Divorced/widowed	10 (9.4)	1 (3.8)	2 (7.2)	3 (11.5)	4 (15.4)	
Educational Level, *n* (%)						0.917
Junior high school and below	8 (7.5)	2 (7.7)	2 (7.1)	2 (7.7)	2 (7.7)	
High school	16 (15.1)	3 (11.5)	3 (10.7)	4 (15.4)	6 (23.1)	
Bachelor's degree and above	82 (77.4)	21 (80.8)	23 (82.1)	20 (76.9)	18 (69.2)	
Average Monthly Income, *n* (%)						0.528
<3000 CNY	18 (17.0)	4 (15.4)	4 (14.2)	5 (19.2)	5 (19.2)	
3000~8000 CNY	42 (39.6)	11 (42.3)	12 (42.9)	6 (23.1)	13 (50.0)	
≥8000 CNY	46 (43.4)	11 (42.3)	12 (42.9)	15 (57.7)	8 (30.8)	
Smoking, *n* (%)						0.021
Never smoked	99 (93.4)	23 (88.5)	24 (85.7)	26 (100.0)	26 (100.0.)	
Former smoker	4 (3.8)	3 (11.5)	1 (3.6)	0 (0.0)	0 (0.0)	
Current smoker	3 (2.8)	0 (0.0)	3 (10.7)	0 (0.0)	0 (0.0)	
Alcohol Consumption, *n* (%)						0.273
Never drinks	94 (88.7)	23 (88.5)	22 (78.5)	24 (92.3)	25 (96.2)	
Former drinker	9 (8.5)	3 (11.5)	5 (17.9)	1 (3.9)	0 (0.0)	
Current drinker	3 (2.8)	0 (0.0)	1 (3.6)	1 (3.8)	1 (3.8)	
Frequent Tea Drinking, *n* (%)						0.744
Yes	49 (46.2)	12 (46.2)	15 (53.6)	12 (46.2)	10 (38.5)	
No	57 (53.8)	14 (53.8)	13 (46.4)	14 (53.8)	16 (61.5)	
Medical History, *n* (%)						
T2DM	84 (79.2)	22 (84.6)	23 (82.1)	17 (65.4)	22 (84.6)	0.100
Hypertension	71 (67.0)	16 (61.5)	19 (67.9)	18 (69.2)	18 (69.2)	0.756
Dyslipidemia	65 (61.3)	17 (65.4)	14 (50)	15 (57.7)	19 (73.1)	0.526
Heart disease	98 (92.5)	22 (84.6)	26 (92.9)	25 (96.2)	25 (96.2)	0.526

[a] Continuous variables are presented as mean ± SD; categorical variables are shown as frequencies (*n*) and percentages (%); SD, standard deviation; CNY, Chinese Yuan; T2DM, Type 2 diabetes.

3.2. Effects of LPT with Different Aged Years on Metabolic Parameters

Table 2 presented the intervention effects of LPT with different aged years on metabolic parameters. After 90 days of tea intervention, SBP significantly reduced in all four groups. In the 10-year-aged group, SBP dropped from 131.32 ± 16.08 mmHg to 121.16 ± 15.16mmHg ($p < 0.001$). DBP showed a significant decline in the 1-year-aged, 4-year-aged, and 10-year-aged groups ($p < 0.05$). TC in the 1-year-aged group showed a marked reduction ($p < 0.001$), from 5.10 ± 1.08mmol/L to 4.87 ± 0.88 mmol/L. LDL-C decreased significantly in the 1-year-aged, 4-year-aged, and 10-year-aged groups ($p < 0.05$), while HDL-C notably increased in the 7-year-aged group ($p < 0.05$). APOA1 was significantly lower in all four groups ($p < 0.05$). FBG levels in the 4-year-aged group were significantly reduced, from 5.75 ± 1.19 mmol/L to 5.50 ± 1.07mmol/L ($p < 0.05$). Additionally, LPT had no significant effect on most liver health biomarkers (T-BIL, ALT, AST, ALP, CK, LDH), indicating minimal impact on liver function and confirming its good safety profile (Table S2).

Table 2. The intervention effects of LPT with different years of aging on metabolic parameters [a].

Metabolic Parameters [b]	1-Year-Aged Group (n = 26)		4-Year-Aged Group (n = 28)		7-Year-Aged Group (n = 26)		10-Year-Aged Group (n = 26)	
	Baseline	Follow-Up	Baseline	Follow-Up	Baseline	Follow-Up	Baseline	Follow-Up
Metabolic parameters								
SBP, mmHg	129.84 ± 22.35	122.92 ± 21.47 *	130.70 ± 14.48	121.37 ± 17.15 *	131.38 ± 13.55	115.30 ± 23.93 *	131.32 ± 16.08	121.16 ± 15.16 **
DBP, mmHg	76.28 ± 12.79	73.08 ± 12.67 *	78.41 ± 8.54	73.56 ± 11.59 *	78.54 ± 10.75	74.08 ± 15.91	81.72 ± 14.56	72.00 ± 10.94 **
TC, mmol/L	5.10 ± 1.08	4.87 ± 0.88 *	5.01 ± 1.12	4.83 ± 1.10	5.19 ± 0.91	5.38 ± 0.85	5.26 ± 0.79	4.96 ± 0.86
TG, mmol/L	1.26 ± 0.82	1.35 ± 0.88	1.28 ± 1.09	1.08 ± 0.51	1.63 ± 0.87	1.66 ± 1.16	1.73 ± 1.86	1.63 ± 1.61
HDL-C, mmol/L	1.34 ± 0.36	1.36 ± 0.31	1.30 ± 0.23	1.36 ± 0.21	1.29 ± 0.29	1.37 ± 0.25 *	1.34 ± 0.22	1.37 ± 0.18
LDL-C, mmol/L	2.91 ± 0.68	2.63 ± 0.55 *	2.90 ± 0.83	2.65 ± 0.76 *	3.08 ± 0.64	2.94 ± 0.50	2.95 ± 0.52	2.63 ± 0.55 *
APOA1, g/L	1.33 ± 0.23	1.25 ± 0.21 *	1.30 ± 0.16	1.22 ± 0.14 *	1.31 ± 0.16	1.24 ± 0.14 *	1.36 ± 0.15	1.25 ± 0.12 **
APOB, g/L	0.85 ± 0.17	0.82 ± 0.14	0.84 ± 0.21	0.83 ± 0.24	0.88 ± 0.16	0.88 ± 0.11	0.83 ± 0.15	0.82 ± 0.14
HbA1c, %	5.59 ± 1.00	5.68 ± 0.99 *	5.60 ± 0.68	5.63 ± 0.68	5.87 ± 1.21	5.92 ± 1.12	5.79 ± 1.29	5.87 ± 1.37
FBG, mmol/L	5.98 ± 1.87	5.88 ± 2.09	5.75 ± 1.19	5.50 ± 1.07 *	8.38 ± 11.56	6.05 ± 1.69	6.05 ± 2.47	5.90 ± 1.97
INS, pmol/L	75.93 ± 33.55	71.70 ± 44.12	73.31 ± 31.06	62.21 ± 28.87	72.68 ± 35.16	67.78 ± 35.83	83.19 ± 50.47	74.28 ± 46.00
Body composition								
Weight, kg	69.70 ± 13.81	68.21 ± 13.19 **	69.00 ± 11.96	66.70 ± 10.77 **	67.36 ± 10.62	65.33 ± 10.92 *	67.06 ± 11.3	65.57 ± 11.90 **
BFM, kg	21.78 ± 5.24	20.28 ± 4.77 **	21.88 ± 5.64	19.54 ± 5.59 **	21.13 ± 4.66	19.15 ± 4.44 **	22.09 ± 6.08	20.12 ± 5.43 **
LBM, kg	46.34 ± 9.17	46.92 ± 9.77 *	47.12 ± 9.02	47.63 ± 9.44	46.61 ± 8.07	46.83 ± 8.59	46.33 ± 9.93	47.01 ± 11.38
MM, kg	42.48 ± 8.49	43.09 ± 9.09 *	42.08 ± 11.06	43.43 ± 9.03	42.76 ± 7.52	43.06 ± 8.00	42.46 ± 9.21	43.18 ± 10.61
BMI, kg/m^2	25.90 ± 3.24	25.44 ± 3.12 *	26.22 ± 2.99	25.38 ± 2.76 **	25.19 ± 2.65	24.47 ± 2.80 *	25.47 ± 3.17	24.93 ± 3.29 *
WC, cm	84.51 ± 8.43	82.60 ± 7.33 **	85.55 ± 8.21	82.04 ± 6.71 **	84.66 ± 7.91	82.20 ± 7.74 **	84.60 ± 9.65	79.37 ± 17.19
BFP, %	31.54 ± 4.98	30.21 ± 4.56 *	31.62 ± 6.37	29.10 ± 7.07 **	31.18 ± 5.16	29.02 ± 4.97 **	32.20 ± 5.62	30.07 ± 5.66 **
VFA, cm^2	99.13 ± 40.25	86.52 ± 33.63 **	100.85 ± 36.11	80.56 ± 31.05 **	93.96 ± 29.61	79.29 ± 27.78 **	101.60 ± 39.97	85.00 ± 32.52 **

** $p < 0.001$, * $p < 0.05$. [a] Values are expressed as mean ± standard. [b] Abbreviations: SBP: systolic blood pressure; DBP: diastolic blood pressure; TC: total cholesterol; TG: triglycerides; HDL-C: high-density lipoprotein cholesterol; LDL-C: low-density lipoprotein cholesterol; APOA1: apolipoprotein A1; APOB: apolipoprotein B; HbA1c: hemoglobin A1c; FBG: fasting blood glucose; INS: insulin; BFM: body fat mass; LBM: lean body mass; MM: muscle mass; BMI: body mass index; WC: waist circumference; BFP: body fat percentage; VFA: visceral fat area.

3.3. Effects of LPT with Different Years of Aging on Body Weight and Composition

After 90 days of intervention, the 1-year-aged group, 4-year-aged group, 7-year-aged group and 10-year-aged group exhibited significant reductions in weight and BFM ($p < 0.05$). In the 4-year-aged group, weight was reduced from 69.00 ± 11.96 kg to 66.70 ± 10.77 kg ($p < 0.001$), and BFM dropped from 21.88 ± 5.64 kg to 19.54 ± 5.59 kg ($p < 0.001$). The 1-year-aged group showed significant increases in LBM and MM ($p < 0.05$). BMI was significantly lower in the 1-year, 4-year, 7-year and 10-year-aged groups ($p < 0.05$), and WC was significantly reduced post-intervention in the first three groups ($p < 0.001$). Additionally, BFP and VFA were significantly reduced in the 1-year, 4-year, 7-year, and

10-year-aged groups ($p < 0.001$) (Table 2). Overall, LPT demonstrated significant effects in reducing body fat, visceral fat, and abdominal fat accumulation.

3.4. Comparison of the Effects of LPT with Different Years of Aging on Metabolic and Body Composition Parameters

The comparison of the effects of 1-year, 4-year, 7-year, and 10-year-aged LPT groups on metabolic parameters and body composition was conducted by calculating the differences between follow-up and baseline levels. The results revealed no significant differences among the groups in reducing blood pressure (SBP, DBP), lipid metabolism (TC, TG, HDL-C, LDL-C), glucose metabolism (HbA1c, FBG), or changes in body weight and composition (weight, BFM, BMI, BFP), with *p*-values for intergroup differences exceeding 0.05 (Figure S1).

3.5. Characteristics of Gut Microbiota in LPT Groups with Different Years of Aging

The principal coordinate analysis (PCoA) of beta diversity revealed no significant differences in gut microbiota structure between baseline and follow-up across all four LPT groups with different years of aging (Figure 2a). Specifically, no significant differences were observed in either PCoA1 or PCoA2 for any of the groups, with *p*-values ranging from 0.343 to 0.865. Similarly, the analysis of alpha diversity using the Shannon, Simpson, and Chao1 indices showed no significant changes in microbial diversity before and after the intervention in all four groups (Figure 2b). These results indicate that the gut microbiota structure and diversity remained stable over time across all experimental groups.

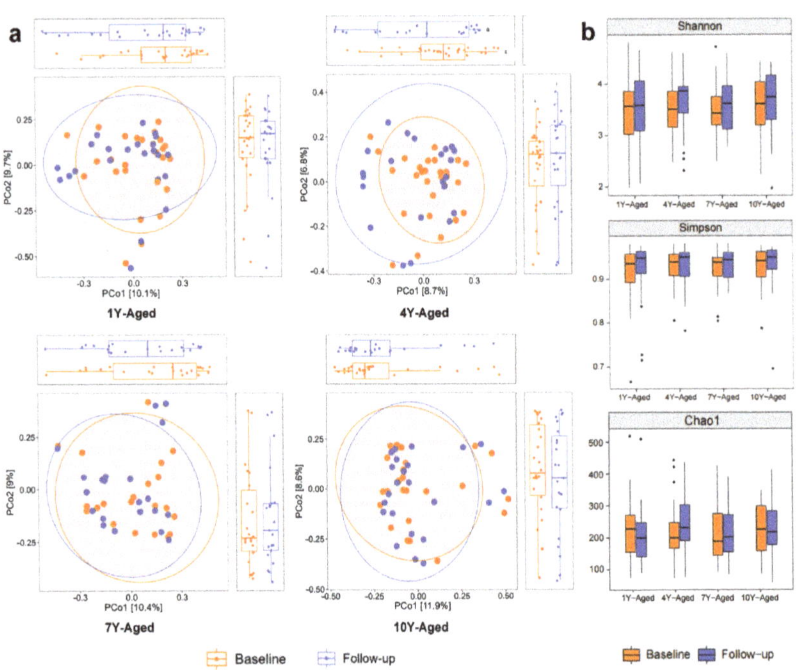

Figure 2. Principal coordinate analysis (PCoA) and alpha diversity of gut microbiota across four LPT groups with different years of aging. (**a**) PCoA plots showed the differences in gut microbiota structure before and after intervention in the 1-year, 4-year, 7-year, and 10-year-aged groups. (**b**) Box plots of Shannon, Simpson, and Chao1 indices measuring alpha diversity in the four groups.

At the phylum level, the gut microbiota composition remained highly consistent across the 1-year, 4-year, 7-year, and 10-year-aged groups, with Firmicutes, Bacteroidota, Proteobacteria, Actinobacteriota, and Verrucomicrobiota being the predominant phyla (Figure 3a). However, a notable reduction in the relative abundance of Firmicutes was observed post-intervention in all four groups. In contrast, Bacteroidota exhibited a decrease in relative abundance in the 1-year, 4-year, and 7-year-aged groups, while the 10-year-aged group showed a significant increase in Bacteroidota abundance, accompanied by a rise in the Bacteroidetes/Firmicutes ratio. At the genus level, the gut microbiota composition remained consistent across all four groups, with Bacteroides being the dominant genus and exhibiting the highest relative abundance in the 1-year, 4-year, 7-year, and 10-year-aged groups. Prevotella_9 and Faecalibacterium followed as the next most abundant genera. Post-intervention, the 1-year-aged group showed a significant increase in the relative abundances of Escherichia−Shigella and Bifidobacterium compared to baseline levels. Beyond the top three dominant genera, the 4-year-aged group exhibited elevated abundances of Bifidobacterium and Megamonas, whereas the 10-year-aged group had higher relative abundances of Blautia and Roseburia (Figure 3b).

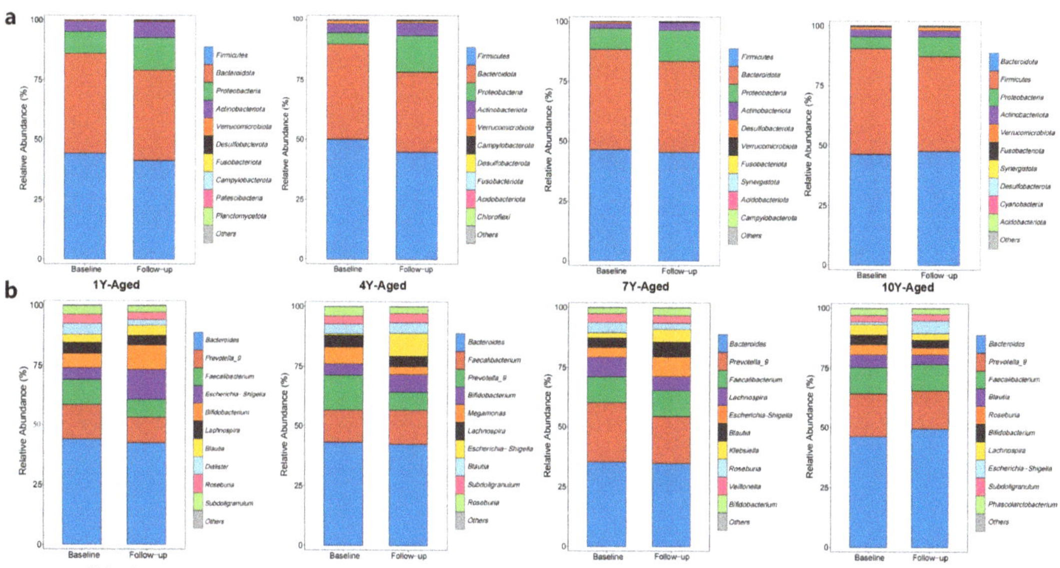

Figure 3. Changes in the relative abundance of gut microbiota at the phylum (**a**) and genus (**b**) levels before and after intervention with LPT of different years of aging.

At the genus level, significant differences were observed in the gut microbiota composition across the four LPT groups with different years of aging before and after the intervention. In the 1-year-aged group, the relative abundance of Stenotrophomonas, Rothia, and Delftia decreased significantly, while Intestinibacter and Eubacterium halli_group increased notably (Figure 4a). In the 4-year-aged group, there were significant increases in Peptostreptococcus, Dubosiella Faecalibaculum, Allobaculum, Eubacterium_xylanophilum_group, Monoglobus, Eubacterium_ventriosum_group, and Escherichia-Shigell after the intervention (Figure 4b). As for the 7-year-aged group, Pseudomonas showed a significant decrease, while UCG.007 and Oxalobacter, displayed notable increases (Figure 4c). The 10-year-aged group exhibited a significant reduction in Erysipelotrichaceae_UCG-003,

FukuN18_freshwater_group, and Polynuceobacter (Figure 4d). Overall, the LPT intervention exhibited varying effects across the different years of aging, significantly altering the relative abundance of certain genera in the gut microbiota. Functional prediction analysis using PICRUSt2 revealed no significant differences in the top ten functional features among the differently aged LPT groups (Table S4).

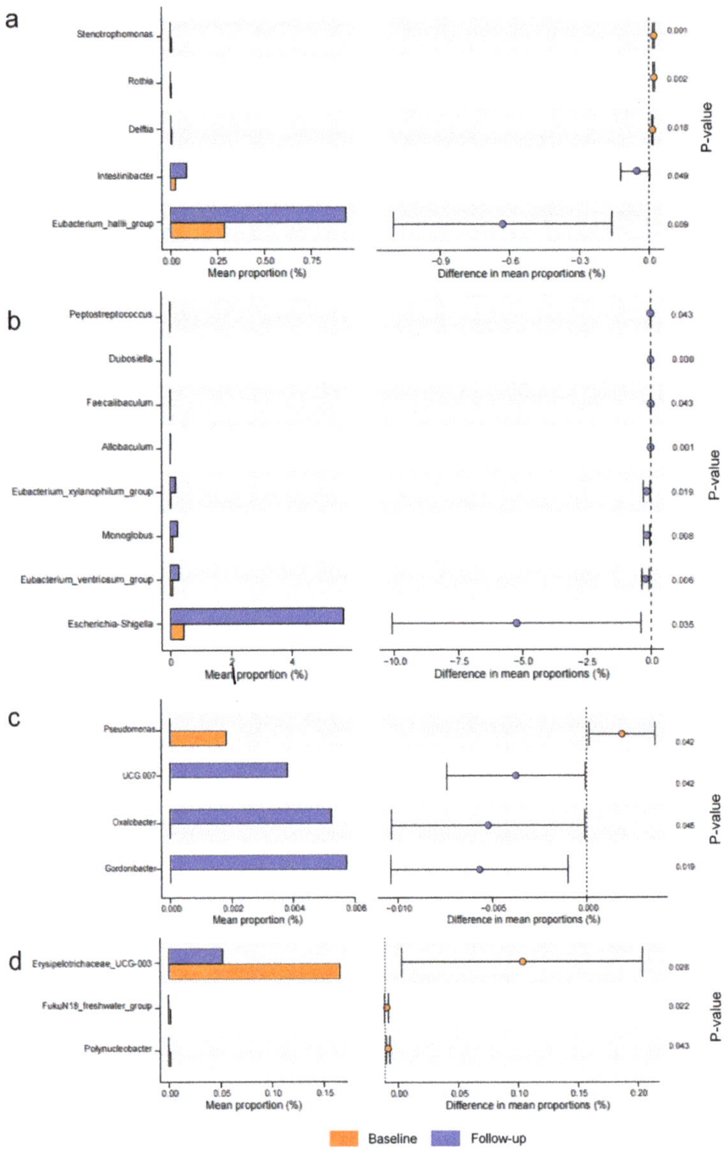

Figure 4. Analysis of the gut bacteria at the genus level in LPT groups with different years of aging (only showing results with statistical significance). (**a**) 1-year-aged group; (**b**) 4-year-aged group; (**c**) 7-year-aged group; (**d**) 10-year-aged group.

4. Discussion

4.1. Main Findings and Comparison with Prior Studies

This study systematically assessed the effects of LPT aged for various years on metabolic parameters, body composition, and gut microbiota, revealing for the first time its beneficial roles in regulating blood pressure, lipids, glucose, and body weight in a human population.

Similar to our study, epidemiological studies associated black tea consumption with reductions in SBP of 1.79–5.31 mmHg and DBP of 0.47–1.02 mmHg [26]. Li et al. demonstrated in a rat model that Pu-erh tea extract enhanced the antihypertensive effects of low-dose nifedipine, achieving blood pressure reductions comparable to high-dose nifedipine alone [27]. LPT is rich in secondary metabolites, particularly polyphenols, which significantly increase the number of functionally active circulating angiogenic cells, thereby counteracting the impairment of flow-mediated dilation caused by fat intake [28,29].

LPT also positively influenced lipid and glucose metabolism. TC levels significantly decreased in the 1-year-aged group, and HDL-C levels significantly increased in the 7-year-aged group, consistent with animal studies. Huang et al., using a hyperlipidemia Sprague–Dawley rat model, found that TC and TG levels in various dosage LPT groups were significantly lower than in the control group, with LDL-C levels also reduced. LPT was more effective than green tea in lowering TC, TG, and LDL-C levels [16]. Wu et al. reported that LPT extract reduced body weight in obese mice, lowered liver TC, TG, and LDL-C levels, and increased HDL-C levels, indicating that LPT alleviates lipid metabolic disorders induced by a high-fat diet [30]. Tannic acid, catechins, and naringenin are identified as key active components of LPT in mitigating hyperlipidemia [24]. Yang et al., using a non-alcoholic fatty liver disease mouse model, found that LPT extract significantly reduced the number and size of hepatic lipid droplets, with therapeutic effects comparable to simvastatin [31]. Regarding blood glucose, FBG significantly decreased in the 4-year-aged group. This aligns with Wu et al.'s findings in animal models, where LPT intervention accelerated the reduction in blood glucose following oral glucose administration, confirming that aged LPT improves glucose tolerance in high-fat diet-induced obese mice [32]. Ding et al. discovered that high-dose LPT extract treatment for six weeks significantly reduced 2 h postprandial blood glucose (2h-PBG) and FBG levels in diabetic rats, exhibiting a dose-dependent hypoglycemic effect [17]. As a post-fermented tea, LPT undergoes long-term microbial fermentation, during which polyphenolic compounds such as tea polyphenols and catechins oxidize and polymerize to form higher molecular weight derivatives like theaflavins. These polyphenols activate the AMP-activated protein kinase (AMPK) pathway, promoting β-oxidation of fatty acids, increasing fat decomposition, and enhancing energy expenditure [33]. Additionally, AMPK activation inhibits fat synthesis-related enzymes (e.g., fatty acid synthase and acetyl-CoA carboxylase), reducing fat synthesis and accumulation, and lowering serum TC and LDL-C levels [34].

All LPT groups exhibited significant reductions in body weight and BFM post-intervention, with the 4-year-aged group showing the most pronounced decreases. BMI, BFP and VFA significantly declined across all LPT groups, and WC also significantly decreased in the 1-year-aged, 4-year-aged, and 7-year-aged groups. Wu et al. found that LPT extract reduced body weight and significantly alleviated liver damage in mice [30]. Li et al. reported that Sichuan black tea significantly reduced weight gain in high-fat diet mice and markedly lowered serum lipid profiles and atherosclerosis indices [35]. Additionally, studies have shown that black tea reduces weight gain and fat accumulation, improves glucose tolerance, alleviates metabolic endotoxemia, and regulates the mRNA expression levels of lipid metabolism-related genes, resulting in weight loss of 15–16% and body fat reduction of 38–44% [36]. Notably, the 1-year-aged group also showed significant increases in LBM and MM, suggesting that LPT may not only reduce fat but also promote muscle growth and maintenance. Previous research identified epigallocatechin as a major bioactive compound in black tea extract that alleviates palmitic acid-induced muscle atrophy by regulating mitochondrial function in C2C12 cells [37]. Catechins in tea maintain the dynamic balance between protein synthesis and degradation and promote mitochondrial energy metabolism, thereby mitigating muscle atrophy and maintaining muscle homeostasis [38].

Regarding gut microbiota, all groups showed a significant decrease in the relative abundance of Firmicutes post-intervention, while Bacteroidota decreased in the 1-year-aged, 4-year-aged, and 7-year-aged groups, with only the 10-year-aged group showing a significant increase in Bacteroidota and Bacteroidetes/Firmicutes ratio. The Bacteroidetes/Firmicutes ratio is a crucial indicator of gut microbiota health. Ding et al. found that LPT extract beneficially modulated diabetes-induced gut microbiota dysbiosis by increasing the Bacteroidetes/Firmicutes ratio [17]. Zhou et al., using a rat model of irritable bowel syndrome, observed that LPT increased the Bacteroidota/Firmicutes ratio and significantly reconstructed the microbial community to alleviate irritable bowel syndrome [39]. At the genus level, Bacteroides, the predominant genus, had the highest relative abundance in all aged LPT groups, followed by Prevotella_9 and Faecalibacterium. Similar to the 10-year-aged group, previous studies showed that LPT extract increases Bacteroides abundance in high-fat diet mouse models [40]. Polysaccharides in LPT more effectively promote butyrate production and increase Bacteroides abundance [41]. The Bacteroides genus plays a vital role in decomposing complex polysaccharides, synthesizing vitamins, and regulating the host immune system [42].

Our study also found that the 1-year-aged LPT intervention significantly increased the relative abundance of Intestinibacter and the Eubacterium_hallii group in the gut microbiota, consistent with Yang et al.'s findings. Eubacterium_hallii is an efficient butyrate and propionate producer, and its increased abundance is closely associated with improvements in metabolic parameters, such as reduced blood glucose, improved lipid profiles, and decreased body fat [43]. In the 4-year-aged group, several gut genera significantly increased post-intervention, and gut microbial diversity also improved. Notably, the Eubacterium_xylanophilum and Eubacterium_ventriosum groups exhibited the most significant changes; these bacteria produce short-chain fatty acids by digesting dietary fiber, exerting anti-obesity effects [44]. Additionally, Dubosiella abundance significantly increased in the 4-year-aged group. Studies have shown that Dubosiella, as a next-generation probiotic, can inhibit the mTOR pathway, thereby reducing blood lipids and enhancing insulin sensitivity [45]. We also observed a significant increase in Allobaculum abundance in the 4-year-aged group. Zheng et al. found that Allobaculum abundance positively correlates with the expression of ANGPTL4, a key regulator of lipid metabolism [46]. In the 10-year-aged group, the abundance of Erysipelotrichaceae UCG-003 significantly decreased. Previous studies indicate that increased Erysipelotrichaceae abundance is associated with a higher risk of developing irritable bowel syndrome and is related to intestinal inflammation processes [47]. The top ten functional features by relative abundance did not show significant differences, which may be related to the insufficient sample size and the instability of functional predictions. However, these limitations highlight the need for future studies to employ shotgun metagenomic sequencing, which can provide a more comprehensive understanding of the microbial community's functional potential by directly sequencing the total DNA of the microbiota.

Despite traditional Chinese beliefs that longer-aged LPT possesses superior quality, our results indicate that LPT samples aged for different durations do not exhibit significant differences in improving blood pressure, blood lipids, blood glucose, body weight, or body composition parameters. Previous research has found that the antioxidant capacity of LPT follows a parabolic trend with storage time, peaking at five years of storage. Beyond this threshold, the antioxidant activity of LPT may decline, potentially due to reductions in catechins and glycosylation of flavonoids [24]. While Wu et al. found that the 10-year-aged group had better anti-obesity effects than the 1-year-aged group in obese mice, our study did not observe significant differences in the health effects of LPT across the different aging durations [32]. This may be related to the dynamic changes in LPT's components. As the aging time extends, antioxidant components such as tea polyphenols and catechins gradually decrease, while flavonoids and soluble dietary fiber show dynamic changes or increasing trends. The synergistic effects of these components may, to some extent, balance the impact of the reduction in individual components on metabolic health. Additionally, individual differences and limitations in experimental design may also mask potential differences in the effects of aging time on metabolic health. While aging time can greatly affect the composition and flavor of LPT, its influence on human metabolic health is likely more complex and not determined solely by aging duration. Future studies should employ larger sample sizes and extended intervention periods to comprehensively explore how these chemical changes impact metabolic health.

4.2. The Strengths and Limitations of the Study

The strength of this study lies in its comprehensive assessment of the impact of LPT on metabolic health and gut microbiota based on a population experiment. By including LPT aged for different numbers of years, the study provides a unique perspective for analyzing the variations in their health effects. This study has several limitations. Firstly, due to practical and ethical considerations, the study design lacked a control group, as all participants received LPT interventions of different aging durations. This absence makes it impossible to rule out time effects or other external factors influencing the results, thereby limiting our ability to establish the independent health effects of LPT. Consequently, it remains uncertain whether the observed metabolic improvements are solely attributable to LPT consumption. Secondly, the study did not thoroughly investigate and record participants' dietary and physical activity habits. Although participants were instructed to maintain their usual diet and exercise routines, the lack of systematic measurement may have allowed minor changes in dietary patterns or physical activity levels, potentially confounding the true effects of LPT. These two limitations may impact the internal validity and generalizability of the findings. Future research should incorporate a control group and meticulously monitor lifestyle factors to more accurately evaluate the independent effects of LPT on metabolic health.

4.3. Clinical Implications and Future Directions

This study demonstrates that LPT as a natural, non-pharmacological intervention, has a broad clinical application prospect in the prevention and management of metabolic diseases. Future studies should focus on addressing the current limitations of the research, such as increasing the sample size, extending the intervention period, incorporating a control group, and systematically monitoring participants' lifestyle factors, in order to more accurately assess the independent health effects of LPT.

5. Conclusions

LPT significantly regulates blood pressure, lipid and glucose metabolism, reduces body weight, and improves body composition parameters, while actively modulating gut microbiota. These effects do not show significant differences across LPT aged for different durations, challenging the traditional belief that longer-aged LPT possesses superior health advantages. The study results indicate that LPT, as a functional beverage, has the potential to promote metabolic health, warranting further research into its mechanisms and long-term impacts.

Supplementary Materials: The following supporting information can be downloaded at: https://www.mdpi.com/article/10.3390/foods14050866/s1, Table S1. The content of main water-soluble components in Liupao tea with different aged years. Table S2. The effects of Liupao tea intervention on liver health biomarkers. Table S3. Intention-to-treat analysis of the intervention effects of Liupao tea with different aging years on metabolic parameters. Table S4. Relative abundance and changes of PICRUSt2-predicted KO functions in Liupao tea of different aging years. Figure S1. Differences in the effects of Liupao tea with different aging years on improving metabolic parameters and body composition.

Author Contributions: Y.W. and D.M. conducted the investigation, developed the methodology, performed formal analysis, curated the data, and contributed to writing the original draft. Q.H. and B.C. participated in the investigation and data curation. D.M. led the conceptualization and supervision of the project, and contributed to the review and editing of the manuscript. All authors have read and agreed to the published version of the manuscript.

Funding: This research received no external funding.

Institutional Review Board Statement: This study was approved by the Biomedical Ethics Committee of Peking University (NO.IRB00001052-23203) on 10 March 2024, and registered with the Chinese Clinical Trial Registry (Trial Registration number: ChiCTR2400089348).

Informed Consent Statement: Written informed consent was secured from all subjects before their enrollment in the study.

Data Availability Statement: The data presented in this study are available on request from the corresponding author due to privacy concerns.

Acknowledgments: The authors thank Guangxi Zhaoping County General Mountain Agricultural Technology Co., Ltd. for providing the Liupao tea samples used in this study.

Conflicts of Interest: The authors declare no conflicts of interest.

Abbreviations

APOA1: apolipoprotein A1; APOB: apolipoprotein B; BMI: body mass index; BFP: body fat percentage; BFM: body fat mass; DBP: diastolic blood pressure; FBG: fasting blood glucose; HbA1: hemoglobin A1c; HDL-C: high-density lipoprotein cholesterol; INS: insulin; LBM: lean body mass; LDL-C: low-density lipoprotein cholesterol; LPT: Liupao tea; MM: muscle mass; SBP: systolic blood pressure; TC: total cholesterol; TG: triglycerides; VFA: visceral fat area; WC: waist circumference.

References

1. Balakrishnan, V.S. Europe's obesity burden on the rise: WHO report. *Lancet Diabetes Endocrinol.* **2022**, *10*, 488. [CrossRef] [PubMed]
2. Ji, C.; Ge, X.; Zhang, J.; Tong, H. The Stroke Burden in China and Its Long-Term Trends: Insights from the Global Burden of Disease (GBD) Study 1990–2021. *Nutr. Metab. Cardiovasc. Dis.* **2025**, 103848, *In Press*. [CrossRef]
3. Collaboration, N.C.D.R.F. Worldwide trends in underweight and obesity from 1990 to 2022: A pooled analysis of 3663 population-representative studies with 222 million children, adolescents, and adults. *Lancet* **2024**, *403*, 1027–1050. [CrossRef]
4. Athanasakis, K.; Bala, C.; Kokkinos, A.; Simonyi, G.; Karoliova, K.H.; Basse, A.; Bogdanovic, M.; Kang, M.; Low, K.; Gras, A. The economic burden of obesity in 4 south-eastern European countries associated with obesity-related co-morbidities. *BMC Health Serv. Res.* **2024**, *24*, 354. [CrossRef]

5. Mendenhall, E.; Singer, M. The global syndemic of obesity, undernutrition, and climate change. *Lancet* **2019**, *393*, 741. [CrossRef] [PubMed]
6. Broussard, J.L.; Garfield, A.; Zarini, S.; Brozinick, J.T.; Perreault, L.; Newsom, S.A.; Kahn, D.; Kerege, A.; Berry, K.Z.; Bui, H.H.; et al. Combined diet and exercise training decreases serum lipids associated with insulin resistance. *Obesity* **2024**, *32*, 2334–2344. [CrossRef]
7. Papakonstantinou, E.; Oikonomou, C.; Nychas, G.; Dimitriadis, G.D. Effects of Diet, Lifestyle, Chrononutrition and Alternative Dietary Interventions on Postprandial Glycemia and Insulin Resistance. *Nutrients* **2022**, *14*, 823. [CrossRef]
8. Castro-Barquero, S.; Ruiz-Leon, A.M.; Sierra-Perez, M.; Estruch, R.; Casas, R. Dietary Strategies for Metabolic Syndrome: A Comprehensive Review. *Nutrients* **2020**, *12*, 2983. [CrossRef]
9. Gao, X.; Xie, Q.; Kong, P.; Liu, L.; Sun, S.; Xiong, B.; Huang, B.; Yan, L.; Sheng, J.; Xiang, H. Polyphenol- and Caffeine-Rich Postfermented Pu-erh Tea Improves Diet-Induced Metabolic Syndrome by Remodeling Intestinal Homeostasis in Mice. *Infect. Immun.* **2018**, *86*, e00601-17. [CrossRef]
10. Liu, J.Y.; He, D.; Xing, Y.F.; Zeng, W.; Ren, K.; Zhang, C.; Lu, Y.; Yang, S.; Ou, S.J.; Wang, Y.; et al. Effects of bioactive components of Pu-erh tea on gut microbiomes and health: A review. *Food Chem.* **2021**, *353*, 129439. [CrossRef]
11. Wang, S.; Qiu, Y.; Gan, R.Y.; Zhu, F. Chemical constituents and biological properties of Pu-erh tea. *Food Res. Int.* **2022**, *154*, 110899. [CrossRef]
12. Yuan, Y.; He, J.; Tang, M.; Chen, H.; Wei, T.; Zhang, B.; Liang, D.; Nie, X. Preventive effect of Ya'an Tibetan tea on obesity in rats fed with a hypercaloric high-fat diet revealed by gut microbiology and metabolomics studies. *Food Res. Int.* **2023**, *165*, 112520. [CrossRef] [PubMed]
13. Costa, M.A.C.; Vilela, D.L.S.; Fraiz, G.M.; Lopes, I.L.; Coelho, A.I.M.; Castro, L.C.V.; Martin, J.G.P. Effect of kombucha intake on the gut microbiota and obesity-related comorbidities: A systematic review. *Crit. Rev. Food Sci. Nutr.* **2023**, *63*, 3851–3866. [CrossRef]
14. Qin, H.; Huang, L.; Teng, J.; Wei, B.; Xia, N.; Ye, Y. Purification, characterization, and bioactivity of Liupao tea polysaccharides before and after fermentation. *Food Chem.* **2021**, *353*, 129419. [CrossRef]
15. Wang, H.; Teng, J.; Huang, L.; Wei, B.; Xia, N. Determination of the variations in the metabolic profile and sensory quality of Liupao tea during fermentation through UHPLC-HR-MS metabolomics. *Food Chem.* **2023**, *404*, 134773. [CrossRef] [PubMed]
16. Huang, S.; Chen, H.; Teng, J.; Wu, Z.; Huang, L.; Wei, B.; Xia, N. Antihyperlipidemic effect and increased antioxidant enzyme levels of aqueous extracts from Liupao tea and green tea in vivo. *J. Food Sci.* **2022**, *87*, 4203–4220. [CrossRef] [PubMed]
17. Ding, Q.; Zhang, B.; Zheng, W.; Chen, X.; Zhang, J.; Yan, R.; Zhang, T.; Yu, L.; Dong, Y.; Ma, B. Liupao tea extract alleviates diabetes mellitus and modulates gut microbiota in rats induced by streptozotocin and high-fat, high-sugar diet. *Biomed. Pharmacother.* **2019**, *118*, 109262. [CrossRef]
18. Ma, L.; Hu, Y.; Alperet, D.J.; Liu, G.; Malik, V.; Manson, J.E.; Rimm, E.B.; Hu, F.B.; Sun, Q. Beverage consumption and mortality among adults with type 2 diabetes: Prospective cohort study. *BMJ* **2023**, *381*, e073406. [CrossRef]
19. Ni, D.; Ai, Z.; Munoz-Sandoval, D.; Suresh, R.; Ellis, P.R.; Yuqiong, C.; Sharp, P.A.; Butterworth, P.J.; Yu, Z.; Corpe, C.P. Inhibition of the facilitative sugar transporters (GLUTs) by tea extracts and catechins. *FASEB J.* **2020**, *34*, 9995–10010. [CrossRef]
20. Perkin, O.J.; Chen, Y.C.; Johnson, D.A.; Thomas, J.E.; Atkinson, G.; Betts, J.A.; Gonzalez, J.T. Postprandial Metabolic Mesponses to High-fat Feeding in Healthy Adults Following Ingestion of Oolong Tea-Derived Polymerized Polyphenols: A Randomized, Double-blinded, Placebo-controlled Crossover Study. *Am. J. Clin. Nutr.* **2023**, *118*, 132–140. [CrossRef]
21. Xia, H.M.; Wang, J.; Xie, X.J.; Xu, L.J.; Tang, S.Q. Green tea polyphenols attenuate hepatic steatosis, and reduce insulin resistance and inflammation in high-fat diet-induced rats. *Int. J. Mol. Med.* **2019**, *44*, 1523–1530. [CrossRef] [PubMed]
22. Wei, L.; Huang, L.; Du, L.; Sun, Q.; Chen, C.; Tang, J.; Teng, J.; Wei, B. Structural Characterization and In Vitro Antioxidant, Hypoglycemic and Hypolipemic Activities of a Natural Polysaccharide from Liupao Tea. *Foods* **2023**, *12*, 2226. [CrossRef]
23. Long, J.; Yu, L.; Huang, L.; Xia, N.; Teng, J.; Wei, B. Isolation, identification, and community diversity of microorganisms during tank fermentation of Liupao tea. *J. Food Sci.* **2023**, *88*, 4230–4246. [CrossRef]
24. Zhou, H.; Wang, X.; She, Z.; Huang, L.; Wei, H.; Yang, S.; Wei, Z.; Chen, H.; Yang, B.; Hu, Z.; et al. Combining bioinformatics and multiomics strategies to investigate the key microbiota and active components of Liupao tea ameliorating hyperlipidemia. *J. Ethnopharmacol.* **2024**, *333*, 118438. [CrossRef]
25. Chu, S.L.; Fu, H.; Yang, J.X.; Liu, G.X.; Dou, P.; Zhang, L.; Tu, P.F.; Wang, X.M. A randomized double-blind placebo-controlled study of Pu'er tea extract on the regulation of metabolic syndrome. *Chin. J. Integr. Med.* **2011**, *17*, 492–498. [CrossRef] [PubMed]
26. Zhao, Y.; Tang, C.; Tang, W.; Zhang, X.; Jiang, X.; Duoji, Z.; Kangzhu, Y.; Zhao, X.; Xu, X.; Hong, F.; et al. The association between tea consumption and blood pressure in the adult population in Southwest China. *BMC Public Health* **2023**, *23*, 476. [CrossRef] [PubMed]

27. Li, Y.; Li, X.; Zhou, W.; Li, C.; Ma, H. Puer tea aqueous extract combined with nifedipine for hypotensive effect study. *Tea Sci.* **2015**, *35*, 165–170. [CrossRef]
28. Guo, J.; Li, K.; Lin, Y.; Liu, Y. Protective effects and molecular mechanisms of tea polyphenols on cardiovascular diseases. *Front. Nutr.* **2023**, *10*, 1202378. [CrossRef]
29. Tang, W.; Yuan, M.; Li, Z.; Lin, Q.; Zhen, Y.; Li, Z.; Zhou, H.; Xia, F. Polyphenol-Rich Liupao Tea Extract Prevents High-Fat Diet-Induced MAFLD by Modulating the Gut Microbiota. *Nutrients* **2022**, *14*, 4930. [CrossRef]
30. Wu, Y.; Sun, H.; Yi, R.; Tan, F.; Zhao, X. Anti-obesity effect of Liupao tea extract by modulating lipid metabolism and oxidative stress in high-fat-diet-induced obese mice. *J. Food Sci.* **2021**, *86*, 215–227. [CrossRef]
31. Yang, S.; Wei, Z.; Luo, J.; Wang, X.; Chen, G.; Guan, X.; She, Z.; Liu, W.; Tong, Y.; Liu, H.; et al. Integrated bioinformatics and multiomics reveal Liupao tea extract alleviating NAFLD via regulating hepatic lipid metabolism and gut microbiota. *Phytomedicine* **2024**, *132*, 155834. [CrossRef]
32. Wu, W.; Hu, Y.; Zhang, S.; Liu, D.; Li, Q.; Lin, Y.; Liu, Z. Untargeted metabolomic and lipid metabolism-related gene expression analyses of the effects and mechanism of aged Liupao tea treatment in HFD-induced obese mice. *RSC Adv.* **2021**, *11*, 23791–23800. [CrossRef] [PubMed]
33. Cai, X.; Liu, Z.; Dong, X.; Wang, Y.; Zhu, L.; Li, M.; Xu, Y. Hypoglycemic and lipid lowering effects of theaflavins in high-fat diet-induced obese mice. *Food Funct.* **2021**, *12*, 9922–9931. [CrossRef] [PubMed]
34. Desjardins, E.M.; Steinberg, G.R. Emerging Role of AMPK in Brown and Beige Adipose Tissue (BAT): Implications for Obesity, Insulin Resistance, and Type 2 Diabetes. *Curr. Diabetes Rep.* **2018**, *18*, 80. [CrossRef]
35. Li, L.; Shi, M.; Salerno, S.; Tang, M.; Guo, F.; Liu, J.; Feng, Y.; Fu, M.; Huang, Q.; Ma, L.; et al. Microbial and metabolomic remodeling by a formula of Sichuan dark tea improves hyperlipidemia in apoE-deficient mice. *PLoS ONE* **2019**, *14*, e0219010. [CrossRef]
36. Qu, J.; Ye, M.; Wen, C.; Cheng, X.; Zou, L.; Li, M.; Liu, X.; Liu, Z.; Wen, L.; Wang, J. Compound dark tea ameliorates obesity and hepatic steatosis and modulates the gut microbiota in mice. *Front. Nutr.* **2023**, *10*, 1082250. [CrossRef]
37. Yoo, A.; Ahn, J.; Seo, H.D.; Hahm, J.H.; Jung, C.H.; Ly, S.Y.; Ha, T.Y. Fuzhuan brick tea extract ameliorates obesity-induced skeletal muscle atrophy by alleviating mitochondrial dysfunction in mice. *J. Nutr. Biochem.* **2024**, *125*, 109532. [CrossRef]
38. Luk, H.Y.; Appell, C.; Chyu, M.C.; Chen, C.H.; Wang, C.Y.; Yang, R.S.; Shen, C.L. Impacts of Green Tea on Joint and Skeletal Muscle Health: Prospects of Translational Nutrition. *Antioxidants* **2020**, *9*, 1050. [CrossRef] [PubMed]
39. Zhou, D.; Liu, X.; Lan, L.; Yu, W.; Qiu, R.; Wu, J.; Teng, C.; Huang, L.; Yu, C.; Zeng, Y. Protective effects of Liupao tea against high-fat diet/cold exposure-induced irritable bowel syndrome in rats. *Heliyon* **2023**, *9*, e16613. [CrossRef]
40. Pan, W.; Li, W.; Wu, H.; Xie, X.; Xie, M.; Nie, Q.; Liu, Z.; Cai, S. Aging-Accelerated Mouse Prone 8 (SAMP8) Mice Experiment and Network Pharmacological Analysis of Aged Liupao Tea Aqueous Extract in Delaying the Decline Changes of the Body. *Antioxidants* **2023**, *12*, 685. [CrossRef]
41. Qiu, S.; Huang, L.; Xia, N.; Teng, J.; Wei, B.; Lin, X.; Khan, M.R. Two Polysaccharides from Liupao Tea Exert Beneficial Effects in Simulated Digestion and Fermentation Model In Vitro. *Foods* **2022**, *11*, 2958. [CrossRef] [PubMed]
42. Zafar, H.; Saier, M.H., Jr. Gut Bacteroides species in health and disease. *Gut Microbes* **2021**, *13*, 1848158. [CrossRef] [PubMed]
43. Kumari, M.; Singh, P.; Nataraj, B.H.; Kokkiligadda, A.; Naithani, H.; Azmal Ali, S.; Behare, P.V.; Nagpal, R. Fostering next-generation probiotics in human gut by targeted dietary modulation: An emerging perspective. *Food Res. Int.* **2021**, *150*, 110716. [CrossRef]
44. Freitas, R.; Vasques, A.C.J.; da Rocha Fernandes, G.; Ribeiro, F.B.; Solar, I.; Shivappa, N.; Hebert, J.R.; de Almeida-Pititto, B.; Geloneze, B.; Ferreira, S.R.G. Gut bacterial markers involved in association of dietary inflammatory index with visceral adiposity. *Nutrition* **2024**, *122*, 112371. [CrossRef] [PubMed]
45. Gao, Y.; Mo, S.; Cao, H.; Zhi, Y.; Ma, X.; Huang, Z.; Li, B.; Wu, J.; Zhang, K.; Jin, L. The efficacy and mechanism of Angelica sinensis (Oliv.) Diels root aqueous extract based on RNA sequencing and 16S rDNA sequencing in alleviating polycystic ovary syndrome. *Phytomedicine* **2023**, *120*, 155013. [CrossRef]
46. Zheng, Z.; Lyu, W.; Ren, Y.; Li, X.; Zhao, S.; Yang, H.; Xiao, Y. Allobaculum Involves in the Modulation of Intestinal ANGPTLT4 Expression in Mice Treated by High-Fat Diet. *Front. Nutr.* **2021**, *8*, 690138. [CrossRef]
47. Gryaznova, M.; Smirnova, Y.; Burakova, I.; Morozova, P.; Lagutina, S.; Chizhkov, P.; Korneeva, O.; Syromyatnikov, M. Fecal Microbiota Characteristics in Constipation-Predominant and Mixed-Type Irritable Bowel Syndrome. *Microorganisms* **2024**, *12*, 1414. [CrossRef]

foods

Article

Characteristic Changes and Potential Markers of Flavour in Raw Pu-Erh Tea with Different Ageing Cycles Analysed by HPLC, HS-SPME-GC-MS, and OAV

Jiayi Xu [1,†], Xiujuan Deng [1,†], Yamin Wu [1], Miao Zhou [1], Cen Du [2], Qiaomei Wang [1], Yuxin Xia [1], Junjie He [1], Wenxia Yuan [1], Wendou Wu [3], Hongxu Li [1], Yankun Wang [1], Tong Li [4,*] and Baijuan Wang [1,*]

[1] College of Tea Science, Yunnan Agricultural University, Kunming 650201, China; 18088111511@163.com (J.X.); 15808869561@163.com (X.D.); m1678447405@163.com (Y.W.); 18724979985@163.com (M.Z.); wqm19850127@163.com (Q.W.); 15140964046@163.com (Y.X.); junjie19991207@163.com (J.H.); yuanwenxia2023@163.com (W.Y.); 13085315910@163.com (H.L.); wyk18787295901@163.com (Y.W.)

[2] Tea Distribution Association in Wenshan Prefecture, Wenshan 663000, China; 13577059654@163.com

[3] Yunnan Organic Tea Industry Intelligent Engineering Research Center, Yunnan Agricultural University, Kunming 650201, China; wuwd2004@126.com

[4] Yunnan Key Laboratory of Crop Production and Smart Agriculture, Kunming 650201, China

* Correspondence: tli@ynu.edu.cn (T.L.); wangbaijuan2023@163.com (B.W.); Tel.: +86-138-8808-6872 (T.L.); +86-138-8897-6094 (B.W.)

† These authors contributed equally to this work.

Abstract: To investigate the flavour evolution mechanism of raw Pu-erh tea (RPT) during storage, the volatile and non-volatile compounds of RPT with different storage years (1–10 years) from the same raw material origin, manufacturer, and storage location in Wenshan Prefecture, Yunnan Province, were systematically analysed by HPLC, HS-SPME-GC-MS, and OAV. The results showed that both cluster analyses based on non-volatile and volatile compounds could classify RPT of different storage years into three ageing cycles, with key turning points in the third and eighth years of storage, which is also accompanied by the colour changing from green to orange or brown, the aroma changing from a faint scent to woody and ageing, the astringency diminishing, and the sweet and mellow increasing. Theophylline was identified as the potential marker of RPT stored 1–3 years, while (−)-catechin gallate, (−)-gallocatechin gallate, quercetin, and rutin as those for a storage of 9–10 years. The volatile compounds indicate a general trend of an initial increase followed by a decrease. Forty-four key aroma compounds (OAV ≥ 1) were identified. Eucalyptol, β-Caryophyllene, 2-Amylfuran, Copaene, Estragole, and α-Terpinene originated as potential markers for RPT stored 1–3 years, while (Z)-Linalool oxide (furanoid), α-Terpineol, Terpinen-4-ol, and cis-Anethol were for RPT stored 8–10 years. This study revealed the flavour characteristics and quality changes of RPT over the course of storage, and constructed a sensory flavour wheel, providing theoretical underpinnings for the quality control and assessment of RPT.

Keywords: raw Pu-erh tea; ageing cycles; quality; flavour; OAV

1. Introduction

Tea drinking in China boasts a history exceeding 1000 years, with Yunnan Province recognised as the origin of the tea plant [1]. Pu-erh tea, recognised with Protected Geographical Indication status, is derived from sun-dried green tea made from the large-leaf varietals of the Yunnan province. According to different processing techniques and quality characteristics, Pu-erh tea is categorised into two varieties: raw Pu-erh tea and ripe Pu-erh tea [2,3]. Pu-erh tea is popular for its weight and fat reduction, sugar reduction, uric acid reduction, anti-fatigue, antioxidant, anti-cancer, and free radical scavenging effects [4–7]. At the same time, the characteristic of "the more it ages, the more fragrant it is" adds unique enjoyment to the Pu-erh tea and makes it a good collector's item [8]. Raw Pu-erh tea is produced by steaming and pressing sun-dried green tea from large-leafed varieties in Yunnan Province. Its quality is characterised by a pure and long-lasting aroma, as well as a rich and sweet taste. In scientific warehousing and with ageing time, its taste will evolve towards being mellow, sweet, thick, and smooth. The aroma will transform from a clear fragrance to a woody, sweet, and unique aged aroma [9]. The ageing of raw Pu-erh tea involves the interaction of various factors, including microorganisms, enzymes, humidity, temperature, and oxidation. These factors contribute to a series of chemical changes in the tea under specific environmental conditions, giving rise to the qualities characteristic of raw Pu-erh tea [10]. The quality of Pu-erh tea is intricately linked to both the geographical origin of its raw materials and the conditions of its storage, in addition to the duration of storage. Therefore, it is essential to investigate the effect of storage years on the quality characteristics of raw Pu-erh tea, along with its implications for both drinking and commercial value.

At present, numerous studies have explored the internal mechanisms of raw Pu-erh tea ageing across different storage years and analysed the reasons for the quality differences in raw Pu-erh tea. In terms of sensory quality, Jiao [11] studied the regular changes of quality indicators during the storage process of raw Pu-erh tea. The results showed that, with the increase of storage time, the colour of the soup of raw Pu-erh tea deepened gradually, the aroma changed from fresh to aged, the taste changed from mellow to thick, and the colour of the bottom of the leaves became deeper and darker. The sensory quality improved after 10 years of storage and above but began to decline after 18 years of storage. Shifting to non-volatile compounds, Liu [12] then discovered that as the ageing period extended, free amino acids and tea polyphenols decreased, whereas water-soluble proteins and tea pigments increased in raw Pu-erh tea. Duan [13] reported an upward trend in the water extract of raw Pu-erh tea with lengthening storage durations, while caffeine content exhibited a downward trend. Furthermore, Zhou [14] determined that storing raw Pu-erh tea under hot and humid conditions promoted the degradation of ester catechins and the concurrent accumulation of flavonoids and GA. With respect to volatile compounds, Rong [15] utilised an integrated approach involving GC-E-Nose, GC-MS, and GC-IMS to analyse the change of the volatile components of Pu-erh tea over different storage periods; nine volatile constituents, including Linalool and (E)-2-Hexenal, were selected as key variables for differentiating Pu-erh teas based on storage duration. Ma [16] utilised multivariate statistical methods to analyse the aroma composition of raw Pu-erh tea from Lincang City; a clear distinction could be observed between samples originating from the southern and northern regions of Lincang City. Southern samples were represented by relatively high concentrations of alcohols and esters, whereas northern samples displayed higher levels of aldehydes, ketones, and other substances, the altitude was hypothesised as a potentially significant factor contributing to the observed regional variations in aroma profiles [17]. Xu et al. concluded that the aroma of raw Pu-erh tea is affected by the storage environment. In a dry and cold storage environment, raw Pu-erh tea is more likely to produce floral and sweet aromas. In contrast, in a hot and humid environment, raw Pu-erh tea is more likely to produce woody and aged aromas [18].

Wenshan Prefecture is located in the southeastern part of Yunnan Province, 103°35′~106°12′ E, 22°48′~24°28′ N, with an altitude of 107.02–2991.20 m. It has a subtropical humid monsoon climate. Moreover, Wenshan Prefecture is within the protected area of the Pu-erh Tea Geographical Indication, boasting abundant tea tree planting resources and a wide distribution. It also features significant regional characteristics, such as a rich and fragrant aroma and a fresh and mellow taste [19]. This study focused on raw Pu-erh tea sourced from the same origin, manufacturer, and storage location in Wenshan Prefecture, due to the scarcity of research on tea quality characteristics in the region. It subsequently conducted a thorough and systematic analysis of the impact of storage duration on the quality characteristics of raw Pu-erh tea.

2. Materials and Methods

2.1. Tea Samples

Samples stored for 1–10 years were all offered by Longli Chun Tea Industry Co., Ltd., located in Wenshan Zhuang and Miao Autonomous Prefecture, Yunnan Province, in China. All samples were produced from Yunnan large-leaf varietals, utilising one bud and two to three leaves as fresh leaf raw material. The production followed the processing method stipulated in GB/T22111-2008 [20], and the products were then uniformly stored in a dry and well-ventilated warehouse located in Wenshan Prefecture, with a relative humidity of ≤70% and an indoor temperature of ≤25 °C. Before further processing, each raw Pu-erh tea sample was ground and sieved through 60-mesh; tea samples were pulverised and kept at −20 °C. Table 1 presents detailed sample information.

Table 1. Tea samples of raw Pu-erh tea different storage years in Wenshan Prefecture.

No.	Storage Time	Year of Production	Place of Origin	Manufacturer	Storage Location
Y1	1 year	2023	Pengzhai Village, Wenshan Prefecture	Longli Chun Tea Industry Co.	Wenshan Zhuang and Miao Autonomous Prefecture
Y2	2 years	2022			
Y3	3 years	2021			
Y4	4 years	2020			
Y5	5 years	2019			
Y6	6 years	2018			
Y7	7 years	2017			
Y8	8 years	2016			
Y9	9 years	2015			
Y10	10 years	2014			

2.2. Chemicals

The main reagent standards (purity ≥ 98%) employed in this analysis included theophylline, caffeine, gallic acid, (+)-catechin, (−)-catechin gallate, (−)-epicatechin, (−)-epicatechin gallate, (−)-epigallocatechin, (−)-epigallocatechin gallate, (+)-gallocatechin, (−)-gallocatechin gallate, quercetin, rutin, kaempferol, myrcene, and ethyl decanoate, procured from Shanghai Yuanye Biotechnology Co. Anthrone (ACS reagent) (Shanghai, China) and ninhydrin (ACS reagent) were sourced from Sigma-Aldrich Company (St. Louis, MO, USA). Methanol and acetonitrile (all chromatographic grade) were purchased from Thermo Fisher Company (Waltham, MA, USA). All remaining reagents were of analytical grade and obtained from Tianjin Damao Chemical Reagent Factory.

2.3. Sensory Evaluation

The sensory evaluation of raw Pu-erh tea samples from different storage years in Wenshan Prefecture was conducted following China's national standard GB/T23776-2018 "Methodology for Sensory Evaluation of Tea" [21]. Five experienced members (3 females and 2 males, aged between 22 and 55 years old) from the Tea College of Yunnan Agricultural University performed the assessment. Prior to the assessment, the group members received professional training in the evaluation of raw Pu-erh tea in accordance with the "Methodology for Sensory Evaluation of Tea". For each evaluation, 3 g of tea were placed in a professional evaluation cup, and 150 mL of boiling water was added twice, steeping for 2 min on the first and 5 min on the second. The results were primarily based on the second steeping while also taking into account the first steeping. The members assessed various characteristics of the tea, including appearance (a), aroma (b), liquor colour (c), taste (d), and tea residue (e). Each sample underwent three rounds of evaluation. Ultimately, the group's assessment data were compiled; subsequently, the quality of the tea samples was quantitatively evaluated using the prescribed formula. The formula is as follows:

$$\text{PRT (total score)} = 20\% \times (a) + 30\% \times (b) + 10\% \times (c) + 35\% \times (d) + 5\% \times (e)$$

2.4. Detection of Non-Volatile Compounds

Tea polyphenol content was determined according to GB/T8313-2018, "Determination of Total Polyphenols and Catechins Content in Tea" [22]. Free amino acid content was determined according to GB/T8314-2013, "Tea–Determination of Free Amino Acids Content" [23]. The content of water extracts was determined in accordance with GB/T 8305-2013, "Tea–Determination of Water Extracts Content" [24]. The content of soluble sugar was determined by the anthrone-sulfuric acid colorimetric method, and the contents of theaflavins, thearubigins, and theabrownins were measured by the system analysis method [25]. The measurements using the above methods were all carried out on a UV-2102PC spectrophotometer (Element Analytical Instrument Co., Ltd., Shanghai, China).

The determination of catechin, flavonoid, and purine alkaloid fractions was carried out by the high-performance liquid chromatography (HPLC) with reference to the method described by Nian [26]. The instrument used in this study is a 1200 high-speed liquid chromatograph equipped with a C18 column (4.6 mm × 100 mm, 2.7 μm, Agilent (Santa Clara, CA, USA)). This method employed a mobile phase A consisting of 0.261% phosphoric acid and 5% acetonitrile, and a mobile phase B of 80% methanol. The elution gradient proceeded as follows: mobile phase B increased linearly from 10% to 45% between 0 and 16 min; from 16 to 22 min, mobile phase B increased linearly from 45% to 65%; mobile phase B was held constant at 65% from 22 to 25.9 min; from 25.9 to 29 min, mobile phase B increased linearly to 100%; and mobile phase B was maintained at 100% from 29 to 30 min. The column temperature was maintained at 35 °C. Each sample was extracted and analysed in triplicate.

2.5. Detection of Volatile Compounds

Headspace solid-phase microextraction (HS-SPME) combined with gas chromatography–mass spectrometry (GC-MS) was utilised to separate and identify the volatile compounds of raw Pu-erh tea from Wenshan Prefecture with different storage years. The instruments used in this study include a 7890A-5975C headspace solid-phase microextraction GC-MS (Agilent, USA), a DB-WAX column (30 m × 0.25 mm × 0.25 μm, Agilent, USA), and a 65 μm solid-phase microextraction head (PDMS/DVB, Supelco (Bellefonte, PA, USA)).

Headspace solid-phase microextraction: 1 g of tea sample was placed in a 20 mL headspace vial. A total of 1 μg of ethyl acetate (decanoic acid ethyl ester) was added as the internal standard, followed by the addition of 6 mL of boiling water. The mouth of the vial was then sealed. CTC autosampler: 60 °C, stabilised for 10 min; 65 μm polydimethylsiloxane/divinylbenzene (PDMS/DVB) extraction head; 60 °C extraction for 30 min at a rotational speed of 250 rpm. Gas chromatography–mass spectrometry (GC-MS) coupled detection, with an inlet port temperature of 230 °C and a desorption time of 5 min.

GC conditions were as follows: column: DB-WAX (30 m × 0.25 mm × 0.25 μm); carrier gas: He; oven: 50 °C (5 min) to 230 °C (7 min) at 6 °C/min; column temperature: 50 °C (5 min) to 230 °C (7 min), temperature rising rate: 6 °C/min; split ration: no split; MS conditions: mass spectrometry conditions; ion source: EI; gas interface temperature: 280 °C; and ion source temperature: 230 °C; quadrupole temperature: 150 °C.

Qualitative and quantitative techniques were employed to identify and measure volatile compounds, following the approach outlined by Deng et al. [27,28]. The preliminary detection of volatile compounds was performed utilising the NIST 17 mass spectral library from the National Institute of Standards and Technology. Furthermore, the identity of each compound was confirmed by matching its retention index (RI). The relative mass concentration of each compound was calculated using the internal standard semi-quantitative method. Furthermore, the odour activity value (OAV) is calculated by the ratio of the relative mass concentration of the volatile component to its olfactory threshold in water.

2.6. Data Analysis

Data standardisation was performed with Microsoft Excel 2021, with standard deviation presented as (±). Compound content significance was determined ($p < 0.05$), along with a fold change calculation. Principal component analysis (PCA) and orthogonal partial least squares discriminant analysis (OPLS-DA) were conducted using SIMCA 14.1. Data visualisation and representation were achieved with Origin 2021, MetaboAnalyst 6.0, TBtools V2.1, and Chiplot (www.chiplot.online). Online, Excel 2019, and PowerPoint 2019.

3. Results and Discussion

3.1. The Evolution of Sensory Flavour of RPT During Storage

The sensory flavour wheel of RPT from Wenshan Prefecture across different storage durations is presented in Figure 1. The sensory evaluation scoring outcomes are presented in Table 2. It was noted that, as storage duration extends, the elements such as the colour, aroma, and taste of the RPT have experienced enriching changes. Appearance colour transitions from green to yellow and finally to auburn, with the auburn hue intensifying over time. Soup colour follows a similar progression from green to yellow and then to orange, indicating a gradual increase in chroma. The aroma, initially a faint and sweet scent, develops woody notes and eventually exhibits ageing characteristics. The ageing aroma intensifies with extended storage, with Y1 and Y9 indicating particularly robust and persistent aromatic qualities. Taste evolves from an initial astringency to sweetness, culminating in a mellow character. The astringent flavour gradually dissipates as mellowness becomes more clear. The colour of the tea residue similarly shifts from green to yellow and finally to auburn, its intensity deepening with age.

Table 2. The sensory evaluation scoring outcomes of raw Pu-erh tea with different storage years in Wenshan Prefecture, Yunnan Province.

Sample	Appearance	s	Aroma	s	Soup Colour	s	Taste	s	Tea Residue	s	Total
Y1	Dark green	94	Faint scent with sweet, strong and lasting (+)	96	Yellowish-green, bright	92	Strong and mellow, astringent	88	Yellowish-green	92	92.20
Y2	Auburnish-green	94	Faint scent with sweet (+), strong	92	greenish yellow, bright	92	Strong and mellow, little astringent	90	Greenish-yellow	92	91.70
Y3	Greenish-auburn	94	Sweet, less pure	78	Light yellow, bright (+)	94	Nearly strong and mellow, little astringent	86	Greenish-yellow (+)	94	86.40

Table 2. *Cont.*

Sample	Appearance	s	Aroma	s	Soup Colour	s	Taste	s	Tea Residue	s	Total
Y4	Brownish-yellow	92	Woody, approach pure	80	Yellow, bright (+)	94	Approach mellow, little astringent	78	Auburnish-green	92	83.70
Y5	Yellowish-brown	92	Woody, pure, lasting	92	Yellow (+), bright (+)	94	Approach sweet and mellow, little astringent	78	Auburnish-green (+)	94	87.40
Y6	Auburnish-yellow	92	Ageing with sweet, strong and lasting	94	Deep yellow, bright	92	Nearly sweet and mellow, little astringent	86	Auburnish-yellow	92	90.50
Y7	Yellowish-auburn	92	Ageing with sweet, strong	92	Orange, bright	92	Mellow and sweet, little astringent	90	Auburnish-yellow (+)	94	91.40
Y8	Yellowish-auburn (+)	94	Ageing, more strong	90	Orange, bright	92	Mellow (+) and sweet	94	Yellowish-auburn	92	92.50
Y9	Auburnish-brown	92	Ageing with sweet, strong and lasting (+)	96	Orange, bright	92	Mellow (++) and sweet	95	Yellowish-auburn (+)	94	94.35
Y10	Brownish-auburn	92	Ageing with sweet, strong and lasting	94	Orange (+)	90	Mellow and thick and sweet	92	Yellowish-auburn (++)	95	92.55

Notes. (+), (++) indicate the degree of the presence of this characteristic.

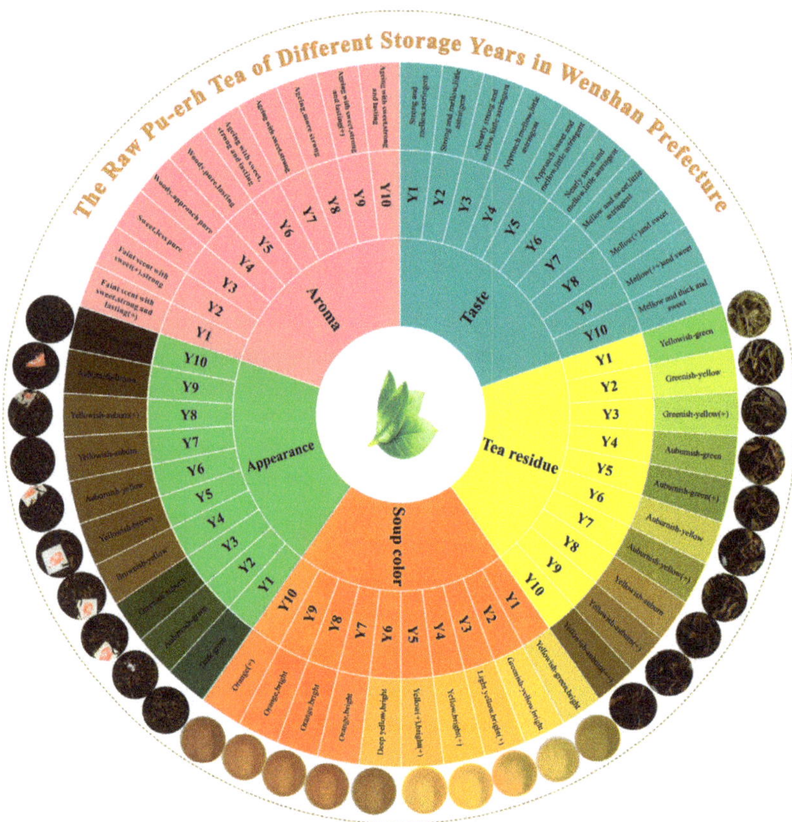

Figure 1. The flavour wheel of raw Pu-erh tea from different storage years in Wenshan Prefecture, Yunnan Province. (+), (++) indicate the degree of the presence of this characteristic.

3.2. Dynamic Changes of Non-Volatile Compounds in RPT Stored for Different Years

According to Figure 2A,B, which presents PCA and clustered heat map analyses of 22 non-volatile compounds, three ageing cycles for RPT from the Wenshan Prefecture stored for 1–10 years are indicated: Group A1 (stored for 1–3 years), Group B1 (stored for 4–8 years), and Group C1 (stored for 9–10 years). With the prolongation of storage duration, the contents of 22 non-volatile compounds in the three groups of RPT samples can be categorised into four changing trends. The first trend of change showed a gradual decrease, including the content of TP, ECG, and Theo (Figure S1A–C). The second trend of change showed a fluctuating decrease, including the content of FAA, SS, CA, EGCG, EGC, EC, GC, and Myr (Figure S1D–G). According to previous authors [24,25], changes in the content of amino acids, tea polyphenols, catechins, flavonoids, purine alkaloids, etc., affect the taste characteristics of tea broth, including freshness, bitterness, astringency, and sweetness. Therefore, the reduction in the content of TP, ECG, EGCG, EC, GC, Theo, and other bitter compounds will lead to a decrease in bitterness and astringency, as well as a reduction in the intensity of the taste of RPT from different storage years in Wenshan Prefecture. On the contrary, the third trend showed a gradual increase, including the contents of WE, TF, TR, TB, CG, GCG, Que, and Rut (Figure S1H–K). The fourth trend showed a fluctuating increase, including the contents of Kea, C, and GA (Figure S1L,M). The inherent instability of catechins renders them susceptible to degradation, oxidation, polymerisation, and isomerisation during storage. These reactions, potentially interacting in competitive or synergistic relationships, likely occur simultaneously, causing varying degrees of change in the catechin content of RPT from Wenshan Prefecture across different storage durations [29,30]. Such changes include the isomerisation of EGCG and ECG, resulting in higher CG and GCG levels. Lv [2] explains that TF, TR, and TB, arising from the hydrolysis, oxidation, and polymerisation of catechins, give the tea liquor its yellow, red, and auburn colours, respectively. These pigments, in conjunction with proteins, influence the appearance of the tea leaves. The progressive darkening of the leaves, liquor, and residue during extended storage can be partly attributed to increasing tea pigment content. Simultaneously, elevated TR and TB levels enhance the harmonious, sweet, and mellow flavour of tea, balancing its bitterness and astringency [31,32]. Potential markers (VIP > 1, FC ≥ 1.5, or FC ≤ 0.8) were further analysed by OPLS-DA and fold change calculations for three groups of RPT with different ageing cycles, yielding that different ageing cycles have different potential markers (Figure 2C). Among them, Theo is a potential marker for distinguishing Group B1 and Group C1 from Group A1. FAA, EGC, and GC are potential markers for distinguishing Group B1 from Group A1. CG, GCG, Que, and Rut are potential markers for distinguishing Group A1 and Group B1 from Group C1. TF and TR are potential markers for distinguishing Group A1 from Group C1. TB is a potential marker for distinguishing Group A1 from Group C1, and C is a potential marker for distinguishing Group C1 from Group B1. These evolving characteristics likely contribute to the unique regional flavour and allow for differentiation based on the storage duration of RPT (Table S1).

3.3. The Evolution of Volatile Compounds in RPT Stored for Different Years

3.3.1. The Volatile Profile of RPT

Based on HS-SPME-GC-MS, a total of 184 volatile compounds were detected in RPT from Wenshan Prefecture across a range of storage periods. According to Figure 3A, the total concentration of these compounds exhibited a general trend of increasing and then decreasing with longer storage. More specifically, the total volatile compound concentration rose sharply, reaching its peak in samples stored for 3 years. A sharp decline was observed in samples stored for 4–5 years, followed by a slight rebound in samples stored for 6–8 years, and a further decrease in samples stored for 9–10 years, creating a wave-like pattern of decline. This indicates that the third year of storage represents a critical point in the evolution of volatile compounds in RPT from Wenshan Prefecture. It can be seen that the third year of storage is the key turning point in the change of volatile compounds in raw Pu-erh tea from different years in Wenshan Prefecture. This is consistent with the conclusion that 3–4 years of storage ageing is an important turning point for volatile compounds in RPT, as reported by Guo [33]. In addition, the eighth year of storage was the second key turning point in the changes of volatile compounds in RPT from different years in Wenshan Prefecture. The volatile compounds fell into nine distinct categories, including 44 alkenes, 34 alkanes, 24 esters, 20 alcohols, 22 heterocycles, 14 aldehydes, 14 aromatic hydrocarbons, nine ketones, and three other compounds. Their relative mass concentrations accounted for 18.12%, 12.97%, 14.97%, 15.84%, 4.19%, 12.57%, 8.41%, 10.72%, and 2.21% of the total, respectively (Figure 3B) (Table S2). Among them, olefins had the highest relative mass concentrations (211.10–1533.84 μg/L), followed by alcohols (233.12–1410.99 μg/L). The remaining six categories of volatile compounds also exhibit varying degrees of distribution, and they show different levels of fluctuation changes with the increase of storage years. Figure 3C illustrates that cluster analysis, based on 188 volatile compounds, allows for the categorisation of RPT from Wenshan Prefecture into three groups according to storage duration: Group A2 (Y1, Y2, Y3, 1–3 years of storage), Group B2 (Y4, Y5, Y6, Y7, 4–7 years of storage), and Group C2 (Y8, Y9, Y10, 8–10 years of storage).

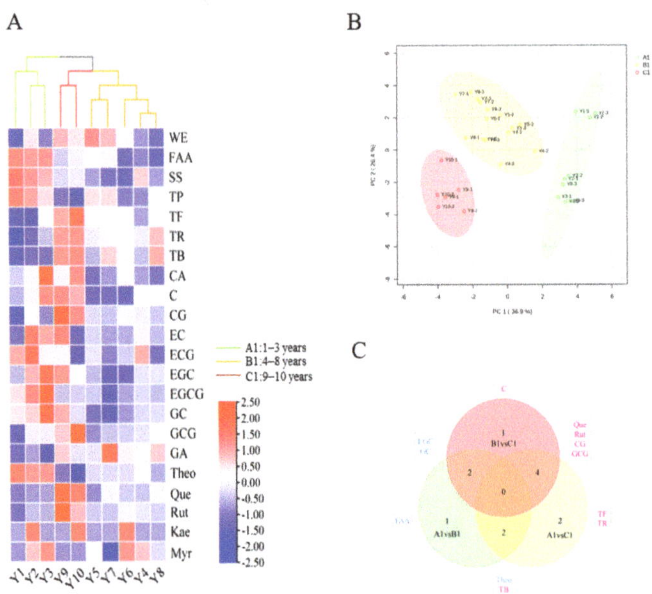

Figure 2. WE: water extracts, FAA: free amino acids, SS: soluble sugars, TP: tea polyphenol, TF: theaflavins, TR: thearubigins, TB: theabrownins, CA: caffeine, C: (+)-catechin, CG: (−)-catechin gallate, EC: (−)-epicatechin, ECG: (−)-epicatechin gallate, EGC: (−)-epigallocatechin, EGCG: (−)-epigallocatechin gallate, GC: (+)-gallocatechin, GCG: (−)-gallocatechin gallate, GA: gallic acid, Theo: theophylline, Que: quercetin, Rut: rutin, Kae: kaempferol, Myr: myrcene. (**A**) Heat diagram of the non-volatile compounds. (**B**) Scatter plot for the PCA model (total). (**C**) Wayne diagrams of potential makers, filter conditions are VIP > 1, FC > 1.5 or <0.8. Blue-labelled compounds were highest in group A1 and purple labelled compounds were highest in group C1.

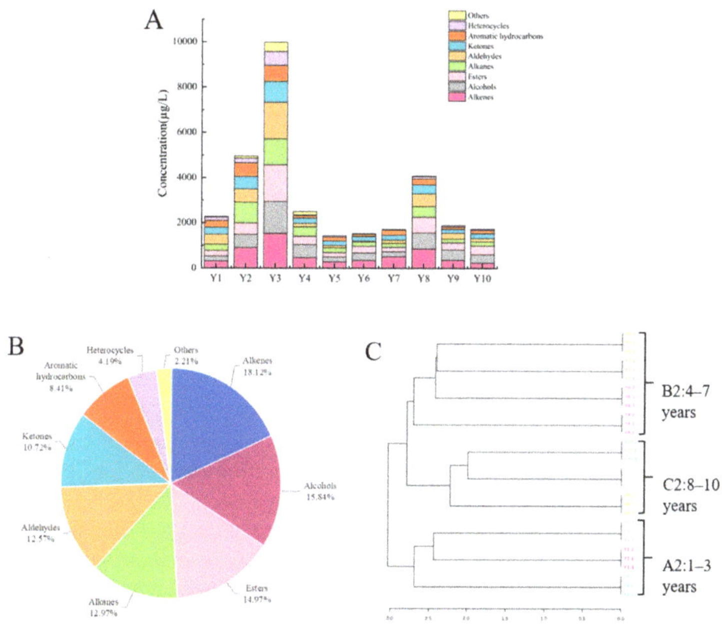

Figure 3. (**A**) The volatile compounds' concentration histogram in different storage years. (**B**) Percentage of content of volatile compounds by types. (**C**) Cluster dendrogram of VOCs in RPT with different storage years in Wenshan Prefecture.

3.3.2. Screening of Aroma-Active Compounds in RPT from Different Storage Years

A total of 44 aroma-active compounds (OAV $\geq$ 1) were identified based on their odour thresholds and aroma profiles (Figure 4). Of these, 42 aroma-active compounds exhibited OAV $\geq$ 10, while two aroma-active compounds had 10 $\geq$ OAV $\geq$ 1 (Figure 5) (Table S3). The aroma characteristics of these 44 compounds fell into nine descriptive categories: floral, fruity, woody, sweet, green, herbal, nuts, fatty, and roasted (Figure 4). These characteristics have been previously utilised to characterise the aroma of RPT. Moreover, Linalool (OAV = 106,598.48–3875.95, floral), β-Ionone (OAV = 74,251.90–12,348.77, violet, floral, sweet, fruity, woody), D-Limonene (OAV = 14,880.11–2078.29 fruity, sweet, lemon, orange), (Z)-3,7-Dimethylocta-1,3,6-triene (OAV = 2024.92–129.29, floral, herbal, sweet), γ-Terpinene (OAV = 174.85–8.57, citrus, lemon), Cedrol (OAV = 704.85–125.46, woody), (±)-Dihydroactinidiolide (OAV = 971.25–18.20, fruity, ripe apricot, oily, woody), 2,2,6-Trimethyl-cyclohexanone (OAV = 1834.61–120.21, honey, black pepper), Naphthalene (OAV = 481.36–59.70 camphor, tar) and nine other aroma substances were the key aroma active substances (OAV $\geq$ 1) common to all groups of different storage years samples, which together formed the basis of the overall aroma styles of the different storage years of RPT in Wenshan Prefecture.

From Figure 5 and Table S3, it can be seen that some aroma-active compounds appeared only at specific storage stages. 2-Amylfura (OAV = 4201.53–15,656.58, roasted, sweet fruity) was exclusively found in tea samples aged 1–3 years and reached the highest value (OAV = 15,656.58) in tea samples from the storage of 3 years. 2-Amylfura is a notable component of the 'honey-orchid' scent, produced via the Maillard reaction during the pressing and drying of tea [34,35]. Cis-Anethol (OAV: 421.88–3719.92, aniseed, herbaceous, sweet aroma) appeared only in tea samples from 2–4 years of storage and reached its highest value in tea samples from the storage of 3 years (OAV = 3719.92). Cis-Anethol is also the main aroma component of RPT after the addition of exogenous amino acid treatment [36]. (Z)-Linalool oxide (furanoid) (OAV = 66.90–538.16, strong woody and floral, camphor) appeared only in tea samples from 5–10 years of storage and reached its highest value in tea samples from the storage of 8 years (OAV = 538.16). According to the previous authors, (Z)-Linalool oxide (furanoid) is not only the distinguishing aroma-active compound of PRT in different storage environments but also the characteristic aroma-active compound of white tea, black tea, rock tea, and other tea types in different storage years [37–40]. At the same time, there are also some aroma-active compounds that are only present in separate storage years, including 3-Nonen-2-one (OAV = 11.61, fruity, spicy, liquorice odour) in the 1-year storage tea sample, 3-Carene (OAV = 107.62, pine odour) in the 2-year storage tea sample, Estragole (OAV = 4146.70, herbal aroma), Limonene (OAV = 4146.70, fruity, berry), (+)-α-Pinene (OAV = 2462.64, strong woody, piney odour), and Dihydrocarvone (OAV = 3.96, herbal, lingonberry-like odour) in the storage 8-year tea sample. Moreover, 28 aroma-active compounds reached their maximum values at the first critical turning point, i.e., tea samples from 3 years of storage, including Linalool (OAV = 106,598.48), β-Cyclocitral (OAV = 87,428.53), and β-Ionone (OAV = 74,251.90) with OAV ≥ 10,000, Safranal (OAV = 73,552.80), Copaene (OAV = 28,492.76), Eucalyptol (OAV = 23,073.88), 1-Octen-3-ol (OAV = 16,220.49), 2-Amylfuran (OAV = 15,656.58), D-Limonene (OAV = 14,880.11), Nonanal (OAV = 10,023.89), and other aroma-active compounds with floral, sweet, fresh, fruity, and roasted aroma characteristics. However, eight aroma-active compounds reached their maximum values at the second critical turning point, i.e., tea samples from the storage year 8, and they were Limonene with OAV ≥ 1 (OAV = 3383.37), (+)-α-Pinene (OAV = 2462.64), (E)-Linalool oxide (furanoid) (OAV = 541.16), (Z)-Linalool oxide (furanoid) (OAV = 538.16), α-Terpineol (OAV = 319.06), Terpinolene (OAV = 319.06), Terpinen-4-ol (OAV = 24.89) Dihydrocarvone (OAV = 3.96) and other aroma-active compounds with fruity, woody and herbal aroma characteristics.

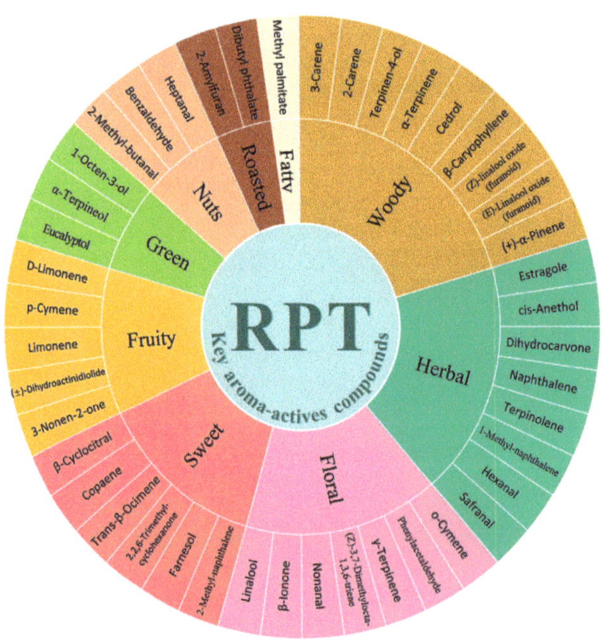

Figure 4. The wheel of key aroma-actives compounds in RPT from different storage years (OAV ≥ 1).

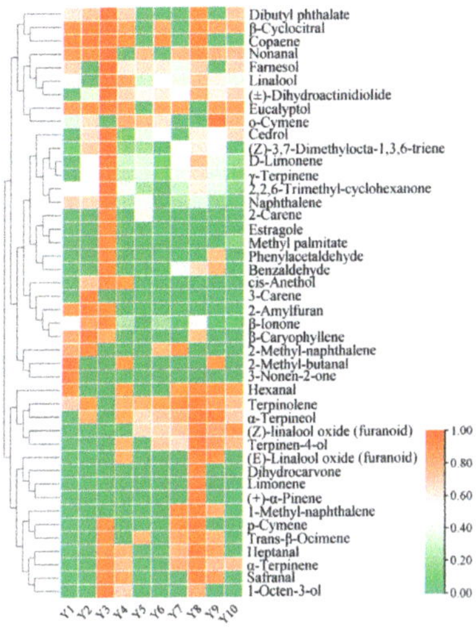

Figure 5. OAV heat map of key aroma-active compounds in PRT of different storage years (OAV $\geq$ 1).

3.3.3. Changes of Aroma-Active Compounds in RPT with Different Storage Ageing Cycles

With increasing storage duration, the aroma-active compounds and their corresponding OAV values shift across different ageing stages, resulting in differences in the aroma type characteristics of RPT from Wenshan Prefecture in each storage group. In group A2 (1–3 years of storage), 33 aroma-active compounds exhibited higher levels compared to group B2 and C2. Among the most potent aroma-active compounds, β-Ionone (OAV = 52280.17, violet, floral, sweet, fruity, woody) presented the highest OAV, succeeded by β-Cyclocitral (OAV = 52,259.92, sweet, fruity, herbal), Linalool (OAV = 41,921.32, floral), Safranal (OAV = 24,517.6, saffron, herbal, woody), Copaene (OAV = 20,053.18, honey, spicy, woody), Eucalyptol (OAV = 12,582.03, mint), 2-Amylfuran (OAV = 9981.23, roasted aroma, sweet fruity), D-Limonene (OAV = 7291.68, sweet fruity, lemon, orange), Nonanal (OAV = 5678.43, light and sweet rose aroma, citrus), and 1-Octen-3-ol (OAV = 5406.83, greasy, mushroom, floral) (Figure 6) (Table S4). These compounds, contributing floral, green, sweet, fruity, herbal, and roasted aromatic notes, help form the aroma of 'faint scent and sweet' in the A2 tea samples.

In group B2 (4–7 years of storage), Linalool (OAV = 26,077.19, floral) presented the highest odour activity value, succeeded by β-Ionone (OAV = 16019.8, violet, floral, sweet, fruity, woody) and β-Cyclocitral (OAV = 52,259.92, sweet, fruity, herbaceous). These findings align with the top aroma-active compounds observed in A2. Moreover, aroma-active compounds arose from B2 compared to A2: Terpinolene (OAV = 135.85, herbal aroma, woody), 1-Methyl-Naphthalene (OAV = 815.36, camphor, herbal aroma, chemical), Terpinen-4-ol (OAV = 7.57, woody), o-Cymene (OAV = 1.58 floral), (Z)-Linalool oxide (furanoid) (OAV = 58.57, strong woody and floral aroma, camphor), and (E)-Linalool oxide (furanoid) (OAV = 52.97 woody, floral); whereas, Naphthalene (OAV = 100.78, camphor, tar), β-Ionone (OAV = 16,019.8, violet, floral, sweet, fruity, woody), β-Cyclocitral (OAV = 52,259.92 sweet, fruity, herbaceous), and Nonanal (OAV = 563.68, fresh sweet roses, citrus) demonstrated lower OAVs in later B2 relative to A2 ($p < 0.05$). In addition, 2-Methyl-Naphthalene (OAV = 2729.03, sweet, floral, woody) and cis-Anethol (OAV = 578.85, herbal, aniseed, sweet) displayed higher OAVs in B2 compared to the tea samples from group C2 (Figure 6) (Table S4). Overall, relative to the A2 group, the B2 group demonstrated a decrease in aroma-active compounds associated with sweet, fruity, floral, and green aromas, with an increase in compounds contributing to herbal and woody aromas. This shift resulted in the development of a 'woody with sweet' aroma profile characteristic of the B2 tea samples. The woody aroma is characteristic of Pu-erh tea and serves as a significant aromatic attribute of Pu-erh tea stored in hot and humid circumstances [18].In group C2 (8–10 years of storage), Linalool (OAV = 29,107.94, floral) presented the highest odour activity value among aroma-active compounds, followed by β-Ionone (OAV = 29,107.94, violet, floral, sweet, fruity, woody) and Safranal (OAV = 18,132.77, saffron, herbal, woody). Furthermore, several other aroma-active compounds attained notable OAVs: Terpinolene (OAV = 140.31, herbal aroma, woody), 1-Methyl-Naphthalene (OAV = 1227.6, camphor, herbal aroma, chemical), Terpinen-4-ol (OAV = 16.8, woody), o-Cymene (OAV = 1.96, flowery), (Z)-Linalool oxide (furanoid) (OAV = 323.2, strong woody and floral odour, camphor), (E)-Linalool oxide (furanoid) (OAV = 306.41, woody, floral odour), (+)-α-Pinene (OAV = 820.88, pine, coniferous, resinous, strong woody and piney odour), Limonene (OAV = 1127.79, fruity, berries), Dihydrocarvone (OAV = 1.32, herbaceous, such as spearmint aroma), α-Terpineol (OAV = 203.73, orchid, light, fruity, banana), and Hexanal (OAV = 2485.41, herbaceous, fruity). These compounds reached their peak OAVs in the three tea sample groups. Terpinen-4-ol (OAV = 16.8, woody) exhibited a significantly greater OAV compared to group B2, while (Z)-Linalool oxide (furanoid) (OAV = 323.2, woody, floral) exhibited a significantly greater OAV compared to both A2 and B2 ($p < 0.05$) (Figure 6) (Table S4). The increased presence of woody aroma-active compounds, coupled with the diverse array of other aroma-active compounds, contributed to the characteristic 'ageing' aroma profile of the C2 tea sample. Moreover, no methoxy aroma-active compounds, commonly associated with 'Ageing aroma' characteristics in Pu-erh tea from previous studies, were detected in C2. However, Shen [37] stated that the Maillard reaction, or the transformation of certain aroma-active compounds during storage ageing, leads to the formation and accumulation of the aroma of herbals, wood, and medicinal herbs. Therefore, both the foundational aroma-active compounds of RPT and those from the later transformation process originate the 'Ageing aroma' characteristic of RPT.

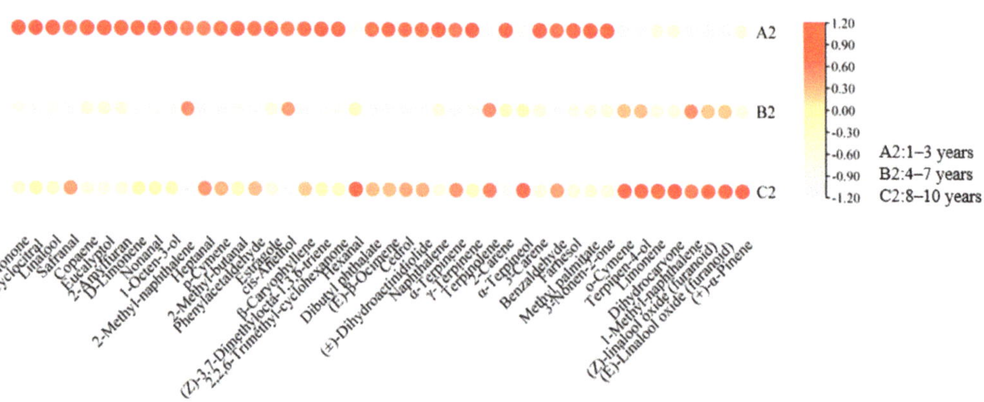

Figure 6. Clustering heat map of key aroma compounds from different storage ageing cycles (OAV $\geq$ 1).

3.3.4. Potential Markers Responsible for Aroma Differences of RPT in Three Ageing Cycles

From the above, it can be seen that the types and contents of aroma active compounds varied greatly among RPT from Wenshan Prefecture at different cycles of storage and ageing times. Based on OPLS-DA analysis combined with fold change for volatile compounds, firstly, a total of 67 significantly different volatile compounds (VIP > 1, FC ≥ 1.5, or FC ≤ 0.8) were screened (Figure 7) (Table S4). In Group B2, relative to Group A2, 40 volatile compounds exhibited down-regulation, while 2 demonstrated up-regulation. Similarly, compared to group A2, 40 volatile compounds were down-regulated and 2 volatile compounds were up-regulated in group B2. In Group C2, compared to Group A2, 37 volatile compounds were down-regulated and 6 were up-regulated. In addition, contrasting Group C2 with Group B2 indicated down-regulation of 6 volatile compounds and up-regulation of 29. Specifically, 12 volatile compounds demonstrated significant differences across all three groups. The relative mass concentrations of both Azulene and Pentanoic acid, 2,2,4-trimethyl-3-carboxyisopropyl, isobutyl ester, decreased significantly with extended storage; whereas, Nonanal, β-Ionone, β-Cyclocitral, Diisobutyl phthalate, Dibutyl phthalate, Phytone, 1-Methylene-1H-indene, 2,2,6-Trimethyl-cyclohexanone, 2,4,4-trimethylpentane-1,3-diyl bis(2-methylpropanoate), and 1-Chloro-heptacosane exhibited an initial decrease followed by an increase in relative mass concentration, reaching their lowest levels in Group B2 compared to Groups A2 and C2. As shown in Figure 8, Eucalyptol, β-Caryophyllene, 2-Amylfuran, Copaene, Estragole, and α-Terpinene were the potential aroma markers for group A2 to differentiate between group B2 and group C2. (Z)-Linalool oxide (furanoid), α-Terpineol, and Terpinen-4-ol were potential aroma markers for group C2 to differentiate group A2 from group B2. The aroma potential markers for group A2 to differentiate group B2 were Naphthalene, (Z)-3,7-Dimethylocta-1,3,6-triene, and Cedrol. The aroma potential markers for group A2 to differentiate group C2 were Linalool and 2-Carene. The aroma potential markers for group C2 to distinguish group B2 were (E)-Linalool oxide (furanoid) and Limonene. Additionally, β-Ionone, β-Cyclocitral, Nonanal, 2,2,6-Trimethyl-Cyclocitral, 2,2,6-Trimethyl-Cyclohexanone, and Dibutyl phthalate were the characteristic difference markers shared by all three groups. Moreover, β-Ionone, Linalool, β-Cyclocitral, (E)-Linalool oxide (furanoid), and Limonene have been shown to be potential markers of aroma in Pu-erh tea [27,41], and 2,2,6-Trimethyl-cyclohexanone is an intermediate that can be used to synthesise β-Ionone [42]. In previous studies, the variety of tea, the technology of processing, and the conditions of storage, including pathways such as oxidation and degradation of compounds caused by the environment, affect the formation of potential markers of aroma [43,44].

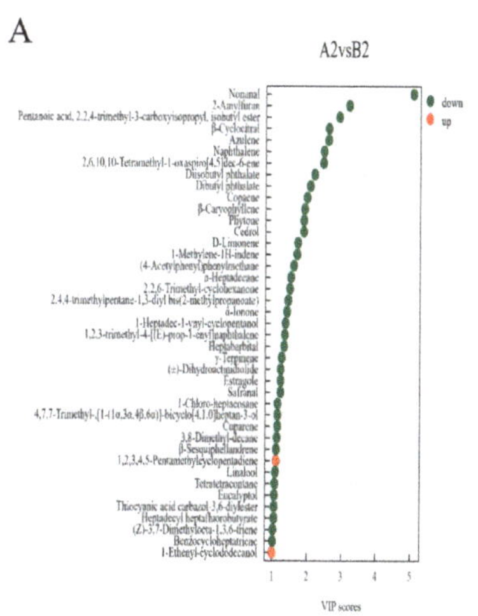

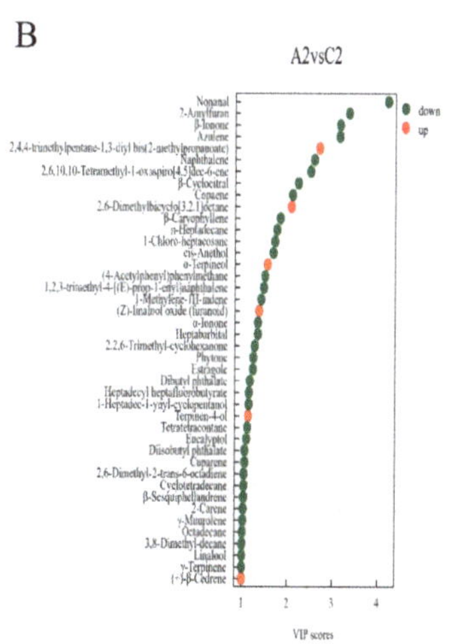

Figure 7. *Cont.*

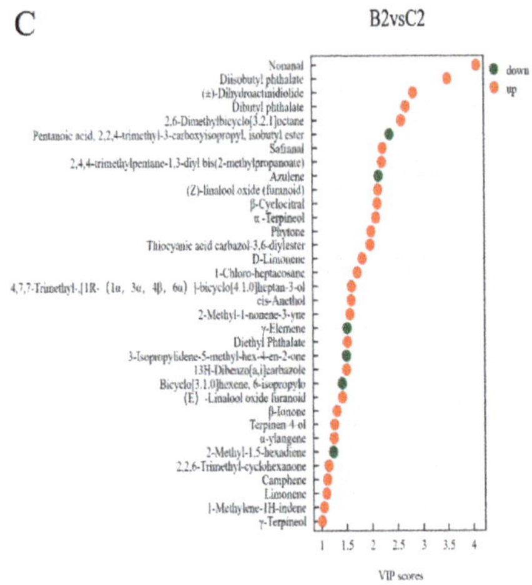

Figure 7. Key differential volatile compounds of RPT in three ageing cycles; filter conditions are VIP > 1, FC > 1.5, or <0.8. (**A**) A2vsB2, green-labelled compounds down-regulated by B2 and red-labelled compounds up-regulated by B2; (**B**) A2vsC2, green-labelled compounds down-regulated by C2 and red-labelled compounds up-regulated by C2; (**C**) B2vsC2, green-labelled compounds down-regulated by C2 and red-labelled compounds up-regulated by C2.

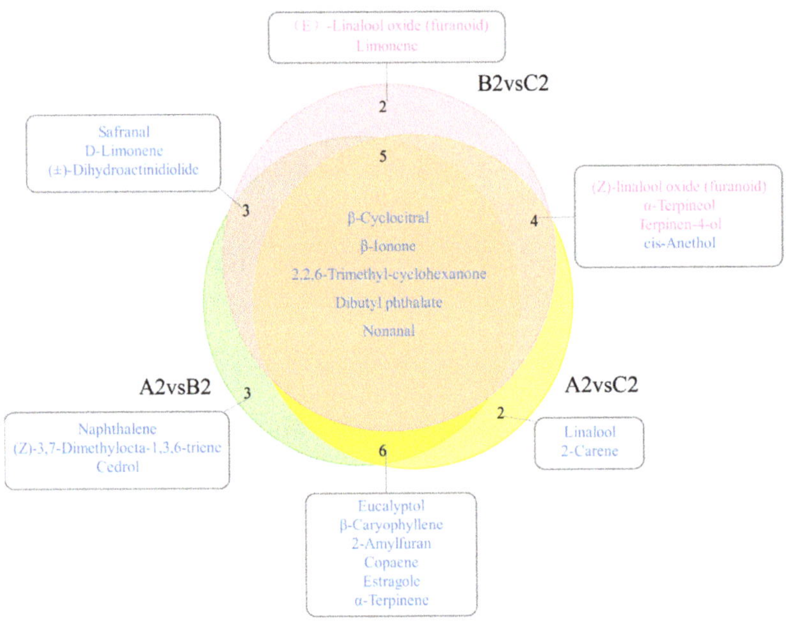

Figure 8. Wayne diagrams of potential differential aroma markers of RPT in three ageing cycles; screening conditions were VIP > 1, FC > 1.5, or <0.8. Blue-labelled compounds were highest in group A2 and purple-labelled compounds were highest in group C2.

3.3.5. Metabolic Evolutionary Pathways of Major Aroma-Active Compounds

The above studies have shown that the aroma characteristics of RPT change during storage due to the influence of various factors. Generally, glycosides, fatty acids, amino acids, carbohydrates, and carotenoids are significant aroma precursors for the synthesis of volatile chemicals in tea [45]. Leveraging KEGG data, key aroma compound metabolic routes were investigated further. Through the results of the study, it was found that the aroma-active compounds of RPT from Wenshan Prefecture were mainly synthesised as monoterpenes, which are usually woody, floral and fruity, and favorably impacts RPT's scent profile [46], including Linalool, β-Ionone, β-Cyclocitral, and Safranal. Previous research indicates that the synthesis of these compounds in tea occurs primarily via the metabolic transformation of two principal precursors: isopentenyl diphosphate and di-methylallyl pyrophosphate. Furthermore, the 2-C-methyl-D-erythritol-4-phosphate pathway and the carotenoid cleavage enzyme play critical roles in the production of monoterpenes and the oxidation of carotenoids [47]. Among them, Linalool oxides will be produced due to Linalool through an oxidoreductase reaction, including (Z)-Linalool oxide (furanoid) and (E)-Linalool oxide (furanoid) [48], and the production of these Linalool oxides helps to promote the wood flavour of RPT. β-primeverosides and β-glucopyranosides are the main precursors for the biosynthesis of Linalool and Linalool oxides in tea, which are produced by degradation and hydrolysis of the corresponding enzymes [48]. β-Ionone, terpene aldehyde, and terpene ketone are carotenoids that are produced by enzymatic and thermo-chemical reactions, and β-Cyclocitral is also an important intermediate in the synthesis of Safranal [44].

4. Conclusions

This study utilised sensory evaluation, HPLC, and HS-SPME-GC-MS, coupled with multivariate statistical analysis, to elucidate the evolving quality and characteristic profiles of RPT from different storage years in Wenshan Prefecture. This combined analytical approach enabled the classification of samples into three groups reflecting different ageing cycles based on storage duration. In this way, it illustrates that during the storage process, the sensory characteristics, non-volatile components, and aroma-active compounds of RPT will undergo differential changes over time.

Among them, changes in catechins, theophyllines, flavonoids, purine alkaloids, and free amino acids were potential markers for distinguishing the three groups of raw Pu-erh tea based on their ageing cycle. In addition, CG, GCG, Que, and Rut were potential markers in group C1 as storage time increased. A total of 184 volatile chemicals were identified using HS-SPME-GC-MS, with their overall quantity exhibiting an initial increase followed by a subsequent decrease across tea samples of varying storage years. Combined with OAV analyses of Linalool, β-Ionone, D-Limonene, (Z)-3,7-Dimethylocta-1,3,6-triene, γ-Terpinene, Cedrol, (±)-Dihydroactinidiolide, 2,2,6-Trimethyl-cyclohexanone, and Naphthalene, a total of nine key aroma-active compounds (OAV $\geq$ 10) together constitute the overall aroma profiles of RPT from different storage years in Wenshan Prefecture. These analyses also identified three sets of aroma-differentiating active compounds associated with different ageing cycles and highlighted (Z)-Linalool oxide (furanoid), α-Terpineol, Terpinen-4-ol, and cis-Anethol as potential markers for group C2. Moreover, β-Ionone, β-Cyclocitral, Nonanal, 2,2,6-Trimethyl-cyclohexanone, and Dibutyl phthalate were identified as common potential markers across the three groups of PRT with varying storage durations from Wenshan Prefecture. This research offers a certain theoretical basis and data support for an in-depth understanding of the quality characteristics of RPT across different storage years and the evolutionary paths of those characteristics. Furthermore, RPT quality is affected by a variety of factors, including harvest season, raw material grade, processing methods, and storage conditions; future studies with larger sample sizes and enhanced variable control are warranted. Integrating metabolomics and microbiological analysis could enhance the investigation of the transformation pathways and processes of compounds during storage ageing. At the same time, advanced technologies such as GC-IMS and GC-O were applied to conduct aroma reconstitution and deletion experiments. These experiments further clarify the characteristic markers, thereby providing more objective bases for the quality assessment of RPT, its flavour characteristics, and the improvement of processing and storage technologies.

Supplementary Materials: The following supporting information can be downloaded at https://www.mdpi.com/article/10.3390/foods14050829/s1: Table S1: Content of non-volatile compounds, *p*-value, VIP-value and fold change of PRT with different ageing cycles. Figure S1: Changes in the content of non-volatile compounds in RPT with different ageing cycles. Table S2: Sum of the absolute contents of volatile compounds in PRT with different ageing years. Table S3: OAV of key aroma-active compounds of RPT with different ageing years. Table S4: OAV, *p*-value, VIP-value and fold change of key aroma-active compounds with different ageing cycles.

Author Contributions: Conceptualization, investigation, methodology, and writing—original draft preparation, J.X. and X.D.; methodology, Y.W. (Yamin Wu) and M.Z.; formal analysis, Q.W. and Y.X.; resources, C.D. and Y.W. (Yankun Wang); data curation, W.W. and H.L.; visualization, J.H. and W.Y.; writing—review and editing, T.L.; funding acquisition, B.W. All authors have read and agreed to the published version of the manuscript.

Funding: This research was funded by the Yunnan Tea Industry Artificial Intelligence and Big Data Application Innovation Team (202405AS350025); the research and development and demonstration project of data sensing technology and equipment for high-altitude mountain smart agriculture (202302AE09002001); the study on the screening mechanism of phenotypic plasticity characteristics of large-leaf tea plants in Yunnan driven by Abased on data fusion (202301AS070083); the Yunnan Menghai County Smart Tea Industry Science and Technology Mission (202304BI090013); and the Yunnan Provincial Department of Science and Technology, Basic Research Special-Youth Project (202301AU070108).

Institutional Review Board Statement: Not applicable.

Informed Consent Statement: Informed consent was obtained from all subjects involved in the study.

Data Availability Statement: The original contributions presented in the study are included in the article, further inquiries can be directed to the corresponding authors.

Conflicts of Interest: The authors declare no conflicts of interest.

References

1. Lewis, R. Gascoyne, Kevin & others. Tea: History, Terroirs, Varieties. *Lib. J.* **2012**, *137*, 127–129.
2. Lv, H.P.; Zhang, Y.J.; Li, Z. Processing and chemical constituents of Pu-erh tea: A review. *Food Res. Int.* **2013**, *53*, 608–618. [CrossRef]
3. Zhang, L.; Zhang, Z.Z.; Zhou, Y.B. Chinese dark teas: Postfermentation, chemistry and biological activities. *Food Res. Int.* **2013**, *53*, 600–607. [CrossRef]
4. Ye, J.; Zhao, Y.; Chen, X.M. Pu-erh tea ameliorates obesity and modulates gut microbiota in high fat diet fed mice. *Food Res. Int.* **2021**, *144*, 110360. [CrossRef] [PubMed]
5. Wu, E.K.; Zhang, T.; Zhang, T.; Tan, C.; Peng, C.; Chisti, Y.; Wang, Q.; Gong, J. Theabrownin from Pu-erh tea together with swinging exercise synergistically ameliorates obesity and insulin resistance in rats. *Eur. J. Nutr.* **2020**, *59*, 1937–1950. [CrossRef]
6. Wu, S.; Yen, G.; Wang, B.; Chiu, C.; Yen, W.; Chang, L.; Duh, P. Antimutagenic and antimicrobial activities of pu-erh tea. *LWT-Food Sci. Technol.* **2007**, *40*, 506–512. [CrossRef]
7. Qian, Z.; Guan, J.; Yang, F.; Li, S. Identification and quantification of free radical scavengers in Pu-erh tea by HPLC-DAD-MS coupled online with 2, 2′-azinobis (3-ethylbenzthiazolinesulfonic acid) diammonium salt assay. *J. Agric. Food Chem.* **2008**, *56*, 11187–11191. [CrossRef]
8. Wang, T.; Li, X.; Yang, H.; Wang, F.; Kong, J.; Qiu, D.; Li, Z. Mass spectrometry-based metabolomics and chemometric analysis of Pu-erh teas of various origins. *Food Chem.* **2018**, *268*, 271–278. [CrossRef]

9. Wang, M.Q.; Sheng, Y.B.; Fan, Y.H. Analysis of key components and quality evaluation of Pu-erh tea under different storage conditions. *Food Res. Dev.* **2022**, *43*. (In Chinese)
10. Lv, S.; Wu, Y.; Wei, J.; Lian, M.; Wang, C.; Gao, X.; Meng, Q. Application of gas chromatography-mass spectrometry and chemometrics methods for assessing volatile profiles of Pu-erh tea with different processing methods and ageing years. *RSC Adv.* **2015**, *5*, 87806–87817. [CrossRef]
11. Jiao, T.T.; Shi, M.N.; Luo, H. Analysis of regular changes in quality indexes during storage of raw Pu-erh tea. *Mod. Food Sci. Technol.* **2024**, *40*. (In Chinese)
12. Liu, X.Y.; Shao, J.L.; Chen, X.L. HPLC fingerprints and vintage discrimination of Puerh tea with different Ageing ages. *Chin. J. Anal. Lab.* **2015**, *34*, 1159–1163. (In Chinese)
13. Duan, H.X.; Zhou, H.; Hu, C.M. Study on the Changes of Internal Composition of Puerh Tea with Different Storage Times. *Southwest China J. Agric. Sci.* **2012**, *25*, 111–114. (In Chinese)
14. Zhou, B.; Ma, C.; Wu, T.; Xu, C.; Wang, J.; Xia, T. Classification of raw Pu-erh teas with different storage time based on characteristic compounds and effect of storage environment. *LWT* **2020**, *133*, 109914. [CrossRef]
15. Rong, Y.; Xie, J.; Yuan, H.; Wang, L.; Liu, F.; Deng, Y.; Jiang, Y.; Yang, Y. Characterization of volatile metabolites in Pu-erh teas with different storage years by combining GC-E-Nose, GC–MS, and GC-IMS. *Food Chem. X* **2023**, *18*, 100693. [CrossRef] [PubMed]
16. Ma, L.L.; Luo, Z.F.; Liu, Y.L. Analysis of aroma components of Lincang Puerh raw tea based on multivariate statistical methods. *Sci. Technol. Food Ind.* **2023**, *44*, 331–337. (In Chinese)
17. Wang, C.; Li, J.; Wu, X.; Zhang, Y.; He, Z.; Zhang, Y.; Zhang, X.; Li, Q.; Huang, J.; Liu, Z. Pu-erh tea unique aroma: Volatile components, evaluation methods and metabolic mechanism of key odour-active compounds. *Trends Food Sci. Technol.* **2022**, *124*, 25–37. [CrossRef]
18. Xu, S.; Zeng, X.; Wu, H.; Shen, S.; Yang, X.; Deng, W.; Ning, J. Characterizing volatile metabolites in raw Pu'er tea stored in wet-hot or dry-cold environments by performing metabolomic analysis and using the molecular sensory science approach. *Food Chem.* **2021**, *350*, 129186. [CrossRef]
19. Ma, L.; Jiang, H.B.; He, Q.Y. Survey and Analysis of Ancient Tea Tree Resources in Wenshan Prefecture, Yunnan Province, China. *Southwest China J. Agric. Sci.* **2017**, *30*. (In Chinese)
20. *GB/T22111-2008*; Product of Geographical Indication–Puer Tea. General Administration of Quality Supervision, Inspection and Quarantine of the People's Republic of China: Beijing, China, 2008.
21. *GB/T23776-2018*; Methodology for Sensory Evaluation of Tea. General Administration of Quality Supervision, Inspection and Quarantine of the People's Republic of China: Beijing, China, 2018.
22. *GB/T8313-2018*; Determination of Total Polyphenols and Catechins Content in Tea. National Standard of the People's Republic of China: Beijing, China, 2018.
23. *GB/T8314-2013*; Tea–Determination of Free Amino Acids Content. Administration of Quality Supervision, Inspection and Quarantine of the People's Republic of China: Beijing, China, 2013.
24. *GB/T 8305-2013*; Tea–Determination of Water Extracts Content. Administration of Quality Supervision, Inspection and Quarantine of the People's Republic of China: Beijing, China, 2013.
25. Wang, Q.; Peng, C.; Gong, J. Effects of enzymatic action on the formation of theabrownin during solid state fermentation of Pu-erh tea. *J. Sci. Food Agric.* **2011**, *91*, 2412–2418. [CrossRef]
26. Nian, B.; Chen, L.; Yi, C.; Shi, X.; Jiang, B.; Jiao, W.; Liu, Q.; Lv, C.; Ma, Y.; Zhao, M. A high performance liquid chromatography method for simultaneous detection of 20 bioactive components in tea extracts. *Electrophoresis* **2019**, *40*, 2837–2844. [CrossRef]
27. Deng, X.; Huang, G.; Tu, Q.; Zhou, H.; Li, Y.; Shi, H.; Wu, X.; Ren, H.; Huang, K.; He, X. Evolution analysis of flavour-active compounds during artificial fermentation of Pu-erh tea. *Food Chem.* **2021**, *357*, 129783. [CrossRef] [PubMed]
28. Nie, C.; Zhong, X.; He, L.; Gao, Y.; Zhang, X.; Wang, C.; Du, X. Comparison of different aroma-active compounds of Sichuan Dark brick tea (*Camellia sinensis*) and Sichuan Fuzhuan brick tea using gas chromatography–mass spectrometry (GC–MS) and aroma descriptive profile tests. *Eur. Food Res. Technol.* **2019**, *245*, 1963–1979. [CrossRef]
29. Chen, Z.; Dai, W.; Xiong, M.; Gao, J.; Zhou, H.; Chen, D.; Li, Y. Metabolomics investigation of the chemical variations in white teas with different producing areas and storage durations. *Food Chem. X* **2024**, *21*, 101127. [CrossRef] [PubMed]
30. Ananingsih, V.K.; Sharma, A.; Zhou, W. Green tea catechins during food processing and storage: A review on stability and detection. *Food Res. Int.* **2013**, *50*, 469–479. [CrossRef]
31. Wang, Y.; Zhang, M.; Zhang, Z.; Lu, H.; Gao, X.; Yue, P. High-theabrownins instant dark tea product by Aspergillus niger via submerged fermentation: α-glucosidase and pancreatic lipase inhibition and antioxidant activity. *J. Sci. Food Agric.* **2017**, *97*, 5100–5106. [CrossRef] [PubMed]
32. Long, P.; Rakariyatham, K.; Ho, C.T.; Zhang, L. Thearubigins: Formation, structure, health benefit and sensory property. *Trends Food Sci. Technol.* **2023**, *133*, 37–48. [CrossRef]

33. Guo, J.; Yu, Z.; Liu, M.; Guan, M.; Shi, A.; Hu, Y.; Li, S.; Yi, L.; Ren, D. Analysis of volatile profile and aromatic characteristics of raw Pu-erh tea during storage based on GC-MS and odour activity value. *Foods* **2023**, *12*, 3568. [CrossRef]
34. Ma, X.M.; Li, H.Y.; Zeng, B. Analysis of Key odourants Responsible for Honey Orchid-like Aroma of Black Teas Made by Tea Cultivar 'Gougunao No.2'. *Mod. Food Sci. Technol* **2024**, *3*, 0008. (In Chinese)
35. Yang, Y.; Xie, J.; Chen, J.; Deng, Y.; Shen, S.; Hua, J.; Wang, J.; Zhu, J.; Yuan, H.; Jiang, Y. Characterization of N, O-heterocycles in green tea during the drying process and unraveling the formation mechanism. *Food Control* **2022**, *139*, 109079. [CrossRef]
36. Zhang, X. Effect of Amino Acids and Sugars on the Qualityof pu-er Tea. Master's Thesis, South China University of Technology, Guangzhou, China, 2012.
37. Shen, S.; Huang, J.; Li, T.; Wei, Y.; Xu, S.; Wang, Y.; Ning, J. Untargeted and targeted metabolomics reveals potential marker compounds of an tea during storage. *LWT* **2022**, *154*, 112791. [CrossRef]
38. Cui, J.; Wu, B.; Wang, J.; Jing, T.; Jin, J.; Zhao, M.; Hu, Y.; Wu, Y.; Yu, F.; Zhang, N. Effect of Storage Time on Aroma Profiles of Wuyi Rock Tea. *LWT* **2024**, *203*, 116367. [CrossRef]
39. Qi, Z.; Huang, W.; Liu, Q.; Ning, J. Variation in the Aroma Composition of Jasmine Tea with Storage Duration. *Foods* **2024**, *13*, 2524. [CrossRef]
40. Huang, Y. Analysis of aroma characteristics and volatile components of Zhenghe white tea with different storage years. *J. Tea Sci.* **2023**, *43*, 667–680.
41. Feng, T.; Sun, J.; Wang, K.; Song, S.; Chen, D.; Zhuang, H.; Lu, J.; Li, D.; Meng, X.; Shi, M. Variation in volatile compounds of raw Pu-Erh tea upon steeping process by gas chromatography–ion mobility spectrometry and characterization of the aroma-active compounds in tea infusion using gas chromatography–olfactometry–mass spectrometry. *J. Agric. Food Chem.* **2022**, *70*, 13741–13753. [CrossRef] [PubMed]
42. Breining, T.; Schmidt, C.; Polos, K. A palladium-catalyzed synthesis of β-ionone from 2, 2, 6-trimethylcyclohexanone. *Synth. Commun.* **1987**, *17*, 85–88. [CrossRef]
43. Guo, X.; Ho, C.; Schwab, W.; Song, C.; Wan, X. Aroma compositions of large-leaf yellow tea and potential effect of theanine on volatile formation in tea. *Food Chem.* **2019**, *280*, 73–82. [CrossRef] [PubMed]
44. Su, D.; He, J.; Zhou, Y.; Li, Y.; Zhou, H. Aroma effects of key volatile compounds in Keemun black tea at different grades: HS-SPME-GC-MS, sensory evaluation, and chemometrics. *Food Chem.* **2022**, *373*, 131587. [CrossRef]
45. Feng, Z.; Li, Y.; Li, M.; Wang, Y.; Zhang, L.; Wan, X.; Yang, X. Tea aroma formation from six model manufacturing processes. *Food Chem.* **2019**, *285*, 347–354. [CrossRef] [PubMed]
46. Ma, C.; Gao, C.; Li, Y.; Zhou, X.; Fan, G.; Tian, D.; Huang, Y.; Li, Y.; Zhou, H. The Characteristic Aroma Compounds of GABA Sun-Dried Green Tea and Raw Pu-Erh Tea Determined by Headspace Solid-Phase Microextraction Gas Chromatography–Mass Spectrometry and Relative odour Activity Value. *Foods* **2023**, *12*, 4512. [CrossRef] [PubMed]
47. Schwab, W.; Davidovich Rikanati, R.; Lewinsohn, E. Biosynthesis of plant-derived flavour compounds. *Plant J.* **2008**, *54*, 712–732. [CrossRef] [PubMed]
48. Zheng, X.; Li, Q.; Xiang, L.; Liang, Y. Recent advances in volatiles of teas. *Molecules* **2016**, *21*, 338. [CrossRef]

foods

MDPI

Article

Influence of Different Shaping Techniques on the Aroma Quality and Volatile Metabolites of Green Tea Revealed by Gas Chromatography Electronic Nose and Gas Chromatography–Tandem Mass Spectrometry

Jiahao Tang [1,2], Jiajing Hu [2], Xianxiu Zhou [2], Qiwei Wang [2], Yongwen Jiang [2], Haibo Yuan [2], Yujie Wang [1,*] and Yanqin Yang [2,*]

[1] State Key Laboratory of Tea Plant Biology and Utilization, Anhui Agricultural University, Hefei 230036, China

[2] National Key Laboratory for Tea Plant Germplasm Innovation and Resource Utilization, Tea Research Institute, Chinese Academy of Agricultural Sciences, Hangzhou 310008, China

* Correspondence: wangyj@ahau.edu.cn (Y.W.); yangyq@tricaas.com (Y.Y.)

Abstract: The shaping process is recognized as a crucial step in the manufacturing of green tea. However, its influence on aroma quality remains unclear. In this study, the effects of four shaping techniques, including flat green tea (FGT), straight green tea (SGT), phoenix green tea (PGT), and curled green tea (CGT), on the aroma quality and volatile metabolites of green tea were investigated by gas chromatography electronic nose (GC-E-Nose) and gas chromatography–tandem mass spectrometry (GC-MS/MS). The findings indicated that distinct shaping processes significantly influenced the development of the aroma quality and aroma components of green tea. The PGT processing facilitated the attainment of superior aroma quality of green tea. In total, 60 volatile components were identified by GC-MS/MS, with 54 of these compounds being consistently detected across four different shaping techniques. In particular, the PGT processing method was effective in yielding elevated levels of alcohols, esters and ketones. Moreover, 20 key odorants were screened out, with (*E*,*E*)-2,4-decadienal, (*E*,*E*)-2,4-nonadienal, phenylethyl alcohol, and benzeneacetaldehyde proven to be substantial contributors to the overall aromas of green tea under diverse shaping procedures. These key odorants were primarily derived from lipid degradation and the Maillard reaction. GC-E-Nose served as a significant adjunct to sensory evaluation, enabling the swift differentiation of green tea samples that have undergone various shaping processes. These findings offer both theoretical and technical perspectives that may guide the creation of innovative green tea products distinguished by their unique shapes.

Keywords: green tea; shaping technique; volatile metabolites; GC-E-Nose; GC-MS/MS

1. Introduction

Tea is globally acknowledged as one of the three most renowned non-alcoholic beverages, celebrated universally amongst people from all over the world [1–3]. Green tea, holding a preeminent position in China's traditional six major categories, is the most abundant type in China [4,5]. Green tea possesses distinct characteristics such as a high aroma, a mellow taste, and an appealing appearance [6,7]. The processing procedure typically involves the stages of fixation, rolling (or shaping), and drying [8,9]. Currently, investigations into green tea processing technology predominantly concentrate on the fixation and drying stages. For example, the aroma undergoes a progressive transformation, evolving from a faint scent to a chestnut-like scent, and ultimately to a high-fire aroma as the drying temperature increases [10]. Shaping is a crucial step in the manufacturing of premium green tea. This procedure employs either a manual or mechanical external force to guide the tea leaves into corresponding shapes and concurrently dry the tea under specific temperature conditions [11]. Currently, there are scarce reports on the shaping technology of green tea. Different shaping technologies and parameters significantly influence both the external shape and internal quality of tea. Therefore, it is important to clarify the influence of shaping technology on the volatile substances of green tea, which is conducive to the improvement of the processing technology of green tea.

Aroma plays a pivotal role in determining tea flavor, exerting significant influence on its quality [12,13]. It is closely linked to the superiority or inferiority of tea quality and influences consumer preferences. The essence of tea aroma is a complex blend of numerous volatile components in varying concentrations, primarily derived from the transformations of precursor substances such as glycosides, carotenoids, fatty acids, and amino acids during processing [10]. Up to now, more than 700 volatile substances have been isolated and identified from tea [14]. Nevertheless, not all volatile substances contribute directly to the formation of tea aroma due to the distinctive odor characteristics and odor thresholds of various compounds. Only a portion of odor substances with high aroma intensities or high aroma activity values play a vital role in tea aroma quality [15,16].

Advanced instrumental analysis techniques have progressively emerged as indispensable tools for the evaluation of tea aroma [17–20]. These techniques are characterized by a minimal sampling volume, high sensitivity, excellent repeatability, rapid analytical speed, and high degree of automation [21]. They facilitate the separation and identification of targets from complex mixtures and the quantification of trace and ultra-trace substances. The predominant analytical technology for tea aroma substances primarily encompasses gas chromatography–mass spectrometry (GC-MS), nuclear magnetic resonance, and gas chromatography–tandem mass spectrometry (GC-MS/MS) [22–27]. Additionally, it also includes analytical technologies based on sensors such as electronic nose [28–30]. Given that tea contains numerous trace or ultra-trace aroma components, it is challenging to achieve the comprehensive characterization of tea aroma components with a single analysis method.

In this study, the influence of different shaping techniques on the aroma quality and volatile metabolites of green tea was comprehensively evaluated through the integration of gas chromatography electronic nose (GC-E-Nose) and GC-MS/MS. The results are expected to provide a theoretical and technical foundation for the processing of premium green tea.

2. Materials and Methods

2.1. Sample Preparation

Freshly harvested tea leaves from Shengzhou City in Zhejiang Province in April 2024 were processed into four distinct types of green tea: flat green tea (FGT), straight green tea (SGT), phoenix green tea (PGT), and curled green tea (CGT). For example, FGT is characterized by its flat appearance, with Xihu Longjing being exemplar of this type, whereas PGT is noted for its resemblance to a phoenix tail, with Anji white tea serving as its representative. The comprehensive manufacturing procedure and specific parameters for the production of FGT, SGT, PGT, and CGT are detailed as follows: the freshly plucked tea leaves from the *Camellia sinensis* cultivar 'Zhongcha 125', comprising a single bud and a single leaf, were spread at a temperature of 20 ± 2 °C for a duration of about 14 h. Upon completion of the spreading process, the tea leaves were divided into four equal portions to facilitate the preparation of the four distinct shapes of green tea. A simplified overview of the processing steps is illustrated in Figure 1. For FGT, the spread tea leaves underwent fixation using an automatic flat tea stir-frying machine (Zhejiang Hengfeng technology development Co., Ltd., Shaoxing, China) at 260 °C, followed by a second fixation at 160 °C, and they were subsequently dried using a pot-drying machine (6CH-3.0A, Zhejiang Hengfeng technology development Co., Ltd.) at 130 °C until the moisture content reached 6.15%. For SGT, the spread tea leaves were subjected to fixation using a strip machine (6CL-80-12D, Zhejiang red five ring tea equipment Co., Ltd., Quzhou, China) at 280 °C with a vibration frequency of 30 and then dried in a hot air-drying machine (6CH-6, Zhejiang red five ring tea equipment Co., Ltd.) at 90 °C for 30 min. For PGT, the spread tea leaves were subjected to fixation using a roller fixation machine (6CST-50, Zhejiang red five ring tea equipment Co., Ltd.) at 300 °C, followed by treatment in a strip machine at 210 °C with a vibration frequency of 30, and they were then dried in a hot air-drying machine at 90 °C for 30 min. For CGT, the spread tea leaves underwent fixation in a roller fixation machine at 300 °C. This was succeeded by a rolling process in a rolling machine (6CR-30, Zhejiang red five ring tea equipment Co., Ltd.) for 45 min. Subsequently, the leaves underwent a first round of drying utilizing a hot air-drying machine at 110 °C for 20 min, followed by a secondary drying phase at 90 °C for 30 min. The final tea samples, produced through various shaping techniques, were stored at −20 °C prior to further analysis.

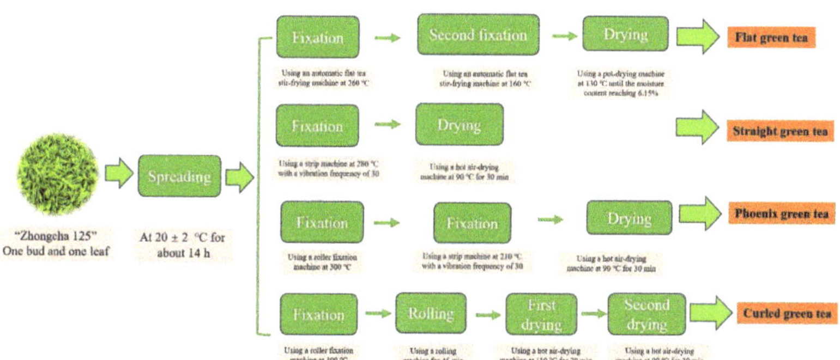

Figure 1. Flowcharts of the manufacturing processes of green tea with different shaping techniques. Independent batches (n = 3) were performed.

2.2. Materials and Reagents

The 20 mL headspace vials and magnetic PTFE/silicone caps were purchased from Agilent Technologies Inc. (Palo Alto, CA, USA). Ethyl decanoate was purchased from TCI Chemical Industry Development Co., Ltd. (Shanghai, China). A blend of n-alkanes spanning C7 to C40 was obtained from Sigma-Aldrich (Shanghai, China), and purified water was obtained from Hangzhou Wahaha Group Co., Ltd. (Hangzhou, China). Detailed information regarding the standards of volatile compounds is presented in Table S1.

2.3. Sensory Evaluation

The aroma quality of four distinct green tea samples was evaluated by a panel consisting of six specialists with extensive experience, adhering to the Chinese national standard (GB/T 23776-2018) [31]. The specialists consensually agreed to participate and use their data, and ethical permission was not required. Specifically, 3 g of tea samples were introduced into an evaluation cup and infused with 150 mL of boiling water for 4 min. Subsequently, the tea infusion was filtered and transferred into an evaluation bowl for aroma assessment. The aroma types, intensity, and persistence were confirmed through hot sniffing, warm sniffing, and cold sniffing. All 6 participants reached a consensus on the evaluation of aroma attributes. The aroma quality score was calculated using a hundred-point system.

2.4. GC-MS/MS Analysis

The headspace solid-phase microextraction technique, adapted from earlier studies [15,32], was employed for the extraction of samples, which underwent various shaping techniques. The manually controlled sampling handle and divinylbenzene/carboxen/polydimethylsiloxane fiber (DVB/CAR/PDMS, 50/30 μm, 2 cm) obtained from Supelco (Bellefonte, PA, USA) were utilized in this experiment. Specifically, 0.5 g of the tea samples and 5 mL of purified water were placed into a 20 mL headspace vial. Following this, the DVB/CAR/PDMS fiber was inserted into the vial prior to incubation at 60 °C for one hour. Finally, the fiber was directed towards the injection port at a temperature of 250 °C, where it underwent five-minute thermal desorption.

A 7890B gas chromatograph system, integrated with a 7000C series triple-quadrupole detector (both manufactured by Agilent Technologies, located in Palo Alto, CA, USA), was employed for analyzing the volatiles, featuring a DB-5MS column (30 m × 0.25 mm × 0.25 μm). Initially, the column's operational temperature was initiated at 40 °C for a duration of 5 min, progressively increasing to 160 °C at 4 °C/min, maintaining this elevated temperature for another 2 min, and culminating in a final increase to 270 °C at a rate of 10 °C/min (maintained for 2 min). Helium was selected as the carrier gas, flowing at a rate of 1.0 mL/min. The splitless mode was adopted for analysis, with the temperatures of the ion source and transfer line being maintained consistently at 230 °C and 270 °C, respectively. Mass spectrometric analysis was conducted in electron ionization mode at 70 eV, with the mass scanning range spanning from 40 m/z to 450 m/z.

The retention index (RI) was calculated by employing normal alkanes from C7 to C40, which served as a basis for comparison with theoretical values. The volatile metabolites were qualitatively evaluated based on a comparison between their distinct mass spectra against those of the standardized references and NIST11 library. Additionally, the RIs were referenced from the databases available at https://webbook.nist.gov/chemistry/ and https://www.flavornet.org/flavornet.html (accessed on 11 September 2024). The accurate quantification of the volatile compounds was achieved via the external standard approach, with the corresponding quantitative information presented in Table S2. For the aroma compounds that lack commercially available standards, quantitative analysis was carried out using the internal standard method with ethyl decanoate.

2.5. GC-E-Nose Analysis

A GC-E-Nose instrument (Alpha M.O.S., Toulouse, France) was utilized for the detailed analysis of tea samples subjected to diverse shaping techniques, with replicate analyses performed thrice. The adopted parameters were acquired from the earlier reported literature [32]. Typically, 0.5 g of the tea samples was precisely measured into a 20 mL headspace vial before being exposed to 60 °C for a duration of 20 min. Then, 5000 μL of headspace gas underwent injection into the GC-E-Nose injection port, operated at 200 °C. Volatile substances were adsorbed for 27 s via the utilization of an odor concentration trap (Tenax TA) maintained at 20 °C. The temperature profiles of the MXT-5 column and MXT-1701 column, exhibiting distinct polarity, were as follows: starting from 50 °C (holding for 5 s), ascending progressively to 80 °C at 0.1 °C/s, followed by a sharp transition to 250 °C at 0.4 °C/s (holding for 10 s). Both flame ionization detectors (FIDs) were set to operate at 260 °C.

2.6. Odor Activity Value Analysis

The contribution of volatile compounds to the overall aroma was assessed using the odor activity value (OAV) method. Compounds featuring an OAV exceeding one are recognized as vital contributors to the aroma profile of the analyzed samples [15,33]. The mathematical equation employed for this analysis is OAV = C/OT, wherein C (μg/L) denotes the concentration level of each volatile compound, and OT indicates its odor threshold in water.

2.7. Statistical Analysis

Three independent batches of green tea processing, comprising both experimental and validation batches, were conducted in accordance with the same methodology. Partial least squares discrimination analysis (PLS-DA) was implemented via the application of SIMCA-P 13.0 (Umetrics, Umea, Sweden). The peak areas obtained from GC-E-Nose and the contents of volatile compounds obtained from GC-MS/MS were designated as the independent variables, respectively, whereas the results of the sensory evaluation were defined as the dependent variables. Significance analysis was evaluated on the contents of volatile compounds in green tea samples by means of one-way analysis of variance (ANOVA) using SPSS 20.0 (Chicago, IL, USA), followed by Duncan's test. Venn diagrams were constructed by employing the robust Venny software (https://jvenn.toulouse.inrae.fr/app/example.html (accessed on 3 September 2024)). Bar graphs were constructed using Origin software 2024 (Northampton, MA, USA). Hierarchical cluster analysis (HCL) was performed on 21 significant differential volatile compounds utilizing Metware Cloud (https://cloud.metware.cn (accessed on 20 September 2024)), adhering to the criteria of $p < 0.05$ and VIP > 1. Red denotes higher than the average concentration, while green signifies lower than the average concentration. The darker the color, the higher the concentration.

3. Results and Discussion

3.1. Analysis of the Effects of Different Shaping Techniques on the Aroma Quality of Green Tea by Sensory Evaluation

In this study, green tea samples obtained from four distinctive shaping methods, namely PGT, FGT, CGT, and SGT, were subjected to sensory evaluation. The results indicated that significant variations in the overall aroma quality of green tea were observed under diverse shaping procedures (Table S3). Specifically, the PGT sample attained the highest score of 89.6, manifesting a delightful chestnut-like aroma. The FGT sample exhibited a cooked chestnut-like aroma, attaining a score of 88.17. The CGT presented a fresh scent, scoring 87.33, whereas the SGT exhibited a neutral tea aroma, earning a score of 86.4. In summary, the PGT processing proved beneficial to obtain high-quality aroma of green tea. It is possible that during the production process of PGT, various high-temperature steps (such as fixation and drying) were performed with multiple types of tea-manufacturing apparatus (such as a roller fixation machine, strip machine, and hot air-drying machine) to prompt the occurrence of lipid oxidation degradation and the Maillard reaction, thereby enhancing the aroma quality [2,16].

3.2. Analysis of the Effects of Different Shaping Techniques on the Volatile Profiles of Green Tea Using GC-E-Nose

Traditional sensor-based electronic noses face challenges such as sensor instability and odor contamination, which reduce the response values. As an innovative electronic nose technology, GC-E-Nose integrates two gas chromatography columns with disparate polarities to efficiently analyze the volatile profiles of target samples. With the aid of advanced algorithm models such as PLS-DA incorporated within GC-E-Nose, it allows for the rapid differentiation of samples with diverse odor characteristics. In this study, the volatile profiles of green tea subjected to diverse shaping methods were analyzed by GC-E-Nose. As illustrated in Figure S1, the volatile profiles collected from the MXT-5 and MXT-1701 columns with different shaping methods exhibited a distinct disparity.

To further investigate the distinctions in volatile profiles among tea samples subjected to different shaping methods, a supervised PLS-DA model was employed for analysis. As shown in Figure 2A, a distinct discrimination was achieved in the score plots of PLS-DA, indicating that there were substantial differences in their aroma profiles. The model's parameters indicated an R^2Y of 0.987, with a Q^2 of 0.909. Generally, the closer the R^2Y value is to 1, the stronger the explanatory power that the model possesses. Moreover, a higher Q^2 value indicates better predictive accuracy. Based on the aforementioned results, the model exhibited robust explanatory and predictive capabilities. To further verify the reliability of the model, 200 iterations of permutation testing were conducted. The results for R^2 = (0.0, 0.561) and Q^2 = (0.0, −0.375) demonstrated that the model was reliable and not overfitted (Figure 2B). Variable importance in projection (VIP) could quantify the contribution of each volatile compound to the classification ability of the model. Typically, the higher the VIP value of a volatile compound, the greater its contribution to the classification. In this study, a total of 23 variables with VIP greater than 1 were screened out (Figure S2). In conclusion, the GC-E-Nose functioned as an important complement to the sensory evaluation, and could rapidly differentiate tea samples obtained from diverse processing methods.

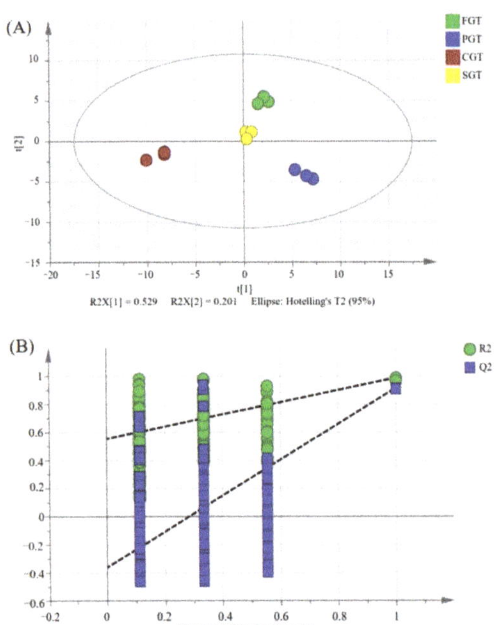

Figure 2. PLS-DA performed on the volatile profiles in tea samples with four distinct shaping techniques obtained by GC-E-nose. (**A**) Score plots of PLS-DA (R^2Y = 0.987, Q^2 = 0.909); (**B**) permutation test conducted 200 times with R^2 = (0.0, 0.561) and Q^2 = (0.0, −0.375). FGT represents flat green tea. PGT represents phoenix green tea. CGT represents curled green tea. SGT represents straight green tea.

3.3. Analysis of the Effects of Different Shaping Techniques on the Volatile Metabolites of Green Tea Using GC-MS/MS

3.3.1. Comparative Analysis of Volatile Metabolites

Clarifying the effect of diverse shaping techniques on volatile components will facilitate the improvement of the manufacturing process of green tea. In this study, a total of 60 volatile compounds were identified, as detailed in Table S4. These volatile compounds primarily comprised 16 aldehydes, 15 alcohols, 10 alkenes, 8 ketones, 5 esters, 3 heterocyclic compounds, 2 aromatic hydrocarbons, and 1 phenol, with aldehydes (26.67%) and alcohols (25.00%) representing the primary aroma categories (Figure 3A). Moreover, 54 common volatile compounds were detected across four shaping techniques (Figure 3B). Notably, 1-heptanol and citral were exclusively detected in the PGT, while 1-nonanal was solely detected in the FGT and PGT. Additionally, *cis*-2-penten-1-ol was detected in the FGT, SGT, and PGT, and acetic acid, butyl ester was present in all tea samples except for the SGT. Nerol was detected in all samples except for the FGT.

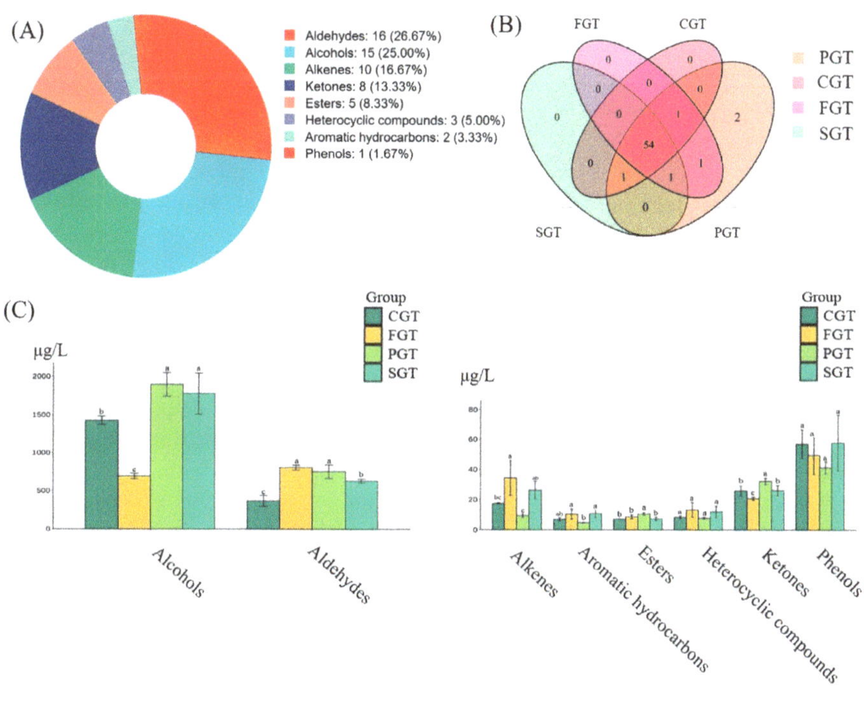

Figure 3. Volatile compounds in tea samples with four distinct shaping techniques analyzed by GC-MS/MS. (**A**) Pie chart of the proportion of various compounds; (**B**) Venn diagrams of volatile compounds; (**C**) Comparison of different categories of volatile compounds. Different letters indicate significant differences ($p < 0.05$). FGT represents flat green tea. PGT represents phoenix green tea. CGT represents curled green tea. SGT represents straight green tea.

Distinct shaping procedures exert a considerable influence on the contents of volatile compounds in green teas. As illustrated in Figure 3C, the contents of various categories of volatile compounds exhibited obvious differences under distinct shaping processes. Notably, alcohols attained the highest contents among the aroma components, endowing green tea with floral and fruity fragrances. Prior studies indicate that alcohols were the predominant compounds produced under various light irradiation spreading conditions, aligning with our findings [32]. Specifically, the content of alcohols was the highest under the PGT, reaching 1895.89 μg/L, whereas it was the lowest under the FGT, merely 691.21 μg/L. The variations in heat treatment processes significantly influence the Maillard reaction and the oxidative degradation of lipids, which in turn affects the production of alcohols during various shaping processes. For instance, phenylethyl alcohol, a notable alcohol compound, is produced through the thermal degradation of phenylalanine, whereas 1-octen-3-ol is generated from the oxidative degradation of the unsaturated fatty acid linoleic acid [2]. Aldehydes play an important role in the formation of green tea aroma. The content of aldehydes was the maximum in the FGT (806.68 μg/L) and the minimum in the CGT (368.11 μg/L). Olefins typically exhibit distinctive aromatic odors, possessing the highest content in the FGT (34.40 μg/L). The contents of olefin compounds for other shaping methods were ordered as follows: SGT (26.26 μg/L) > CGT (17.47 μg/L) > PGT (9.45 μg/L). The content of ketones was the highest in the PGT (32.33 μg/L) and the lowest in the FGT (20.72 μg/L), with minimal disparity between the CGT and SGT samples. The content of heterocyclic substances reached the highest level in the FGT (13.24 μg/L) and the lowest in the PGT (7.71 μg/L). The levels of aromatic hydrocarbons in the SGT (10.76 μg/L) and FGT (10.39 μg/L) were not significantly different, with the contents in CGT and PGT being 7.13 μg/L and 4.73 μg/L, respectively. The contents of esters under the four shaping methods were as follows: PGT (10.58 μg/L), FGT (8.68 μg/L), SGT (7.33 μg/L), and CGT (7.18 μg/L). The content of phenols reached the highest level in the SGT (57.96 μg/L) and the lowest in the PGT (41.40 μg/L). The aforementioned results underscored that different shaping procedures have a substantial impact on the number and contents of the aroma components of green tea.

3.3.2. Multivariate Statistical Analysis

To investigate the key differential components in tea samples under different shaping methods, PLS-DA was conducted. As depicted in Figure 4A, excellent discrimination was achieved. The PGT samples were located in the fourth quadrant of the score plots, whereas the CGT and SGT were distributed in the second quadrant. The FGT samples were mainly positioned in the third quadrant. The parameters of the PLS-DA model are presented as follows: the cumulative $R^2Y = 0.971$ and $Q^2 = 0.818$, indicating that the model displayed commendable explanatory and predictive capabilities. To further verify the stability of the model, 200 iterations of permutation testing were conducted. The obtained parameters were $R^2 = (0.0, 0.561)$ and $Q^2 = (0.0, -0.496)$ (Figure 4B), demonstrating that the model was dependable and not overfitted. Volatile compounds with VIP values greater than 1 are typically recognized as significant volatile compounds. As illustrated in Figure 4C, a total of 22 volatile compounds with VIP values exceeding 1 were identified. Adhering to the screening principles of $p < 0.05$ and VIP > 1, a total of 21 volatile compounds were selected as pivotal compounds for distinguishing tea samples under different shaping methods, with aldehydes, esters, and alcohols being the predominant volatiles. Furthermore, hierarchical cluster analysis was performed on the 21 key differential volatile compounds. As can be seen from Figure 5, volatiles such as methyl salicylate, γ-terpinene, and styrene possessed higher contents in the SGT compared to the other treatment groups. Nerolidol, acetic acid, butyl ester, dihydroactinidiolide, 1-heptanol, and 1-nonanol exhibited higher contents in the PGT than the other treatment groups. 2-Ethylfuran, α-calacorene, benzeneacetaldehyde, 1-nonanal, and α-muurolene exhibited higher contents in the FGT than the other treatment groups. Representative compounds such as 3-methylbutanal β-ionone, (*E*)-3-hexen-1-ol, and methyl geranate displayed higher contents in the CGT than the other treatment groups. Due to the disparity in technological parameters and so forth, diverse shaping processes exhibit distinct thermophysical and chemical effects, thereby inducing considerable variations in the levels of volatile metabolites.

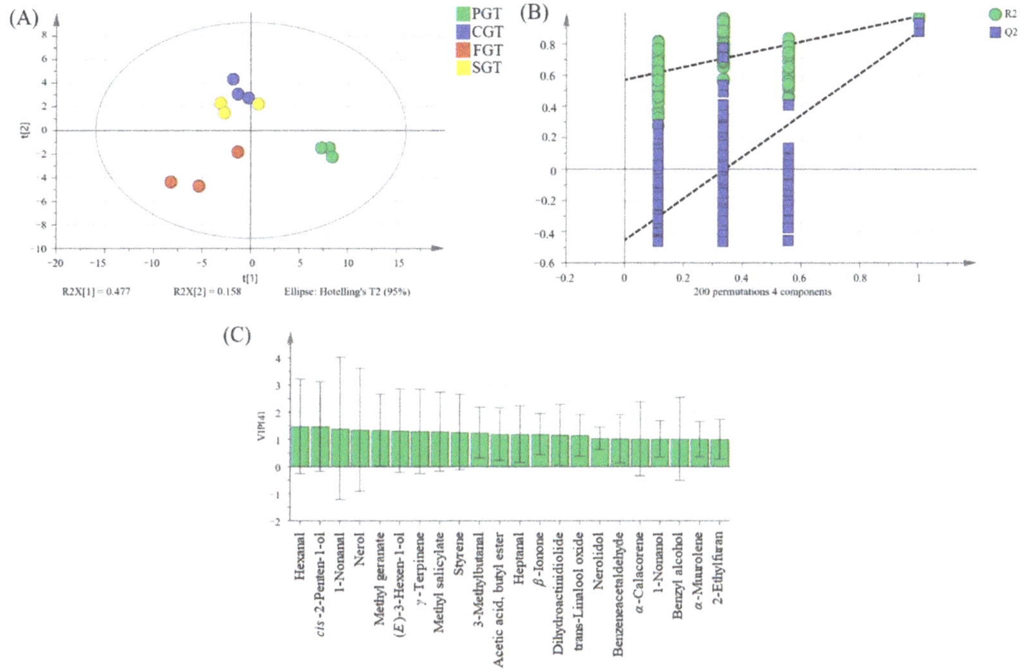

Figure 4. PLS-DA performed on the volatile compounds in tea samples with four distinct shaping techniques obtained from GC-MS/MS. (**A**) Score plot of PLS-DA ($R^2Y = 0.971$, $Q^2 = 0.818$); (**B**) permutation test conducted 200 times with $R^2 = (0.0, 0.561)$ and $Q^2 = (0.0, -0.496)$; (**C**) a total of 22 key differential volatile substances with VIP greater than 1. FGT represents flat green tea. PGT represents phoenix green tea. CGT represents curled green tea. SGT represents straight green tea.

3.3.3. OAV Analysis

Not all volatile compounds necessarily contribute to the fragrance of green tea. Hence, OAV analysis is indispensable to further ascertain those volatile constituents that significantly contribute to the overall aroma of diversely shaped samples. Generally speaking, when the OAV of a volatile compound is greater than 1, it can be deduced that this compound makes an important contribution to the overall aroma. The higher the OAV, the greater its impact on the aroma quality [34]. In this study, the odor thresholds and calculated OAVs of the volatile components of green tea with four different shaping techniques are listed in Table S5. A total of 20 volatile compounds with OAVs greater than 1 were screened out from the tea samples produced via the four shaping methods (Figure 6). Based on the magnitude of the OAVs, these compounds were primarily divided into two distinct groups. The first group encompassed volatiles, including 3-methylbutanal, 2-methylbutanal, heptanal, 1-octen-3-ol, 6-methyl-5-hepten-2-one, benzeneacetaldehyde, 1-octanol, 2-methoxyphenol, β-cyclocitral, nerol, geraniol, citral, *cis*-jasmone, and (*Z*)-2-heptenal, characterized by OAVs below 100. The second group typically featured volatile compounds such as β-ionone, hexanal, (*E*,*E*)-2,4-nonadienal, (*E*,*E*)-2,4-decadienal, linalool, and phenylethyl alcohol, which exhibited OAVs exceeding 100.

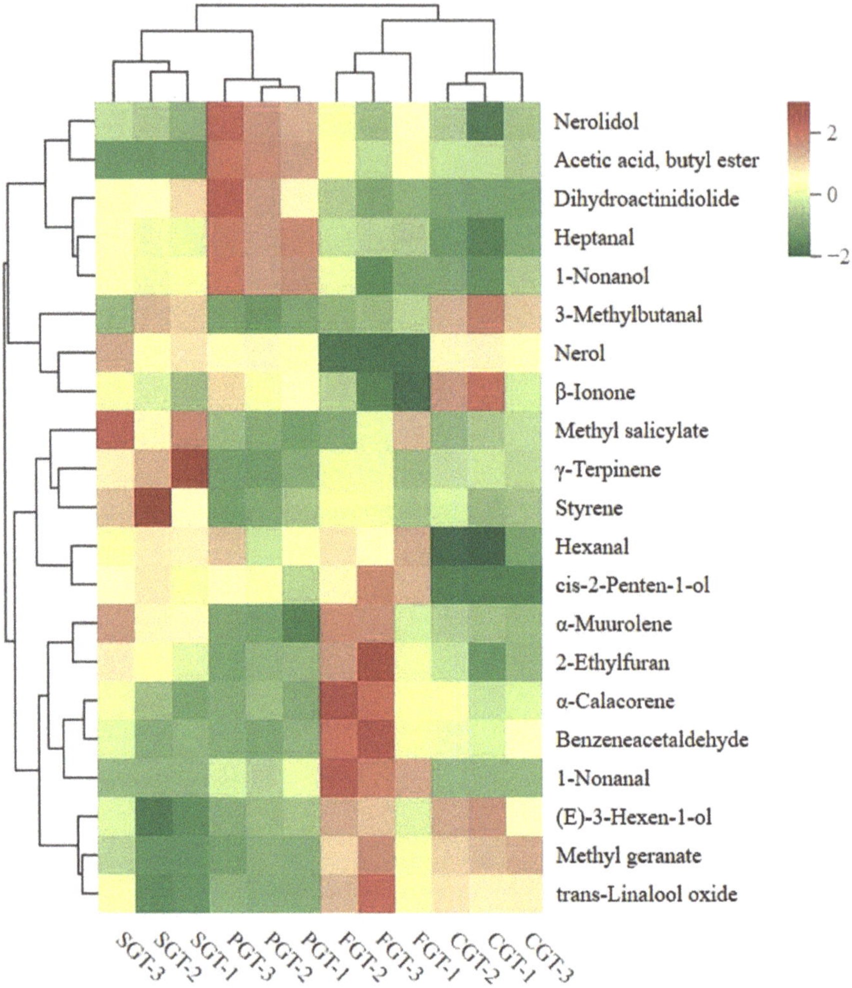

Figure 5. Hierarchical cluster analysis of 21 volatile compounds in tea samples with four distinct shaping techniques according to $p < 0.05$ and VIP > 1. FGT represents flat green tea. PGT represents phoenix green tea. CGT represents curled green tea. SGT represents straight green tea.

Among the volatiles with OAVs surpassing 100, the OAVs of (*E*,*E*)-2,4-nonadienal and (*E*,*E*)-2,4-decadienal displayed minimal fluctuations under the four distinct shaping methods. The OAV of hexanal in the PGT, FGT, and SGT samples was notably higher compared to CGT. β-Ionone is a pivotal ketone compound that affects the aroma profile of green tea and typically exhibits a floral fragrance [14]. In this study, its OAV reached the highest value in the CGT (115.05) and the lowest in the FGT (41.45). As a representative monoterpene alcohol in tea, linalool primarily presents citrus and floral scents. Its OAV attained the highest value in the SGT (211.39) and the lowest in the PGT (151.61). Phenylethyl alcohol also presents an appealing fruity and floral fragrance [2]. Its OAV surpassed 500 in all four shaping methods, attaining the maximum in the PGT (1654.82). To sum up, the diverse shaping techniques significantly influenced the OAVs of the aroma compounds, thereby influencing the manifestation of the fragrance style.

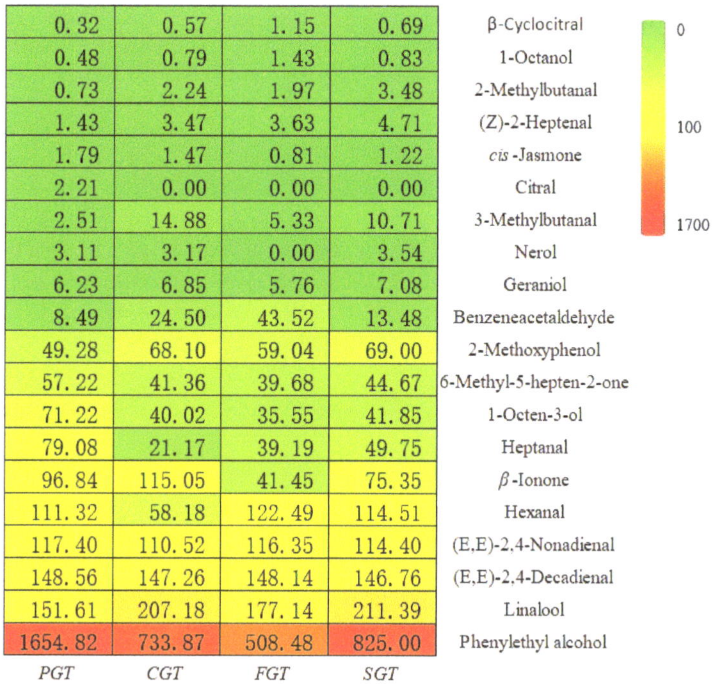

Figure 6. Comparison of 20 key odorants with OAV > 1 in tea samples with four distinct shaping techniques.

3.4. Discussion on the Potential Formation Pathways of Key Aroma Components in Green Tea with Different Shaping Techniques

The volatile compounds present in tea are intricately blended in specific proportions to generate its distinctive aroma. The mechanisms underlying the formation of these volatile compounds are highly complex. Following the metabolic pathways, the aforementioned 20 essential aroma components were primarily divided into four categories: carotenoid degradations (3), lipid oxidation degradations (8), glycosides as precursors (4), and Maillard reaction products (5) (Figure 7). The lipid degradation and Maillard reaction were the primary formation pathways for key aroma components of green teas subjected to four distinct shaping methods.

To be specific, β-cyclocitral, β-ionone, and 6-methyl-5-hepten-2-one were derived from the oxidative degradation of carotenoids [2]. β-Cyclocitral and β-ionone originated from the dioxygenase oxidation and cleavage of β-carotene. β-Cyclocitral exhibited the highest content in the FGT (3.44 μg/L) and the lowest in the PGT (0.95 μg/L) (Figure 7A). β-Ionone exhibited the lowest content in the FGT (0.41 μg/L), yet it displayed comparable levels in the PGT (0.97 μg/L) and CGT (1.15 μg/L). 6-Methyl-5-hepten-2-one originated from ξ-carotene, and it exhibited the highest content in the PGT (9.15 μg/L).

Volatiles such as heptanal, hexanal, 1-octen-3-ol,1-octanol, (Z)-2-heptenal, (*E*,*E*)-2,4-decadienal, (*E*,*E*)-2,4-nonadienal, and *cis*-jasmone originated from the oxidation degradation of unsaturated fatty acids such as α-linolenic acid, linoleic acid, and palmitoleic acid [34–36]. As shown in Figure 7B, (*E*,*E*)-2,4-decadienal, *cis*-jasmone, (*Z*)-2-heptenal, and hexanal originated from the precursor of α-linolenic acid, with the content of (*E*,*E*)-2,4-decadienal being comparable across the four shaping methods and the content of *cis*-jasmone exhibiting the highest level in the PGT and the lowest level in the FGT. The concentration of (Z)-2-heptenal was notably higher in the SGT, FGT, and CGT when compared to that in the PGT. Hexanal exhibited the highest content in the FGT (551.20 μg/L), followed by SGT (515.30 μg/L), PGT (500.94 μg/L), and CGT (261.81 μg/L). Both heptanal and 1-octanol originated from the oxidation degradation of palmitoleic acid. The content of heptanal in the PGT (71.17 μg/L), SGT (44.77 μg/L), and FGT (35.27 μg/L) was significantly higher than that in the CGT (19.05 μg/L), whereas 1-octanol exhibited the maximum content in the FGT (4.29 μg/L) and the minimum in the PGT (1.43 μg/L). The synthetic precursor of 1-octen-3-ol and (*E*,*E*)-2,4-nonadienal was linoleic acid, with the content of 1-octen-3-ol in the PGT (71.22 μg/L) being significantly higher than in the CGT (40.02 μg/L), FGT (35.55 μg/L), and SGT (41.85 μg/L) and the content of (*E*,*E*)-2,4-nonadienal exhibiting the highest level in the PGT (7.04 μg/L).

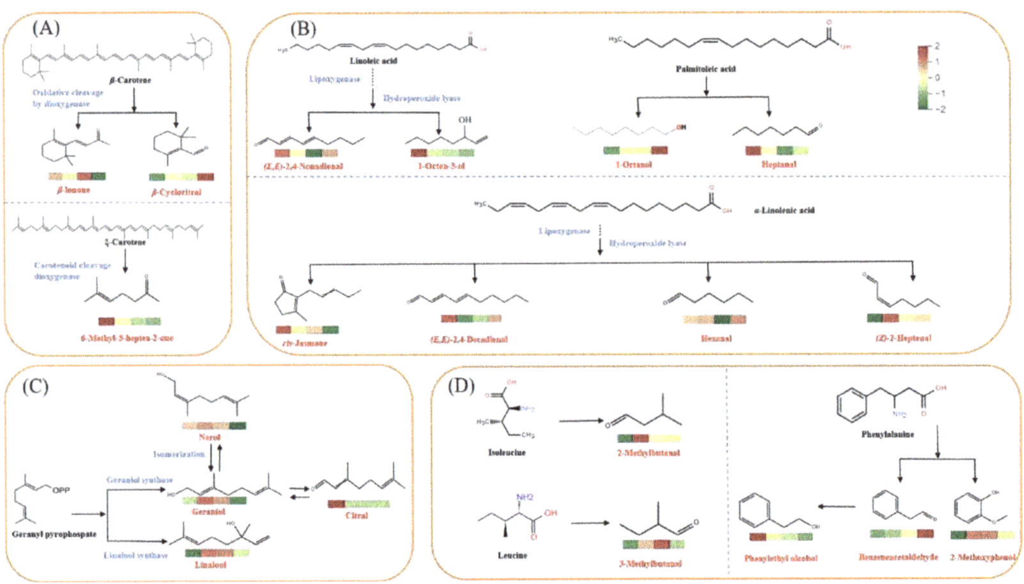

Figure 7. Content variations and key metabolic pathways of important compounds in tea samples with four distinct shaping techniques. (**A**) Carotenoid degradations; (**B**) Lipid oxidation degradations; (**C**) Glycosides as precursors; (**D**) Maillard reaction pathway. Different colored boxes represent the contents of metabolites in PGT, SGT, CGT, and FGT from left to right. Red and green colors indicated metabolite levels above and below the average, respectively.

Geraniol, linalool, nerol, and citral are glycoside-derived volatiles (Figure 7C). By employing geranyl pyrophosphate as the precursor substrate, geraniol and linalool were synthesized via the catalysis of geraniol synthase and linalool synthase, respectively [2,7]. Geraniol and linalool are prevalent volatile compounds in tea, exerting a pivotal influence on the development of tea aroma [19]. The content of linalool fluctuated slightly across the four shaping methods, around approximately 100 μg/L, whereas the content of geraniol varied around 48 μg/L. In addition, citral was generated by the oxidation of geraniol [37]. It is noteworthy that citral was exclusively identified in the PGT sample. Nerol was isomerized from geraniol and exhibited the highest level among the vital odorants, with its content being around 950 μg/L in the PGT, CGT, and SGT samples.

Benzeneacetaldehyde, phenylethyl alcohol, 3-methylbutanal, 2-methoxyphenol, and 2-methylbutanal were derived from the degradation of amino acids, with benzeneacetaldehyde, phenylethyl alcohol, and 2-methoxyphenol being derived from phenylalanine [2,7,38]. These volatiles play pivotal roles in generating fruity and floral odors. Benzeneacetaldehyde was produced from phenylalanine through Strecker degradation, and its content in the FGT (52.23 μg/L) was markedly higher than that for the other three shaping techniques (Figure 7D). Phenylethyl alcohol displayed the highest content in the PGT (579.19 μg/L), followed by the SGT (288.75 μg/L), CGT (256.86 μg/L), and FGT (177.97 μg/L) samples. 2-Methoxyphenol was a prevalent phenolic compound in tea, and its content showed no significant variations among the four shaping methods. 3-Methylbutanal was derived from the degradation of leucine, and its content was significantly higher in the CGT and SGT samples compared to that in the PGT and FGT samples. 2-Methylbutanal was derived from the degradation of isoleucine and possessed a content comparable with that of 3-methylbutanal in the PGT, FGT, and SGT samples.

4. Conclusions

This study investigated the effect of diverse shaping techniques on the aroma quality and volatile metabolites of green tea, employing advanced analytical techniques such as GC-E-Nose and GC-MS/MS. Our findings demonstrated that distinct shaping procedures notably influenced the overall aroma quality. The specific procedure designated as PGT yielded a superior aroma quality, characterized by a pleasant chestnut-like aroma. Through the application of GC-MS/MS, 60 volatile components were identified, of which 21 were selected as key differentiators among tea samples subjected to various shaping techniques, based on VIP > 1.0 and $p < 0.05$. Furthermore, 20 critical odorants with OAVs > 1.0 were pinpointed. Notably, (*E*,*E*)-2,4-decadienal, (*E*,*E*)-2,4-nonadienal, phenylethyl alcohol, and benzeneacetaldehyde were confirmed as substantial contributors responsible for the aroma of green tea. Lipid degradation and Maillard reactions were the primary pathways contributing to the generation of these pivotal odorants under diverse shaping conditions. Additionally, GC-E-Nose served as an effective tool for the rapid differentiation of green tea samples subjected to various shaping methods, complementing sensory evaluation. Future research will involve the use of stable isotope tracer technology to explore the mechanisms underlying the formation of these crucial aroma components across different shaping processes. Moreover, a molecular docking technique will be employed to elucidate the perceptual mechanisms of these vital aroma components at the molecular level. These findings provide a theoretical foundation for enhancing processing technologies related to green tea.

Supplementary Materials: The following supporting information can be downloaded at https://www.mdpi.com/article/10.3390/foods14050816/s1: Table S1. The standard information of volatile compounds used in this study; Table S2. The quantification information of the volatile compounds in tea samples with four distinct shaping techniques obtained from GC-MS/MS; Table S3. The aroma evaluation of green tea with different shaping techniques, Table S4. The volatile compounds in tea samples with four distinct shaping techniques obtained from GC-MS/MS; Table S5. The OAVs of volatile compounds in tea samples with four distinct shaping techniques [14,22,39,40]. Figure S1. The volatile fingerprints of green tea with different shaping techniques obtained from the MXT-5 and MXT-1701 columns in parallel; Figure S2. A total of 23 variables with VIP greater than 1obtained from GC-E-Nose.

Author Contributions: Conceptualization, Y.W. and Y.Y.; Methodology, J.T. and Y.Y.; Software, J.T., J.H. and X.Z.; Validation, J.T. and J.H.; Formal analysis, J.T., J.H., X.Z. and Q.W.; Resources, Q.W. and Y.Y.; Data curation, J.T.; Writing—original draft, J.T.; Writing—review & editing, Y.Y.; Supervision, Y.J., H.Y., Y.W. and Y.Y.; Project administration, Y.J. and H.Y.; Funding acquisition, H.Y. All authors have read and agreed to the published version of the manuscript.

Funding: This research was funded by the Science and Technology Innovation Project of the Chinese Academy of Agricultural Sciences, grant number CAAS-ASTIP-TRICAAS, Guizhou Provincial Science and Technology Support Project, grant number Qiankehe Support [2023] General 069.

Institutional Review Board Statement: Not applicable.

Informed Consent Statement: Informed consent was obtained from all subjects involved in the study.

Data Availability Statement: The original contributions presented in this study are included in the article/Supplementary Materials. Further inquiries can be directed to the corresponding authors.

Conflicts of Interest: The authors declare that they have no known competing financial interests or personal relationships that could have appeared to influence the work reported in this paper.

References

1. Yang, Y.; Qian, M.C.; Deng, Y.; Yuan, H.; Jiang, Y. Insight into aroma dynamic changes during the whole manufacturing process of chestnut-like aroma green tea by combining GC-E-Nose, GC-IMS, and GC × GC-TOFMS. *Food Chem.* **2022**, *387*, 132813. [CrossRef]
2. Ho, C.-T.; Zheng, X.; Li, S. Tea aroma formation. *Food Sci. Hum. Wellness* **2015**, *4*, 9–27. [CrossRef]
3. Zhang, J.; Feng, W.; Xiong, Z.; Dong, S.; Sheng, C.; Wu, Y.; Deng, G.; Deng, W.-W.; Ning, J. Investigation of the Effect of Over-Fired Drying on the Taste and Aroma of Lu'an Guapian Tea Using Metabolomics and Sensory Histology Techniques. *Food Chem.* **2023**, *437*, 137851. [CrossRef]
4. Zeng, L.; Fu, Y.-Q.; Gao, Y.; Wang, F.; Liang, S.; Yin, J.-F.; Fauconnier, M.-L.; Ke, L.; Xu, Y.-Q. Dynamic changes of key metabolites in Longjing green tea during processing revealed by widely targeted metabolomic profiling and sensory experiments. *Food Chem.* **2024**, *450*, 139373. [CrossRef] [PubMed]
5. Ge, Y.; Wang, L.; Huang, Y.; Jia, L.; Wang, J. Characteristic flavor compounds in Guizhou green tea and the environmental factors influencing their formation: Investigation using stable isotopes, electronic nose, and headspace-gas chromatography ion migration spectrometry. *LWT* **2024**, *196*, 115887. [CrossRef]
6. Wang, Q.; Xie, J.; Wang, L.; Jiang, Y.; Deng, Y.; Zhu, J.; Yuan, H.; Yang, Y. Comprehensive investigation on the dynamic changes of volatile metabolites in fresh scent green tea during processing by GC-E-Nose, GC–MS, and GC × GC-TOFMS. *Food Res. Int.* **2024**, *187*, 114330. [CrossRef] [PubMed]
7. Yin, P.; Kong, Y.-S.; Liu, P.-P.; Wang, J.-J.; Zhu, Y.; Wang, G.-M.; Sun, M.-F.; Chen, Y.; Guo, G.-Y.; Liu, Z.-H. A critical review of key odorants in green tea: Identification and biochemical formation pathway. *Trends Food Sci. Technol* **2022**, *129*, 221–232. [CrossRef]
8. Yang, Y.; Xie, J.; Chen, J.; Deng, Y.; Shen, S.; Hua, J.; Wang, J.; Zhu, J.; Yuan, H.; Jiang, Y. Characterization of N,O-heterocycles in green tea during the drying process and unraveling the formation mechanism. *Food Control* **2022**, *139*, 109079. [CrossRef]
9. Yang, Y.; Zhang, M.; Hua, J.; Deng, Y.; Jiang, Y.; Li, J.; Wang, J.; Yuan, H.; Dong, C. Quantitation of pyrazines in roasted green tea by infrared-assisted extraction coupled to headspace solid-phase microextraction in combination with GC-QqQ-MS/MS. *Food Res. Int.* **2020**, *134*, 109167. [CrossRef]
10. Yang, Y.; Chen, J.; Jiang, Y.; Qian, M.C.; Deng, Y.; Xie, J.; Li, J.; Wang, J.; Dong, C.; Yuan, H. Aroma dynamic characteristics during the drying process of green tea by gas phase electronic nose and gas chromatography-ion mobility spectrometry. *LWT* **2022**, *154*, 112691. [CrossRef]
11. Wang, Q.; Shi, D.; Hu, J.; Tang, J.; Zhou, X.; Wang, L.; Xie, J.; Jiang, Y.; Yuan, H.; Yang, Y. Unraveling the dynamic changes of volatile compounds during the rolling process of Congou black tea via GC-E-nose and GC–MS. *Front. Sustain. Food Syst.* **2024**, *8*, 1436542. [CrossRef]
12. Yang, Y.; Xie, J.; Wang, Q.; Deng, Y.; Zhu, L.; Zhu, J.; Yuan, H.; Jiang, Y. Understanding the dynamic changes of volatile and non-volatile metabolites in black tea during processing by integrated volatolomics and UHPLC-HRMS analysis. *Food Chem.* **2024**, *432*, 137124. [CrossRef]
13. Zhu, J.; Cao, X.; Niu, Y.; Xiao, Z. Investigation of Lactone Chiral Enantiomers and Their Contribution to the Aroma of Longjing Tea by Odor Activity Value and S-Curve. *J. Agric. Food Chem.* **2023**, *71*, 6691–6698. [CrossRef] [PubMed]

14. Zhai, X.; Zhang, L.; Granvogl, M.; Ho, C.-T.; Wan, X. Flavor of tea (*Camellia sinensis*): A review on odorants and analytical techniques. *Compr. Rev. Food Sci. Food Saf.* **2022**, *21*, 3867–3909. [CrossRef] [PubMed]
15. Yang, Y.; Xie, J.; Wang, Q.; Wang, L.; Shang, Y.; Jiang, Y.; Yuan, H. Volatolomics-assisted characterization of the key odorants in green off-flavor black tea and their dynamic changes during processing. *Food Chem. X* **2024**, *22*, 101432. [CrossRef]
16. Jiang, G.; Xue, R.; Xiang, J.; Wang, Y.; Liu, B.; Yuan, Y.; Pu, Q.; Fang, X.; Hu, X.; Liu, X.; et al. Dynamic changes in the aroma profiles and volatiles of Enshi Yulu tea throughout its industrial processing. *Food Chem.* **2024**, *458*, 140145. [CrossRef] [PubMed]
17. Wang, L.; Xie, J.; Wang, Q.; Hu, J.; Jiang, Y.; Wang, J.; Tong, H.; Yuan, H.; Yang, Y. Evaluation of the quality grade of Congou black tea by the fusion of GC-E-Nose, E-tongue, and E-eye. *Food Chem. X* **2024**, *23*, 101519. [CrossRef] [PubMed]
18. Wang, L.; Xie, J.; Miao, Y.; Wang, Q.; Hu, J.; Jiang, Y.; Wang, J.; Tong, H.; Yuan, H.; Yang, Y. Exploration of the effects of geographical regions on the volatile and non-volatile metabolites of black tea utilizing multiple intelligent sensory technologies and untargeted metabolomics analysis. *Food Chem. X* **2024**, *23*, 101634. [CrossRef]
19. Wei, Y.; Zhang, J.; Li, T.; Zhao, M.; Song, Z.; Wang, Y.; Ning, J. GC–MS, GC–O, and sensomics analysis reveals the key odorants underlying the improvement of yellow tea aroma after optimized yellowing. *Food Chem.* **2024**, *431*, 137139. [CrossRef]
20. Sun, Z.; Lin, Y.; Yang, H.; Zhao, R.; Zhu, J.; Wang, F. Characterization of honey-like characteristic aroma compounds in Zunyi black tea and their molecular mechanisms of interaction with olfactory receptors using molecular docking. *LWT* **2024**, *191*, 115640. [CrossRef]
21. Wang, S.; Chen, H.; Sun, B. Recent progress in food flavor analysis using gas chromatography-ion mobility spectrometry (GC-IMS). *Food Chem.* **2020**, *315*, 126158. [CrossRef] [PubMed]
22. Xie, J.; Wang, L.; Deng, Y.; Yuan, H.; Zhu, J.; Jiang, Y.; Yang, Y. Characterization of the key odorants in floral aroma green tea based on GC-E-Nose, GC-IMS, GC-MS and aroma recombination and investigation of the dynamic changes and aroma formation during processing. *Food Chem.* **2023**, *427*, 136641. [CrossRef] [PubMed]
23. Liang, Y.; Wang, Z.; Zhang, L.; Dai, H.; Wu, W.; Zheng, Z.; Lin, F.; Xu, J.; Huang, Y.; Sun, W. Characterization of volatile compounds and identification of key aroma compounds in different aroma types of Rougui Wuyi rock tea. *Food Chem.* **2024**, *455*, 139931. [CrossRef] [PubMed]
24. Fang, X.; Xu, W.; Jiang, G.; Sui, M.; Xiao, J.; Ning, Y.; Niaz, R.; Wu, D.; Feng, X.; Chen, J.; et al. Monitoring the dynamic changes in aroma during the whole processing of Qingzhuan tea at an industrial scale: From fresh leaves to finished tea. *Food Chem.* **2024**, *439*, 137810. [CrossRef] [PubMed]
25. Jin, G.; Zhu, Y.; Cui, C.; Yang, C.; Hu, S.; Cai, H.; Ning, J.; Wei, C.; Li, A.; Hou, R. Tracing the origin of Taiping Houkui green tea using 1H NMR and HS-SPME-GC–MS chemical fingerprints, data fusion and chemometrics. *Food Chem.* **2023**, *425*, 136538. [CrossRef] [PubMed]
26. Fang, X.; Liu, Y.; Xiao, J.; Ma, C.; Huang, Y. GC–MS and LC-MS/MS metabolomics revealed dynamic changes of volatile and non-volatile compounds during withering process of black tea. *Food Chem.* **2023**, *410*, 135396. [CrossRef]
27. Zhan, S.; Liu, Z.; Su, W.; Lin, C.-C.; Ni, L. Role of roasting in the formation of characteristic aroma of wuyi rock tea. *Food Control* **2023**, *147*, 109614. [CrossRef]
28. Chen, J.; Yang, Y.; Deng, Y.; Liu, Z.; Xie, J.; Shen, S.; Yuan, H.; Jiang, Y. Aroma quality evaluation of Dianhong black tea infusions by the combination of rapid gas phase electronic nose and multivariate statistical analysis. *LWT* **2022**, *153*, 112496. [CrossRef]
29. Wang, S.; Zhang, Q.; Liu, C.; Wang, Z.; Gao, J.; Yang, X.; Lan, Y. Synergetic application of an E-tongue, E-nose and E-eye combined with CNN models and an attention mechanism to detect the origin of black pepper. *Sens. Actuators A Phys.* **2023**, *357*, 114417. [CrossRef]
30. Xu, M.; Wang, J.; Zhu, L. The qualitative and quantitative assessment of tea quality based on E-nose, E-tongue and E-eye combined with chemometrics. *Food Chem.* **2019**, *289*, 482–489. [CrossRef] [PubMed]
31. *GB/T 23776-2018*; Methodology for Sensory Evaluation of Tea. Chinese National Standard: Beijing, China, 2018.
32. Xie, J.; Wang, Q.; Hu, J.; Wang, L.; Yu, X.; Yuan, H.; Jiang, Y.; Yang, Y. Uncovering the effects of spreading under different light irradiation on the volatile and non-volatile metabolites of green tea by intelligent sensory technologies integrated with targeted and non-targeted metabolomics analyses. *Food Chem.* **2025**, *463*, 141482. [CrossRef]
33. Wang, C.; Yue, Y.; Yuan, B.; Deng, Q.; Liu, Y.; Zhou, Q. Identification of the key aroma compounds in flaxseed milk using stir bar sorptive extraction, aroma recombination, and omission tests. *Food Chem.* **2024**, *446*, 138782. [CrossRef]
34. Wang, C.; Li, J.; Wu, X.; Zhang, Y.; He, Z.; Zhang, Y.; Zhang, X.; Li, Q.; Huang, J.; Liu, Z. Pu-erh tea unique aroma: Volatile components, evaluation methods and metabolic mechanism of key odor-active compounds. *Trends Food Sci. Technol.* **2022**, *124*, 25–37. [CrossRef]
35. Yang, Z.; Baldermann, S.; Watanabe, N. Recent studies of the volatile compounds in tea. *Food Res. Int* **2013**, *53*, 585–599. [CrossRef]
36. Huang, F.-F.; Yang, P.-D.; Bai, S.-L.; Liu, Z.-H.; Li, J.; Huang, J.-A.; Xiong, L.-G. Lipids: A noteworthy role in better tea quality. *Food Chem.* **2024**, *431*, 137071. [CrossRef]

37. Yang, Y.; Wang, Q.; Xie, J.; Deng, Y.; Zhu, J.; Xie, Z.; Yuan, H.; Jiang, Y. Uncovering the Dynamic Alterations of Volatile Components in Sweet and Floral Aroma Black Tea during Processing. *Foods* **2024**, *13*, 728. [CrossRef]
38. Wang, H.; Chen, L.; Xu, A.; Zhao, Y.; Wang, Y.; Liu, Z.; Xu, P. Thermochemical reactions in tea drying shape the flavor of tea: A review. *Food Res. Int* **2024**, *197*, 115188. [CrossRef] [PubMed]
39. Van Gemert, L.J. *ODOUR THRESHOLDS—Compilations of Odour Threshold Values in Air, Water and Other Media*; Oliemans Punter & Partners BV: Zeist, The Netherlands, 2011.
40. Guo, X.; Ho, C.-T.; Wan, X.; Zhu, H.; Liu, Q.; Wen, Z. Changes of volatile compounds and odor profiles in Wuyi rock tea during processing. *Food Chem.* **2021**, *341*, 128230. [CrossRef]

foods

MDPI

Article

The Combination of Shaking and Yellow-Light Withering Promote the Volatile Aroma Components and the Aroma Quality of Black Tea

Zeyi Ai, Shuangming Hu, Lingfei Ji, Bing Mu and Yiyang Yang *

Jiangsu Key Laboratory for Horticultural Crop Genetic Improvement, Institute of Leisure Agriculture, Jiangsu Academy of Agricultural Sciences, Nanjing 210014, China; aizeyi@jaas.ac.cn (Z.A.); lingfeiji@jaas.ac.cn (L.J.)

* Correspondence: yiyangyang@jaas.ac.cn; Tel./Fax: +86-025-84391693

Abstract: The application of shaking during the withering process has been shown to significantly enhance the floral aroma of black tea. However, prior to this study, there was limited research on the effects of shaking combined with other withering treatments on the aroma components of black tea. In this study, the aroma attributes of black teas processed with shaking combinations of yellow-light or high-temperature withering (YLS, HTS, and HYS) were evaluated through sensory evaluation, and the volatile composition and identification of key aroma compounds in black teas and in-process withered tea leaves were analyzed by gas chromatography–mass spectrometry (GC-MS). The results indicated that black teas subjected to different shaking combinations exhibited a distinct floral aroma with higher volatile compound content, with the YLS treatment showing the most significant aroma improvement. Eight volatile compounds with relative odor activity values (rOAV) > 1 were identified from 53 differential volatile compounds in black tea under different shaking combinations as the most important contributors to aroma quality. Linalool, trans-β-Ionone, α-cedrene, and nonanal were identified as key floral volatiles with high rOAVs. Their concentrations in YLS were notably higher compared to those in S, both in withered tea leaves (with the exception of trans-β-Ionone) and in the final dry black teas, suggesting that there may be a synergistic effect between the yellow-light withering and the shaking process in enhancing these key floral compounds. Overall, this study found that shaking combined with yellow-light withering can improve the aroma composition and quality of black tea, providing a theoretical basis and practical guidance for the production and optimization of high-aroma black tea.

Keywords: black tea; shaking combinations; yellow-light withering; volatile substance; aroma improvement

1. Introduction

Black tea, renowned globally for its distinct, diverse, and rich aroma, is the most widely produced and consumed tea type worldwide [1]. The aroma of black tea is characterized by sweet, floral, honey, and caramel-like notes [2,3]. Aroma performance is constituted by the mixture of multiple volatile flavor compounds (VFCs) [4,5], making the study of VFCs a focal point in tea research. To date, over 600 aroma components have been identified in black tea, primarily consisting of aldehydes, alcohols, esters, alkanes, and ketones [6]. Among these, geraniol (rose floral), trans-β-ionone (floral), methyl salicylate (floral), (Z/E)-linalool oxides (floral), linalool (sweet floral), phenylacetaldehyde (honey), and β-damascenone (sweet, honey-like) amongst others are recognized as the predominant aroma-active compounds [7,8], which are primarily generated during tea processing.

The processing of black tea involves several stages: withering, rolling/rolling and cutting, fermentation, and drying. Withering, the initial step, plays a crucial role in shaping the overall aroma quality of black tea [9,10]. Appropriate withering techniques can promote the favorable transformation of volatile aroma compounds in fresh tea leaves, thereby enhancing the tea's aroma quality [11]. With continuous innovation in withering technology, many black-tea-producing regions in China have incorporated shaking into the withering process to enrich flavor, resulting in what is known as shaken black tea (SBT) [12]. Shaking involves rotating withered leaves using a water sieve or shaker machine. Studies have shown that shaking facilitates friction-induced damage along the leaf edges, stimulating enzyme activities such as oxidase, hydrolase, and reactive oxygen species metabolism enzymes [13], thereby promoting the synthesis of floral compounds like indole, methyl jasmonate, and β-ionone, and significantly enhancing the floral flavor of black tea [14,15].

Additionally, previous research demonstrated that yellow, orange, and red-light withering can significantly improve the aroma and taste quality of black tea, with yellow light particularly effective in increasing the content of VFCs such as linalool and geraniol [16]. Numerous studies also highlighted the role of light quality in tea spreading and withering processes. For instance, yellow light combined with low-temperature treatment markedly enhanced the synthesis and accumulation of methyl salicylate and indole during green tea spreading [17], while red-light withering primarily influenced the accumulation of terpene volatile components in the later stages of black tea withering by upregulating key synthase gene expression [18]. Furthermore, warm-air withering is commonly used in black tea production to accelerate water loss and shorten withering time [19], though there is limited research on the effects of high-temperature withering on black tea aroma components.

While existing research has primarily focused on the impact of individual technologies on black tea quality, there is a lack of systematic investigation into the combined effects of multiple withering technologies, particularly on aroma quality. The influence of different combinations of yellow light, high temperature, and shaking on the metabolism of volatile compounds in black tea remains unclear. Therefore, this study explores the effects of different shaking combinations of yellow-light or high-temperature withering on the volatile aroma compounds of black tea, aiming to provide theoretical and practical insights for the application of various withering techniques in black tea production and to lay the groundwork for optimizing processing technologies for high-quality black tea.

2. Materials and Methods

2.1. Reagents and Equipment

Ethyl decanoate (≥99% purity) was purchased from Macklin Biochemical Technology Co., Ltd. (Shanghai, China). A C3-C25 N-alkane mixture was obtained from o2si smart solutions International LLC (North Charleston, SC, USA). The TSQ 8000 EVO gas chromatograph-mass spectrometer (GC-MS) was purchased from Thermo Fisher Scientific, Inc. (Waltham, MA, USA). Supelco SPME fiber (PDMS/DVB 65 μm) was purchased from Merck KGaA (Darmstadt, Germany). The LCWD-5 withering machine, 6CWL-90 shaker machine, and 6CFJ-0.7 fermentation machine were obtained from Fujian Jiayou Tea Machinery Intelligent Technology Co., Ltd. (Anxi, China). Yellow light-emitting diode (LED) lamps (cold light source, 585–590 nm, peak wavelength 587.9 nm) and the LI-250A light meter were purchased from Zhongshan Light of Sea Emperor Co., Ltd. (Zhongshan, China) and LI-COR Biosciences Co., Ltd. (Lincoln, NE, USA).

2.2. Tea Sample Processing

Fresh tea leaves with one bud and two leaves were harvested from the Lj43 cultivar (*Camellia sinensis var. sinensis cv. Lj43*) in modern tea plantations in Pukou County at 32.101216° N, 118.657506° E (Nanjing City, Jiangsu Province, China) in July 2023. The tea shoots were subjected to five different withering treatments: indoor natural withering (CK), shaking (S), yellow light plus shaking (YLS), high temperature plus shaking (HTS), and high temperature with yellow light plus shaking (HYS). For the CK treatment, tea leaves were withered in a withering machine (LCWD-5) for 12 h at 25 °C and 60% relative humidity (RH) with natural light intensity below 200 lux. For the S treatment, after 12 h of indoor natural withering, the tea leaves were shaken three times using a shaker machine (6CWY-90) for 10 min each, with 15 min intervals between shakes. For the other treatments, tea leaves were withered under yellow-light irradiation (YLS, 25 °C, 60% RH, 844–864 lux), at a high room temperature (HTS, 32 °C, 60% RH, <200 lux), or at a high room temperature with yellow-light irradiation (HYS, 32 °C, 60% RH, 844–864 lux) before shaking. The shaking parameters were consistent with those of the S treatment. During the 12 h withering process, the leaves were stirred every hour to ensure uniform light exposure. After five different treatments, the tea leaves underwent further withering by fan in withering machine (LCWD-5) until their moisture content decreased to 60.0 ± 1.0%, respectively. Subsequently, half of the withered leaves were frozen in liquid nitrogen and stored at −80 °C for subsequent experiments, while the other half were processed into black tea through rolling and cutting (2 times), fermentation (30 °C, 90% RH for 3 h), and drying (110 °C for 30 min, then 80 °C until the moisture content reached approximately 6%). Each treatment was repeated three times, and the experimental processing flowchart is shown in Figure 1.

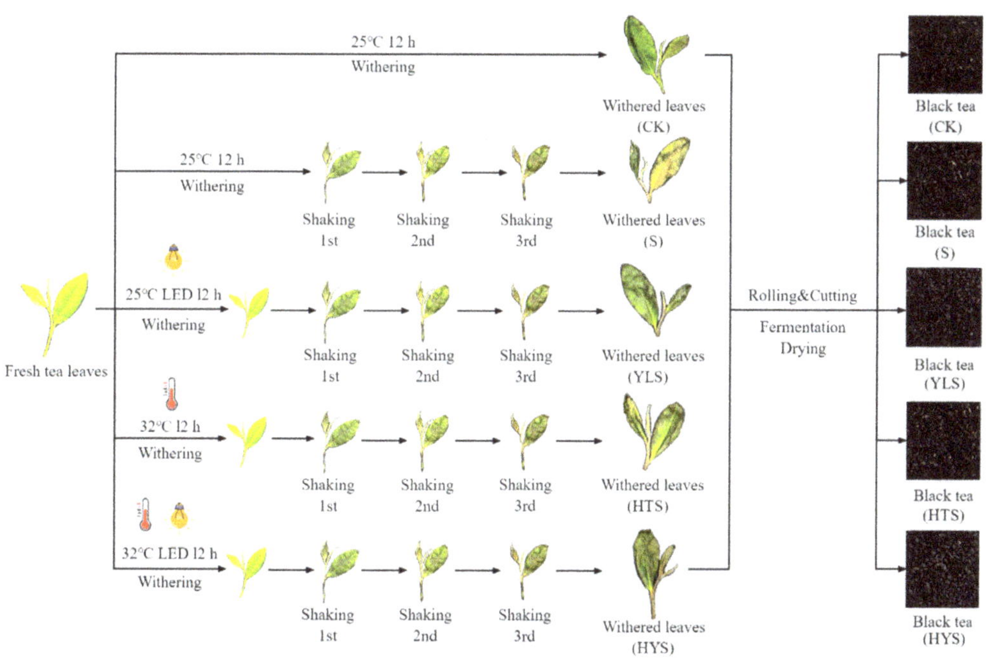

Figure 1. The experimental processing flowchart of black tea.

2.3. Sensory Evaluation

Black tea samples were prepared from tea leaves subjected to the five postharvest treatments (CK, S, YLS, HTS, and HYS). Sensory evaluation was conducted by five professional tea tasters with Senior Tea Assessor qualifications (China National Professional Qualification Level 3). Following the Chinese Tea Evaluation Standard Procedure (GB/T 23776-2018), the appearance of dry tea was evaluated first. After brewing 3.0 g of dry tea with 150 mL of boiling water for 5 min, the brew color, aroma, taste, and infused leaves were assessed. Scores were based on a 100-point scale, with appearance accounting for 20%, infusion color for 10%, aroma for 30%, taste for 30%, and infused leaves for 10% [20]. Sensory assessments were conducted blindly and independently by the five panelists.

2.4. Extraction and Analysis of Volatile Compounds

Volatile compounds were extracted using the headspace solid-phase microextraction (HS-SPME) method. For withered tea leaves, 2.000 g of freeze-dried powder was weighed into a 20 mL headspace bottle, and 100 μL of ethyl decanoate (1 μL/L) was added as an internal standard. The bottle was sealed, and an SPME fiber (65 μm PDMS/DVB) was inserted and heated in a 60 °C water bath for 60 min to adsorb the aroma components. For black tea, 2.000 g of crushed tea was mixed with 10 mL of boiling distilled water and 100 μL of ethyl decanoate (4 μL/L, internal standard) and extracted using a 50/30 μm CAR/DVB/PDMS SPME fiber for 60 min in a 60 °C water bath [18]. Each sample was analyzed in triplicate.

Volatile compounds were analyzed using a GC-MS. After extraction, the fiber was immediately inserted into the GC injector for thermal desorption at 250 °C for 5 min. The analysis was performed under the following conditions: chromatographic column: TG-5MS (30 m × 0.25 mm × 0.25 μm, Thermo Fisher Scientific, USA); injection mode: splitless; carrier gas: high-purity helium (purity $\geq$ 99.99%); column flow rate: 1.2 mL/min; inlet temperature: 250 °C; heating program: initial temperature of 35 °C held for 2 min, increased to 180 °C at 4 °C/min and held for 2 min, then increased to 280 °C at 15 °C/min and held for 5 min; ion source (EI) electron energy: 70 eV; ion source temperature: 300 °C; mass scanning range (m/z):35–400 amu. Volatiles were identified by matching the National Institute of Standards and Technology mass spectral database (NIST 17, USA) and retention index (RI) [21]. The RI for each compound was calculated using the linear formula for C3-C25 N-alkanes. Compounds were further screened based on the retention index difference compared to the NIST database.

2.5. Quantification and Relative Odor Activity Value (rOAV) Calculation

Xcalibur software (Thermo Fisher Scientific, Waltham, MA, USA) was used for GC-MS data analysis, including peak extraction, baseline filtering, calibration, peak alignment, deconvolution, and peak area integration. Volatile compounds were semi-quantified based on the peak areas of the internal standard. rOAVs were calculated by dividing the content of each volatile compound by its odor threshold in water [22]. Odor thresholds (OTs) in water were based on the literature [5,15,23,24].

2.6. Data Processing Method and Statistical Analysis

All data are presented as means ± standard deviation (SD). Significant differences were determined using one-way analysis of variance (ANOVA) in SPSS 20.0 (SPSS, Inc., Chicago, IL, USA) based on Duncan's multiple-range test ($p < 0.05$). Partial least squares discriminant analysis (PLS-DA) was performed using MetaboAnalyst 3.0 (https://www.metaboanalyst.ca/ accessed on 18 October 2023). Figures were generated using Processon (https://www.processon.com/ accessed on 23 August 2024), GraphPad Prism 9 (GraphPad Software company, Boston, MA, USA) and Adobe Illustrator CC 2019 (Adobe Systems Incorporated, San Jose, CA, USA).

3. Results and Discussion

3.1. Sensory Quality of Black Tea Under Different Shaking Combinations

Sensory evaluation revealed that shaking and shaking-involved combinations significantly influenced the aroma, taste, and infusion color of black tea, with less impact on appearance and infused leaf quality. As shown in Table 1, the brightness of infusion color under high-temperature withering combinations (HTS and HYS) were both lower than that of CK, suggesting that high-temperature withering may negatively affect infusion color. Compared to CK, all shaking combinations positively influenced black tea taste, except for HYS, which exhibited a less mellow taste. Regarding aroma quality, all shaking combinations resulted in a distinct floral aroma alongside a sweet aroma. In contrast to a slightly floral and sweet aroma of HTS and HYS, the floral and sweet aroma of YLS and S was stronger and more persistent. It indicated that shaking is crucial for forming the floral aroma of black tea, while additional withering treatments (yellow light or high temperature) can influence aroma intensity and persistence instead of the aroma type. The combination of yellow-light withering and shaking (YLS) achieved the highest aroma score, suggesting that yellow-light withering may further enhance the aroma quality of shaken black tea.

Table 1. Sensory evaluation of black tea under different shaking combinations.

Sample	Appearance	Infusion Color	Aroma	Taste	Infused Leaf	Total Score
CK	90.00 ± 0.00 a	90.23 ± 0.21 a	85.00 ± 0.41 d	88.23 ± 0.21 c	90.00 ± 0.00 a	88.24 ± 0.13 d
S	90.00 ± 0.00 a	90.32 ± 0.13 a	88.23 ± 0.21 b	90.17 ± 0.12 b	90.00 ± 0.00 a	89.64 ± 0.06 b
YLS	90.00 ± 0.00 a	90.67 ± 0.24 a	90.57 ± 0.33 a	92.23 ± 0.21 a	90.00 ± 0.00 a	90.88 ± 0.09 a
HTS	90.00 ± 0.00 a	88.50 ± 0.41 b	87.33 ± 0.24 c	90.07 ± 0.09 b	90.00 ± 0.00 a	89.20 ± 0.11 c
HYS	90.00 ± 0.00 a	84.23 ± 0.21 c	88.07 ± 0.09 b	87.13 ± 0.09 d	90.00 ± 0.00 a	88.08 ± 0.07 d

The numerical values presented in this table represent the evaluation scores (mean ± SD, n = 5 panelists)., which are graded on a 100-point scale. The total scores were computed with the following weightings: appearance constitutes 20%, infusion color accounts for 10%, aroma makes up 30%, taste represents 30%, and infused leaves contribute 10% [21]. Treatments: CK represents indoor natural withering, S represents shaking, YLS represents yellow light plus shaking, HTS represents high temperature plus shaking, HYS represents high temperature with yellow light plus shaking. Different letters in the same column indicate a significant difference between different withering treatments ($p < 0.05$).

3.2. Volatile Compound Analysis of Black Tea Under Different Shaking Combinations

3.2.1. Qualitative and Quantitative Analysis of Volatile Compounds

A total of 53 volatile components were identified across different black tea samples (Table S1), including fourteen alcohols, twelve olefins, eleven esters, nine aldehydes, two ketones, one ether, one organosulfur compound, one aromatic compound, one heterocyclic compound, and one acid. The addition of shaking or shaking combinations during withering increased the number of volatile substances in black tea, with the YLS treatment group showing the highest number of volatile compounds (Figure 2A). Venn diagram analysis revealed 30 common volatile compounds across all samples, with six overlapping volatiles in four shaking-involved treatments, including benzyl alcohol, cis-3-hexenyl isovalerate, cedryl acetate, β-copaene, D-limonene, and 2-pentyl-furan (Figure 2B). Five components, such as β-guaiene, α-alaskene, 10-undecenyl-3-phenylpropanoate, 2-ethyl-1-hexanol, and geranyl vinyl ether, were unique to YLS, while geranyl isovalerate and trans-2-hexenoic acid were unique to HTS. No unique substances were detected in S and HYS. These findings suggest that shaking increases the variety of volatile compounds, and additional withering treatments (YLS and HTS) can further enhance this effect, thus changing the volatile composition of black tea.

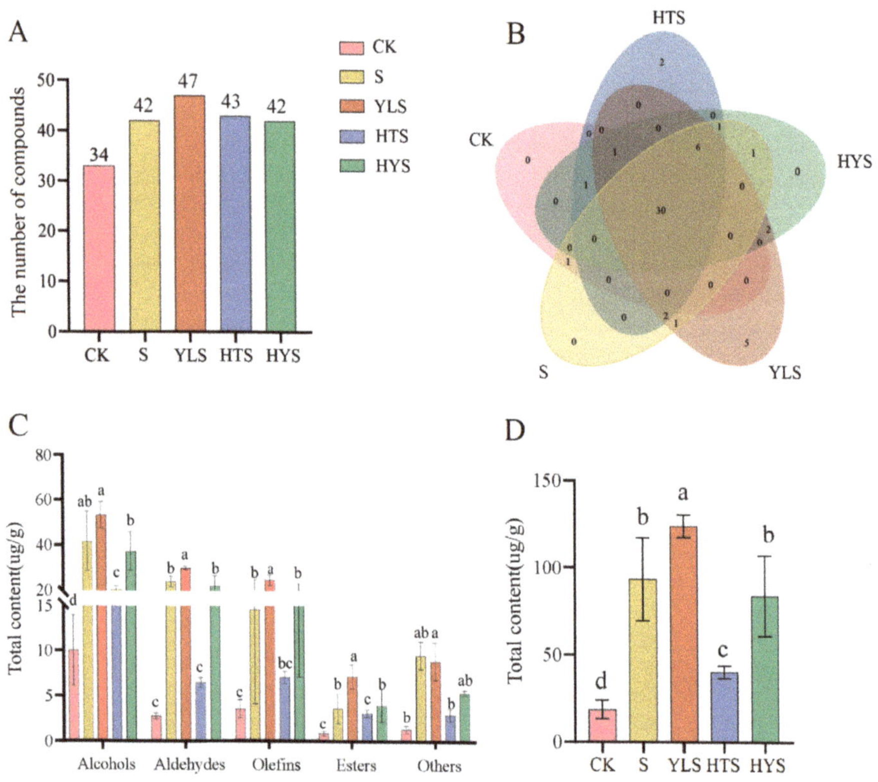

Figure 2. The distribution of volatile compounds in black tea under different withering combinations. (**A**) The number of volatile compounds identified in each withering treatment. (**B**) Venn diagram showing the overlap and unique volatile compounds among the five withering treatments. The numbers in the diagram represent the count of volatile compounds in each overlapping or unique region. (**C**) Total volatile contents of the five groups categorized by chemical class. (**D**) Total contents of all the volatiles. The total content of volatiles was presented as mean ± SD (n = 3). Different lowercase letters above the bars in the same volatile category indicate significant differences at $p < 0.05$. Treatments: CK represents indoor natural withering, S represents shaking, YLS represents yellow light plus shaking, HTS represents high temperature plus shaking, and HYS represents high temperature with yellow light plus shaking.

Quantitative analysis showed that different shaking combinations improved the content of various volatile categories (Figure 2C). YLS had the highest content of alcohols, aldehydes, olefins, esters, and other compounds, leading to the highest level in total volatile content (Figure 2D). Compared to CK, S and HYS also promoted the accumulation of these categories, while HTS only significantly increased alcohol content. Major alcohols such as linalool, cedrol, and trans-linalool oxide (furanoid) increased significantly after shaking combinations, contributing to the overall increase in alcohol content. Similarly, S, YLS, and HYS increased the content of main aldehydes (e.g., 3-methylbutyraldehyde), olefins (e.g., β-cedrene and α-cedrene), esters (e.g., α-terpinyl acetate), and other compounds (e.g., dimethyl sulfide). The total content of these categories exhibited the following order: YLS > S ≈ HYS > HTS > CK, indicating that YLS is most beneficial for volatile compound formation in black tea, thereby enhancing aroma quality, which was consistent with the aroma scores of sensory evaluations.

3.2.2. PLS-DA Analysis of Volatile Compounds

The partial least squares discriminant analysis (PLS-DA) was shown in Figure 3. The model parameters (R2X(cum) = 0.934, R2Y(cum) = 0.986, Q2(cum) = 0.925) indicated high credibility and predictive ability. Principal component 1 (57.1%) and principal component 2 (9.2%) effectively separated and predicted black tea samples under different shaking combinations (Figure 3A). Apparently, the tea samples were clearly classified into five groups, with CK on the far left and YLS on the far right showed the largest difference, which validated the results of sensory evaluation. HTS is just next to CK, and they are both on the left side of the horizontal axis. HYS approaches the origin and S is positioned between HYS and YLS. These results confirm that different shaking combinations significantly affect the composition and content of volatile components in black tea.

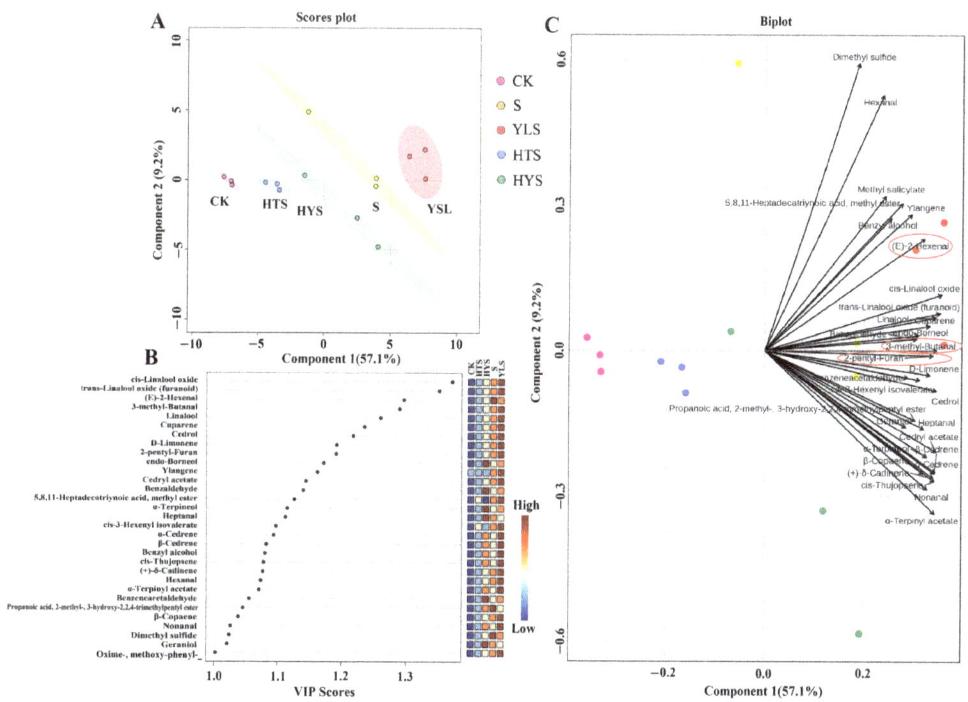

Figure 3. PLS-DA analysis of main components in black teas under different shaking combinations. (**A**) Score plot showing the distribution of samples based on the first two principal components. Each point represents a sample, and the ellipses indicate confidence intervals for each group. (**B**) VIP (Variable Importance in Projection) plot displaying the key volatile compounds (VIP scores > 1.0) contributing to the separation of groups with. (**C**) Biplot illustrating the relationship between samples (points) and volatile compounds (vectors) in the PLS-DA model. Treatments: CK represents indoor natural withering, S represents shaking, YLS represents yellow light plus shaking, HTS represents high temperature plus shaking, and HYS represents high temperature with yellow light plus shaking.

According to the value of variable importance factor (VIP > 1, $p < 0.05$), thirty-one compounds were identified as key components differentiating black teas under different shaking combinations, with their contents significantly promoted in YLS (Figure 3B). The biplot (Figure 3C) further illustrated the contributions of these volatiles to classification, with most of the key differentiating compounds are in the right quadrant near S and YLS. It was noted that 3-methyl-butanal, (E)-2-hexenal and 2-pentyl-furan were closer to YLS, indicating their significant influence on aroma differentiation.

3.2.3. Key Aroma Compounds of Black Teas

The relative odor activity values (rOAVs) were calculated to evaluate the contribution of specific volatiles to black tea aroma (Table S2). Eight compounds (rOAV > 1) were identified as the most important contributors to black tea aroma under different shaking combinations (Table 2), including linalool, cedrol, 3-methyl-butanal, trans-β-ionone, dimethyl sulfide, benzeneacetaldehyde, α-cedrene, and nonanal. Among these, linalool, cedrol, 3-methyl-butanal, and trans-β-ionone had average rOAV values > 10, indicating their substantial contributions to black tea aroma.

Table 2. Relative odor activity values (rOAVs) of 8 key aroma compounds (rOAV> 1) in black teas under different shaking combinations.

Name of Odorants	OT ($\mu g \cdot L^{-1}$)	Odor Characteristic	Average rOAV				
			CK	S	YLS	HTS	HYS
Linalool	0.22	Floral, sweet, grape-like, woody	4.72 ± 0.75 c	30.34 ± 11.08 b	54.78 ± 0.82 a	22.14 ± 1.71 b	23.20 ± 7.45 b
Cedrol	1.00	Cedarwood aroma	6.08 ± 2.65 c	19.16 ± 7.02 a	21.29 ± 2.96 a	7.50 ± 0.33 c	14.33 ± 4.30 ab
3-methyl-Butanal	0.50	Apple-like and chocolate-like flavor	2.53 ± 0.16 d	18.14 ± 5.07 ab	25.63 ± 4.82 a	8.19 ± 0.75 c	16.87 ± 3.13 b
Trans-β-Ionone	0.007	Violet-like, floral, and raspberry-like	16.29 ± 6.28 c	8.51 ± 3.60 c	65.13 ± 7.99 a	37.66 ± 3.77 b	78.50 ± 32.76 a
Dimethyl sulfide	0.33	Corn, sulfurous flavor	0.88 ± 0.22 c	12.35 ± 4.39 a	7.64 ± 0.61 a	2.91 ± 0.36 b	4.16 ± 1.00 b
Benzeneacetaldehyde	1.20	Hawthorne, honey, sweet	0.20 ± 0.04 c	1.02 ± 0.36 b	1.77 ± 0.03 a	0.82 ± 0.10 b	1.79 ± 0.24 a
α-Cedrene	2.13	Floral	0.89 ± 0.19 c	4.66 ± 3.20 abc	5.67 ± 0.46 a	1.66 ± 0.13 bc	3.68 ± 1.57 ab
Nonanal	1.10	Rose aroma, aroma of citrus flower, creamy aroma	0.27 ± 0.05 c	1.14 ± 0.69 ab	1.90 ± 0.07 a	0.80 ± 0.03 b	1.21 ± 0.76 ab

rOAV: calculated by the relative concentration of odorant in tea with OTs in water (mean ± SD, n = 3). OTs: odor threshold in water based on the literature [5,23,24]. Treatments: CK represents indoor natural withering, S represents shaking, YLS represents yellow light plus shaking, HTS represents high temperature plus shaking, and HYS represents high temperature with yellow light plus shaking. Different lowercase letters in a row indicate a significant difference between different treatments ($p < 0.05$).

Compared to CK, the rOAV values of key aroma components increased significantly after shaking combinations, particularly in YLS, which had the highest rOAVs for all aroma-active compounds, suggesting superior aroma quality, further corroborated the results of sensory evaluation. Trans-β-ionone in HYS (78.50 ± 32.76) and dimethyl sulfide in S (12.35 ± 4.39) also showed high rOAVs. Referring to the previous literature for the aroma description (Table 2), linalool, trans-β-ionone, α-cedrene, and nonanal were identified as key floral aroma substances, with their highest rOAV values in YLS, indicating that the increase in floral aroma compounds is crucial for aroma quality improvement under YLS treatment.

3.3. Volatile Compounds Analysis of Withered Leaves under Different Shaking Combinations

3.3.1. Overall Analysis of Volatile Compounds

In total, 52 volatile components were obtained from in-process withered leaves (Table S3). The total volatile content in withered leaves increased significantly after S and HYS treatments compared to CK, while YLS showed a non-significant increase (Figure 4A). HTS exhibited a decrease in volatile content, suggesting that high-temperature withering may hinder volatile compound formation in withered leaves. Moreover, PLS-DA analysis of withered leaves (Figure 4B) showed distinct clustering of samples under different treatments, with S and YLS in the first quadrant, HTS in the second, CK in the third, and HYS in the fourth, indicating significant differences in the volatile composition of withered tea leaves.

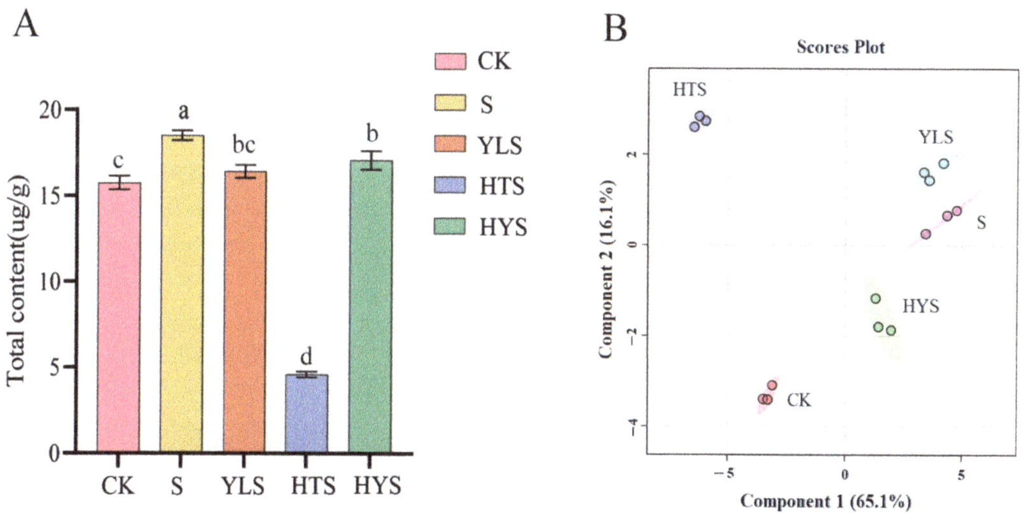

Figure 4. Effect of different shaking combinations on volatile components in withered leaves. (**A**) The total content of volatile components in withered leaves under different shaking combinations (mean ± SD, n = 3). Different lowercase letters above the bars indicate significant differences at $p < 0.05$. (**B**) Score plot of PLS-DA (partial least squares discriminant analysis) showing the distribution of samples based on the first two principal components. Each point represents a sample, and the ellipses indicate confidence intervals for each group. Treatments: CK represents indoor natural withering, S represents shaking, YLS represents yellow light plus shaking, HTS represents high temperature plus shaking, and HYS represents high temperature with yellow light plus shaking.

3.3.2. Analysis of Key Aroma Components

The key aroma components of withered leaves were compared with those in black teas (Figure 5). Key floral aroma compounds in withered leaves, such as linalool, α-cedrene, and nonanal, increased significantly after S, YLS, and HYS treatments (Figure 5A–C), while trans-β-ionone was not detected in withered leaves (Figure 5D). The content of these floral aroma compounds in both withered leaves and dry black teas was highest in YLS, suggesting that yellow-light withering further enhances the accumulation of floral substances on the basis of shaking, thereby improving black tea aroma quality. Benzeneacetaldehyde, 3-methyl-butanal, dimethyl sulfide, and cedrol also increased significantly under different shaking combinations in both withered leaves and dry black teas (Figure 5E–H), with YLS promoting the highest accumulation of benzeneacetaldehyde 3-methyl-butanal and cedrol and S promoting the highest accumulation of dimethyl sulfide and cedrol. The accumulation of these key aroma components in the withering process lays the foundation for the formation of tea aroma in the subsequent processing, thus facilitating the improvement in tea aroma quality.

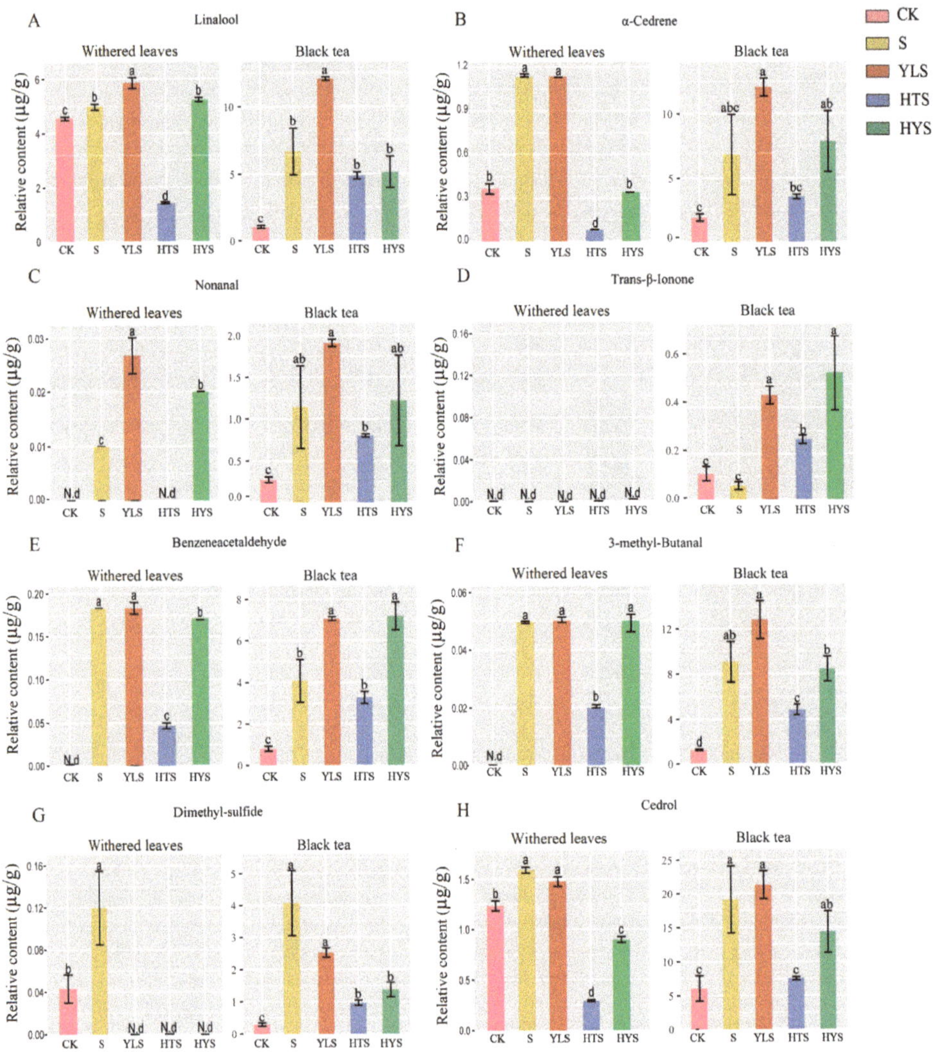

Figure 5. The relative contents of key aroma components in withered leaves and black teas: (**A**) linalool, (**B**) α-cedrene, (**C**) nonanal, (**D**) trans-β-ionone, (**E**) benzeneacetaldehyde, (**F**) 3-methyl-butanal, (**G**) dimethyl sulfide, and (**H**) cedrol. The relative contents were presented as mean ± SD (n = 3). Different lowercase letters above the bars in the same volatile category indicate significant differences at $p < 0.05$. N.d denotes that the component was not detected in the sample. Treatments: CK represents indoor natural withering, S represents shaking, YLS represents yellow light plus shaking, HTS represents high temperature plus shaking, and HYS represents high temperature with yellow light plus shaking.

4. Discussion

Aroma is a pivotal factor in tea evaluation, directly influencing tea pricing and consumer preference. Enhancing aroma quality through innovative processing techniques is one of the most effective strategies in tea production. Withering, the first and indispensable step in black tea processing, plays a crucial role in flavor formation, as volatile components are significantly generated during this stage [11,20]. During withering, photosynthesis, respiration, and most metabolic activities in tea leaves continue [9,17], and these processes can be modulated by postharvest treatments such as shaking and light irradiation. Previous research has demonstrated that introducing shaking procedures can significantly improve black tea quality, particularly its aroma [14,15,25]. Consistent with these findings, this study revealed that shaking and its combinations increased the content of alcohols, aldehydes, olefins, esters, and other compounds in black tea, promoting the enrichment of floral aroma volatiles and enhancing overall aroma quality. However, the effects varied across different shaking combinations.

Among the treatments, yellow-light withering combined with shaking (YLS) exhibited the most significant improvement in volatile compound formation, with the highest content of floral aroma substances in both withered leaves and dry black teas. This finding underscores the critical role of light quality in enhancing tea aroma and volatile metabolites [18,26–28]. It also suggests that yellow-light withering can further amplify the floral aroma quality of shaken black tea, aligning with previous studies that supplementary LED light during the standing stage of shaking promotes volatile formation in oolong tea [29]. A synergistic effect between yellow-light withering and shaking is hypothesized. During the yellow-light withering of the YLS treatment, yellow light may upregulate genes associated with volatile synthesis, such as linalool synthase genes (CsLIS2, CsLIS3, and CsLIS4) and farnesene synthase genes (CsFS5 and CsFS10), which have been reported to significantly increase under yellow light [30] and, thus, promote an increase in the content of related volatiles. As a cold light source, yellow light minimally affects leaf temperature, allowing tea leaves to remain fresh and responsive to external stresses like dehydration and mechanical damage during subsequent shaking. The dehydration and mechanical stress from shaking activate genes in volatile-related metabolic pathways [31], alters protein expression, and enhances enzyme activities that modulate volatile formation [32,33], thereby enriching the aroma components of black tea. Thus, yellow light may play a role in the early withering stage, while shaking contributes to the later stage, collectively promoting the accumulation of aroma substances to a greater extent.

In contrast, high-temperature withering combined with shaking (HTS) seemed to show an antagonistic effect. Although the aroma scores and total aroma content of HTS were higher than those of the control (CK), the content of many volatile constituents, including key aroma components such as linalool and α-cedrene, decreased after withering. Moreover, the total aroma content of HTS was lower than that of S (shaking alone), suggesting that high-temperature withering may counteract the positive effects of shaking.

High-temperature withering accelerates water loss in tea leaves, potentially disrupting the activities of oxidase, hydrolase, and other enzymes involved in aroma anabolism [34]. For instance, higher air temperatures have been shown to decrease polyphenol oxidase activity while enhancing peroxidase activity in tea leaves [35]. Therefore, changes in enzyme activity may be the key to antagonism between shaking and high-temperature withering. Interestingly, the volatile content in HYS (high-temperature with yellow-light withering plus shaking) was similar to that of S, likely due to the counteracting effects of yellow light (which increases volatiles) and high temperature (which decreases volatiles).

However, the precise mechanisms by which multiple treatments synergistically enhance black tea aroma quality require further investigation.

The concentration of volatile compounds alone is not sufficient to determine their significant contribution to the overall aroma. However, our study did not conduct specialized olfactory measurements for individual compounds. In order to evaluate the contribution of specific volatiles to the overall aroma of black tea, relative odor activity values (rOAVs) were calculated based on the odor threshold and the descriptions of specific compound odor characteristics are based on references from previous literature [28]. In this study, eight volatiles were identified as the most important aroma components (rOAVs > 1). To better understand the mechanisms underlying volatile changes during different withering treatments, these eight key compounds were classified into four categories based on their sources: volatile terpenes (VTs) (e.g., linalool, cedrol, and α-cedrene), amino acid-derived volatiles (AADVs) (e.g., benzeneacetaldehyde, 3-methyl-butanal, and dimethyl sulfide), carotenoid-derived volatiles (CDVs) (e.g., trans-β-ionone), and others (e.g., nonanal). Linalool can also be derived from the hydrolysis of glycosidically bound volatiles (GBVs) [36]. The YLS treatment showed the highest content of key VTs and AADVs, suggesting that YLS may enhance tea aroma by regulating terpenoid metabolism and amino acid degradation. VTs in tea leaves are synthesized via the methyl-erythritol phosphate (MEP) and mevalonic acid (MVA) pathways. Previous research has shown that light quality (e.g., red and yellow), and shaking can significantly upregulate genes associated with VT formation [18,26]. The higher content of VTs in YLS compared to S indicates that yellow light and shaking may synergistically regulate terpenoid metabolism, with yellow light promoting VT accumulation in the early withering stage and shaking enhancing it in the later stage. Similarly, amino acids undergo Strecker degradation under thermal conditions, producing key Strecker aldehydes such as 3-methyl-butanal, dimethyl sulfide, and benzeneacetaldehyde, with precursors that include leucine (Leu), methionine (Met), and phenylalanine (Phe), respectively [36]. Yellow-light withering [16,27] and shaking [25] have been shown to increase the content of free amino acids, which will provide abundant precursors for AADV formation and ultimately leading to higher concentrations in black tea.

The floral aroma is not only a hallmark of high-quality black tea but also a critical indicator of its excellence [37]. With the growing application of molecular sensory science in tea research, numerous studies have identified key aroma components responsible for the floral notes in black tea. For instance, indole and methyl jasmonate had been recognized as the most critical compounds determining the floral aroma intensity of Hunan black tea [15], while linalool, (E)-β-ionone, geraniol, β-myrcene, and (E)-2-hexenal, with high OAVs, have been identified as key floral volatiles in summer black tea [25]. Consistent with these findings, this study also identified linalool and trans-β-ionone as key floral components based on odor descriptions from **previous literature** [5]. Additionally, α-cedrene and nonanal, which contribute to floral aromas [23], were found to play significant roles in enhancing the aroma quality of black tea. The contents of linalool, α-cedrene, and nonanal were significantly upregulated under different shaking combinations, particularly in the YLS treatment, in both withered tea leaves and dry black teas. Trans-β-ionone, with its distinctive "violet-like" aroma [5], is a carotenoid-derived compound that can be produced either through enzymatic reactions during fermentation or thermal degradation during drying [36]. As a result, it was not detected in withered leaves, but its content in black tea increased remarkably in YLS-treated samples. Consistent with sensory evaluations, YLS, which exhibited the highest content of key floral components, also achieved the highest aroma score, characterized by a long-lasting floral aroma.

In conclusion, our findings highlight the significant impact of shaking and yellow-light withering on the aroma quality of black tea. The YLS treatment, in particular, demonstrated the most pronounced enhancement in floral aroma components, suggesting its potential as a valuable technique in black tea processing. Future research should integrate olfactory measurements and molecular sensory science to further elucidate the specific contributions of individual compounds to black tea aroma. This approach will provide a more comprehensive understanding of aroma formation and enable the development of targeted strategies for producing high-quality black tea.

5. Conclusions

This study evaluated the aroma attributes of black teas processed with different shaking combinations through sensory evaluation and GC-MS analysis. Compared to CK, all shaking combinations resulted in a distinct floral aroma, with YLS showing the highest floral aroma intensity, volatile compound content, and number. Eight key aroma compounds, including linalool, cedrol, 3-methyl-butanal, trans-β-ionone, dimethyl sulfide, benzeneacetaldehyde, α-cedrene, and nonanal, were identified as the most important contributors to black tea aroma, with their content significantly increased after shaking combinations in withered leaves (except trans-β-ionone). YLS exhibited the highest content of floral volatiles with high rOAVs, indicating a synergistic effect between yellow light and shaking. In summary, the findings of this study enhance our understanding of the aroma formation of black tea under different shaking combinations and provide a theoretical basis and practical guidance for producing high-aroma black tea through optimized shaking and withering techniques.

Supplementary Materials: The following supporting information can be downloaded at: https://www.mdpi.com/article/10.3390/foods14050758/s1. Table S1: Volatile components and their content in black tea under different shaking combinations; Table S2: Relative Odor activity values (rOAV) and Odor characteristics of Volatile compounds in black teas under different withering combinations; Table S3: Volatile components and their content in withered tea leaves under different shaking combinations; Figure S1: EI mass spectra of key aroma compounds and their corresponding mass spectra from database; Figure S2: Figure S2 Total ion chromatogram of black tea samples under different withering combinations; Figure S3: Figure S3 Total ion chromatogram of withered leaves under different withering combinations.

Author Contributions: Z.A.: Conceptualization, Methodology, Investigation, Data curation, Writing. S.H.: Formal analysis, Investigation. Visualization. B.M.: Writing—review and editing. L.J.: Visualization, Writing—review and editing. Y.Y.: Conceptualization, Writing—review and editing, Funding acquisition. All authors have read and agreed to the published version of the manuscript.

Funding: This work was funded by Jiangsu Agricultural Science and Technology Innovation Fund (CX[22]2008); Jiangsu Modern Seed Industry Revitalization Project (JBGS[2021]085); Jiangsu Provincial Key Research and Development Program (BE2023364).

Institutional Review Board Statement: This study was approved by the Ethics Committee of Jiangsu Academy of Agricultural Sciences (Approval Date: 20 August 2024). In the process of ethical review for scientific research projects, within our unit's management procedures, the approval code is not an element that must be generated or provided.

Informed Consent Statement: Informed consent was obtained from all subjects involved in the study.

Data Availability Statement: The original contributions presented in the study are included in the article/supplementary material; further inquiries can be directed to the corresponding author.

Conflicts of Interest: The authors declare that they have no known competing financial interests or personal relationships that could have appeared to influence the work reported in this paper.

References

1. Huang, W.; Fang, S.; Wang, J.; Zhuo, C.; Luo, Y.; Yu, Y.; Li, L.; Wang, Y.; Deng, W.-W.; Ning, J. Sensomics Analysis of the Effect of the Withering Method on the Aroma Components of Keemun Black Tea. *Food Chem.* **2022**, *395*, 133549. [CrossRef] [PubMed]
2. Kumazawa, K.; Wada, Y.; Masuda, H. Characterization of Epoxydecenal Isomers as Potent Odorants in Black Tea (Dimbula) Infusion. *J. Agric. Food Chem.* **2006**, *54*, 4795–4801. [CrossRef]
3. Wang, Q.; Qin, D.; Huang, G.; Jiang, X.; Fang, K.; Wang, Q.; Ni, E.; Li, B.; Pan, C.; Li, H.; et al. Identification and Characterization of the Key Volatile Flavor Compounds in Black Teas from Distinct Regions Worldwide. *J. Food Sci.* **2022**, *87*, 3433–3446. [CrossRef]
4. Flaig, M.; Qi, S.C.; Wei, G.; Yang, X.; Schieberle, P. Characterisation of the Key Aroma Compounds in a Longjing Green Tea Infusion (Camellia Sinensis) by the Sensomics Approach and Their Quantitative Changes during Processing of the Tea Leaves. *Eur. Food Res. Technol.* **2020**, *246*, 2411–2425. [CrossRef]
5. Guo, X.; Schwab, W.; Ho, C.-T.; Song, C.; Wan, X. Characterization of the Aroma Profiles of Oolong Tea Made from Three Tea Cultivars by Both GC–MS and GC-IMS. *Food Chem.* **2022**, *376*, 131933. [CrossRef]
6. Chen, Z.; Li, Z.; Zhao, Y.; Zhu, M.; Li, J.; Wang, K. A Meta-Analysis of Dynamic Changes of Key Aroma Compounds during Black Tea Processing. *Food Biosci.* **2024**, *58*, 103784. [CrossRef]
7. Kang, S.; Yan, H.; Zhu, Y.; Liu, X.; Lv, H.-P.; Zhang, Y.; Dai, W.-D.; Guo, L.; Tan, J.-F.; Peng, Q.-H.; et al. Identification and Quantification of Key Odorants in the World's Four Most Famous Black Teas. *Food Res. Int.* **2019**, *121*, 73–83. [CrossRef] [PubMed]
8. Jin, L.; Lian, X.; Chen, L.; Lei, Y.; Li, J.; Yang, Z.; Li, D. Characteristic Aroma Analysis and Interaction Study of Key Aroma Compounds of Chuanhong Congou Black Tea. *Eur. Food Res. Technol.* **2024**, *250*, 441–454. [CrossRef]
9. Deb, S.; Jolvis Pou, K.R. A Review of Withering in the Processing of Black Tea. *J. Biosyst. Eng.* **2016**, *41*, 365–372. [CrossRef]
10. Wang, Y.; Zheng, P.-C.; Liu, P.-P.; Song, X.-W.; Guo, F.; Li, Y.-Y.; Ni, D.-J.; Jiang, C.-J. Novel Insight into the Role of Withering Process in Characteristic Flavor Formation of Teas Using Transcriptome Analysis and Metabolite Profiling. *Food Chem.* **2019**, *272*, 313–322. [CrossRef] [PubMed]
11. Fang, X.; Liu, Y.; Xiao, J.; Ma, C.; Huang, Y. GC–MS and LC-MS/MS Metabolomics Revealed Dynamic Changes of Volatile and Non-Volatile Compounds during Withering Process of Black Tea. *Food Chem.* **2023**, *410*, 135396. [CrossRef] [PubMed]
12. Xue, J.; Liu, P.; Yin, J.; Wang, W.; Zhang, J.; Wang, W.; Le, T.; Ni, D.; Jiang, H. Dynamic Changes in Volatile Compounds of Shaken Black Tea during Its Manufacture by GC × GC–TOFMS and Multivariate Data Analysis. *Foods* **2022**, *11*, 1228. [CrossRef]
13. Zeng, L.; Zhou, X.; Su, X.; Yang, Z. Chinese Oolong Tea: An Aromatic Beverage Produced under Multiple Stresses. *Trends Food Sci. Technol.* **2020**, *106*, 242–253. [CrossRef]
14. Wang, J.; Ouyang, W.; Zhu, X.; Jiang, Y.; Yu, Y.; Chen, M.; Yuan, H.; Hua, J. Effect of Shaking on the Improvement of Aroma Quality and Transformation of Volatile Metabolites in Black Tea. *Food Chem. X* **2023**, *20*, 101007. [CrossRef]
15. Yin, X.; Xiao, Y.; Wang, K.; Wu, W.; Huang, J.; Liu, S.; Zhang, S. Effect of Shaking Manners on Floral Aroma Quality and Identification of Key Floral-Aroma-Active Compounds in Hunan Black Tea. *Food Res. Int.* **2023**, *174*, 113515. [CrossRef]
16. Ai, Z.; Zhang, B.; Chen, Y.; Yu, Z.; Chen, H.; Ni, D. Impact of Light Irradiation on Black Tea Quality during Withering. *J. Food Sci. Technol.* **2017**, *54*, 1212–1227. [CrossRef] [PubMed]
17. Yu, X.; Li, Y.; He, C.; Zhou, J.; Chen, Y.; Yu, Z.; Wang, P.; Ni, D. Nonvolatile Metabolism in Postharvest Tea (*Camellia sinensis* L.) Leaves: Effects of Different Withering Treatments on Nonvolatile Metabolites, Gene Expression Levels, and Enzyme Activity. *Food Chem.* **2020**, *327*, 126992. [CrossRef] [PubMed]
18. Li, Y.; He, C.; Yu, X.; Zhou, J.; Ntezimana, B.; Yu, Z.; Chen, Y.; Ni, D. Study on Improving Aroma Quality of Summer-Autumn Black Tea by Red-Light Irradiation during Withering. *LWT* **2022**, *154*, 112597. [CrossRef]
19. Huang, W.; Lu, G.; Deng, W.-W.; Ning, J. Effects of Different Withering Methods on the Taste of Keemun Black Tea. *LWT* **2022**, *166*, 113791. [CrossRef]
20. Zhou, J.; Yu, X.; He, C.; Qiu, A.; Li, Y.; Shu, Q.; Chen, Y.; Ni, D. Withering Degree Affects Flavor and Biological Activity of Black Tea: A Non-Targeted Metabolomics Approach. *LWT* **2020**, *130*, 109535. [CrossRef]
21. Qin, M.; Zhou, J.; Luo, Q.; Zhu, J.; Yu, Z.; Zhang, D.; Ni, D.; Chen, Y. The Key Aroma Components of Steamed Green Tea Decoded by Sensomics and Their Changes under Different Withering Degree. *Food Chem.* **2024**, *439*, 138176. [CrossRef]
22. Wang, M.-Q.; Ma, W.-J.; Shi, J.; Zhu, Y.; Lin, Z.; Lv, H.-P. Characterization of the Key Aroma Compounds in Longjing Tea Using Stir Bar Sorptive Extraction (SBSE) Combined with Gas Chromatography-Mass Spectrometry (GC–MS), Gas Chromatography-Olfactometry (GC-O), Odor Activity Value (OAV), and Aroma Recombination. *Food Res. Int.* **2020**, *130*, 108908. [CrossRef] [PubMed]

23. Hou, Z.-W.; Wang, Y.-J.; Xu, S.-S.; Wei, Y.-M.; Bao, G.-H.; Dai, Q.-Y.; Deng, W.-W.; Ning, J.-M. Effects of Dynamic and Static Withering Technology on Volatile and Nonvolatile Components of Keemun Black Tea Using GC-MS and HPLC Combined with Chemometrics. *LWT* **2020**, *130*, 109547. [CrossRef]
24. Xie, J.; Wang, L.; Deng, Y.; Yuan, H.; Zhu, J.; Jiang, Y.; Yang, Y. Characterization of the Key Odorants in Floral Aroma Green Tea Based on GC-E-Nose, GC-IMS, GC-MS and Aroma Recombination and Investigation of the Dynamic Changes and Aroma Formation during Processing. *Food Chem.* **2023**, *427*, 136641. [CrossRef] [PubMed]
25. Wang, J.; Qu, L.; Yu, Z.; Jiang, Y.; Yu, C.; Zhu, X.; Lin, Q.; Niu, L.; Yu, Y.; Lin, Q.; et al. Targeted Quantitative Metabolomic and Flavor Objective Quantification Technique Reveal the Impact Mechanism of Shaking on Black Tea Quality and Non-Volatile Metabolites. *Food Chem.* **2024**, *458*, 140226. [CrossRef]
26. Hua, J.; Zhu, X.; Ouyang, W.; Yu, Y.; Chen, M.; Wang, J.; Yuan, H.; Jiang, Y. Non-Target and Target Quantitative Metabolomics with Quantitative Aroma Evaluation Reveal the Influence Mechanism of Withering Light Quality on Tea Aroma and Volatile Metabolites Evolution. *Food Res. Int.* **2024**, *192*, 114773. [CrossRef] [PubMed]
27. Fu, X.; Chen, Y.; Mei, X.; Katsuno, T.; Kobayashi, E.; Dong, F.; Watanabe, N.; Yang, Z. Regulation of Formation of Volatile Compounds of Tea (Camellia Sinensis) Leaves by Single Light Wavelength. *Sci. Rep.* **2015**, *5*, 16858. [CrossRef] [PubMed]
28. He, Y.; Li, J.; Mei, H.; Zhuang, J.; Zhao, Z.; Jeyaraj, A.; Wang, Y.; Chen, X.; Li, X.; Liu, Z. Effects of Leaf-Spreading on the Volatile Aroma Components of Green Tea under Red Light of Different Intensities. *Food Res. Int.* **2023**, *168*, 112759. [CrossRef] [PubMed]
29. Ni, Z.; Yang, Y.; Zhang, Y.; Hu, Q.; Lin, J.; Lin, H.; Hao, Z.; Wang, Y.; Zhou, J.; Sun, Y. Dynamic Change of the Carotenoid Metabolic Pathway Profile during Oolong Tea Processing with Supplementary LED Light. *Food Res. Int.* **2023**, *169*, 112839. [CrossRef] [PubMed]
30. Yu, X.; He, C.; Li, Y.; Zhou, J.; Chen, Y.; Yu, Z.; Wang, Y.; Ni, D. Effects of Different Spreading Treatments on the Formation of Aroma Quality in Green Tea. *Beverage Plant Res.* **2021**, *1*, 14. [CrossRef]
31. Zheng, X.-Q.; Li, Q.-S.; Xiang, L.-P.; Liang, Y.-R. Recent Advances in Volatiles of Teas. *Molecules* **2016**, *21*, 338. [CrossRef] [PubMed]
32. Wu, L.; Wang, Y.; Liu, S.; Sun, Y.; Li, C.; Lin, J.; Wei, S. The Stress-Induced Metabolites Changes in the Flavor Formation of Oolong Tea during Enzymatic-Catalyzed Process: A Case Study of Zhangping Shuixian Tea. *Food Chem.* **2022**, *391*, 133192. [CrossRef]
33. Zhou, Z.-W.; Wu, Q.-Y.; Yang, Y.; Hu, Q.-C.; Wu, Z.-J.; Huang, H.-Q.; Lin, H.-Z.; Lai, Z.-X.; Sun, Y. The Dynamic Change in Fatty Acids during the Postharvest Process of Oolong Tea Production. *Molecules* **2022**, *27*, 4298. [CrossRef]
34. Chen, Y.; Zeng, L.; Liao, Y.; Li, J.; Zhou, B.; Yang, Z.; Tang, J. Enzymatic Reaction-Related Protein Degradation and Proteinaceous Amino Acid Metabolism during the Black Tea (Camellia Sinensis) Manufacturing Process. *Foods* **2020**, *9*, 66. [CrossRef] [PubMed]
35. Das, S.; Samanta, T.; Datta, A.K. Improving Black Tea Quality through Optimization of Withering Conditions Using Artificial Neural Network and Genetic Algorithm. *J. Food Process. Preserv.* **2021**, *45*, e15273. [CrossRef]
36. Ho, C.-T.; Zheng, X.; Li, S. Tea Aroma Formation. *Food Sci. Hum. Wellness* **2015**, *4*, 9–27. [CrossRef]
37. Yao, H.; Su, H.; Ma, J.; Zheng, J.; He, W.; Wu, C.; Hou, Z.; Zhao, R.; Zhou, Q. Widely Targeted Volatileomics Analysis Reveals the Typical Aroma Formation of Xinyang Black Tea during Fermentation. *Food Res. Int.* **2023**, *164*, 112387. [CrossRef]

foods

MDPI

Article

Insight into the Volatile Profiles and Key Odorants of Rizhao Green Tea by Application of SBSE-GC-MS, OAVs and GC-O Analysis

Mengqi Wang [1], Dapeng Song [1], Hongxu Yin [2], Fengxiang Fang [1], Yali Shi [3], Hui Wang [1], Jiyan Li [1], Kunpeng Wang [1], Yin Zhu [4], Haipeng Lv [4,*] and Shibo Ding [1,*]

[1] Tea Research Institute, Rizhao Academy of Agricultural Sciences, Rizhao 276800, China; wmqrznky@163.com (M.W.); sdp20073882@163.com (D.S.); sdrzffx@163.com (F.F.); wh2009tea@126.com (H.W.); rzljy2663@163.com (J.L.); wkpahj888@163.com (K.W.)

[2] Rizhao Donggang District Agriculture and Rural Bureau, Rizhao 276800, China; sdjnyinhongxu@163.com

[3] Tea Research Institute, Shandong Academy of Agricultural Sciences, Jinan 250100, China; shiyali97@163.com

[4] Key Laboratory of Biology, Genetics and Breeding of Special Economic Animals and Plants, Ministry of Agriculture and Rural Affairs, Tea Research Institute, Chinese Academy of Agricultural Sciences, Hangzhou 310008, China; zhuy_scu@tricaas.com

* Correspondence: lvhaipeng@caas.cn (H.L.); rznynky@rz.shandong.cn (S.D.); Tel.: +86-571-86653154 (H.L.); +86-633-8221315 (S.D.)

Abstract: Rizhao green tea (RZT), a renowned green tea, is cultivated in China's northernmost tea region. Its unique environment endows it with a strong chestnut- and seaweed-like aroma. This study sought to explore the volatile profiles of RZT and pinpoint its key odorants by employing stir bar sorptive extraction (SBSE) coupled with gas chromatography–mass spectrometry (GC-MS), determining the odor activity value (OAV), and performing gas chromatography–olfactometry (GC-O). A total of 112 volatiles were identified, and the major volatile compounds were esters (2035.25 μg/kg), alcohols (1799.02 μg/kg), alkanes (991.88 μg/kg), and ketones (691.96 μg/kg), comprising 74.91% of the total. A molecular aroma wheel was preliminarily established based on these key odorants. These insights might contribute to the scientific elucidation of the flavor chemical basis of RZT.

Keywords: Rizhao green tea (RZT); SBSE-GC-MS; odor activity values; aroma quality; key odorants; aroma wheel

1. Introduction

Green tea is the most consumed and produced tea product in China, eith widespread appreciation from consumers around the globe for its delightful flavor. The quality of tea is intricately linked to factors such as the planting environment, climate, soil and geography [1]. Consequently, green teas from China's northern regions exhibit distinctive flavor profiles that set them apart from their southern counterparts [2,3]. Rizhao green tea (RZT) is among the few highly esteemed green tea products cultivated in the northernmost tea region of China (35°04′ N–36°04′ N) (Figure 1). By 2023, Rizhao's total tea planting area had grown to 20,000 hectares, producing over 19,800 tons of Rizhao green tea with an output value of CNY 3.8 billion. The unique growing conditions at this latitude, characterized by a humid maritime monsoon climate, low-temperature stress, and pronounced diurnal temperature fluctuations, result in the slower but more robust growth of RZT tea leaves and buds. Accordingly, these unique environments provide it with some special flavor characteristics different from those of the green tea grown in other tea regions in China, especially reflected in the RZT aroma quality: a rich chestnut- and seaweed-like aroma.

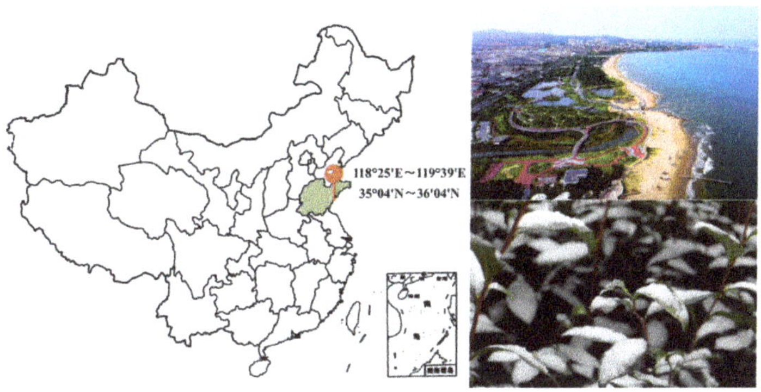

Figure 1. Geographical location and natural growth conditions of Rizhao green tea.

Aroma is a crucial parameter for assessing tea quality, primarily arising from the complex interaction of various volatile compounds in distinct proportions [4,5]. Research into the key odorants present in tea and the environmental factors influencing their synthesis can provide valuable insights for guiding tea cultivation and enhancing tea quality. RZT possesses special aroma quality that make it a valuable subject of study. However, the specific key odorants that shape its distinctive aroma and the biological processes behind their development remain unclear, which hinders the development of a comprehensive scientific understanding of how the aroma profile of RZT is linked to its cultivation environment.

It is widely recognized that different extraction methods can significantly influence the detection of tea volatiles in terms of both composition and quantity [6]. With high sensitivity and recovery, SBSE has been established as a practical and highly effective aroma enrichment technique that demonstrates high sensitivity and recovery. Presently, SBSE has gained significant traction in analytical applications, with widespread use in the analysis of tea, wine, and fruit [7–9]. In order to figure out the volatile profiles and clarify the key odorants of the green tea products produced in the Chinese northern tea regions, the volatile compounds of the representative RZT samples were analyzed in this study using SBSE in combination with GC-O and OAVs. The findings could enhance the understanding of green tea's aroma quality across different cultivation conditions, as well as bridge the current research gap regarding the aroma quality of RZT, offering a theoretical framework for the quality assessment of RZT.

2. Materials and Methods

2.1. Tea Samples

Thirty-one Rizhao green tea samples (Table A1), commercially available and harvested from one bud and two leaves of fresh tea leaves (*Camellia sinensis* cv. Huangshan zhong), were sourced from the northern tea region, specifically Rizhao city in Shandong Province, China, and were gathered in extensive amounts in May 2021. Additionally, sensory evaluation of the above 31 RZT samples was performed according to the "Tea Sensory Evaluation Method" (GB/T 23776–2018) [10] and "Tea Sensory Evaluation Terminology" (GB/T 14487–2017) [11]. The sensory evaluation was conducted by 5 tea experts who are qualified as senior tea assessors, all of whom have more than 10 years of experience in tea sensory evaluation. Experts rated the tea samples for appearance, soup color, aroma, taste and the bottom of the leaves, and those with sensory scores <89 were removed. Ultimately, six tea samples that exhibited the most representative and quintessential flavor of RZT were finally selected as the subjects of this study. All tea samples were preserved in aluminum foil bags, shielded from light at −24 °C. Prior to analysis, the tea samples were ground for 40 s into powder at a speed of 24,000 rpm using a tube mill (CS025, IKA, Staufen, Germany). Additionally, a composite sample of the aforementioned six RZT samples was prepared for subsequent GC-O analysis.

2.2. Chemicals

Ethyl decanoate (≥98%) was purchased from Sigma-Aldrich (Shanghai, China). n-alkanes (C3–C7, C8–C40) were purchased from J&K Scientific (Beijing, China). Distilled water was purchased from Wahaha Group Company (Hangzhou, China). Aroma standards were purchased from J&K Scientific (Beijing, China) or Sigma-Aldrich (Shanghai, China), including hexanal (97%), 3-hexen-1-ol (98%), heptanal (98%), hexanoic acid, methyl ester (99%), 1-octen-3-ol (98%), 2-pentyl-furan (98%), octanal (98%), (E)-β-ocimene (≥95%), 3,5-octadien-2-one (95%), linalool (98%), nonanal (98%), phenethyl alcohol (98%), (Z)-3-nonen-1-ol (95%), methyl salicylate (98%), β-cyclocitral (95%), geraniol (98%), citral (99%), (E)-β-ionone (≥98%), indole (99%), (Z)-3-hexenyl hexanoate (99%), hexanoic acid, hexyl ester (99%), (Z)-Jasmone (99.0%), α-ionone (≥98%), geranylacetone (97%), phenylethyl alcohol (99.0%), dihydroactinidiolide (≥98%), and methyl jasmonate (95%).

2.3. SBSE Procedure

The volatiles were extracted using the SBSE technique, and the extraction conditions were based on those in the literature with slight modifications [12]. In detail, 600 mg of ground tea powder together with 500 mg of NaCl was weighed into a 20 mL sample bottle and brewed in 10 mL of boiling water. The PDMS twister (10 mm length, 1 mm thickness, 24 μL capacity, Gerstel, Germany) was immersed into the tea infusion. Then, the infusion was stirred and absorbed for 30 min at 80 °C and 1250 rpm on a multiposition hotplate stirrer (SP200-2 T; Miu Instruments Co., Ltd., Hangzhou, China). Finally, the twister was rinsed with purified water, wiped by nonwovens after extraction, and transferred to a thermal desorption tube for subsequent GC-MS analysis.

2.4. Thermal Desorption

The parameter settings for thermal desorption were based on those in the literature with small modifications [13,14]. For the GC–MS and GC–O analysis, the twister was inserted into the thermal desorption unit (TDU, Gerstel, Mülheim an der Ruhr, Germany) after extraction. The parameters used for the TDU were as follows: the desorption program was held at 30 °C for 1 min and raised to 240 °C (held for 5 min) at a rate of 100 °C/min, in splitless mode. The cooled injection system (Gerstel CIS-4 PTV injector) was maintained at −100 °C using liquid nitrogen (99.99%), and then raised to 280 °C (held for 3 min) at a rate of 12 °C/min sec from −100 °C (held for 1 min) after the desorption of the aroma volatile compounds.

2.5. GC-MS Analysis

The identification and quantification of volatiles were analyzed using an Agilent 7890B GC system coupled with an Agilent 5977B MSD mass spectrometer (Agilent, Santa Clara, CA, USA), equipped with an HP–5MS capillary column (30 m × 0.25 mm × 0.25 μm). The analytical method of GC-MS was based on previous studies [13,14]. The GC oven temperature program was follows: it started from 50 °C for 2 min, increased at 4 °C/min to

170 °C for 5 min, increased at 10 °C/min to 265 °C, and maintained for 5 min. The carrier gas was helium (>99.99%) with a constant flow rate of 1.6 mL/min, in solvent vent mode. Mass spectra were recorded in electron impact (EI) ionization mode at 70 eV. The ion source temperature was 220 °C. The interface temperature was set to 280 °C. The mass scan range was 30–600 AMU.

2.6. Identification and Qualification of the Volatile Compounds in RZT Samples

The volatile compounds were identified based on the NIST 2014 database library. The retention indices (RIs) of the compounds obtained from the experiment were compared with the theoretical values of the n-alkane (C3–C7, C8–C40) series under the same GC conditions. The quantification of the compounds was conducted based on their peak abundance in GC-MS analysis, with the relative content of the volatile compounds expressed in relation to that of ethyl decanoate, utilized as an external standard. For all the aroma volatile compounds extracted via SBSE, quantification was performed by comparing their peak areas to those of the external standard. In this study, the equation defining the external standard curve is as follows:

$$Y = 12{,}766.062X - 111{,}889.342\ (R^2 = 0.996)$$

where $R^2 = 0.996$ indicates a very high degree of correlation between X and Y. Y represents the abundance of volatiles' chromatographic peaks, while X signifies the concentration of these volatiles, expressed in micrograms per kilogram (μg/kg).

2.7. OAVs Calculation

The odor activity value (OAV) is a standard metric used to assess the contributions of aroma compounds. The OAV is calculated as the ratio of the concentration of the compound in the tea infusion (C) to its odor threshold in water (OT). A compound is generally considered to have a vital contributing role to tea aroma quality, when the OAV is $\geq$1.

2.8. GC-O Analysis

The procedure for the GC-O analysis was based on that described in the literature [15]. The GC-O analysis was performed using GC-MS equipped with an ODP-2 olfactory detection port (Gerstel GmbH & Co. KG, Mülheim an der Ruhr, Germany). The volatiles extracted were split between the olfactory detection port and MS in a 1:1 ratio. The temperature of the GC-O injector was set at 230 °C and that of the transfer line was set at 250 °C. The carrier gas was high-purity nitrogen (99.99%). The temperature ramping procedure was the same as that described in Section 2.5.

GC-O analysis was conducted by a panel of experienced assessors, comprising five healthy, non-smoking individuals (three females and two males, all aged between 30 and 40). The GC-O analysis, in conjunction with the detection frequency method, was utilized to evaluate the aroma characteristics and odor intensities on a 4-point scale ranging from 1 to 4. This scale is defined as follows: 1 represents a weak intensity, 2 indicates a moderate intensity, 3 signifies a strong intensity, and 4 denotes an extremely strong intensity.

2.9. Statistical Analysis

All values were presented in the form of means. Figures were drawn using GraphPad Prism 9.0.0 software (GraphPad Software, San Diego, CA, USA), ChemDraw professional 16.0, and the OriginPro 2021 and online website (https://www.chiplot.online/, accessed on 25 August 2024).

3. Results and Discussion

3.1. Characterization of Aroma Compounds in RZT Identified by SBSE-GC-MS

As displayed in Table 1, a total of 112 volatile compounds from six representative and typical RZT samples were detected and identified by SBSE-GC-MS (Figure 2a). According to the differences in chemical structure, these volatiles can be further divided into 12 categories, including aldehydes, alkenes, esters, alcohols, oxygen heterocycles, nitrogen heterocycles, aromatics, phenols, alkanes, organic acids, and amines. Figure 2b illustrates the distribution of quantities and contents for a variety of volatile compounds in RZT. RZT had the greatest number of esters (22 types), followed by alkanes (21 types), alcohols (18 types), ketones (13 types), and so on. In addition, the primary volatile categories of RZT were esters (2035.25 μg/kg), alcohols (1799.02 μg/kg), alkanes (991.88 μg/kg) and ketones (691.96 μg/kg), which together accounted for 74.91% of the total volatile contents. In comparison, phenols (94.73 μg/kg) and amines (155.96 μg/kg) were present at the lowest concentration. As Figure 2c illustrates, twenty volatiles with high abundance (>100 μg/kg) were identified in RZT samples. Notably, (Z)-3-hexanoic acid hexenyl ester (433.71 μg/kg) and geraniol (406.46 μg/kg) were the most abundant. The former has an apple- and pear-like scent and was previously found in high abundance in XinYang MaoJian green tea [16]. It was discovered that (Z)-3-hexanoic acid hexenyl ester could increase the expression of the essential enzyme genes in the synthesis pathway, *CsADH1*, *CsADH3*, and *CsLOX3*, to strengthen the tea plant's tolerance to the cold [17]. Geraniol, a colorless, oily terpene with a rose-like fragrance, is reported to be the most prevalent volatile in pan-fired green teas [12]. Linalool (292.88 μg/kg) was the third most abundant in RZT, sharing the same precursor (geranyl pyrophosphate, GPP) with geraniol [18]. It exhibits a distinct floral and sweet aroma, differing from that of geraniol, and is also abundant in various premium green tea products [5,19,20]. According to earlier research, cold stress can also cause a rise in the amounts of geraniol and linalool [21]. Although the exact process by which geraniol and linalool are inducted in tea is still unknown, it is hypothesized that the rise in these volatiles is connected to an alteration in the tea tree's metabolic pathway at low temperatures.

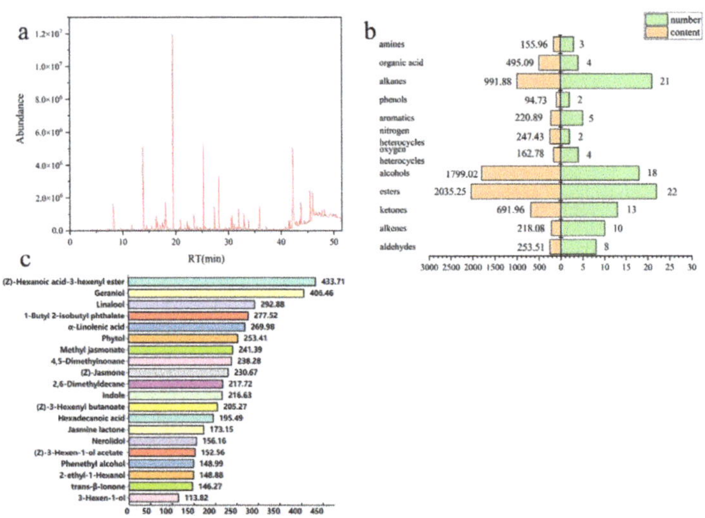

Figure 2. Result of volatile compounds in RZT identified by SBSE-GC-MS. (**a**) A representative total ion flow diagram of the RZT sample. (**b**) Volatile profiles of RZT samples. The left part of the figure shows the content of each volatile category, and the right part shows the number of volatiles in each category. (**c**) Details of volatile compounds with high contents (>100 μg/kg).

Table 1. Volatile compounds of RZT identified by SBSE-GC-MS.

No	RT [a]	Compounds	CAS	RI [b]	Mean Content of RZT (μg/kg)	Content Range of RZT (μg/kg)	OT [c] (μg/kg)	OAV	Odor Type
Aldehydes									
1	7.74	Hexanal	66-25-1	800	51.70	47.62–55.80	4.5	11.49	Grass
2	10.79	Heptanal	111-71-7	903	39.00	34.72–44.76	3	13.00	Green
3	14.33	Octanal	124-13-0	1005	18.88	17.24–20.26	0.7	26.97	Aldehydic
4	16.14	2,6-Dimethyl-5-heptenal	106-72-9	1052	12.50	12.30–12.60	10	1.25	Melon
5	17.96	Nonanal	124-19-6	1106	56.35	49.11–61.05	1	56.35	Aldehydic
6	21.51	Decanal	112-31-2	1208	16.75	15.19–18.75	0.1	167.53	Aldehydic
7	22.18	β-Cyclocitral	432-25-7	1223	34.93	33.36–36.99	3	11.64	Tropical
8	23.83	Citral	5392-40-5	1276	23.40	21.65–24.42	62	<1	Citrus
SUM					253.51				
Alkenes									
1	8.81	2,4-Dimethylhept-1-ene	19549-87-2	843	20.12	18.71–23.94	NF	-	-
2	11.90	α-Pinene	80-56-8	940	16.58	14.94–19.14	180	<1	Herbal
3	13.87	α-Myrcene	123-35-3	993	22.65	20.78–24.60	44.5	<1	Spicy
4	14.10	(+)-4-Carene	29050-33-7	1002	11.39	9.94–13.04	NF	-	-
5	15.32	D-Limonene	138-86-3	1032	34.82	30.67–39.75	4	8.71	Fruity
6	15.94	(E)-β-Ocimene	3779-61-1	1051	22.03	20.48–23.30	18.7	1.18	Floral
7	17.55	4-Methyl-1-undecene	74630-39-0	1085	37.78	29.61–46.95	NF	-	-
8	18.37	(E)-4,8-Dimethylnona-1,3,7-triene	19945-61-0	1117	27.01	24.21–28.85	NF	-	-
9	26.40	α-Cubebene	17699-14-8	1353	13.52	12.99–14.64	NF	-	Herbal
10	27.72	β-cubebene	13744-15-5	1391	12.18	11.71–12.85	NF	-	Citrus
SUM					218.08				
Ketones									
1	7.81	4-methyl-3-Penten-2-one	141-79-7	802	50.31	46.90–54.60	NF	-	Vegetable
2	10.57	2-Heptanone	110-43-0	893	30.74	30.15–31.49	140	4.52	Cheesy
3	13.31	2,2-Dimethyl-3-heptanone	19078-97-8	967	22.53	19.36–29.29	NF	-	-
4	15.52	2,2,6-Trimethyl-cyclohexanone	2408-37-9	1041	28.45	27.17–29.90	NF	-	Spicy Cedar
5	17.02	3,5-Octadien-2-one	38284-27-4	1090	49.28	47.86–50.74	0.5	98.57	Fruity
6	22.93	2-Isopropyl-5-methyl-2-cyclohexen-1-one	5113-66-6	1251	9.98	9.74–10.40	NF	-	-
7	24.46	2-Undecanone	112-12-9	1294	23.43	21.09–24.67	7	3.35	Fruity
8	28.02	(Z)-Jasmone	488-10-8	1397	230.67	219.63–246.44	7	32.95	Floral
9	28.82	α-Ionone	127-41-3	1430	22.10	20.54–23.58	0.4	55.26	Woody, Floral
10	29.45	Geranylacetone	3796-70-1	1455	43.78	38.65–47.43	60	<1	Floral
11	30.52	4-(2,6,6-Trimethylcyclohexa-1,3-dienyl)but-3-en-2-one	1203-08-3	1485	17.29	16.80–17.94	NF	-	-
12	30.59	(E)-β-Ionone	79-77-6	1490	146.27	138.46–159.99	0.09	1625.22	Floral
13	30.74	2-Tridecanone	593-08-8	1501	17.13	16.11–17.80	10,000	<1	Waxy
SUM					691.96				
Esters									
1	11.61	Hexanoic acid, methyl ester	106-70-7	928	22.16	20.11–23.88	4	5.54	Fruity
2	14.40	3(Z)-Hexen-1-ol acetate,	3681-71-8	1007	152.56	144.25–163.41	NF	-	Green

Table 1. *Cont.*

No	RT [a]	Compounds	CAS	RI [b]	Mean Content of RZT (μg/kg)	Content Range of RZT (μg/kg)	OT [c] (μg/kg)	OAV	Odor Type
3	20.76	(Z)-3-Hexenyl butanoate	16491-36-4	1189	205.27	186.02–219.25	20,000	<1	Green
4	21.05	(E)-Butanoic acid- 2-hexenyl ester,	53398-83-7	1196	19.15	17.63–19.98	NF	-	Green
5	21.47	Methyl salicylate	119-36-8	1194	85.68	77.20–93.51	40	2.14	Minty
6	22.32	(Z)-3-Hexenyl 2-methylbutanoate	53398-85-9	1234	76.61	69.87–82.52	NF	-	Green
7	26.48	Propanoic acid, 2-methyl-, 2-ethyl-3-hydroxyhexyl ester	74367-31-0	1373	37.78	29.61–46.95	NF	-	-
8	27.17	Propanoic acid, 2-methyl-, 3-hydroxy-2,2,4-trimethylpentyl ester	77-68-9	1380	68.00	54.05–88.04	NF	-	-
9	27.19	(Z)-Hexanoic acid-3-hexenyl ester,	31501-11-8	1383	433.71	425.88–438.43	NF	-	Green
10	27.33	Hexanoic acid, hexyl ester	6378-65-0	1386	19.49	17.90–20.15	6400	<1	Green
11	27.34	cis-3-Hexenyl cis-3-hexenoate	61444-38-0	1389	35.96	27.91–52.94	NF	-	Green
12	27.43	(E)-Hexanoic acid-2-hexenyl ester	53398-86-0	1391	21.68	20.33–22.65	NF	-	Waxy
13	29.23	β-Phenylethyl butyrate	103-52-6	1450	16.30	15.35–17.04	NF	-	Floral
14	30.99	Jasmine lactone	25524-95-2	1518	173.15	163.00–188.20	2	86.58	Creamy
15	32.22	Dihydroactinidiolide	17092-92-1	1542	61.41	58.56–62.81	NF	-	Musk Coumarin
16	33.04	Acetaminophen	1068-90-2	1602	21.46	19.72–22.63	NF	-	-
17	35.83	Methyl jasmonate	1211-29-6	1655	241.39	230.21–252.00	5700	<1	Floral
18	40.18	Isoamyl laurate	6309-51-9	1848	11.33	10.91–11.65	NF	-	Waxy
19	41.92	1-Butyl 2-isobutyl phthalate	17851-53-5	1933	277.52	240.18–320.20	NF	-	-
20	42.80	Hexadecanoic acid, 15-methyl-, methyl ester	6929-04-0	1984	29.11	19.01–43.86	NF	-	-
21	45.07	9,12-Octadecenoic acid, methyl ester	2462-85-3	2091	11.48	10.61–12.22	NF	-	-
22	45.15	Linolenic acid, methyl ester	301-00-8	2095	14.05	12.28–15.49	NF	-	-
SUM					2035.25				
Alcohols									
1	9.62	(Z-)3-Hexen-1-ol	928-96-1	857	111.82	103.84–123.67	NF	-	Green
2	10.06	1-Hexanol	111-27-3	872	17.14	15.73–20.01	200	<1	Green
3	13.63	1-Octen-3-ol	3391-86-4	982	64.78	61.84–67.61	14	4.63	Earthy
4	15.34	2-Ethyl-1-hexanol	104-76-7	1033	148.88	142.35–158.97	270,000	<1	Citrus
5	16.89	2-Furanmethanol, 5-ethenyltetrahydro-5-trimethyl-, cis-	104188-13-8	1066	69.08	64.60–73.70	NF	-	-
6	17.92	Linalool	78-70-6	1101	292.88	281.05–305.56	4	73.22	Floral
7	18.89	Phenethyl alcohol	60-12-8	1121	148.99	142.27–155.66	1200	<1	Floral
8	19.85	(Z)-3-Nonen-1-ol	10340-23-5	1159	12.02	11.69–12.50	NF	-	Waxy
9	20.44	1-Nonanol	143-08-8	1175	16.47	16.31–16.62	50	<1	Floral, Citrus
10	21.30	α-Terpineol	98-55-5	1191	21.69	20.63–23.67	280	<1	Terpenic
11	23.28	Geraniol	106-24-1	1258	406.46	402.09–415.01	75	5.42	Floral
12	30.15	1-Dodecanol	112-53-8	1477	15.23	13.75–16.69	73	<1	Waxy
13	31.02	(3S,3aR,3bR,4S,7R,7aR)-4-Isopropyl-3,7-dimethyloctahydro-1H	23445-02-5	1515	21.85	20.26–23.35	NF	-	Spicy
14	31.63	Bicyclo [3.1.0]hexan-2-ol, 5-[(1R)-1,5-dimethyl-4-hexen-1-yl]-2-methyl	58319-05-4	1540	11.84	11.10–12.57	NF	-	

Table 1. *Cont.*

No	RT [a]	Compounds	CAS	RI [b]	Mean Content of RZT (μg/kg)	Content Range of RZT (μg/kg)	OT [c] (μg/kg)	OAV	Odor Type
15	32.82	Nerolidol	7212-44-4	1568	156.16	146.93–169.98	10	15.62	Floral
16	34.58	(-)-Torreyol	19435-97-3	1642	17.85	16.55–19.40	NF	-	Herbal
17	35.31	α-Cadinol	481-34-5	1653	12.47	12.31–12.73	NF	-	Herbal
18	45.31	Phytol	150-86-7	2109	253.41	220.29–292.77	0.64	395.95	Floral
SUM					1799.02				
Oxygen heterocycles									
1	20.60	(E)-linalool oxide (pyranoid)	39,028-58-5	1180	48.98	45.76–51.61	3000	<1	Woody
2	13.93	2-Pentyl-furan	3777-69-3	995	33.43	23.28–37.09	6	5.57	Fruity
3	13.94	(2R,5R)-2-Methyl-5-(prop-1-en-2-yl)-2-vinyltetrahydrofuran	54750-70-8	994	17.46	16.20–19.83	NF	-	-
4	17.46	(E)-Linalool oxide (furanoid)	34995-77-2	1091	62.91	58.10–70.62	190	<1	Woody, Floral
SUM					162.78				
Nitrogen heterocycles									
1	16.27	Tea pyrrole	2167-14-8	1053	30.80	29.06–32.38	65,000	<1	Roasted
2	25.35	Indole	120-72-9	1302	216.63	200.18–237.29	140	1.55	Floral
SUM					247.43				
Aromatics									
1	7.01	Toluene	108-88-3	771	82.69	75.05–94.57	NF	-	-
2	9.88	1,3-Dimethyl-benzene	108-38-3	873	65.94	59.29–82.55	41	1.61	Benzene-like
3	15.19	1-Methyl-3-(1-methylethyl)-benzene	535-77-3	1029	23.16	19.86–27.59	NF	-	-
4	31.34	Butylated Hydroxytoluene	128-37-0	1518	34.99	31.34–39.13	NF	-	Phenolic
5	28.49	1,6-Dimethylnaphthalene	575-43-9	1428	14.11	12.82–15.34	NF	-	-
SUM					220.89				
Phenols									
1	31.46	2,4-Bis(1,1-dimethylethyl)-phenol	96-76-4	1523	83.33	74.51–92.86	NF	-	-
2	36.42	Juniper camphor	473-04-1	1700	11.40	10.27–12.14	NF	-	-
SUM					94.73				
Alkanes									
1	8.22	2,4-Dimethyl-heptane	2213-23-2	822	90.53	85.76–98.94	NF	-	-
2	9.25	2,3-dimethyl-heptane	3074-71-3	856	15.57	14.41–19.23	NF	-	-
3	9.48	4-Methyl-octane	2216-34-4	864	59.15	54.37–67.28	NF	-	-
4	14.93	2,6-Dimethyl-nonane	17302-28-2	1020	31.60	23.80–42.38	NF	-	-
5	16.20	4,5-Dimethylnonane	17302-23-7	1057	238.28	220.13–254.99	NF	-	-
6	18.01	2,6-Dimethyldecane	13,150-81-7	1112	217.72	209.81–227.89	NF	-	-
7	18.71	3,7-Dimethyl-decane,	17312-54-8	1126	13.93	12.86–16.16	NF	-	-
8	21.24	4,7-Dimethyl-undecane,	17301-32-5	1208	33.16	25.34–41.45	NF	-	-
9	23.91	2,6,11-Trimethyl-dodecane	31295-56-4	1275	47.10	35.96–59.10	NF	-	-
10	24.64	2,3,5,8-Tetramethyl-decane,	192823-15-7	1318	21.34	17.30–25.02	NF	-	-
11	25.42	4,6-Dimethyl-dodecane,	61141-72-8	1325	35.34	28.16–46.36	NF	-	-
12	25.75	2,2,4,4,6,8,8-Heptamethylnonane	4390-04-9	1326	22.78	19.32–27.97	NF	-	-
13	26.08	1-Cyclohexylheptane	5617-41-4	1346	11.48	11.00–12.06	NF	-	-
14	26.30	2,6,10-Trimethyl-dodecane,	3891-98-3	1364	13.60	11.79–15.46	NF	-	-

Table 1. *Cont.*

No	RT [a]	Compounds	CAS	RI [b]	Mean Content of RZT (μg/kg)	Content Range of RZT (μg/kg)	OT [c] (μg/kg)	OAV	Odor Type
15	26.87	3-Methyl-tridecane,	6418-41-3	1372	12.97	11.87–14.27	NF	-	-
16	27.78	6-Ethyl-2-methyldecane	62108-21-8	1390	24.05	18.99–28.36	NF	-	-
17	30.79	5-Propyl-tridecane,	55045-11-9	1502	20.56	18.50–22.65	NF	-	-
18	32.40	Nonylcyclohexane	2883-02-5	1558	15.46	14.25–16.47	NF	-	-
19	37.66	Phytane	638-36-8	1789	16.75	16.07–17.58	NF	-	-
20	40.48	2,6,10,14-Tetramethylheptadecane	18344-37-1	1872	17.79	15.65–21.63	NF	-	-
21	41.30	1,2-Epoxyoctadecane	7390-81-0	1900	32.72	23.14–43.23	NF	-	-
SUM					991.88				
Organic acid									
1	26.26	2-Methoxy-, methyl ester benzoic acid,	579-75-9	1347	11.93	11.68–12.28	NF	-	-
2	39.93	Pentadecylic acid	1002-84-2	1861	17.69	13.95–21.24	NF	-	Waxy
3	43.66	Hexadecanoic acid	57-10-3	1975	195.49	173.03–210.17	10,000	<1	Waxy
4	45.87	α-Linolenic acid	463-40-1	2119	269.98	241.30–291.66	NF	-	Fatty
SUM					495.09				
Amines									
1	24.97	N,N-dibutyl-formamide	761-65-9	1310	51.24	47.65–56.54	NF	-	-
2	42.25	5-Methyl-2-benzylhydrazide-3-isoxazolecarboxylic acid,	59-63-2	1945	14.81	14.00–15.23	NF	-	-
3	47.14	Hexadecanamide	629-54-9	2182	89.91	80.21–98.10	NF	-	-
SUM					155.96				

[a] Retention time. [b] The retention index (RI) from the published literature and an online library (https://webbook.nist.gov/chemistry/cas-ser.html, accessed on accessed on 5 July 2024); [c] odor threshold, all the odor thresholds were obtained from the literature (Fenaroli's handbook of flavor ingredients). -, no odor description information was found in the literature. NF, not found in the literature.

In addition to the aforementioned, RZT contains a variety of other volatiles in higher concentrations, such as 1-butyl-2-isobutyl phthalate, α-linolenic acid, phytol, methyl jasmonate, (Z)-jasmone, indole, (Z)-3-hexenyl butanoate, hexadecanoic acid, jasmine lactone, (Z)-3-hexen-1-ol acetate, phenethyl alcohol, 2-ethyl-1-hexanol, (E)-β-ionone, (Z)-3-hexen-1-ol and others. Phytol, found in high abundance in Jingmai and Wuliang pu-erh green teas, has been reported to contribute to the aroma characteristics of green odor [22]. (E)-β-ionone, found in the majority of teas, possesses a scent reminiscent of violets. Both 3-hexen-1-ol and (Z)-3-hexenyl butanoate exhibit green flavor profiles; the former is crucial to the aroma of Longjing and Xinyang Maojian [16,23], while the latter is the predominant volatile and characteristic aroma compound of white tea [24]. Phenethyl alcohol has been detected in Laoshan green tea and imparts a rose-like scent [25]. 2-ethyl-1-hexanol contributes a citrus flavor and indole has floral aroma in a highly diluted form; both have been identified as key contributors to the chestnut-like aroma of Taiping Houkui green tea [26], and it is hypothesized that these two aforementioned volatiles may also have been factors contributing to the chestnut-like flavor of RZT samples in this research. α-linolenic acid is a representative unsaturated fatty acid in tea, and emits a waxy and fatty odor. Additionally, the unsaturated fatty acids serve as significant aroma precursors in tea, and their biosynthesis and degradation can be influenced by stress [18].

Figure 3 presents our hypothesis regarding the impact of low-temperature stress on fatty acid derivatives in RZT, based on the volatiles detected in this study. Research has shown that unsaturated fatty acids in tea can be oxidized by lipoxygenase (LOX) and subsequently decomposed into six-carbon aliphatic aromatic compounds. The metabolic pathways diverge based on the fate of hydrogen peroxide, an LOX oxidation product, into two branches: the LOX-AOS (lipoxygenase-allene oxygenase) and LOX-HPL (lipoxygenase-hydroperoxide lyases) pathways [27–29]. Unsaturated fatty acids can be converted into green leaf volatiles such as (Z)-3-hexen-1-ol and (Z)-3-hexen-1-ol acetate through oxidation, reduction, and esterification in the LOX-HPL pathway, and converted to jasmonic acid (JA) through oxidation and reduction in the LOX-AOS pathway [30]. Both pathways are significantly affected by environmental stresses. Low-temperature stress can enhance the accumulation of jasmonic acid in fresh tea leaves, and the concentration of its derivative, methyl jasmonate (MeJA), is also substantially elevated [31]. MeJA, known for its floral, fresh magnolia, and oily waxy scent, acts as a signaling factor that can induce changes in the levels of aroma precursors [32]. According to pertinent studies, the entire JA pathway in tea leaves can be rapidly activated by external MeJA, significantly affecting the biosynthesis of terpenoid backbones and leading to an increase in volatiles like linalool and geraniol [33]. External MeJA can also efficiently promote ROS (reactive oxygen species) scavenging and maintain cell membrane stability under cold stress [34]. Jasmine lactone, a cyclic volatile resulting from fat degradation [4], is characteristic of Oolong tea, and its accumulation in tea can be triggered by external stress such as low temperatures [35]. Therefore, it is hypothesized that the low-temperature stress caused by its high-latitude geographical location could be a major factor affecting the elevated levels of some unsaturated fatty acid derivative volatiles in RZT.

Although 1-butyl-2-isobutyl phthalate was abundant in RZT samples, there are no reports indicating its direct contribution to tea aroma. The same applies to hexadecanoic acid and alkanes; they are presumed to have negligible effects on the flavor of RZT samples due to their undetectable odor thresholds, and they cannot they be detected by GC-O analysis; however, it is hypothesized that these odorless volatiles may interact with other components such as amino acids and tea polyphenols, thereby influencing the flavor of tea.

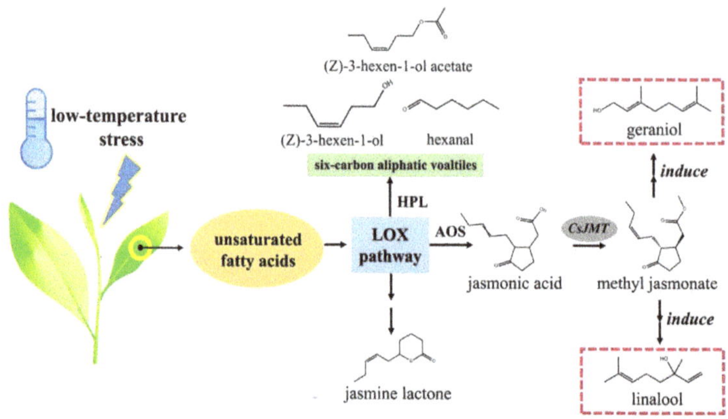

Figure 3. Effects of low-temperature stress on volatiles in fresh leaves of RZT (speculated). LOX: lipoxygenase; HPL: hydroperoxide lyases; AOS: allene oxygenase; JMT: JA carboxyl methyltransferase.

3.2. Key Odorants in RZT Identified by OAV Analysis

As is well acknowledged, not all volatile compounds contribute to the aroma of tea. The quality and type of tea aroma are determined by the complex interaction of various volatile compounds at different concentrations, each with its own aroma contribution [15]. It is therefore insufficient to assess the contribution of aroma compounds to the aroma quality of tea just based on their content solely. The odor activity value (OAV) of a volatile reflects its influence on the overall aroma quality of tea; a compound with an OAV greater than 1 is typically considered a key odorant to the aroma and could help determine the overall aromatic quality [36]. Consequently, this study involved determining odor thresholds and calculating OAVs for 112 volatiles detected by GC-MS in order to screen out the key odorants that substantially influence the aroma quality of RZT. Table 1 presents the final computational and screening results. Out of the 45 volatiles analyzed, 26 were found to have OAVs greater than 1. These volatiles comprised seven aldehydes, two alkenes, six ketones, three esters, five alcohols, one oxygen heterocycle, one nitrogen heterocycle, and one aromatic. In terms of the aroma distribution of volatile compounds with an OAV > 1, the distinctive aroma profiles of RZT were characterized as floral (nine volatiles), fruity (five volatiles), and aldehydic (three volatiles), with cumulative OAVs of 2206.37, 121.74, and 250.85, respectively.

It is a well-established fact that a volatile compound can significantly influence the overall flavor profile of tea when its OAV exceeds 100 [37]. Among the identified volatiles, three volatiles had OAVs > 100, indicating their pivotal roles in the aroma of RZT: (E)-β-ionone (OAV = 1625.22), followed by phytol (OAV = 395.95), and decanal (OAV = 167.53). Notably, decanal is characterized by its fatty odor and is recognized as a key aroma compound in Chinese chestnuts [14,38]. Consequently, decanal is presumed to have a substantial effect on the chestnut-like aroma of RZT. Among other volatiles with OAVs > 10 in RZT, 3,5-octadien-2-one (OAV = 98.57) exhibits a fruity odor and is confirmed as a key odorant of Longjing tea [39]. Hexanal, heptanal, octanal, nonanal, 1-octene-3-ol, (Z)-jasmone, and jasmine lactone are primarily produced through the degradation of fatty acids [6], while β-Cyclocitral, (E)-β-ionone, α-ionone, and nerolidol are aromatic compounds derived from carotenoid cleavage [16]. Additionally, there exists a positive correlation between the quality of the chestnut-like aroma and the levels of (E)-β-ionone and nerolidol [40]. Other volatiles with OAVs > 1 include D-limonene, 1-octen-3-ol, 2-pentylfuran, hexanoic acid, methyl ester, methyl salicylate, indole, and (E)-β-ocimene, which have been identified as characteristic or aroma-active compounds in tea [7,15,41–43]. Additionally, previous research has indicated that D-limonene, 2-pentylfuran, and (E)-β-ocimene are the common aroma compounds of curled and pelleted RZT [44].

While research on the OAVs of tea volatiles has advanced, the limitations of OAV analysis have become increasingly apparent. For instance, certain volatiles exhibit a disproportionate relationship between their intensity and concentrations [7]. Moreover, the OAV alone is inadequate for accurately ranking the contributions of volatiles to a specific flavor profile. Some volatiles with lower OAVs might still be crucial to the overall flavor [45–47]. Consequently, to enhance OAV analysis, GC-O analysis which incorporates instrumental analysis and sensory determinations needs to be employed.

3.3. Key Odorants in RZT Identified by GC-O Analysis

GC-O, which uses the human olfactory system rather than an electronic nose, offers heightened sensitivity and is widely applied in contemporary research to detect a multitude of significant aroma compounds [48–50]. Compared to OAV analysis, GC-O is capable of evaluating the individual contributions of volatiles to the overall aroma, facilitating the identification of aroma-active compounds within complex mixtures [51].

By applying GC-O analysis, the sensory panel detected 37 key odorants from RZT samples, as indicated in Table 2. These odorants were categorized into seven groups (Figure 4a) based on their aroma types: floral and sweet (14 volatiles), fruity, citrus, and tropical (7 volatiles), green and minty (7 volatiles), aldehydic (4 volatiles), herbal (3 volatiles), cheesy (1 volatile), and earthy (1 volatile). Figure 4b shows that the floral and sweet category was the most prevalent in RZT, both in terms of the number of compounds and overall aroma intensities, with linalool exhibiting the highest aroma intensity (AI = 3.67). Followed by the green and minty aroma type, heptanal had the highest aroma intensity value (AI = 2.80). In the fruity, citrus, and tropical category, 2-pentylfuran had the highest aroma intensity (AI = 3.00). Among the aldehydic types, octanal and nonanal were found to have the most pronounced aroma intensity (AI = 3.00). Additionally, two unidentified chemicals (unknown 1 and unknown 2) with high aroma intensities that appeared as coconutty and floral and sweet were included in the samples. However, it is presumed that their concentrations in the tea were too low for structural identification. Therefore, the use of more sensitive instruments with lower detection limits, such as comprehensive two-dimensional gas chromatography coupled with time-of-flight mass spectrometry (GCxGC-TOFMS), is necessary for further systematic identification analysis.

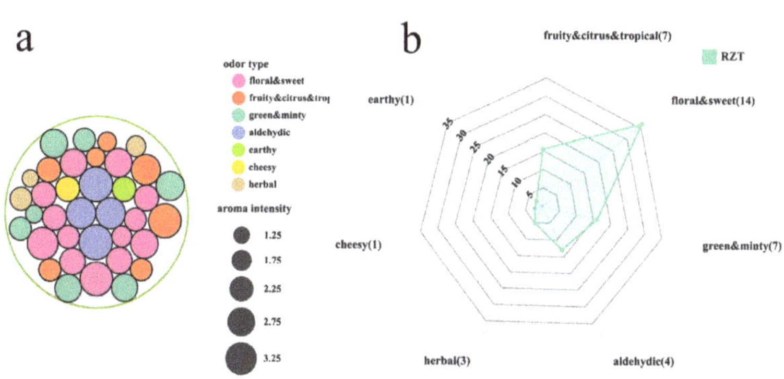

Figure 4. The results of GC-O analysis. (**a**) Circle packing plots of the odorants. (**b**) Radar map of the odorants. The *X*-axis represents the aroma type and number of corresponding compounds in that type, while the *Y*-axis represents the cumulative aroma intensity of different types of odorants.

Table 2. Key odorants in RZT identified by GC-O analysis.

No.	TR (min) [d]	Aroma Compounds	Odor Descriptors	AI [e]	IM [f]
1	7.72–8.01	Hexanal	Grass, leafy	2.67	MS [g], RI [h], A [i], O [j]
2	9.63–9.77	3-Hexen-1-ol	Green, grassy	1.33	MS, RI, A, O
3	10.45–10.57	2-Heptanone	Cheese, coconut	2.25	MS, RI, A
4	10.80–11.23	Heptanal	Fresh, green	2.80	MS, RI, A, O
5	11.50–11.97	Hexanoic acid, methyl ester	Fruity, sweet	2.25	MS, RI, A, O
6	13.45–13.78	1-Octen-3-ol	Green, metallic	2.25	MS, RI, A, O
7	13.75–14.07	2-Pentyl-furan	Fruity, green	3.00	MS, RI, A, O
8	14.32–14.57	Octanal	Aldhydic, waxy	3.50	MS, RI, A, O
9	15.04–15.15	2-Ethyl-1-hexanol	Sweet, herbal, green	1.50	MS, RI, A, O
10	16.19–16.29	(E)-β-Ocimene	Sweet, floral	2.00	MS, RI, A, O
11	16.95–17.23	3,5-Octadien-2-one	Fruity, green	2.00	MS, RI, A, O
12	17.46–17.83	(E)-Linalool oxide (furanoid)	Floral, coffee, baked	2.67	MS, RI, A
13	17.53–18.14	Linalool	Floral, sweet	3.67	MS, RI, A, O
14	17.89–18.14	Nonanal	Waxy aldehydic	3.50	MS, RI, A, O
15	18.89–19.12	Phenethyl alcohol	Sweet, floral	2.50	MS, RI, A, O
16	19.85–20.13	(Z)-3-Nonen-1-ol	Mushroom, green, spicy	3.00	MS, RI, A, O
17	20.49–20.72	(E)-Linalool oxide (pyranoid)	Fresh, woody	3.17	MS, RI, A
18	21.41–21.45	Methyl salicylate	Wintergreen, minty	2.00	MS, RI, A, O
19	22.12–22.46	β-Cyclocitral	Fresh, clean, citrus	1.50	MS, RI, A, O
20	22.38–22.99	(Z)-3-Hexenyl 2-methylbutanoate	Green, apple, sweet	2.00	MS, RI, A
21	23.28–23.49	Geraniol	floral, sweet	2.33	MS, RI, A, O
22	23.93–24.09	Citral	Sweet, citrus, fruity	2.00	MS, RI, A, O
23	25.34–25.56	Indole	Floral, sweet, plastic	2.75	MS, RI, A, O
24	26.50–26.66	α-Cubebene	herbal	2.00	MS, RI, A,
25	26.79–27.05	(Z)-3-Hexenyl hexanoate	Citrus, fruity,	2.60	MS, RI, A, O
26	27.44–27.76	Hexanoic acid, hexyl ester	Fruity, green, sweet	2.60	MS, RI, A, O
27	28.04–28.38	(Z)-Jasmone	Floral, sour, sweet	2.33	MS, RI, A, O
28	28.85–29.04	α-Ionone	Floral, violet	1.75	MS, RI, A, O
29	29.25–29.75	Geranylacetone	Magnolia, sweet	2.00	MS, RI, A, O
30	29.98–30.29	1-Dodecanol	Fatty, waxy, sweet	3.00	MS, RI, A
31	30.51–30.79	(E)-β-Ionone	Floral, orris	3.40	MS, RI, A, O
32	31.06–31.53	Unknown 1	coconut	3.40	-
33	31.87–32.25	Dihydroactinidiolide	Floral, sweet, fresh	2.50	MS, RI, A, O
34	34.46–34.70	(-)-Torreyol	Fresh, sweet, herbal	1.25	MS, RI, A
35	35.33–35.61	α-Cadinol	Herbal. green, mild sweet	1.67	MS, RI, A
36	35.83–36.11	Methyl jasmonate	Floral, sweet, milk	2.25	MS, RI, A, O
37	37.11–37.43	Unknown 2	Floral	2.75	-

[d]: time range. [e]: aroma intensity. [f]: identification method (IM). [g]: mass spectrum (MS), compared with the Nist library. [h]: retention index (RI), compared with the RI in the published literature and an online library (https://webbook.nist.gov/chemistry/cas-ser.html, accessed on 5 July 2024). [i]: aroma descriptors (A). [j]: (O) odor of the authentic standard.

3.4. Comparison of the Key Odorants in RZT Screened by OAV and GC-O

As anticipated, discrepancies exist between the OAV and GC-O results, as evidenced in Tables 1 and 2. Some volatiles with lower OAVs exhibit stronger aroma intensities in GC-O, while certain volatile compounds with higher OAVs might not be detected at all. For example, phytol, with an OAV of approximately 400, was not identified in the GC-O analysis. Similarly, (E)-linalool oxide (furanoid), possessing an aroma intensity of 2.67, had an OAV less than 1. In actuality, the intricate process of volatile compound perception is affected by various factors, namely compound concentration, volatility, and the antagonism and synergy interactions between aroma compounds [52]. Studies on black tea, for instance, have demonstrated that methyl salicylate can exert diverse perceptual interaction effects on volatiles related to floral flavors (a masking effect, additive effect and synergistic effect) [53]. Calculating the OAV for a single chemical may not account for the complex reactions between aroma compounds. Therefore, it is hypothesized that these interaction effects could be responsible for the observed discrepancies in this study.

Despite minor discrepancies between the outcomes of GC-O and OAV analysis, the synergistic application of these two methodologies offers a more holistic portrayal of the aromatic profile of RZT. As depicted in Figure 5a, OAV and GC-O identified 25 and 37 key odorants, respectively, while the two groups mentioned above shared 17 common odorants. Therefore, the above 17 odorants, including linalool, geranial, (Z)-jasmone, α-ionone, (E)-β-ionone, indole, linalool, geranial indole, hexanoic acid methyl ester, 2-pentylfuran, 3,5-octadien-2-one, β-cyclocitral, methyl salicylate, heptanal, hexanal, octanal, nonanal, 1-octen-3-ol, and 2-heptanone, verified by OAV and GC-O methods, were considered the key odorants of RZT. The distinctive aroma quality of RZT arises from the complex interactions among these key odorants, which have thus been characterized as the key contributors to RZT's aroma in this study. It is significant to note that prior research on Rizhao green tea's aroma also identified linalool, hexanal, heptanal, octanal, and nonanal as the predominant compounds responsible for its chestnut-like aroma [54], aligning with our findings. The overall quantity of key odorants and volatiles of RZT found in the aforementioned study, however, slightly diverges from the findings of this investigation, this discrepancy being potentially attributable to differences in the methodologies employed for aroma extraction and detection.

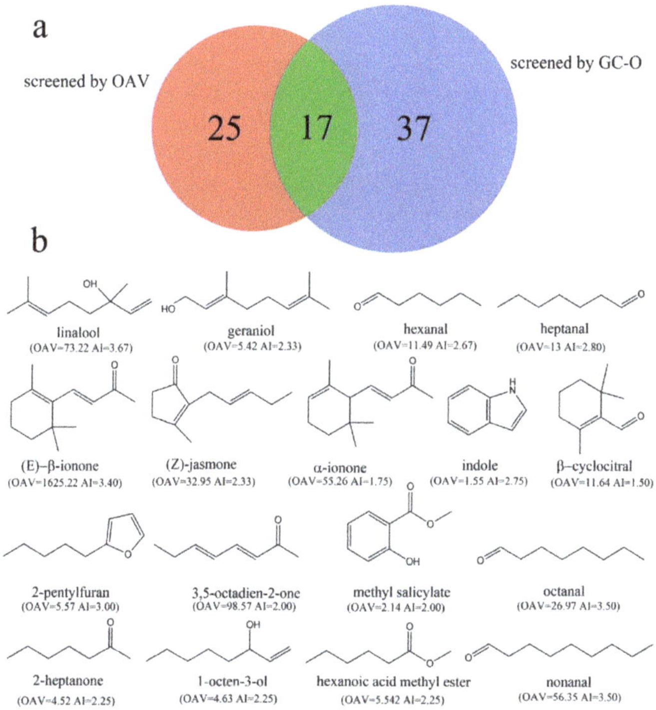

Figure 5. Key odorants based on OAV calculations and GC-O analysis in RZT. (**a**): a venn diagram of key odorants identified by OAV calculations and GC-O analysis. (**b**): chemical structure of 17 key odorants with dual verification of OAV and GC-O (the unknowns in GC-O cannot be categorized as key odorants because their names and chemical structures were not clear).

In addition to the characteristic chestnut-like aroma of RZT, the volatile compounds responsible for its seaweed-like aroma are equally appealing. It has been proven that hexanal contributes to the formation of seafood flavors by providing its seaweed-like aroma [55] and is also identified as one of the principal compounds contributing to the aroma of Japanese matcha [56,57]. Additionally, 2-heptanone, found abundantly in various seafoods such as squid, seaweed, and crabs, is also a key aroma component of many seafood products [58–60]. Nevertheless, it exhibits a cream-like and cheese-like odor, which implies that 2-heptanone may contribute to the seaweed-like aroma of RZT through interactions with other components.

3.5. Establishment of Molecule Aroma Wheel of RZT

The aroma wheel served as a visual and intuitive tool for categorizing diverse types of flavor descriptions [61]. Currently, it is used in flavor research on beverages and condiments such as wine, coffee, and soy sauce [62–64]. The development of an aroma wheel can shed light on the chemical foundation of the distinctive aromatic qualities of RZT.

Building upon the sensory aroma wheel and the findings of previous studies as references [7,65], we refined the classification of key odorants according to their aroma types in order to more accurately define the aroma characteristics of RZT. The final RZT molecular aroma wheel is presented in Figure 6. The aroma quality of RZT was divided into seven groups using the molecular aroma wheel. Notably, the floral and sweet aroma type had the highest number of representative compounds, including linalool, geranial, (Z)-jasmone, α-ionone, (E)-β-ionone, and indole. Typical compounds of the fruity and citrus and tropical categories include hexanoic acid methyl ester, 2-pentylfuran, 3,5-octadien-2-one, and β-cyclocitral. Representative compounds of the green and minty category include methyl salicylate, heptanal, and hexanal. The aldehydic category includes octanal and nonanal, the earthy category contains 1-octen-3-ol, and the cheesy category comprises 2-heptanone. Through the integration of chemical composition and sensory descriptions, the molecular aroma wheel provides a comprehensive and consistent framework for evaluating RZT aromas, which can guide the creation of tea products with specific odor properties in future production.

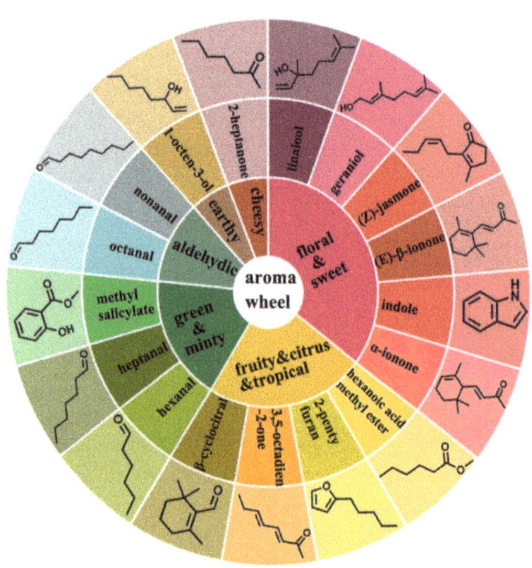

Figure 6. The molecular aroma wheel of RZT.

However, as previously noted, the aroma quality of tea is not solely attributed to the presence of a single volatile; instead, it is the result of the comprehensive actions of multiple volatiles with significant aroma contributions. Moreover, these volatiles interact with one another through various mechanisms, including masking, additive, and synergistic effects. Therefore, the chestnut- and seaweed-like flavor characteristics of RZT do not suggest that the key odorants are confined to those that directly exhibit these flavors. Volatiles with floral, sweet, green, minty, aldehydic, and other odor types might also promote the perception of chestnut- and seaweed-like flavors in RZT. Consequently, future research on the flavor of RZT, including omission testing and other cutting-edge techniques, will be conducted to pinpoint the key odorants and ascertain their respective impacts.

While this study has yielded significant findings, several limitations must be acknowledged. Firstly, commercially available tea samples were used in this study, which may affect result reliability due to the slight differences in the production process. Additionally, the precursors and the biochemical pathways leading to the formation of these key odorants remain obscure, as do the regulatory mechanisms under the specific conditions of high latitude and low temperature. To address these limitations, and to conduct a more comprehensive investigation into RZT's key odorants, future research should adopt a refined sampling strategy by processing the collected fresh tea leaves in a standardized technology. Moreover, sophisticated analysis technologies, including metabonomics and transcriptomics, will be essential to providing deeper insights into the underlying mechanisms.

4. Conclusions

Rizhao green tea (RZT) is renowned for its excellent aroma quality. This is the first comprehensive investigation to demonstrate the key odorants responsible for RZT's unique flavor characteristic by conducting SBSE-GC-MS, OAVs and GC-O analysis. The findings indicate that RZT boasts a rich profile of volatile compounds, with alcohols, esters, alkanes, and ketones emerging as the predominant aromatic categories. Moreover, twenty-six and thirty-seven key odorants were identified by the OAVs and GC-O techniques, respectively, while seventeen key odorants, which were also confirmed through dual verification via OAVs and GC-O assessments, were revealed to be ultimately crucial in shaping RZT's flavor profile. Furthermore, the RZT molecular aroma wheel was established based on these key odorants and their associated odor types. This research offers an in-depth analysis of RZT's aroma characteristics, establishing a basis for the enhancement of processing techniques and the improvement of RZT's aroma quality. Future research will examine how the amount of aroma precursors in tea leaves varies from winter to spring in order to shed light on how temperature and seasonal variations affect RZT's aroma qualities at high latitudes. In addition, we will broaden the tea varieties and sample areas, and strive to provide a more thorough theoretical foundation for the investigation of the aroma characteristics of green tea in northern China.

Author Contributions: Conceptualization, H.L.; data curation, M.W. and D.S.; formal analysis, Y.S.; funding acquisition, S.D.; investigation, H.W. and J.L.; methodology, D.S. and Y.Z.; project administration, H.L.; resources, D.S., K.W. and S.D.; software, M.W. and H.Y.; supervision, H.L. and S.D.; validation, F.F.; visualization, M.W. and Y.S.; writing—original draft, M.W.; writing—review and editing, M.W., D.S. and H.Y. All authors have read and agreed to the published version of the manuscript.

Funding: This work was supported by the earmarked fund for China Agricultural Research System (CARS-19), the Technology System of Modern Agricultural Industry in Shandong Province (SDAIT-

03), the Shandong Tea Industry Quality and Efficiency Technology Integrated Innovation and Demonstration Project (2021TZXD009) and the Rizhao City Natural Science Foundation (RZ2022RZ56).

Institutional Review Board Statement: Not applicable.

Informed Consent Statement: Informed consent was obtained from all subjects involved in the study.

Data Availability Statement: The original contributions presented in this study are included in the article. Further inquiries can be directed to the corresponding authors.

Acknowledgments: We would like to acknowledge the contributions of all collaborators and institutions that supported this research.

Conflicts of Interest: The authors declare no conflicts of interest.

Appendix A

Table A1. Information of commercially available tea samples collected in this research.

No.	Tender	Company	Tea Quality Grade
1.	One bud and two leaves	GuanQing Tea Technology Co., Ltd.	Super Class
2.	One bud and two leaves	HengShan Tianhu Tea Co., Ltd.	Super Class
3.	One bud and two leaves	HengShan Tianhu Tea Co., Ltd.	First class
4.	One bud and two leaves	BiBo Tea Co., Ltd.	Super Class
5.	One bud and two leaves	BiBo Tea Co., Ltd.	First class
6.	One bud and two leaves	BiBo Tea Co., Ltd.	First class
7.	One bud and two leaves	Lin Yuan Tea Industry Co., Ltd.	Super Class
8.	One bud and two leaves	Lin Yuan Tea Industry Co., Ltd.	First class
9.	One bud and two leaves	MaLing Spring Tea Co., Ltd.	Super Class
10.	One bud and two leaves	MaLing Spring Tea Co., Ltd.	First class
11.	One bud and two leaves	KanHai (Lechun Family Farm)	Super Class
12.	One bud and two leaves	KanHai (Lechun Family Farm)	First class
13.	One bud and two leaves	ShengGu Mountain tea Co., Ltd.	Super Class
14.	One bud and two leaves	ShengGu Mountain tea Co., Ltd.	First class
15.	One bud and two leaves	Liu YuanChun Ecological Agriculture Co., Ltd.	Super Class
16.	One bud and two leaves	Liu YuanChun Ecological Agriculture Co., Ltd.	First class
17.	One bud and two leaves	Lukui Tea Industry Co., Ltd.	Super Class
18.	One bud and two leaves	Lukui Tea Industry Co., Ltd.	First class
19.	One bud and two leaves	Lukui Tea Industry Co., Ltd.	First class
20.	One bud and two leaves	FuYuan Spring Tea Co., Ltd.	Super Class
21.	One bud and two leaves	FuYuan Spring Tea Co., Ltd.	First class
22.	One bud and two leaves	FuLaiQing Group Co., Ltd.	Super Class
23.	One bud and two leaves	RiZhao Royal Bay Tea Expo Garden Co., Ltd.	Super Class
24.	One bud and two leaves	RiZhao Royal Bay Tea Expo Garden Co., Ltd.	First class
25.	One bud and two leaves	RiZhao SongChen Tea Industry Trade Co., Ltd.	Super Class
26.	One bud and two leaves	RiZhao SongChen Tea Industry Trade Co., Ltd.	First class
27.	One bud and two leaves	Rizhao Zhulongshan Green Tea Co., Ltd.	First class
28.	One bud and two leaves	Rizhao Zhongtai Tea Co., Ltd.	First class
29.	One bud and two leaves	RiZhao XinTian Tea Leaf Professional Cooperative	Super Class
30.	One bud and two leaves	RiZhao ShengYe Tea Co., Ltd.	First class
31.	One bud and two leaves	RiZhao City JingYang Green Tea Garden	First class

References

1. Lee, J.E.; Lee, B.J.; Chung, J.O.; Hwang, J.A.; Lee, S.J.; Lee, C.H.; Hong, Y.S. Geographical and climatic dependencies of green tea (*Camellia sinensis*) metabolites: A 1H NMR-based metabolomics study. *J. Agric. Food Chem.* **2010**, *58*, 10582–10589. [CrossRef] [PubMed]
2. Gu, H.W.; Yin, X.L.; Peng, T.Q.; Pan, Y.; Cui, H.N.; Li, Z.Q.; Sun, W.; Ding, B.; Hu, X.-C.; Zhang, Z.-H.; et al. Geographical origin identification and chemical markers screening of Chinese green tea using two-dimensional fingerprints technique coupled with multivariate chemometric methods. *Food Control* **2022**, *135*, 108795. [CrossRef]

3. Wang, Q.; Qin, D.; Huang, G.; Jiang, X.; Fang, K.; Wang, Q.; Ni, E.; Li, B.; Pan, C.; Li, H.; et al. Identification and characterization of the key volatile flavor compounds in black teas from distinct regions worldwide. *J. Food Sci.* **2022**, *87*, 3433–3446. [CrossRef]
4. Ho, C.T.; Zheng, X.; Li, S. Tea aroma formation. *Food Sci. Hum. Wellness* **2015**, *4*, 9–27. [CrossRef]
5. Zhu, Y.; Lv, H.; Dai, W.; Guo, L.; Tan, J.; Zhang, Y.; Yu, F.; Shao, C.; Peng, Q.; Lin, Z. Separation of aroma components in Xihu Longjing tea using simultaneous distillation extraction with comprehensive two-dimensional gas chromatography-time-of-flight mass spectrometry. *Sep. Purif. Technol.* **2016**, *164*, 146–154. [CrossRef]
6. Yang, Z.; Baldermann, S.; Watanabe, N. Recent studies of the volatile compounds in tea. *Food Res. Int.* **2013**, *3*, 585–599. [CrossRef]
7. Ma, W.; Zhu, Y.; Ma, S.; Shi, J.; Yan, H.; Lin, Z.; Lv, H. Aroma characterisation of Liu-pao tea based on volatile fingerprint and aroma wheel using SBSE-GC–MS. *Food Chem.* **2023**, *414*, 135739. [CrossRef] [PubMed]
8. Gamero, A.; Wesselink, W.; de Jong, C. Comparison of the sensitivity of different aroma extraction techniques in combination with gas chromatography–mass spectrometry to detect minor aroma compounds in wine. *J. Chromatogr. A* **2013**, *1272*, 1–7. [CrossRef] [PubMed]
9. Madrera, R.R.; Valles, B.S. Determination of volatile compounds in apple pomace by stir bar sorptive extraction and gas chromatography-mass spectrometry (SBSE-GC-MS). *J. Food Sci.* **2011**, *76*, C1326–C1334. [CrossRef]
10. *GB/T 23776–2018*; Methodology for Sensory Evaluation of Tea. Standardization Administration of China: Beijing, China, 2018.
11. *GB/T 14487–2017*; Tea Vocabulary for Sensory Evaluation. Standardization Administration of China: Beijing, China, 2017.
12. Shi, Y.; Zhu, Y.; Ma, W.; Lin, Z.; Lv, H. Characterisation of the volatile compounds profile of Chinese pan-fried green tea in comparison with baked green tea, steamed green tea, and sun-dried green tea using approaches of molecular sensory science. *Cur. Res. Food Sci.* **2022**, *5*, 1098–1107. [CrossRef]
13. Zhu, Y.; Lin, Z.; Shao, C.Y.; Zhang, Y.; Lv, H.; Zhang, Z.; Zeng, J.; Peng, Q.; Zhu, Y.; Lin, Z. Aromatic profiles and enantiomeric distributions of chiral odorants in baked green teas with different picking tenderness. *Food Chem.* **2022**, *15*, 388. [CrossRef]
14. Wang, M.Q.; Ma, W.J.; Shi, J.; Zhu, Y.; Lin, Z.; Lv, H.P. Characterization of the key aroma compounds in Longjing tea using stir bar sorptive extraction (SBSE) combined with gas chromatography-mass spectrometry (GC–MS), gas chromatography-olfactometry (GC-O), odor activity value (OAV), and aroma recombination. *Food Res. Int.* **2020**, *130*, 108908. [CrossRef] [PubMed]
15. Zhu, Y.; Lv, H.P.; Shao, C.Y.; Kang, S.; Zhang, Y.; Guo, L.; Dai, W.; Tan, J.; Peng, Q.; Lin, Z. Identification of key odorants responsible for chestnut-like aroma quality of green teas. *Food Res. Int.* **2018**, *108*, 74–82. [CrossRef]
16. Cui, J.; Zhai, X.; Guo, D.; Du, W.; Gao, T.; Zhou, J.; Schwab, W.G.; Song, C. Characterization of key odorants in xinyang maojian green tea and their changes during the manufacturing process. *J. Agric. Food Chem.* **2022**, *70*, 279–288. [CrossRef] [PubMed]
17. Zhang, X.; Li, J.; Chen, X.; Wang, W.; Li, F.; Ma, Y. Effects of leaf acetate on physiology and biochemistry of cold tolerance of tea plant under low temperature stress. *Jiangsu Agric. Sci.* **2021**, *49*, 127–132. [CrossRef]
18. Zeng, L.; Watanabe, N.; Yang, Z. Understanding the biosyntheses and stress response mechanisms of aroma compounds in tea (*Camellia sinensis*) to safely and effectively improve tea aroma. *Crit. Rev. Food Sci. Nutr.* **2019**, *59*, 2321–2334. [CrossRef]
19. Lee, J.; Chambers, D.H.; Chambers, E., IV; Adhikari, K.; Yoon, Y. Volatile aroma compounds in various brewed green teas. *Molecules* **2013**, *18*, 10024–10041. [CrossRef] [PubMed]
20. Liao, X.; Yan, J.; Wang, B.; Meng, Q.; Zhang, L.; Tong, H. Identification of key odorants responsible for cooked corn-like aroma of green teas made by tea cultivar 'Zhonghuang 1'. *Food Res. Int.* **2020**, *136*, 109355. [CrossRef]
21. Zhao, M.; Wang, L.; Wang, J.; Zhang, J.; Zhang, N.; Lei, L.; Gao, T.; Jing, T.; Zhang, S.; Wu, B.; et al. Induction of priming by cold stress via inducible volatile cues in neighboring tea plants. *J. Int. Plant Biol.* **2020**, *62*, 1461–1468. [CrossRef]
22. Wu, Y.; Lv, S.; Wang, C.; Gao, X.; Li, J.; Meng, Q. Comparative analysis of volatiles difference of Yunnan sun-dried Pu-erh green tea from different tea mountains: Jingmai and Wuliang mountain by chemical fingerprint similarity combined with principal component analysis and cluster analysis. *Chem. Cent. J.* **2016**, *10*, 11. [CrossRef] [PubMed]
23. Lin, J.; Dai, Y.; Guo, Y.N.; Xu, H.R.; Wang, X.C. Volatile profile analysis and quality prediction of Longjing tea (*Camellia sinensis*) by HS-SPME/GC-MS. *J. Zhejiang Uni. Sci. B.* **2012**, *13*, 972–980. [CrossRef] [PubMed]
24. Feng, Z.; Li, Y.; Li, M.; Wang, Y.; Zhang, L.; Wan, X.; Yang, X. Tea aroma formation from six model manufacturing processes. *Food Chem.* **2019**, *285*, 347–354. [CrossRef] [PubMed]
25. Zhu, J.; Niu, Y.; Xiao, Z. Characterization of the key aroma compounds in Laoshan green teas by application of odour activity value (OAV), gas chromatography-mass spectrometry-olfactometry (GC-MS-O) and comprehensive two-dimensional gas chromatography mass spectrometry (GC×GC-qMS). *Food Chem.* **2021**, *339*, 128136. [CrossRef] [PubMed]
26. Liu, N.; Shen, S.; Huang, L.; Deng, G.; Wei, Y.; Ning, J.; Wang, Y. Revelation of volatile contributions in green teas with different aroma types by GC–MS and GC–IMS. *Food Res. Int.* **2023**, *169*, 112845. [CrossRef]
27. Qian, X.; Xu, X.Q.; Yu, K.J.; Zhu, B.Q.; Lan, Y.B.; Duan, C.Q.; Pan, Q.H. Varietal dependence of GLVs accumulation and LOX-HPL pathway gene expression in four *Vitis vinifera* wine grapes. *Int. J. Mol. Sci.* **2016**, *17*, 1924. [CrossRef]

28. De Domenico, S.; Bonsegna, S.; Horres, R.; Pastor, V.; Taurino, M.; Poltronieri, P.; Muhammad, I.; Gunter, K.; Victor, F.; Peter, W.; et al. Transcriptomic analysis of oxylipin biosynthesis genes and chemical profiling reveal an early induction of jasmonates in chickpea roots under drought stress. *Plant Physiol. Biochem.* **2012**, *61*, 115–122. [CrossRef] [PubMed]
29. Xu, T.; Zhou, Q.; Chen, W.; Zhang, G.; He, G.; Gu, D.; Zhang, W. Involvement of jasmonate-signaling pathway in the herbivore-induced rice plant defense. *Chin. Sci. Bull.* **2003**, *48*, 1982–1987. [CrossRef]
30. Qian, X.; Sun, L.; Xu, X.Q.; Zhu, B.Q.; Xu, H.Y. Differential expression of VvLOXA diversifies C6 volatile profiles in some *Vitis vinifera* table grape cultivars. *Int. J. Mol. Sci.* **2017**, *18*, 2705. [CrossRef] [PubMed]
31. Zhang, X.; Yu, Y.; Zhang, J.; Qian, X.; Li, X.; Sun, X. Recent progress regarding jasmonates in tea plants: Biosynthesis, signaling, and function in stress responses. *Int. J. Mol. Sci.* **2024**, *25*, 1079. [CrossRef] [PubMed]
32. Shi, J.; Wang, J.; Lv, H.; Peng, Q.; Schreiner, M.; Baldermann, S.; Lin, Z. Integrated proteomic and metabolomic analyses reveal the importance of aroma precursor accumulation and storage in methyl jasmonate-primed tea leaves. *Hortic. Res.* **2021**, *8*, 95. [CrossRef]
33. Shi, J.; Ma, C.; Qi, D.; Lv, H.; Yang, T.; Peng, Q.; Chen, Z.; Lin, Z. Transcriptional responses and flavor volatiles biosynthesis in methyl jasmonate-treated tea leaves. *BMC Plant Biol.* **2015**, *15*, 233. [CrossRef] [PubMed]
34. Han, Z.; Zhang, C.; Zhang, H.; Duan, Y.; Zou, Z.; Zhou, L.; Zhu, X.; Fang, W.; Ma, Y. CsMYB transcription factors participate in jasmonic acid signal transduction in response to cold stress in tea plant (*Camellia sinensis*). *Plants* **2022**, *11*, 2869. [CrossRef]
35. Zhou, Y.; Zeng, L.; Liu, X.; Gui, J.; Mei, X.; Fu, X.; Dong, F.; Tang, J.; Zhang, L.; Yang, Z. Formation of (E)-nerolidol in tea (*Camellia sinensis*) leaves exposed to multiple stresses during tea manufacturing. *Food Chem.* **2017**, *231*, 78–86. [CrossRef] [PubMed]
36. Wang, B.; Meng, Q.; Xiao, L.; Li, R.; Peng, C.; Liao, X.; Tong, H. Characterization of aroma compounds of Pu-erh ripen tea using solvent assisted flavor evaporation coupled with gas chromatography-mass spectrometry and gas chromatography-olfactometry. *Food Sci. Hum. Wellness* **2022**, *11*, 618–626. [CrossRef]
37. Ge, Y.; Wang, L.; Huang, Y.; Jia, L.; Wang, J. Characteristic flavor compounds in Guizhou green tea and the environmental factors influencing their formation: Investigation using stable isotopes, electronic nose, and headspace-gas chromatography ion migration spectrometry. *LWT* **2024**, *196*, 115887. [CrossRef]
38. Morini, G.; Maga, J.A. Volatile compounds in roasted and boiled Chinese chestnuts (*Castanea molissima*). *LWT—Food Sci. Technol.* **1995**, *28*, 638–640. [CrossRef]
39. Cui, Y.H.; Xu, F.; Wang, F.J.; Wang, J.Z.; Ouyang, J. Effect of processing technology on volatile components of Chinese chestnut. *Food Ferment. Ind.* **2012**, *38*, 99–103. (In Chinese)
40. Ye, G.Z.; Jiang, Y.W.; Yin, J.F.; Yuan, H.B.; Zhang, R.L.; Wang, Z.L.; Shen, D.Y.; Wang, F.; Chen, J.X. Study on the characteristic of aroma components in green tea with chestnut-like aroma. *J. Tea Sci.* **2009**, *29*, 385–394. (In Chinese) [CrossRef]
41. Wang, J.; Shi, J.; Zhu, Y.; Ma, W.; Yan, H.; Shao, C.; Wang, M.; Zhang, Y.; Peng, Q.; Chen, Y.; et al. Insights into crucial odourants dominating the characteristic flavour of citrus-white teas prepared from citrus *reticulata Blanco* 'Chachiensis' and *Camellia sinensis* 'Fudingdabai'. *Food Chem.* **2022**, *377*, 132048. [CrossRef]
42. You, Q.; Shi, Y.; Zhu, Y.; Yang, G.; Yan, H.; Lin, Z.; Lv, H. Effect of Different Processing Technologies on the Key Aroma-Active Compounds of Green Tea. *Food Sci.* **2023**, *44*, 7. (In Chinese) [CrossRef]
43. Shu, C.; She, Y.; Xiao, Z.; Xu, L.; Niu, Y.; Zhu, J. Invetigation on the aroma active compounds in fresh and aged Longjing tea by SPME/GC-MS/GC-O/OAV. *Food Ind.* **2016**, *37*, 279–285. (In Chinese)
44. Liu, R.; Ding, L.; Liang, Q.; Ding, X.; Chen, Z. Comparison of aroma components between Rizhao pelleted green tea and Rizhao curly green tea. *Sci. Technol. Food Ind.* **2015**, *36*, 323–326. [CrossRef]
45. Grosch, W. Evaluation of the key odorants of foods by dilution experiments, aroma models and omission. *Chem. Senses* **2001**, *26*, 533–545. [CrossRef] [PubMed]
46. Reiners, J.; Grosch, W. Odorants of virgin olive oils with different flavor profiles. *J. Agric. Food Chem.* **1998**, *46*, 2754–2763. [CrossRef]
47. Benkwitz, F.; Nicolau, L.; Lund, C.; Beresford, M.; Wohlers, M.; Kilmartin, P.A. Evaluation of key odorants in Sauvignon blanc wines using three different methodologies. *J. Agric. Food Chem.* **2012**, *60*, 6293–6302. [CrossRef]
48. Ma, L.; Gao, M.; Zhang, L.; Qiao, Y.; Li, J.; Du, L.; Zhang, H.; Wang, H. Characterization of the key aroma-active compounds in high-grade Dianhong tea using GC-MS and GC-O combined with sensory-directed flavor analysis. *Food Chem.* **2022**, *378*, 132058. [CrossRef] [PubMed]
49. Barata, A.; Campo, E.; Malfeito-Ferreira, M.; Loureiro, V.; Cacho, J.; Ferreira, V. Analytical and sensorial characterization of the aroma of wines produced with sour rotten grapes using GC-O and GC-MS: Identification of key aroma compounds. *J. Agric. Food Chem.* **2011**, *59*, 2543–2553. [CrossRef]
50. Xiao, Z.; Chen, J.; Niu, Y.; Chen, F. Characterization of the key odorants of fennel essential oils of different regions using GC–MS and GC–O combined with partial least squares regression. *J. Chromatogr. B* **2017**, *1063*, 226–234. [CrossRef]

51. Wardencki, W.; Chmiel, T.; Dymerski, T. Gas chromatography-olfactometry (GC-O), electronic noses (e-noses) and electronic tongues (e-tongues) for in vivo food flavour measurement. In *Instrumental Assessment of Food Sensory Quality*; Woodhead Publishing: Amsterdam, The Netherlands, 2013; pp. 195–229. [CrossRef]
52. Chen, X.; Chen, D.; Jiang, H.; Sun, H.; Zhang, C.; Zhao, H.; Li, X.; Yan, F.; Chen, C.; Xu, Z. Aroma characterization of Hanzhong black tea (*Camellia sinensis*) using solid phase extraction coupled with gas chromatography–mass spectrometry and olfactometry and sensory analysis. *Food Chem.* **2019**, *274*, 130–136. [CrossRef]
53. Niu, Y.; Ma, Y.; Xiao, Z.; Zhu, J.; Xiong, W.; Chen, F. Characterization of the key aroma compounds of three kinds of Chinese representative black tea and elucidation of the perceptual interactions of methyl salicylate and floral odorants. *Molecules* **2022**, *27*, 1631. [CrossRef]
54. Sun, A.; Li, C.; Lv, S.; Gao, J.; Xia, Y.; Geng, Y. Study on the Effects of Processes on Aroma Compounds in Rizhao Green Tea Based on the Gas Chromatography-Ion Mobility Spectrometry. *J. Food Process. Preserv.* **2023**, 3046129. [CrossRef]
55. Vilar, E.G.; O'Sullivan, M.G.; Kerry, J.P.; Kilcawley, K.N. A chemometric approach to characterize the aroma of selected brown and red edible seaweeds/extracts. *J. Sci. Food Agric.* **2021**, *101*, 1228–1238. [CrossRef] [PubMed]
56. Baba, R.; Amano, Y.; Wada, Y.; Kumazawa, K. Characterization of the potent odorants contributing to the characteristic aroma of Matcha by gas chromatography Olfactometry techniques. *J. Agric. Food Chem.* **2017**, *65*, 2984–2989. [CrossRef]
57. Tan, H.R.; Lau, H.; Liu, S.Q.; Tan, L.P.; Sakumoto, S.; Lassabliere, B.; Leong, K.; Sun, J.; Yu, B. Characterisation of key odourants in Japanese green tea using gas chromatography-olfactometry and gas chromatography-mass spectrometry. *LWT* **2019**, *108*, 221–232. [CrossRef]
58. De Alencar Diniz, J.; Rocha, S.; Pires-Cavalcante, K.; Freitas, J.; Nagano, C.; Sampaio, A.; Saker-Sampaio, S. Chemical composition of volatile compounds in two red seaweeds, *Pterocladiella capillacea* and *Osmundaria obtusiloba*, using static headspace gas chromatography mass spectrometry. *J. Appl. Phycol.* **2017**, *29*, 1571–1576. [CrossRef]
59. Rinanda, S.D.; Pratama, R.I.; Riyantini, I.; Rostini, I. Volatile Flavor Compounds Composition of Steamed Squid (*Loligo* sp.). *Asian J. Fish Aquat. Res.* **2021**, *13*, 1–7. [CrossRef]
60. Wu, N.; Gu, S.; Tao, N.; Wang, X.; Ji, S. Characterization of important odorants in steamed male Chinese mitten crab (*Eriocheir sinensis*) using gas chromatography-mass spectrometry-olfactometry. *J. Food Sci.* **2014**, *79*, C1250–C1259. [CrossRef] [PubMed]
61. Deng, X.; Huang, G.; Tu, Q.; Zhou, H.; Li, Y.; Shi, H.; Wu, X.; Ren, H.; Huang, K.; He, X.; et al. Evolution analysis of flavor-active compounds during artificial fermentation of Pu-erh tea. *Food Chem.* **2021**, *357*, 129783. [CrossRef] [PubMed]
62. Capone, S.; Tufariello, M.; Siciliano, P. Analytical characterisation of Negroamaro red wines by "Aroma Wheels". *Food Chem.* **2013**, *141*, 2906–2915. [CrossRef] [PubMed]
63. Spencer, M.; Sage, E.; Velez, M.; Guinard, J.X. Using single free sorting and multivariate exploratory methods to design a new coffee taster's flavor wheel. *J. Food Sci.* **2016**, *81*, S2997–S3005. [CrossRef]
64. Imamura, M. Descriptive terminology for the sensory evaluation of soy sauce. *J. Sens. Stud.* **2016**, *31*, 393–407. [CrossRef]
65. Zhang, Y.B.; Liu, X.; Lu, C.Y. Study on primitive morpheme in sensory terminology and flavor wheel construction of Chinese tea. *J. Tea Sci.* **2019**, *39*, 474–483. [CrossRef]

foods

MDPI

Article

Exploring the Bioaccessibility of Roasted Japanese Green Tea: Impact of Simulated Gastrointestinal Digestion

Wei Qin [1,2,*] and Sunantha Ketnawa [2,3]

1 Faculty of Health Sciences, Hokkaido University, Kita-12, Nishi-5, Kita-Ku, Sapporo 060-0812, Japan
2 Graduate School of Horticulture, Chiba University, 648, Matsudo, Matsudo 271-8510, Japan; sunantha.ketnawa@gmail.com
3 Food Science and Technology Program, School of Agro-Industry, Mae Fah Luang University, Chiang Rai 57100, Thailand
* Correspondence: shini.qinwei@gmail.com; Tel./Fax: +81-11-706-3698

Abstract: In this study, the effects were explored of digestive enzymes and pH on the bioaccessibility of polyphenols, flavonoids, and antioxidant activities in Hojicha (roasted green tea, RT) infusions during simulated in vitro digestion. Roasting modifies its polyphenolic profile and reduces bitterness, making it a popular variation of green tea. In this study, RT was used for assessing how the roasting-induced changes influenced the tea's bioaccessibility and stability under digestive conditions. A two-step gastrointestinal digestion model was applied to mimic real digestion. Total polyphenol content (TPC), total flavonoid content (TFC), and antioxidant activity (DPPH, ABTS, FRAP, and MIC) were measured at different digestion stages. Gastric conditions led to a 2.07-fold reduction in TPC and a 4.27-fold reduction in TFC. Digestive enzymes enhanced bioactive compound stability, with TPC and TFC bioaccessibility reaching 56% and 25% in the simulated digestion with digestive enzymes (MD) group, compared to 52% and 20% in the without digestive enzymes (WOE) group. Antioxidant activities were also better preserved, with antioxidant activity retention at 31% in the MD samples versus 19% in the WOE. These findings emphasize the key role of digestive enzymes in maintaining the antioxidant potential of roasted green tea during digestion, providing insight into future research on roasting methods and tea functionality for product development.

Keywords: roasted green tea; bioactive compounds; antioxidant; in vitro digestion; bioaccessibility

1. Introduction

Green tea has long been recognized for its rich content of bioactive compounds, particularly polyphenols such as catechins, which are known for their antioxidant, anti-inflammatory, and potential health-promoting effects [1–3]. Roasted Japanese green tea (RT), commonly referred to as "Hojicha", is a widely consumed variant of green tea that undergoes a roasting process at high temperatures (160 °C to 180 °C) [4]. This process not only alters the flavor profile of the tea but also impacts its chemical composition, particularly its polyphenol content [5–7]. Roasting enhances the tea's caramel and nutty aroma while reducing the astringency and bitterness associated with catechins [5,8–11], making it more palatable to a broader range of consumers.

Hojicha has become increasingly popular due to its milder flavor and reduced caffeine content, which make it suitable for consumption at any time of day, including by individuals who are sensitive to caffeine [5,8]. The roasting process also contributes to the development of specific volatile compounds, such as pyrazines, pyrroles, and furans, which provide Hojicha with its distinctive aroma [12–14]. However, high roasting temperatures can cause the thermal degradation and transformation of bioactive compounds, such as catechins and flavonoids, which may influence their stability and bioaccessibility during digestion. For example, catechins may undergo oligomerization under roasting conditions, leading to the formation of new polyphenolic compounds with potentially altered functional properties [5,15]. These transformations are critical to the development of roasted tea's flavor but may also impact the stability and bioaccessibility of these compounds during digestion. Similar effects have been observed in other thermally processed foods. For example, in cocoa, high-temperature roasting degrades certain flavan-3-ols while enhancing the formation of stable derivatives, which can influence their functional properties [6]. In coffee, polyphenols can bind to Maillard reaction products, affecting their solubility and bioaccessibility [7]. These findings highlight that thermal processing not only modifies the chemical composition but may also influence the digestive behavior of bioactive compounds.

While studies have explored the sensory properties, volatile profiles, and dry leaf composition of roasted green tea [5,8,15], there remains limited information regarding the bioaccessibility of its bioactive compounds during digestion. Bioaccessibility refers to the fraction of bioactive compounds released from the food matrix during digestion and made available for potential absorption in the gastrointestinal tract [16,17]. This parameter is crucial for understanding the potential health benefits of food products, particularly for polyphenols, whose stability and solubility can be significantly influenced by digestive conditions. Understanding the bioaccessibility of polyphenols and the antioxidant activities of roasted green tea during digestion is essential to better evaluate its functional properties as a health-promoting beverage.

In this study, the hypothesis is that the gastrointestinal digestion process significantly influences the bioaccessibility of bioactive compounds in roasted Japanese green tea, with digestive enzymes enhancing the stability of polyphenols and antioxidant activities under simulated gastric and intestinal conditions. To test this hypothesis, we investigate changes in total polyphenol content (TPC), total flavonoid content (TFC), and antioxidant activities of roasted green tea using assays such as DPPH, ABTS radical scavenging activity, ferric-reducing antioxidant power (FRAP), and metal ion chelating (MIC) activity. By examining the impact of digestion on these parameters, in this study, the aim is to provide insights into the health potential of roasted green tea as a functional beverage.

2. Materials and Methods

2.1. Materials and Preparation of Infusion

Roasted green tea leaves from the commercial brand 'Oi Ocha' (Itoen Ltd., Tokyo, Japan) were sourced from a local supermarket in Matsudo, Japan. For each processed sample, 5 g of tea leaves, with moisture content determined and expressed on a dry basis (d.b.), were placed in a beaker, and hot water at 95 °C was added until the total weight reached 500 g. The mixture was then steeped for 5 min. The resulting infusion was filtered and maintained at 37 °C in a water bath (NTT-20S; Eyela, Tokyo, Japan) to mimic physiological conditions until it was used for simulated in vitro gastrointestinal digestion, which occurred within 30 min.

2.2. In Vitro Gastrointestinal Digestion

The in vitro gastrointestinal digestion model, following a two-phase static simulation, was performed twice, adhering to the procedure described by Qin et al. (2022) [16], with slight modifications. A 150 mL portion of tea infusion was placed in a glass reaction vessel, where it was continuously stirred using a magnetic stirrer while the temperature was held at 37 °C via a water bath throughout the entire digestion process. Prior to the gastric phase (labeled as 'G'), the pH was adjusted to 1.20 using an HCl solution. Afterward, 19 mL of pepsin solution (200 mg of pepsin in 19 mL of 0.1 M HCl) was introduced, and the pH remained at 1.20 ± 0.01 for the duration of the G phase. One hour into the gastric digestion, the process was transitioned to the intestinal phase (denoted as 'I'). The pH was initially adjusted to 4.0 by adding 1M $NaHCO_3$ and subsequently raised to 6.8 using an NaOH solution. At this point, 25 mL of intestinal enzyme solution (625 μM pancreatin solution) for the intestinal stage was added. The mixture was maintained under these conditions for another two hours, with the pH kept stable at 6.80 ± 0.01 throughout the I phase. This digestion protocol was referred to as the mimicked digestion under normal digestive conditions (MD). A control experiment was also conducted without the addition of digestive enzymes (WOE) to assess the effect of pH alone on the bioactive properties of the tea infusion. The procedure followed the same steps as the standard digestion. Samples were collected at various stages, as follows: BF, before the digestion process began; G0, at the start of digestion with a pH of 1.2; G1, after 1 h of the gastric phase, with the pH still at 1.2; G1I0, after the gastric phase with the pH adjusted to 6.8; G1I1, following 1 h of the intestinal phase at a pH of 6.8; and G1I2, after 1 h of gastric digestion and 2 h of intestinal digestion, also at a pH of 6.8.

2.3. Determination of Bioactive Compounds

The bioactive compounds in the tea samples were assessed by measuring total polyphenol content (TPC) and total flavonoid content (TFC). TPC was quantified using the Fast Blue BB (FBBB) assay, as described in a previous study [16]. Then, 0.1 mL of the sample extract was mixed with 0.01 mL of 0.1% FBBB solution, followed by 0.01 mL of 0.5% NaOH in a 96-well plate. After a 90 min incubation at room temperature, absorbance was measured at 420 nm, and results were expressed as milligrams of gallic acid equivalents per gram dry basis samples.

TFC was determined using a colorimetric assay for the 96-well plate. A 250 μL portion of the diluted sample extract was mixed with 0.01 mL of 5% $NaNO_2$, and incubated for 5 min, followed by the addition of 0.01 mL of 10% $AlCl_3$ and a 6 min incubation. The reaction was completed by adding 0.06 mL of 1M NaOH, and absorbance was recorded at 520 nm. TFC was expressed as milligrams of catechin equivalents per gram of dry basis. All samples were analyzed for TPC and TFC in triplicate.

2.4. Antioxidant Activity

The antioxidant potential of the tea samples was evaluated using the following, several assays: DPPH, ABTS, FRAP, and MIC. DPPH radical scavenging activity was measured following a modified version of the method described by Qin et al. (2022) [16]. A 0.02 mL sample extract was combined with 0.30 mL of 60 μM DPPH solution and incubated in the dark for 30 min. Absorbance was recorded at 520 nm, with results expressed as micromoles of Trolox equivalents per gram of dry basis tea leaves.

For the ABTS test, 0.01 mL sample was mixed with 0.32 mL of diluted ABTS solution. The mixture was incubated for 10 min. After incubation, the absorbance was measured at

740 nm. Results were reported as milligrams of ascorbic acid equivalents per gram of dry basis tea leaves.

Ferric reducing antioxidant power (FRAP) was evaluated by mixing 0.04 mL of the sample extract with 0.26 mL of freshly prepared FRAP reagent (a solution of TPTZ, $FeCl_3 \cdot 7H_2O$, and acetate buffer). The mixture was incubated at 37 °C for 30 min in the dark, and absorbance was read at 590 nm. FRAP values were expressed as micromoles of $FeSO_4$ equivalents per gram of dry basis tea leaves.

Metal ion chelating activity (MIC) was assessed by adding 5 μL of 2 mM $FeCl_2 \cdot 3H_2O$ and 0.01 mL of 5 mM FerroZine® to 0.30 mL of the sample extract. After a 10 min incubation at room temperature, absorbance was measured at 560 nm, and MIC activity was expressed as micromoles of EDTA equivalents per gram of dry basis tea leaves.

2.5. Assessment of Bioaccessibility of Bioactive Compounds and Residual Antioxidant Activity

The bioaccessibility and residual antioxidant activity were determined and expressed as percentages (%) using Equations (1) and (2), respectively, as follows:

$$\text{Bioaccessibility (\%)} = (\text{BS}/\text{BD}) \times 100 \quad (1)$$

$$\text{Residual antioxidant activity (\%)} = (\text{AA}/\text{BD}) \times 100 \quad (2)$$

In Equation (1), BS represents the amount of bioactive compounds quantified at each sampling point during the digestion process. In Equation (2), AA refers to the antioxidant activity measured at each sampling point during digestion, following previously described methods. BD, in both cases, refers to the bioactive compound content and antioxidant activity measured in the initial tea infusion before digestion.

Bioaccessibility refers to the proportion of bioactive compounds that remain accessible for potential absorption at various stages of the digestion process, serving as an indicator of their stability under simulated gastrointestinal conditions. Residual antioxidant activity represents the fraction of antioxidant activity retained at each stage of digestion, expressed as a percentage of the antioxidant activity measured in the initial tea infusion before digestion.

2.6. Statistical Analysis

To ensure result accuracy, all experiments were conducted in triplicate. Data analysis was performed using SPSS, Version 26.0 (SPSS Inc., Chicago, IL, USA), with a 95% confidence level ($p < 0.05$). Correlation coefficients for the standard curve were calculated using Microsoft Excel.

3. Results and Discussion

3.1. Change in Bioactive Compounds During In Vitro Digestion

Table 1 illustrates the changes in total polyphenol content (TPC) and total flavonoid content (TFC) of roasted green tea infusions during different stages of simulated in vitro digestion. Initially, the undigested tea infusion (BF) exhibited the highest levels of TPC (46.49 ± 0.37 mg/g d.b.) and TFC (2.05 ± 0.15 mg/g d.b.), reflecting the high concentration of bioactive compounds before exposure to the digestive process. However, a significant reduction in both TPC and TFC was observed at the G0 stage (before enzyme digestion but after pH adjustment to 1.2). TPC decreased by 2.07-fold to 22.42 ± 0.12 mg/g d.b., while TFC saw a more dramatic 4.27-fold decrease, dropping to 0.48 ± 0.02 mg/g d.b. This drastic reduction highlights the sensitivity of polyphenols and flavonoids to acidic conditions, even in the absence of digestive enzymes. These findings are consistent with the results reported by Ferruzzi and Blakeslee (2007) [17]. They found that polyphenols are prone to oxidation and degradation when exposed to acidic environments. Different types of catechins show varying degrees of stability in alkaline conditions, such as EGCG and EGC breaking down quickly, while EC and ECG are more stable. This shows that catechins react differently depending on the pH level. The sensitivity of polyphenols to pH changes shows how complex their behavior is at different pH levels. It also emphasizes how important the digestive environment is for maintaining polyphenol stability [18].

Table 1. Total polyphenol content (TPC) and total flavonoid content (TFC) of roasted green tea infusions at various digestion stages.

Digestion Stage	TPC		TFC	
BF	46.49 ± 0.37 AA		2.05 ± 0.15 AA	
G0	22.42 ± 0.12 EC		0.48 ± 0.02 EE	
	MD	WOE	MD	WOE
G1	26.75 ± 1.80 B *	21.70 ± 0.07 D	0.54 ± 0.02 D *	0.38 ± 0.00 F
G1I0	24.30 ± 0.27 C *	21.95 ± 0.33 D	1.23 ± 0.03 B *	0.92 ± 0.02 B
G1I1	23.67 ± 0.27 D *	21.64 ± 0.25 D	1.12 ± 0.03 C *	0.64 ± 0.02 D
G1I2	25.13 ± 0.33 B *	23.14 ± 0.71 B	1.18 ± 0.02 C *	0.73 ± 0.00 C

Different uppercase letters and underlined characters within the same column represent significant differences ($p < 0.05$) across digestion stages for the digested samples under MD and WOE conditions, respectively. An asterisk (*) denotes a significant difference ($p < 0.05$) between the two digestive conditions at the same digestion stage for TPC and TFC.

During the gastric phase (G1), the MD group showed higher retention of both TPC (26.75 ± 1.80 mg/g d.b.) and TFC (0.54 ± 0.02 mg/g d.b.) compared to the WOE group, which exhibited lower values of TPC (21.70 ± 0.07 mg/g d.b.) and TFC (0.38 ± 0.00 mg/g d.b.). It suggests that digestive enzymes play a significant role in stabilizing bioactive compounds under gastric conditions. Digestive enzymes play a critical role in stabilizing bioactive compounds under gastric conditions by enhancing their solubility and retention. This effect is supported by Bohn et al. (2015) [19], who demonstrated that digestive enzymes improved the stability of plant-based compounds during digestion. Additionally, the interaction between tea polyphenols and digestive enzymes through hydrogen bonds and hydrophobic interactions could explain the observed differences. Tea polyphenols binding to enzymes might affect the activity of these enzymes. This is because the structure of tea polyphenols contains hydroxyl and galloyl groups that can interact with the polar parts of the enzymes, changing how they function [20,21].

In the subsequent intestinal digestion phase (G1I0 to G1I2), TPC and TFC remained relatively stable in both groups, with slight variations. At the final digestion stage (G1I2), TPC was 25.13 ± 0.33 mg/g d.b. in the MD group and 23.14 ± 0.71 mg/g d.b. in the WOE group. TFC was also higher in the MD group (1.18 ± 0.02 mg/g d.b.) compared to the WOE group (0.73 ± 0.00 mg/g d.b.). These results emphasize that digestive enzymes not only enhance the release of bioactive compounds but also protect them from further degradation during the intestinal phase. These results emphasize that digestive enzymes enhance the release of bioactive compounds and protect them from further degradation, as also reported by Tagliazucchi et al. (2010) [22]. The various catechin components of green tea have been previously reported to exhibit different stabilities under alkaline pH conditions. In this context, EGCG and EGC are unstable and degrade rapidly. However, the other major catechins, EC and ECG, are more stable under these conditions. In contrast, under acidic pH conditions, all catechins remain stable for several hours [18,23].

3.2. Variations in Antioxidant Activity Throughout Simulated Digestion

As natural antioxidants, polyphenols help safeguard biomolecules against oxidative damage, which is linked to their role in disease prevention [24]. Assessing antioxidant activity is often the first step in understanding their biological properties since bioactive compounds must be liberated from the food matrix to exhibit their effects [25]. Antioxidant properties of roasted green tea infusions, including DPPH, ABTS, FRAP, and MIC, were evaluated across different stages of digestion (Table 2). The initial antioxidant activities (BF) of the tea infusions were notably high, with DPPH measured at 66.71 ± 1.37 μmol TE/g d.b., ABTS at 216.94 ± 4.63 mg ascorbic acids/g d.b., FRAP at 236.57 ± 1.80 μmol $FeSO_4$/g d.b., and MIC at 7.27 ± 0.11 μmol EDTA/g d.b. These values reflect the robust antioxidant capacity of the roasted tea before any digestion processes. However, upon pH adjustment to 1.2 (G0 stage), a marked reduction in antioxidant activity was observed, with DPPH, ABTS, FRAP, and MIC levels decreasing substantially. These reductions demonstrate the sensitivity of antioxidant compounds to acidic conditions, even before the introduction of digestive enzymes. It is consistent with the significant reduction in polyphenol and flavonoid content at this stage, as noted in Section 3.1, suggesting that the pH adjustment process impacts both bioactive compounds and antioxidant capacity in a similar manner.

Table 2. Antioxidant properties and pH of roasted green tea infusions at each digestion stage reported as DPPH, ABTS, FRAP, and MIC.

Digestion Stage	DPPH	ABTS	FRAP	MIC
BF	66.71 ± 1.37 AA	216.94 ± 4.63 AA	236.57 ± 1.80 AA	7.27 ± 0.11 BB
G0	34.90 ± 0.03 CC	116.54 ± 0.40 BB	141.93 ± 1.31 BD	6.58 ± 0.06 DD
			MD	
G1	39.25 ± 0.01 B *	105.56 ± 4.38 C *	133.85 ± 6.08 C *	6.85 ± 0.04 C
G1I0	30.56 ± 0.01 D *	93.30 ± 2.38 D *	116.06 ± 0.83 D *	7.27 ± 0.03 B *
G1I1	25.25 ± 0.25 E *	94.84 ± 2.11 D *	106.21 ± 2.41 E *	10.51 ± 0.04 A *
G1I2	22.52 ± 0.16 F *	86.06 ± 1.77 E *	96.31 ± 3.27 F *	10.51 ± 0.03 A *
			WOE	
G1	34.39 ± 0.07 C	102.11 ± 4.43 C	123.40 ± 1.63 C	7.73 ± 0.10 B *
G1I0	24.98 ± 0.07 D	82.53 ± 3.71 D	95.98 ± 2.15 D	5.71 ± 0.03 D
G1I1	18.36 ± 0.17 E	66.49 ± 2.08 E	87.83 ± 0.39 E	9.76 ± 0.00 A
G1I2	11.89 ± 0.03 F	68.23 ± 1.35 E	81.41 ± 0.63 F	9.39 ± 0.02 A

Different uppercase letters and underlined characters within the same column represent significant differences ($p < 0.05$) across digestion stages for the digested samples under MD and WOE conditions, respectively. An asterisk (*) denotes a significant difference ($p < 0.05$) between the two digestive conditions at the same digestion stage for TPC and TFC.

During the gastric phase (G1), the MD group exhibited a higher retention of antioxidant activity compared to the WOE group, a trend that parallels the greater stability of polyphenols and flavonoids in the enzyme-treated group noted in Section 3.1, so that digestive enzymes play a key role not only in stabilizing bioactive compounds but also in preserving the associated antioxidant activity. The presence of digestive enzymes helps maintain higher levels of antioxidant compounds in gastric conditions. This finding aligns with the conclusion of Bohn et al. (2015) [19], who stated that digestive enzymes improve the solubility and stability of plant-based compounds. As the digestion progressed through the intestinal phase (from G1I0 to G1I2), antioxidant activity continued to decrease. However, the MD group consistently retained higher antioxidant levels compared to the WOE group. By the end of digestion (G1I2), DPPH, ABTS, and FRAP values dropped more sharply in the WOE group than in the MD group. Even though the MIC values remained more stable in both conditions, the final digestion point showed a higher value than at the beginning of digestion. This result highlights the key role of digestive enzymes in protecting antioxidant compounds from breaking down, especially during the intestinal phase. Since phenolic compounds are known to contribute to antioxidant activity, their retention likely plays a role in the differences in antioxidant capacity observed between the MD and WOE groups. This further confirms that digestive enzymes are important for preserving not only polyphenols and flavonoids but also their related antioxidant effects during digestion.

3.3. Bioaccessibility of TPC and TFC and Associated Residual Antioxidant Activity

Figure 1 illustrates the bioaccessibility (%) of TPC and TFC under MD and WOE conditions, which demonstrate the substantial impact that digestive enzymes have on the bioaccessibility of total polyphenol content (TPC) and total flavonoid content (TFC). The bioaccessibility of TPC during the G phase reached 56% in the MD group compared to 52% in the WOE group. This slight yet significant increase highlights the role of digestive enzymes in enhancing polyphenol release during the early stages of digestion. Qin et al. (2022) [16] also demonstrated that enzymatic activity plays a crucial role in polyphenol stability and release under gastric conditions, corroborating the trend observed here. As digestion continued through the intestinal phase (GI), TPC bioaccessibility slightly decreased to around 50% in both MD and WOE groups, indicating a stabilization of polyphenol release, as also noted by Hollebeeck et al. (2013) [26], who found a similar plateau effect of polyphenol absorption during digestion. The stability in TPC bioaccessibility at the intestinal stage suggests that a portion of the polyphenols remains resistant to further enzymatic breakdown, highlighting their potential for absorption.

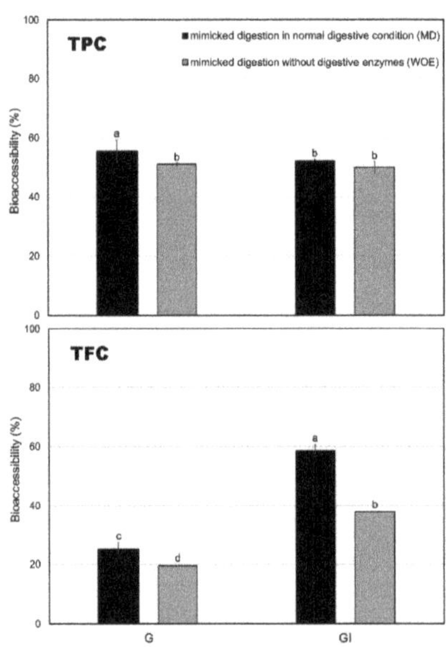

Figure 1. Bioaccessibility (%) of total polyphenol content (TPC) and total flavonoid content (TFC) in roasted green tea infusion during simulated digestion. Bars represent the standard deviation of triplicate determinations. Uppercase letters denote significant differences ($p < 0.05$) between the MD (mimicked digestion) and WOE (without digestive enzymes) treatments within the same compound groups. Lowercase letters indicate significant differences ($p < 0.05$) among different phenolic compound groups within the same digestive condition.

For TFC, the effect of digestive enzymes was even more pronounced. During the G phase, TFC bioaccessibility in the MD group was 25%, compared to only 20% in the WOE group. However, as digestion progressed to the GI phase, the MD group saw a surge in TFC bioaccessibility to nearly 60%, whereas the WOE group increased to only 38%. This significant difference underscores the crucial role of digestive enzymes in flavonoid release, particularly during the intestinal phase. The breakdown of flavonoid glycosides and the release of aglycones through enzymatic hydrolysis may explain the enhanced bioaccessibility observed [27,28]. Nevertheless, it is also important to consider potential limiting factors. Binding with bile salts and interactions with digestive enzymes might lead to the formation of insoluble complexes, partially hindering the release and solubility of flavonoids under intestinal conditions [29,30]. This interplay between enzymatic release and potential bile salt interactions likely determines the overall bioaccessibility of flavonoids. These interactions probably influence the overall bioaccessibility of flavonoids and warrant further investigation.

Figure 2 illustrates the changes in residual antioxidant activity (DPPH, ABTS, FRAP, and MIC) during digestion, revealing a clear trend of decline in antioxidant activity across all assays as digestion progressed, particularly in the WOE group. For DPPH activity, the MD group exhibited significantly higher retention compared to the WOE group by the end of the intestinal phase, maintaining 31% of its initial activity compared to only 19% in the WOE group. This 1.63-fold higher retention highlights the role of digestive enzymes in preserving antioxidant potential. A similar trend was observed for ABTS, where the MD group retained 40% of its activity compared to 31% in the WOE group, further underscoring that digestive enzymes helped maintain polyphenol stability and antioxidant activity in tea during digestion, particularly during the intestinal phase [31]. For FRAP activity, the MD group consistently exhibited higher retention (43%) than the WOE group (33%) by the end of digestion. Enzymatic digestion plays a key role in preserving antioxidant activity and the bioaccessibility of polyphenols during gastrointestinal digestion. The increase in free radical scavenging activity is attributed to the deprotonation of the hydroxyl groups on the aromatic rings of phenolic compounds [22]. The transition from the gastric to the intestinal environment may induce structural changes in phenolic molecules, likely due to the ionization of hydroxyl groups. Moreover, some studies have discussed that the antioxidant activity of chemically extracted food compounds may be underestimated because the solvents used for extraction might be less effective in fully extracting polyphenols [32]. Additionally, due to the impact of pH on the racemization of substances, two chiral enantiomers with different reactivities in respective reagents may be formed [33–35].

In contrast to the other assays, MIC activity showed an increase during digestion, especially in the intestinal phase, where the MD group exhibited a 1.12-fold higher retention of metal chelation capacity compared to the WOE group. This result suggests that digestive enzymes not only stabilize polyphenolic compounds but also enhance the release of the compounds responsible for metal chelation, as observed in the previous research by Tarko et al. (2020) [36]. The increase in MIC activity, particularly in the MD group, indicates that digestive enzymes play a crucial role in activating bioactive compounds that contribute to the metal chelating capacity of green tea during digestion. Overall, digestive enzymes are essential for preserving the antioxidant capacity of roasted green tea infusions throughout the digestive process, particularly during the intestinal phase, where the bioactive compound release and activation are most critical.

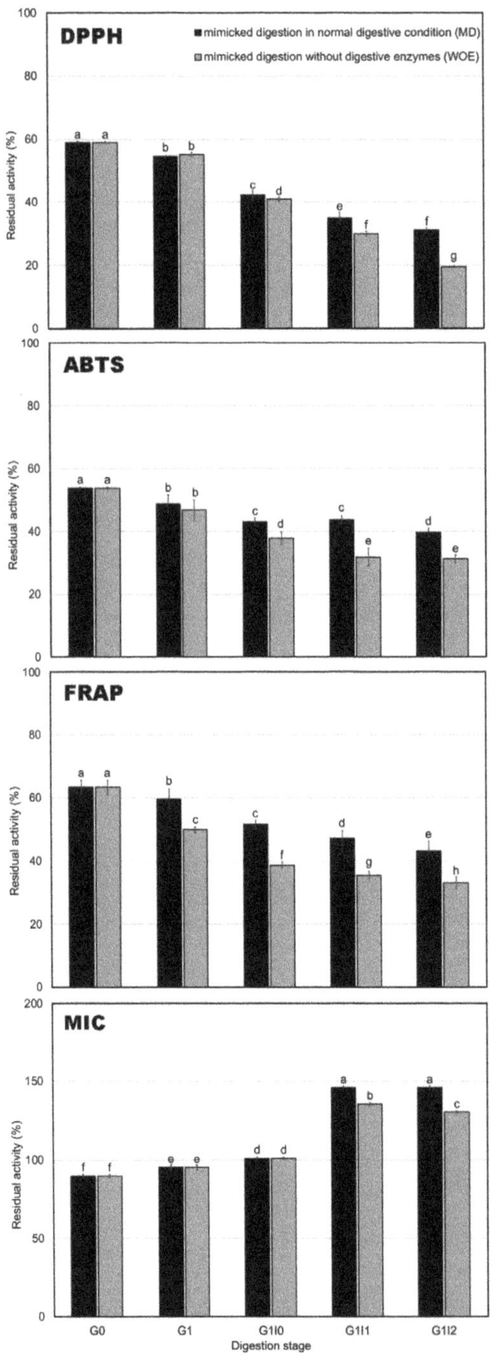

Figure 2. Residual activity of roasted green tea infusion during simulated digestion, evaluated by DPPH, ABTS, FRAP, and MIC assays. Bars represent the standard deviation of triplicate determinations. Lowercase letters indicate significant differences among samples within the digestion stages (G0, G1, G1I0, G1I1, and G1I2) ($p < 0.05$). Uppercase letters denote significant differences between mimicked digestion in MD and WOE within the same digestion stage ($p < 0.05$).

4. Conclusions

In this study, the significant influence is shown of digestive enzymes and pH levels on the bioaccessibility of polyphenols, flavonoids, and antioxidant properties in roasted Japanese green tea infusions during simulated in vitro digestion. The conditions of the gastric phase caused significant reductions in total polyphenol content, total flavonoid content, and antioxidant activities. However, the presence of digestive enzymes notably enhanced the release and stability of these bioactive compounds, particularly during the intestinal phase, where TPC and TFC show higher bioaccessibility, as well as antioxidant activity, were consistently higher in MD group samples compared to the WOE group. The findings indicate that digestive enzymes are essential for maintaining the antioxidant capacity of roasted green tea infusions, particularly in the later stages of digestion, where the metal chelating capacity (MIC) also showed increased activity. The results also suggest that enzyme-mediated digestion contributes to the stability and release of bioactive compounds, shedding light on the digestive behavior of roasted green tea. Future studies should explore how different roasting methods and intensities influence the bioaccessibility and functional properties of roasted tea as well as bioactive compound profiles providing deeper insights as a functional beverage. Such research will further clarify the implications of these findings for dietary recommendations and tea product development.

Author Contributions: Conceptualization, W.Q. and S.K.; methodology, W.Q. and S.K.; formal analysis, W.Q. and S.K.; investigation, W.Q.; resources, W.Q.; data curation, W.Q.; writing—original draft preparation, W.Q.; writing—review and editing, W.Q. and S.K.; supervision, W.Q. and S.K.; project administration, W.Q.; funding acquisition, W.Q. All authors have read and agreed to the published version of the manuscript.

Funding: This work was supported by JSPS KAKENHI Grant Numbers 21J10500, and the Tojuro Iijima Foundation for Food Science and Technology. The APC was funded by the International Journal Paper Submission Support Program of Hokkaido University.

Data Availability Statement: The original contributions presented in this study are included in the article.

Acknowledgments: Special thanks are extended to Yukiharu Ogawa of the Graduate School of Horticulture, Chiba University, for his invaluable guidance and support throughout this study.

Conflicts of Interest: The authors declare no conflict of interest.

References

1. Mao, A.J.; Su, H.; Fang, S.M.; Chen, X.; Ning, J.M.; Ho, C.T.; Wan, X.C. Effects of roasting treatment on non-volatile compounds and taste of green tea. *Int. J. Food Sci. Technol.* **2018**, *53*, 2586–2594. [CrossRef]
2. Wang, L.F.; Kim, D.M.; Lee, C.Y. Effects of heat processing and storage on flavanols and sensory qualities of green tea beverage. *J. Agric. Food Chem.* **2000**, *48*, 4227–4232. [CrossRef] [PubMed]
3. Wang, Y.; Ho, C.T. Polyphenolic chemistry of tea and coffee: A century of progress. *J. Agric. Food Chem.* **2009**, *57*, 8109–8114. [CrossRef] [PubMed]
4. Kurosaka, C.; Tagata, C.; Nakagawa, S.; Kobayashi, M.; Miyake, S. Effects of green tea and roasted green tea on human responses. *Sci. Rep.* **2024**, *14*, 8588. [CrossRef]
5. Morikawa, H.; Okuda, K.; Kunihira, Y.; Inada, A.; Miyagi, C.; Matsuo, Y.; Saito, Y.; Tanaka, T. Oligomerization mechanism of tea catechins during tea roasting. *Food Chem.* **2019**, *285*, 252–259. [CrossRef]
6. De Taeye, C.; Bodart, M.; Caullet, G.; Collin, S. Roasting conditions for preserving cocoa flavan-3-ol monomers and oligomers: Interesting behaviour of Criollo clones. *J. Sci. Food Agric.* **2017**, *97*, 4001–4008. [CrossRef]
7. Moreira, A.S.; Nunes, F.M.; Simões, C.; Maciel, E.; Domingues, P.; Domingues, M.R.M.; Coimbra, M.A. Transglycosylation reactions, a main mechanism of phenolics incorporation in coffee melanoidins: Inhibition by Maillard reaction. *Food Chem.* **2017**, *227*, 422–431. [CrossRef]

8. Zhu, Y.M.; Dong, J.J.; Jin, J.; Liu, J.H.; Zheng, X.Q.; Lu, J.L.; Liang, Y.R.; Ye, J.H. Roasting process shaping the chemical profile of roasted green tea and the association with aroma features. *Food Chem.* **2021**, *353*, 129428. [CrossRef]
9. Chen, M.; Dai, W.D.; Li, P.L.; Zhu, Y.; Chen, Q.C.; Yang, Y.Q.; Tan, J.F.; Lin, Z. Study on the changes of primary metabolites during the manufacturing process of roasted green tea by pre-column derivatization combining with GC-MS. *J. Tea Sci.* **2019**, *39*, 297–308. [CrossRef]
10. Chen, Y.J.; Kuo, P.C.; Yang, M.L.; Li, F.Y.; Tzen, J.T.C. Effects of baking and aging on the changes of phenolic and volatile compounds in the preparation of old Tieguanyin oolong teas. *Food Res. Int.* **2013**, *53*, 732–743. [CrossRef]
11. Fu, Y.Q.; Wang, J.Q.; Chen, J.X.; Wang, F.; Yin, J.F.; Zeng, L.; Shi, J.; Xu, Y.Q. Effect of baking on the flavor stability of green tea beverages. *Food Chem.* **2020**, *331*, 127258. [CrossRef] [PubMed]
12. Kawakami, M.; Yamanishi, T. Formation of aroma components in roasted or pan-fired green tea by roasting or pan-firing treatment. *Nippon Nogeikagaku Kaishi-J. Jpn. Soc. Biosci. Biotechol. Agrochem.* **1999**, *73*, 893–906. [CrossRef]
13. Zheng, X.Q.; Li, Q.S.; Xiang, L.P.; Liang, Y.R. Recent advances in volatiles of teas. *Molecules* **2016**, *21*, 338. [CrossRef] [PubMed]
14. Sasaki, T.; Yuikawa, N.; Tanihiro, N.; Michihata, T.; Enomoto, T. The effects of roasting conditions on the physical appearance traits and aroma and taste components of roasted stem tea. *Food Sci. Technol. Res.* **2020**, *26*, 643–654. [CrossRef]
15. Hara, T.; Kubota, E. Effects of roasting condition on the qualities of roasted green tea (Hojicha). *Nippon Shokuhin Kogyo Gakkaishi* **1969**, *16*, 145–149. [CrossRef]
16. Qin, W.; Ketnawa, S.; Ogawa, Y. Effect of digestive enzymes and pH on variation of bioavailability of green tea during simulated in vitro gastrointestinal digestion. *Food Sci. Hum. Wellness* **2022**, *11*, 669–675. [CrossRef]
17. Ferruzzi, M.G.; Blakeslee, J. Digestion, absorption, and cancer preventative activity of dietary chlorophyll derivatives. *Nutr. Res.* **2007**, *27*, 1–12. [CrossRef]
18. Zhu, Q.Y.; Zhang, A.; Tsang, D.; Huang, Y.; Chen, Z.Y. Stability of green tea catechins. *J. Agric. Food Chem.* **1997**, *45*, 4624–4628. [CrossRef]
19. Bohn, T.; McDougall, G.J.; Alegría, A.; Alminger, M.; Arrigoni, E.; Aura, A.M.; Brito, C.; Cilla, A.; El, S.N.; Karakaya, S. Mind the gap—Deficits in our knowledge of aspects impacting the bioavailability of phytochemicals and their metabolites—A position paper focusing on carotenoids and polyphenols. *Mol. Nutr. Food Res.* **2015**, *59*, 1307–1323. [CrossRef]
20. Dreosti, I.E. Antioxidant polyphenols in tea, cocoa, and wine. *Nutrition* **2000**, *16*, 692–694. [CrossRef]
21. Siebert, K.J.; Troukhanova, N.V.; Lynn, P.Y. Nature of polyphenol−protein interactions. *J. Agric. Food Chem.* **1996**, *44*, 80–85. [CrossRef]
22. Tagliazucchi, D.; Verzelloni, E.; Bertolini, D.; Conte, A. In vitro bio-accessibility and antioxidant activity of grape polyphenols. *Food Chem.* **2010**, *120*, 599–606. [CrossRef]
23. Zhang, A.Q.; Chan, P.T.; Luk, Y.S.; Ho, W.K.K.; Chen, Z.Y. Inhibitory effect of jasmine green tea epicatechin isomers on LDL-oxidation. *J. Nutr. Biochem.* **1997**, *8*, 334–340. [CrossRef]
24. Zhang, Y.; Wu, S.; Qin, Y.; Liu, J.; Liu, J.; Wang, Q.; Ren, F.; Zhang, H. Interaction of phenolic acids and their derivatives with human serum albumin: Structure–affinity relationships and effects on antioxidant activity. *Food Chem.* **2018**, *240*, 1072–1080. [CrossRef]
25. Gullon, B.; Pintado, M.E.; Fernández-López, J.; Pérez-Álvarez, J.A.; Viuda-Martos, M. In vitro gastrointestinal digestion of pomegranate peel (Punica granatum) flour obtained from co-products: Changes in the antioxidant potential and bioactive compounds stability. *J. Funct. Foods* **2015**, *19*, 617–628. [CrossRef]
26. Hollebeeck, S.; Borlon, F.; Schneider, Y.J.; Larondelle, Y.; Rogez, H. Development of a standardised human in vitro digestion protocol based on macronutrient digestion using response surface methodology. *Food Chem.* **2013**, *138*, 1936–1944. [CrossRef]
27. Barak, T.H.; Celep, E.; İnan, Y.; Yesilada, E. Influence of in vitro human digestion on the bioavailability of phenolic content and antioxidant activity of Viburnum opulus L. (European cranberry) fruit extracts. *Ind. Crops Prod.* **2019**, *131*, 62–69. [CrossRef]
28. Chen, G.L.; Chen, S.G.; Zhao, Y.Y.; Luo, C.X.; Li, J.; Gao, Y.Q. Total phenolic contents of 33 fruits and their antioxidant capacities before and after in vitro digestion. *Ind. Crop. Prod.* **2014**, *57*, 150–157. [CrossRef]
29. Czubinski, J.; Wroblewska, K.; Czyzniejewski, M.; Górnaś, P.; Kachlicki, P.; Siger, A. Bioaccessibility of defatted lupin seed phenolic compounds in a standardized static in vitro digestion system. *Food Res. Int.* **2018**, *116*, 1126–1134. [CrossRef]
30. Sęczyk, Ł.; Sugier, D.; Świeca, M.; Gawlik-Dziki, U. The effect of in vitro digestion, food matrix, and hydrothermal treatment on the potential bioaccessibility of selected phenolic compounds. *Food Chem.* **2020**, *344*, 128581. [CrossRef]
31. Wu, Z.; Teng, J.; Huang, L.; Xia, N.; Wei, B. Stability, antioxidant activity and in vitro bile acid-binding of green, black and dark tea polyphenols during simulated in vitro gastrointestinal digestion. *RSC Adv.* **2015**, *5*, 92089–92095. [CrossRef]
32. Anjali, B.; Vinayak, P. Antioxidant activity of garlic using conventional extraction and in vitro gastrointestinal digestion. *Free Radic. Antioxid.* **2013**, *3*, 30–34. [CrossRef]

33. Jamali, B.; Bjørnsdottir, I.; Nordfang, O.; Hansen, S.H. Investigation of racemisation of the enantiomers of glitazone drug compounds at different pH using chiral HPLC and chiral CE. *J. Pharm. Biomed. Anal.* **2008**, *46*, 82–87. [CrossRef] [PubMed]
34. Wootton-Beard, P.C.; Moran, A.; Ryan, L. Stability of the total antioxidant capacity and total polyphenol content of 23 commercially available vegetable juices before and after in vitro digestion measured by FRAP, DPPH, ABTS and Folin–Ciocalteu methods. *Food Res. Int.* **2011**, *44*, 217–224. [CrossRef]
35. Ketnawa, S.; Suwannachot, J.; Ogawa, Y. In vitro gastrointestinal digestion of crisphead lettuce: Changes in bioactive compounds and antioxidant potential. *Food Chem.* **2020**, *311*, 125885. [CrossRef] [PubMed]
36. Tarko, T.; Duda-Chodak, A.; Soszka, A. Changes in phenolic compounds and antioxidant activity of fruit musts and fruit wines during simulated digestion. *Molecules* **2020**, *25*, 5574. [CrossRef] [PubMed]

foods

MDPI

Article

Withering and Fermentation Affect the Transformation and Accumulation of Key Metabolites in Rougui (*Camellia sinensis*) for the Formation of Special Taste Characteristics

Jianghua Ye [1,†], Yangxin Luo [2,†], Yulin Wang [2], Qi Zhang [1], Shuqi Zhang [2], Junbin Gu [2], Yankun Liao [2], Tingting Wang [2], Xiaoli Jia [1] and Haibin Wang [1,2,*]

1 College of Tea and Food, Wuyi University, Wuyishan 354300, China; jhye1998@126.com (J.Y.)
2 College of Life Science, Longyan University, Longyan 364012, China
* Correspondence: 82015018@lyun.edu.cn
† These authors contributed equally to this work.

Abstract: During the production of Wuyi rock tea, withering and fermentation play a crucial role in the primary processing of the tea, greatly influencing the development of its distinct taste characteristics. In this study, Rougui (*Camellia sinensis*) was selected as the research object to investigate the effects of withering and fermentation on metabolites and taste characteristics in tea leaves. The findings revealed that a total of 1249 metabolites were detected in Rougui leaves at various processing stages, of which only 40 key metabolites were significantly altered. The process of withering and fermentation is crucial to increase the content of organic acids, plumerane, alkaloids, nucleotides and derivatives, amino acids and derivatives, and free fatty acids in the leaves of Rougui and to decrease the content of saccharides, phenolic acids, flavonols, flavones, and flavanols, which in turn enhances the mellowness, fresh and brisk taste, and aroma of tea and attenuates the saccharides, bitterness, and astringency. Withering and fermentation had the greatest effect on the bitterness and astringency of Rougui taste characteristics, followed by mellowness. It can be seen that withering and fermentation were extremely important for the development of Rougui's special taste characteristics. The present study provides important support for optimizing Rougui processing and the formation of its special taste characteristics.

Keywords: Wuyi rock tea; withering; fermentation; key metabolites; taste characteristics

1. Introduction

Wuyishan is an influential tea-producing region located at 27°32′36″~27°55′15″ north latitude and 117°24′12″~118°02′50″ east longitude in Fujian Province, China. The unique geological structure of the region, characterized by Danxia geomorphology, is essential for cultivating and producing the renowned Wuyi rock tea, a type of oolong tea with a more intricate process compared to other teas [1]. The primary processing of Wuyi rock tea is divided into two distinctive stages. The first stage involves the fresh leaves' picking, withering, and fermentation, while the second stage mainly involves the green removing, rolling, and drying of the leaves. During the first stage, the chemical composition of the tea leaves undergoes significant changes under the influence of both enzymatic and non-enzymatic reactions, which contribute to the formation of the unique flavor of Wuyi rock tea [2,3]. In the subsequent second stage, after conducting green removing on the tea leaves, the chemical transformation occurs exclusively under non-enzymatic reactions. Therefore, the primary processing stage is the primary determinant of the chemical composition of Wuyi rock tea, with the first stage being particularly critical as it lays the foundation for the second stage and is a crucial factor in the formation of the distinct flavor profile [4–7]. This identifies the significant role that the first stage of the primary processing plays in the production of Wuyi rock tea and the crucial importance of understanding the metabolite change of each node in the production process.

Wuyi rock tea's primary processing is mainly divided into three steps: picking fresh leaves, withering, and fermentation. Tea leaves are carefully picked and then undergo withering, the purpose of which is to induce a moderate loss of water and soften the leaves, and this process encourages the transformation and accumulation of substances within the leaves [8]. For example, Zhou et al. [9] discovered that withering could reduce the total catechin and starch content while increasing the theaflavin, γ-aminobutyric acid, maltose, and soluble sugar content, thereby enhancing the mellow, umami, and sweet taste of white tea. Furthermore, Fang et al. [10] analyzed the effect of withering on black tea metabolites, revealing that withering encouraged the production of alcohols, aldehydes, phenols, hydrocarbons, and halogenated hydrocarbons. Similarly, Shan et al. [11] found that most free amino acids and catechins increased in content after the withering of Longjing green tea, while the content of organic acids, phenolic acids, and their derivatives decreased. This indicates that during withering, the substances inside the tea leaves have already started undergoing transformation and accumulation, and this transformation and accumulation of substances was differential among different types of tea leaves. Once withering is complete, tea leaves are ready for fermentation. During this stage, the enzyme activity within the leaves rapidly increases, intensifying the degree of substance transformation and rapidly raising the content of the total substances [12]. For example, Salman et al. [13] found that fermentation significantly affects the substance transformation of tea, with the higher the degree of fermentation, the higher the content of theaflavins and the lower the content of catechins and total phenols. Wang et al. [14] found that fermentation promotes the oxidative degradation of fatty acids and carotenoids in black tea, thus enhancing the formation and accumulation of terpenoids, ketones, and aldehydes. It can be seen that during the primary processing of tea, withering and fermentation have an important impact on the transformation and accumulation of substances in tea, and for different types of tea, the effect of withering and fermentation on the transformation of substances is obviously different, which leads to different types of tea with different flavors. For instance, Jia et al. [15] concluded that withering is beneficial for enhancing the mellowness, freshness, and brisk taste of oolong tea, and there are 47 characteristic metabolites determining the changes in the taste characteristics, mainly nucleotides and derivatives, terpenoids, and organic acids. Li et al. [16] found that only 71 key metabolites, which are categorized as catechins, flavonoids, phenolic acids, and terpenoids, significantly changed in Fu brick tea after withering and fermentation. Secondly, withering and fermentation reduced the bitterness, astringency, and acidity of Fu brick tea but increased the mellowness of the tea. Cheng et al. [17] found that withering and fermentation are conducive to reducing the bitterness and astringency of Qingzhuan tea and improving the woodiness and mellowness of the tea, and that there were 102 key metabolites determining this change in taste characteristics. It can be seen that the formation of tea taste characteristics was closely linked to changes in key metabolites during processing, and the analysis and acquisition of key metabolites with significant changes during tea processing were of significance for enhancing tea processing and shaping tea flavor.

Rougui (*Camellia sinensis*) is the primary cultivated variety of tea trees in Wuyishan, a representative location of Wuyi rock tea. However, the key metabolites that significantly change after withering and fermentation in Rougui and their effects on tea taste characteristics are still unclear. In-depth investigation into the effect of withering and fermentation on the transformation and accumulation of Rougui metabolites, as well as obtaining the key metabolites and analyzing their effects on the taste characteristics of the tea, are of great significance for optimizing the processing technology of Rougui tea and improving its taste characteristics. Therefore, in the present study, Rougui was chosen as the subject of research. Initially, the tea leaves were processed in accordance with the traditional Wuyi rock tea production method, and fresh leaves, withered leaves, and leaves at different fermentation degrees were collected to assess the effects of different processing processes on the metabolite content of leaves. Meanwhile, key metabolites were screened to obtain significant changes in the different processing nodes and then analyzed for their taste characteristics. The present study analyzed the effect of withering and fermentation on the taste characteristics of Rougui, which is crucial in Rougui processing from a practical production standpoint, providing valuable information for refining Rougui processing and producing superior-quality Rougui tea.

2. Materials and Methods

2.1. Field Experiment and Sample Collection

Located in Fujian Province, Wuyishan City, China, is an important tea-producing region in the subtropical zone. It experiences an average annual temperature of 12~13 °C, with an annual precipitation of more than 2000 mm and a relative humidity of 85%. Wuyi rock tea is a famous traditional Chinese tea because tea trees grow in the cracks of rock, and the tea therefore has a special rock rhyme quality. Rougui tea tree is a highly recognized cultivar native to the renowned Wuyishan region of Fujian Province. This cultivar was formally recognized as a superior provincial variety by the Crop Variety Validation Committee of Fujian Province in 1985. In this study, fresh leaves of the Rougui tea tree (three leaves and one core) were collected, and the tea leaves were primary-processed according to the traditional production mode of Wuyi rock tea, including fresh leaf picking, withering, fermentation, green removing, kneading, and drying [18]. This study focuses on metabolite changes in Rougui leaves at various processing nodes during the first stage of primary processing and their effects on taste characteristics. Therefore, fresh leaves (FL), withered leaves (WD), and leaves after the 1st, 2nd, and 3rd fermentations (FJ1, FJ2, FJ3) of Rougui processing were collected in this study (Figure 1). Meanwhile, the Rougui leaves collected were extracted, enriched, and analyzed for metabolites using ultra-performance liquid chromatography/tandem mass spectrometry (UPLC-MS/MS) with three replicates per sample.

2.2. Metabolite Extraction from Tea Leaves

The tea samples collected were first freeze-dried through a vacuum freeze-dryer (Scientz-100F, Hangzhou, China). The dried samples were ground into powder and subsequently used for the metabolite extraction as described by Zhou et al. [19]. In brief, 60 mg of tea powder was added to 1.5 mL of methanol (70%) and stirred with rapid vortexing and shaking for 45 s. For this process, in a vortex mixer, the prepared sample was agitated for 45 s at 30 min intervals, with the overall number of vortex agitation cycles being 6. Subsequently, the sample was centrifuged in a centrifuge at a speed of 12,000 revolutions per minute for 3 min, and the supernatant was isolated utilizing an ANPEL SCAA-104 0.22 μm filter membrane (Shanghai, China). This isolated supernatant was then utilized for the UPLC-MS/MS (UPLC, Shimadzu's Nexera X2, Kyoto, Japan; MS/MS, Applied Biosystems 4500 QTRAP, MA, USA) analysis. Three independent replicates were set up for each sample.

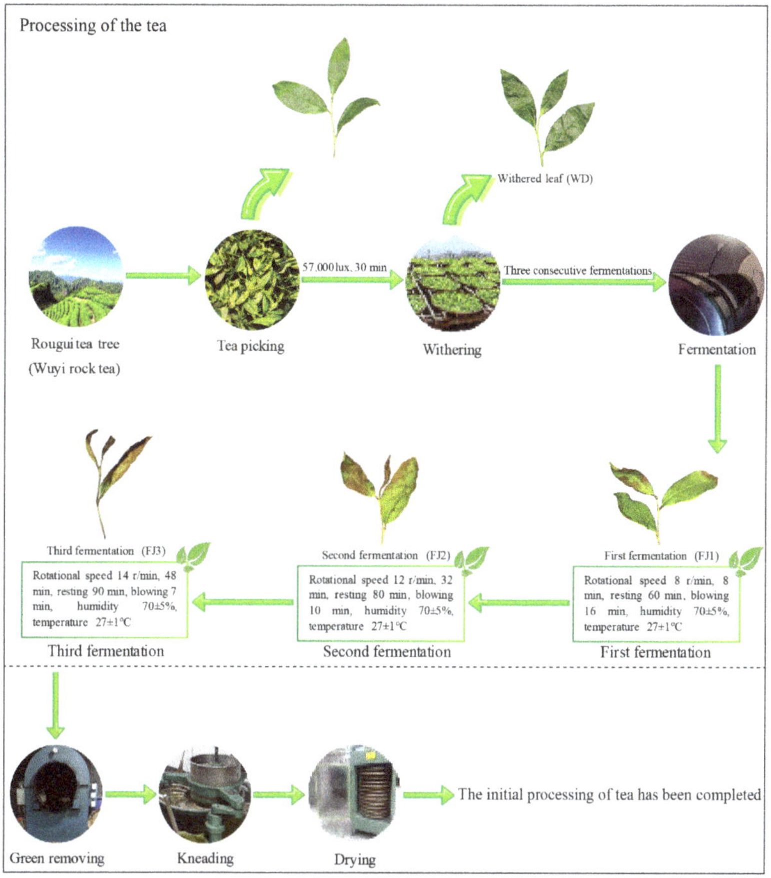

Figure 1. Primary processing of Wuyi rock tea and distribution of sampling points.

2.3. Qualitative and Quantitative Analysis of Metabolites in Tea

Metabolites were determined using UPLC-MS/MS, in which the UPLC was Shimadzu's Nexera X2 (Kyoto, Japan), the column was the Agilent SB-C18 (1.8 µm, 2.1 mm × 100 mm, CA, USA), and the MS/MS was the Applied Biosystems 4500 QTRAP (MA, USA). Metabolite analysis was conducted utilizing ultra-performance liquid chromatography (UPLC) with a gradient elution mode. The mobile phase consisted of ultrapure water with 0.1% formic acid serving as mobile phase A and acetonitrile with 0.1% formic acid functioning as mobile phase B. The UPLC run was carefully designed to achieve a ratio of mobile phase A initially at 95% at 0.00 min, subsequently gradually decreasing to 5% over a period of within 9.00 min and being maintained for an additional 1 min. Subsequently, the proportion of phase A was increased to 95% within 1.1 min and maintained for 2.9 min. The injection volume of the sample during the UPLC assay was kept at 4 µL, the flow rate was set at 0.35 mL/min, and the column temperature was maintained at 40 °C. The MS/MS conditions for metabolite identification were set as follows: the source temperature was set to 550 °C; the ion spray voltage was adjusted to 5500 V for the positive ion mode and −4500 V for the negative ion mode; the collision-activated dissociation was programmed to high; the compositions of the ion sources gas I, gas II, and the curtain gas were adjusted to 50, 60, and 25 psi, respectively; and the spectral scanning was performed by employing a triple series quadrupole mass spectrometer configured for the multiple reaction monitoring mode. Prior to scanning, the instrument was adjusted to an intermediate level of collision gas intensity. It then underwent a series of tuning and mass calibration procedures using 10 and 100 μmol/L of polypropylene glycol solutions as reference substances. The principle applied was to determine the optimal declustering potential and the collision energy of each multiple reaction monitoring ion in order to track specific multiple reaction monitoring ion pairs of eluted metabolites across all subsequent time periods. The characterization of the metabolites was facilitated by the application of advanced mass spectrometry data analysis methods utilizing Analyst (1.6.3) software to obtain retention times and ion current intensities for each metabolite. This approach was further refined by facilitating direct comparisons with NIST20 mass spectrometry database standards. Following the characterization of the metabolites, the characteristic ions and their signal intensities for each metabolite were determined using a triple quadrupole mass spectrometer. Subsequently, the chromatographic peaks corresponding to each metabolite were integrated and corrected using the software package MultiQuant (version 3.0). The peak areas derived from the chromatographic peaks were utilized as a surrogate measure of the metabolite content [20].

2.4. Statistical Analysis

In this study, the raw data of the metabolites were obtained, and, firstly, Excel 2020 was used for the preliminary statistical analysis, including the calculation of the mean value of metabolite content and its classification. Differences in metabolite content across different processing nodes were analyzed using variance analysis and paired Student's t-tests, with $p < 0.05$ being considered statistically significant. After data analysis, graphs were produced uniformly using Rstudio software (v 4.2.3) and different R packages [21]. The R package used for the multi-group box plots was gghalves 0.1.4; for the heat maps, it was pheatmap 1.0.12; for the principal component analysis, it was ggbiplot 0.55; for the Kruskal–Wallis test and LDA analysis, it was microeco; and for the multiple volcano plots, it was ggplot2 and ggnewscale. The R package used for technique for order preference by similarity to ideal solution (TOPSIS) was dplyr 1.1.4, and the R package used for the curvilinear plot was ggplot2.

3. Results and Discussion

3.1. Effects of Withering and Fermentation on Metabolites in Tea Leaves

During processing, substances in tea leaves are transformed and accumulated, and the composition and content of different substances change, which has an important role in the special flavor of tea [2,22]. Wang et al. [23] found that the metabolites in green tea leaves changed significantly during processing, and 527 key metabolites were obtained, which could be categorized into 11 groups. Chen et al. [24] detected the metabolites of oolong tea processing and revealed that the key metabolites that underwent significant changes in oolong tea as a result of processing were catechins and their derivatives, flavonol glycosides, and amino acids. In the present study, it was revealed (Figure 2A) that significant morphological changes occurred during Rougui processing from fresh leaves (FL) to withering leaves (WD) to fermented leaves (FJ1 to FJ3), and that such changes may lead to alterations in the transformation and accumulation of metabolites. Therefore, this study deeply analyzed leaf metabolite changes during Rougui processing through the UPLC-MS/MS technique, and the results revealed (Figure 2B, Table S1) that during the processing of Rougui leaves, a total of 1249 metabolites were detected at each node; moreover, the total amount of metabolites significantly increased with each processing step. Principal component analysis on leaf metabolite content from different processing nodes showed (Figure 2C) that the two principal components could effectively distinguish the different processing nodes. The contribution of principal component 1 was found to be 35.19%, while the contribution of principal component 2 was 13.11%. This resulted in an overall contribution of 48.30%. It is evident that the metabolite content in the leaves significantly altered during the processing of Rougui.

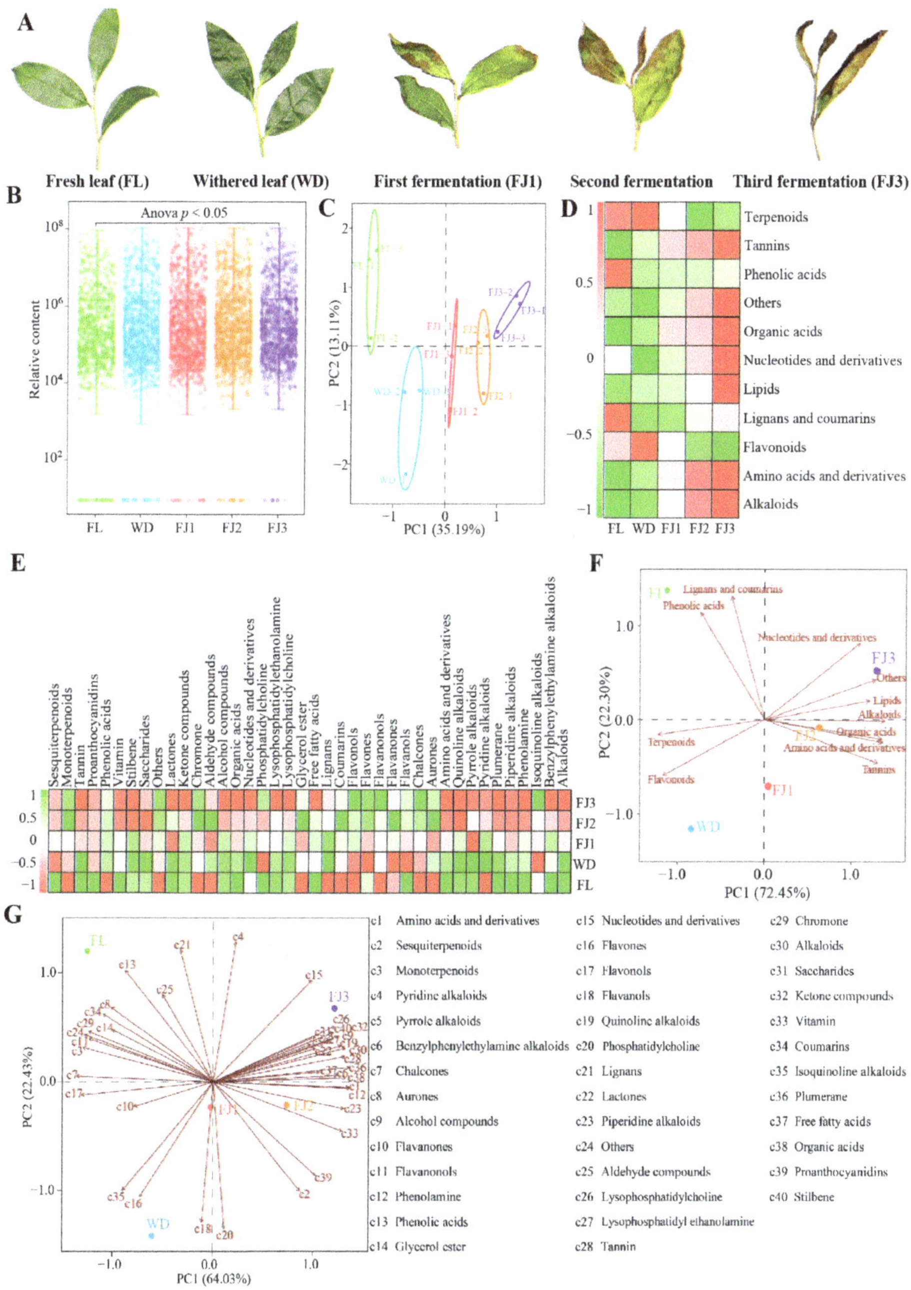

Figure 2. Metabolomics analysis of tea leaves across different processing nodes. Note: FL: fresh leaves; WD: withered leaves; FJ1: first fermentation; FJ2: second fermentation; FJ3: third fermentation; (**A**): tea leaf morphology across different processing nodes; (**B**): metabolite contents in tea leaves across

different processing nodes; (**C**): principal component analysis of metabolite contents in tea leaves across different processing nodes; (**D**): heat map of metabolite contents of primary classification in tea leaves at different processing nodes; (**E**): heat map of metabolite contents of secondary classification in tea leaves across different processing nodes; (**F**): principal component analysis of metabolite contents of primary classification in tea leaves across different processing nodes; (**G**): principal component analysis of metabolite contents of secondary classification in tea leaves across different processing nodes.

The present study continued to analyze the classification of metabolites in Rougui from different processing nodes, and the results showed (Figure 2D) that metabolites were characterized based on the first level of classification, where they could be phylogenetically organized into eleven groups. Five of these groups, encompassing amino acids and derivatives, tannins, alkaloids, organic acids, and lipids, exhibited an increasing trend in their content. Conversely, the content of the four other groups, encompassing phenolic acids, nucleotides and derivatives, lignans and coumarins, and others, displayed a biphasic trend, initially decreasing and then increasing. The secondary classification of metabolites demonstrated (Figure 2E) that they could be categorized into 40 groups. Of these, 10 groups displayed a progressive increase, 4 groups exhibited a progressive decrease, 3 groups exhibited a gradual increase followed by a subsequent decrease, 10 groups demonstrated a gradual decrease followed by a subsequent increase, and 13 groups exhibited fluctuating changes as the tea processing process was executed. The results of the principal component analysis after the first-level categorization of metabolites from different processing nodes showed (Figure 2F) that the contribution of principal component 1 was 72.45%, in which FL, WD, and FJ1 were located at the negative end of principal component 1, with the main contribution coming from four groups of metabolites, while FJ2 and FJ3 were located at the positive end of principal component 1, with the main contribution coming from seven groups of metabolites. The results of the principal component analysis after the secondary categorization of the metabolites from different processing nodes showed (Figure 2G) that the contribution of principal component 1 amounted to 64.03%, in which FL, WD, and FJ1 were still located at the negative end of principal component 1, with the main contribution coming from 16 groups of metabolites, while FJ2 and FJ3 were located at the positive end of principal component 1, with the main contribution coming from 24 groups of metabolites. It is evident that leaf metabolites significantly altered during the processing of Rougui, and significant differences appeared in their contents.

3.2. *Screening of Key Differential Metabolites Across Different Processing Nodes*

There is a large variety of metabolites in tea, and processing leads to changes in their types and contents, while a small number of key metabolites are mainly responsible for determining tea taste characteristics [25]. Shi et al. [26] analyzed the effect of processing on green tea metabolites and found that there were up to 1034 metabolites in green tea, while only 146 key metabolites, mainly flavonoids, amino acids, lipids, and phenolic acid metabolites, were significantly changed during processing. Xue et al. [27] revealed that 368 metabolites were obtained during black tea processing, of which only 93 key metabolites were significantly changed, and key metabolites determined tea taste characteristics. It can be seen that screening and obtaining key metabolites that change significantly during tea processing is important for further analyzing the impact of processing on the formation of tea taste characteristics. Accordingly, this study further screened the key metabolites that were significantly changed during the processing of Rougui. In the present study, the metabolite content in Rougui across different processing nodes was analyzed using the Kruskal–Wallis test, and it was found (Figure 3A) that, out of 1249 metabolites, 492 metabolites were significantly changed. Analysis of 492 metabolites using LDA revealed 144 differential metabolites with LDA values greater than two (Figure 3B). Further analysis of 144 differential metabolites that significantly changed across different processing nodes using multiple volcano plots revealed (Figure 3C) that 71 key metabolites that significantly changed were obtained by screening, and 39 of these metabolites exhibited an increase in content and the remaining 32 presented a decrease as processing progressed. The weight of withering and fermentation effects on the content of 71 metabolites was determined using TOPSIS weight analysis, and it was found (Figure 3D) that only 40 of the 71 metabolites had an effect weight greater than 10%. It can be seen that only 40 key metabolites were significantly changed during the withering and fermentation of Rougui, and the content of key metabolites may affect the taste characteristics of tea.

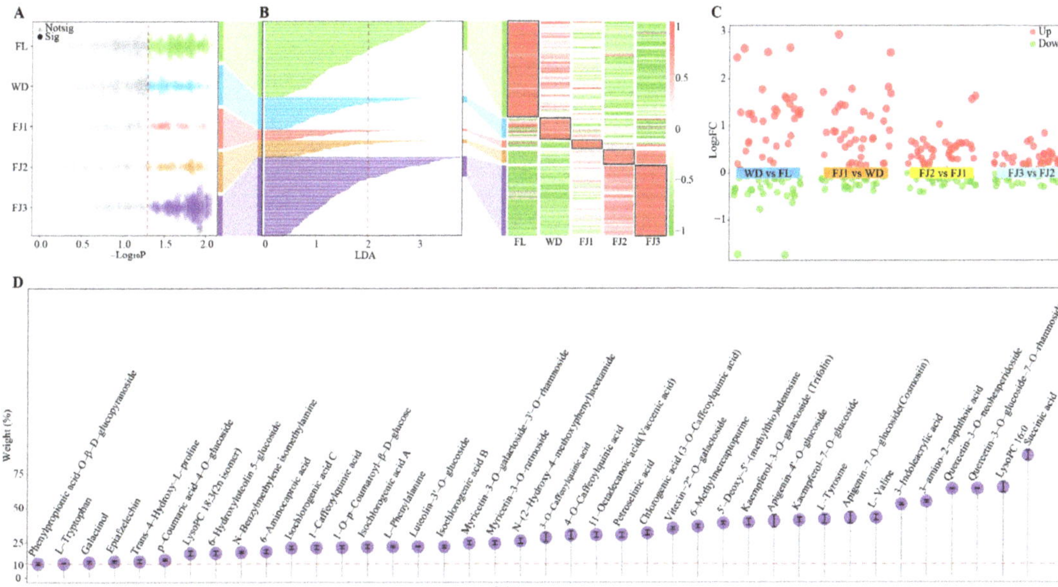

Figure 3. Screening for significantly changed key metabolites in Rougui leaves across different processing nodes. Note: FL: fresh leaves; WD: withered leaves; FJ1: first fermentation; FJ2: second fermentation; FJ3: third fermentation; (**A**): screening for significantly changed metabolites across different processing nodes by Kruskal–Wallis test; (**B**): screening for significantly changed differential metabolites by LDA; (**C**): screening for significantly changed key metabolites using a multiple volcano diagram; (**D**): determination of the weight of the effect of withering and fermentation on key metabolites by TOPSIS.3.3. Taste characteristics of key metabolites.

From the above results, 40 key metabolites obtained in this study were further categorized, and it was found (Figure 4A) that the 40 key metabolites were subsequently classified into 13 groups based on the secondary classification. Five of these groups, including saccharides, phenolic acids, flavonols, flavones, and flavanols, exhibited a significant reduction in content upon processing, while eight groups, including organic acids, plumerane, alkaloids, nucleotides and derivatives, amino acids and derivatives, lysophosphatidyl choline, phenolamine, and free fatty acids, demonstrated a significant increase in content. Reportedly, tea leaves exhibit diverse taste characteristics due to the varying types of metabolites. For example, saccharides mainly present sweetness taste characteristics [28]. Organic acids, plumerane, and alkaloids mainly present mellowness taste characteristics [29,30]. Nucleotides and derivatives and amino acids and derivatives mainly present fresh and brisk taste characteristics [31,32]. Phenolic acids, flavonols, flavones, flavanols, phenolamine, and lysine mainly present bitterness and astringency taste characteristics [33–36]. However, free fatty acids mainly present aroma [37]. It can be seen that, after withering and fermentation, the key metabolites that changed significantly in Rougui mainly exhibited five types of taste characteristics, namely sweetness, mellowness, fresh and brisk taste, bitterness and astringency, and aroma. Further analysis of the trends in the intensity of the five taste characteristics during the processing of Rougui revealed (Figure 4B) that the intensity of the sweetness, bitterness, and astringency of the tea showed a decreasing trend as the processing process progressed, while the intensity of the mellowness, fresh and brisk taste, and aroma showed an increasing trend. TOPSIS analysis showed (Figure 4C) that withering and fermentation had the greatest effect on Rougui taste characteristics in terms of bitterness and astringency, followed by mellowness.

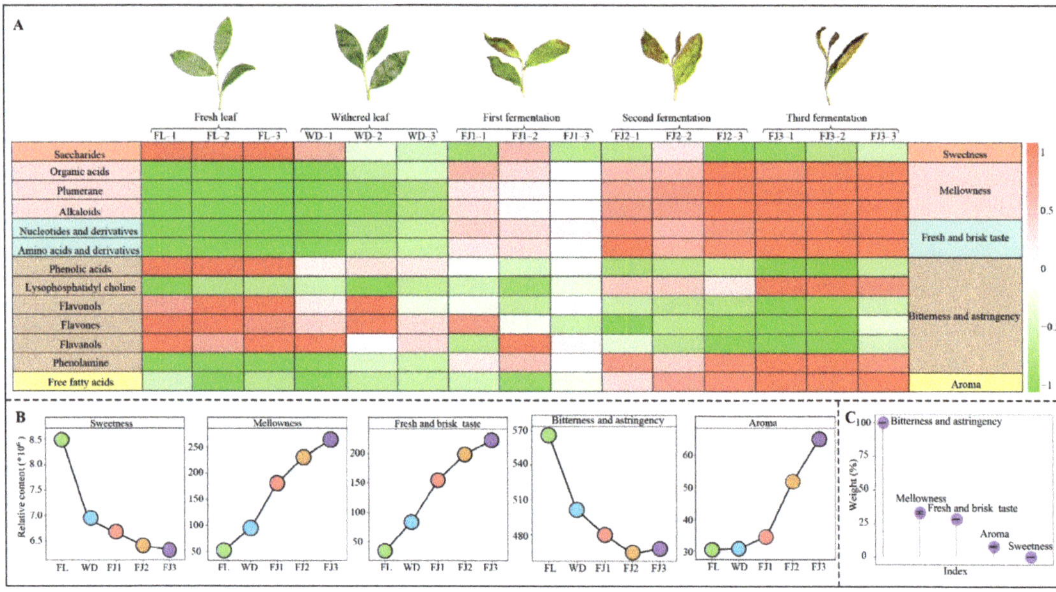

Figure 4. Changes in the content of key metabolites from tea leaves across different processing nodes and their effects on taste characteristics. Note: FL: fresh leaves; WD: withered leaves; FJ1: first fermentation; FJ2: second fermentation; FJ3: third fermentation; (**A**): changes in the content of key metabolites and their taste characteristics at different processing stages; (**B**): trends in the intensity of taste characteristics of tea leaves across different processing nodes; (**C**): TOPSIS determination of the weights of the effects of withering and fermentation on different taste characteristics.

Withering and fermentation are key steps in tea processing, and different teas show different changes in their taste characteristics after undergoing withering and fermentation [38]. For example, after the withering of green tea, the content of polyphenols and flavonoids decreases significantly, and the intensity of the bitterness and astringency of the tea decreases significantly [39]. After the withering and fermentation of black tea, the intensity of the bitterness and astringency of the tea significantly decreases, while the intensity of the sweetness and mellowness significantly increases [10]. The fermentation of Fu brick tea results in a significant decrease in astringency, bitterness, and sourness intensities and a significant increase in mellow intensity [16]. The intensity of bitterness and astringency of Liupao tea significantly increases after withering and fermentation [40]. It is evident that significant differences existed in the impact that withering and fermentation had on the taste characteristics of different teas. Rougui, the subject of this study, belongs to oolong tea, and withering and fermentation were conducive to increasing the intensity of the mellowness, fresh and brisk taste, and aroma while decreasing the intensity of the sweetness, bitterness, and astringency, and withering and fermentation were the most important in significantly reducing the intensity of bitterness and astringency.

4. Conclusions

In the present study, the effects of withering and fermentation on the metabolites and taste characteristics of Rougui were analyzed, and the results revealed (Figure 5) that the metabolite content in tea leaves significantly increased upon processing. Withering and fermentation promote the transformation and accumulation of metabolites in tea leaves and increase the level of metabolites. The content of 40 key metabolites significantly changed during Rougui withering and fermentation, and they were categorized into 13 groups based on secondary classification. It was further found that the content of the five groups of metabolites mentioned above, such as saccharides, phenolic acids, flavonols, flavones, and flavanols, showed a significant decreasing trend with processing, while the content of eight groups, such as organic acids, plumerane, alkaloids, nucleotides and derivatives, amino acids and derivatives, lysophosphatidyl choline, phenolamine, and free fatty acids, showed a significant increase. After withering and fermentation, the key metabolites that changed significantly in Rougui showed five main taste characteristics, namely mellowness, sweetness, fresh and brisk taste, bitterness and astringency, and aroma. For Rougui, withering and fermentation were conducive to increasing the intensity of the mellowness, fresh and brisk taste, and aroma while decreasing the intensity of the sweetness and bitterness and astringency, and withering and fermentation were the most important to significantly reduce the intensity of bitterness and astringency. It is evident that withering and fermentation are pivotal processes in the development of the unique taste characteristics of Wuyi rock tea, Rougui. This research provides a valuable reference for refining the Rougui processing technology and crafting a unique flavor.

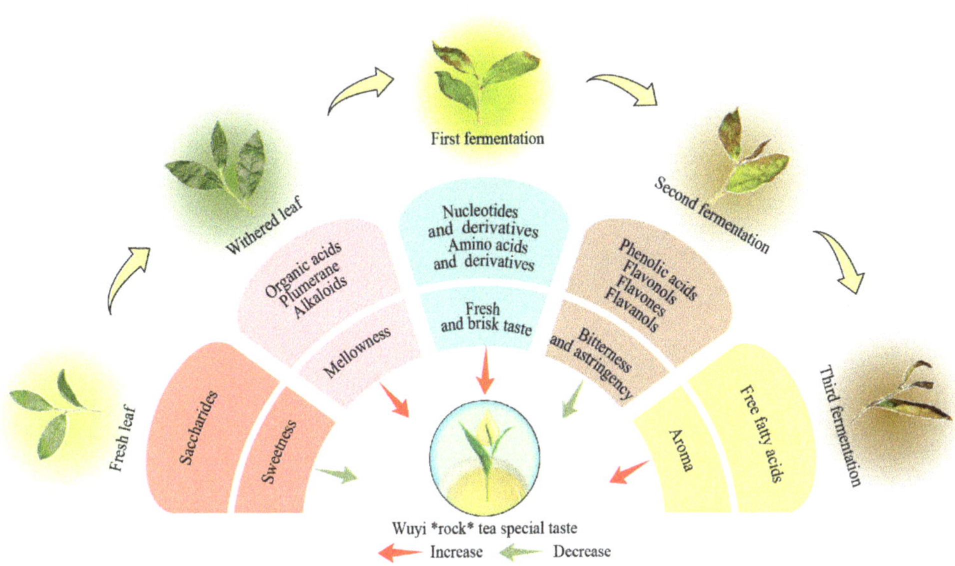

Figure 5. Effect of primary processing on tea quality formation.

Supplementary Materials: The following supporting information can be downloaded at: https://www.mdpi.com/article/10.3390/foods13233955/s1, Table S1: UPLC-MS/MS determination data of tea leaf metabolites at different processing stages.

Author Contributions: J.Y. and Y.L. (Yangxin Luo): conceptualization, formal analysis, methodology, software, writing—original draft. Y.W. and Q.Z.: data curation, methodology, writing reviewing and editing. S.Z. and J.G.: methodology, data curation, writing—original draft. Y.L. (Yankun Liao), T.W. and X.J.: data curation, methodology, writing—review and editing. H.W.: project administration, supervision, funding acquisition, writing—original draft, writing—review and editing. All authors have read and agreed to the published version of the manuscript.

Funding: This work was supported by the Natural Science Foundation of Fujian Province (22024N0009, 2024J01861, 2024J01866); the Nanping City Science and Technology Plan Project (NP2021KTS06, NP2021KTS05, N2021Z012); the Innovation and Entrepreneurship Training Program for College Students (202411312002, S202210397033, 202310397003, 202310397013); the construction of the first-class undergraduate specialty (tea science) in Fujian Province (SJZY2019004); the faculty and students co-creation team of Wuyi University (2021-SSTD-01, 2021-SSTD-05).

Institutional Review Board Statement: Not applicable.

Informed Consent Statement: Not applicable.

Data Availability Statement: The original contributions presented in the study are included in the article/supplementary material, further inquiries can be directed to the corresponding author.

Conflicts of Interest: The authors declare no conflicts of interest.

References

1. Wong, M.; Sirisena, S.; Ng, K. Phytochemical profile of differently processed tea: A review. *J. Food Sci.* **2022**, *87*, 1925–1942. [CrossRef] [PubMed]
2. Liu, L.; Shi, J.; Yuan, Y.; Yue, T. Changes in the metabolite composition and enzyme activity of fermented tea during processing. *Food Res. Int.* **2022**, *158*, 111428. [CrossRef] [PubMed]
3. Wu, H.; Chen, Y.; Feng, W.; Shen, S.; Wei, Y.; Jia, H.; Wang, Y.; Deng, W.; Ning, J. Effects of three different withering treatments on the aroma of white tea. *Foods* **2022**, *11*, 2502. [CrossRef] [PubMed]
4. Aaqil, M.; Peng, C.; Kamal, A.; Nawaz, T.; Zhang, F.; Gong, J. Tea harvesting and processing techniques and its effect on phytochemical profile and final quality of black tea: A review. *Foods* **2023**, *12*, 4467. [CrossRef]
5. Yu, J.Y.; Li, J.Z.; Lin, Z.; Zhu, Y.; Feng, Z.H.; Ni, D.J.; Zeng, S.C.; Zeng, X.H.; Wang, J.M.; Zhang, L.; et al. Dynamic changes and the effects of key procedures on the characteristic aroma compounds of Lu'an Guapian green tea during the manufacturing process. *Food Res. Int.* **2024**, *188*, 114525. [CrossRef]
6. Qiao, D.H.; Zhu, J.Y.; Mi, X.Z.; Xie, H.; Shu, M.T.; Chen, M.Y.; Li, R.; Liu, S.R.; Wei, C.L. Effects of withering time of fresh leaves on the formation of flavor quality of Taiping Houkui tea. *LWT* **2023**, *182*, 114833. [CrossRef]
7. Hao, M.J.; Lai, X.F.; Li, Q.H.; Cao, J.X.; Sun, L.L.; Chen, R.H.; Zhang, Z.B.; Li, Q.; Lai, Z.X.; Sun, S.L. Widely targeted metabolomic analysis reveals metabolite changes induced by incorporating black tea fermentation techniques in oolong tea processing for quality improvement. *Food Chem.* **2024**, *459*, 140433. [CrossRef]
8. Paiva, L.S.; Dias, A.P.; Marcone, M.F.; Baptista, J.A.B. The impact of different withering process conditions on the bioactivity and quality of black tea from Azorean *Camellia sinensis*. *Beverages* **2023**, *9*, 94. [CrossRef]
9. Zhou, C.Z.; Chen, Z.H.U.; Li, X.Z.; Lan, C.H.E.N.; Xie, S.Y.; Chen, G.W.; Zhang, H.; Lai, Z.X.; Lin, Y.L.; Guo, Y.Q. Transcriptome and phytochemical analyses reveal the roles of characteristic metabolites in the taste formation of white tea during the withering process. *J. Integr. Agric.* **2022**, *21*, 862–877. [CrossRef]
10. Fang, X.; Liu, Y.; Xiao, J.; Ma, C.; Huang, Y. GC–MS and LC-MS/MS metabolomics revealed dynamic changes of volatile and non-volatile compounds during withering process of black tea. *Food Chem.* **2023**, *410*, 135396. [CrossRef]
11. Shan, X.J.; Yu, Q.Y.; Chen, L.; Zhang, S.; Zhu, J.Y.; Jiang, Y.W.; Yuan, H.B.; Zhou, Q.H.; Li, J.; Wang, Y.J.; et al. Analyzing the influence of withering degree on the dynamic changes in non-volatile metabolites and sensory quality of Longjing green tea by non-targeted metabolomics. *Front. Nutr.* **2023**, *10*, 1104926. [CrossRef] [PubMed]
12. Zhao, F.; Qian, J.; Liu, H.; Wang, C.; Wang, X.; Wu, W.; Wang, D.; Cai, C.; Lin, Y. Quantification, identification and comparison of oligopeptides on five tea categories with different fermentation degree by Kjeldahl method and ultra-high performance liquid chromatography coupled with quadrupole-orbitrap ultra-high resolution mass spectrometry. *Food Chem.* **2022**, *378*, 132130. [CrossRef] [PubMed]
13. Salman, S.; Öz, G.; Felek, R.; Haznedar, A.; Turna, T.; Özdemir, F. Effects of fermentation time on phenolic composition, antioxidant and antimicrobial activities of green, oolong, and black teas. *Food Biosci.* **2022**, *49*, 101884. [CrossRef]
14. Wang, J.; Ouyang, W.; Zhu, X.; Jiang, Y.; Yu, Y.; Chen, M.; Yuan, H.; Hua, J. Effect of shaking on the improvement of aroma quality and transformation of volatile metabolites in black tea. *Food Chem. X* **2023**, *20*, 101007. [CrossRef]
15. Jia, X.; Zhang, Q.; Chen, M.; Wang, Y.; Lin, S.; Pan, Y.; Cheng, P.; Li, M.; Zhang, Y.; Ye, J.; et al. Analysis of the effect of different withering methods on tea quality based on transcriptomics and metabolomics. *Front. Plant Sci.* **2023**, *14*, 1235687. [CrossRef] [PubMed]
16. Li, Q.; Jin, Y.; Jiang, R.; Xu, Y.; Zhang, Y.; Luo, Y.; Huang, J.; Wang, K.; Liu, Z. Dynamic changes in the metabolite profile and taste characteristics of Fu brick tea during the manufacturing process. *Food Chem.* **2021**, *344*, 128576. [CrossRef]

17. Cheng, L.Z.; Yang, Q.Q.; Chen, Z.Y.; Zhang, J.R.; Chen, Q.; Wang, Y.F.; Wei, X.L. Distinct changes of metabolic profile and sensory quality during Qingzhuan tea processing revealed by LC-MS-based metabolomics. *J. Agric. Food Chem.* **2020**, *68*, 4955–4965. [CrossRef]
18. Wu, W.M.; Jiang, X.Y.; Zhu, Q.; Yuan, Y.; Chen, R.P.; Wang, W.Z.; Liu, A.X.; Wu, C.J.; Ma, C.H.; Li, J.H.; et al. Metabonomics analysis of the flavor characteristics of Wuyi Rock Tea (Rougui) with "rock flavor" and microbial contributions to the flavor. *Food Chem.* **2024**, *450*, 139376. [CrossRef]
19. Zhou, X.; Hu, L.; Hoang, N.H.; Thanh, T.L.; Zhou, C.; Mei, X.; Buensanteai, K. The changes in metabolites, quality components, and antioxidant activity of tea (*Camellia sinensis*) infected with exobasidium vexans by applying UPLC-MS/MS-based widely targeted metabolome and biochemical analysis. *Phytopathology* **2024**, *114*, 164–176. [CrossRef]
20. Wen, O.; Ning, J.; Zhu, X.; Jiang, Y.; Wang, J.; Yuan, H.; Hua, J. UPLC-ESI-MS/MS analysis revealed the dynamic changes and conversion mechanism of non-volatile metabolites during green tea fixation. *LWT* **2024**, *198*, 116010. [CrossRef]
21. Wickham, H. *Advanced R*, 2nd ed.; Chapman and Hall/CRC the R Series; Taylor and Francis: Abingdon, UK, 2019.
22. Gui, A.; Gao, S.; Zheng, P.; Feng, Z.; Liu, P.; Ye, F.; Wang, S.; Xue, J.; Xiang, J.; Ni, D.; et al. Dynamic changes in non-volatile components during steamed green tea manufacturing based on widely targeted metabolomic analysis. *Foods* **2023**, *12*, 1551. [CrossRef] [PubMed]
23. Wang, H.J.; Hua, J.J.; Yu, Q.Y.; Li, J.; Wang, J.J.; Deng, Y.L.; Yuan, H.B.; Jiang, Y.W. Widely targeted metabolomic analysis reveals dynamic changes in non-volatile and volatile metabolites during green tea processing. *Food Chem.* **2021**, *363*, 130131. [CrossRef] [PubMed]
24. Chen, S.; Liu, H.; Zhao, X.; Li, X.; Shan, W.; Wang, X.; Wang, S.; Yu, W.; Yang, Z.; Yu, X. Non-targeted metabolomics analysis reveals dynamic changes of volatile and non-volatile metabolites during oolong tea manufacture. *Food Res. Int.* **2020**, *128*, 108778. [CrossRef] [PubMed]
25. Yu, X.; Xiao, J.; Chen, S.; Yu, Y.; Ma, J.; Lin, Y.; Li, R.; Lin, J.; Fu, Z.; Zhou, Q.; et al. Metabolite signatures of diverse Camellia sinensis tea populations. *Nat. Commun.* **2020**, *11*, 5586. [CrossRef] [PubMed]
26. Shi, Y.; Zhu, Y.; Ma, W.; Shi, J.; Peng, Q.; Lin, Z.; Lv, H. Comprehensive investigation on non-volatile and volatile metabolites in four types of green teas obtained from the same tea cultivar of Longjing 43 (*Camellia sinensis* var. *sinensis*) using the widely targeted metabolomics. *Food Chem.* **2022**, *394*, 133501. [CrossRef] [PubMed]
27. Xue, J.; Liu, P.; Guo, G.; Wang, W.; Zhang, J.; Wang, W.; Le, T.; Yin, J.; Ni, D.; Jiang, H. Profiling of dynamic changes in non-volatile metabolites of shaken black tea during the manufacturing process using targeted and non-targeted metabolomics analysis. *LWT* **2022**, *156*, 113010. [CrossRef]
28. Wei, F.; Luo, L.; Zeng, L. Characterization of key sweet taste compounds in Camellia nanchuanica black tea. *LWT* **2023**, *182*, 114858. [CrossRef]
29. Chen, X.; Lin, L.; Cai, H.; Gao, X. Identification and analysis of metabolites that contribute to the formation of distinctive flavour components of Laoxianghuang. *Foods* **2023**, *12*, 425. [CrossRef]
30. Zhou, J.; Fang, T.; Li, W.; Jiang, Z.; Zhou, T.; Zhang, L.; Yu, Y. Widely targeted metabolomics using UPLC-QTRAP-MS/MS reveals chemical changes during the processing of black tea from the cultivar *Camellia sinensis* (L.) O. Kuntze cv. Huangjinya. *Food Res. Int.* **2022**, *162*, 112169. [CrossRef]
31. Yu, Y.; Zhu, X.; Yuan, B.; Chen, M.; Wang, J.; Zhu, L.; Jiang, Y.; Yuan, H.; Hua, J. Investigation of non-volatile metabolite variations during round green tea processing and effect of pan-frying degree using untargeted metabolomics and objective quantification. *Food Chem.* **2024**, *457*, 140067. [CrossRef]
32. Feng, X.; Wang, H.; Yu, Y.; Zhu, Y.; Ma, J.; Liu, Z.; Ni, L.; Lin, C.; Wang, K.; Liu, Y. Exploration of the flavor diversity of oolong teas: A comprehensive analysis using metabolomics, quantification techniques, and sensory evaluation. *Food Res. Int.* **2024**, *195*, 114868. [CrossRef] [PubMed]
33. Huang, R.; Xu, C. An overview of the perception and mitigation of astringency associated with phenolic compounds. *Compr. Rev. Food Sci. Food Saf.* **2021**, *20*, 1036–1074. [CrossRef] [PubMed]
34. Soares, S.; Brandão, E.; Mateus, N.; de Freitas, V. Sensorial properties of red wine polyphenols: Astringency and bitterness. *Crit. Rev. Food Sci.* **2017**, *57*, 937–948. [CrossRef] [PubMed]
35. Wilson, A.; Johnson, J.B.; Batley, R.; Lal, P.; Wakeling, L.; Naiker, M. Authentication using volatile composition: A proof-of-concept study on the volatile profiles of fourteen Queensland ciders. *Beverages* **2021**, *7*, 28. [CrossRef]
36. Hua, J.; Ouyang, W.; Zhu, X.; Wang, J.; Yu, Y.; Chen, M.; Yang, L.; Yuan, H.; Jiang, Y. Objective quantification technique and widely targeted metabolomic reveal the effect of drying temperature on sensory attributes and related non-volatile metabolites of black tea. *Food Chem.* **2024**, *439*, 138154. [CrossRef]
37. Chen, Q.; Zhu, Y.; Liu, Y.; Liu, Y.; Dong, C.; Lin, Z.; Teng, J. Black tea aroma formation during the fermentation period. *Food Chem.* **2022**, *374*, 131640. [CrossRef]
38. Hu, T.; Shi, S.; Ma, Q. Modulation effects of microorganisms on tea in fermentation. *Front. Nutr.* **2022**, *9*, 931790. [CrossRef]

39. Wei, Y.; Yin, X.; Wu, H.; Zhao, M.; Huang, J.; Zhang, J.; Li, T.; Ning, J. Improving the flavor of summer green tea (*Camellia sinensis* L.) using the yellowing process. *Food Chem.* **2022**, *388*, 132982. [CrossRef]
40. Wang, H.F.; Teng, J.W.; Huang, L.; Wei, B.Y.; Xia, N. Determination of the variations in the metabolic profile and sensory quality of Liupao tea during fermentation through UHPLC–HR–MS metabolomics. *Food Chem.* **2023**, *404*, 134773. [CrossRef]

foods MDPI

Article

Research on Rapid Detection Methods of Tea Pigments Content During Rolling of Black Tea Based on Machine Vision Technology

Hanting Zou, Tianmeng Lan, Yongwen Jiang, Xiao-Lan Yu and Haibo Yuan *

Tea Research Institute, The Chinese Academy of Agricultural Sciences, Hangzhou 310008, China; 13500436167@163.com (H.Z.); ww98317@163.com (T.L.); jiangyw@tricaas.com (Y.J.); yuxiaolan@tricaas.com (X.-L.Y.)

* Correspondence: 192168092@tricaas.com

Abstract: As a crucial stage in the processing of black tea, rolling plays a significant role in both the color transformation and the quality development of the tea. In this process, the production of theaflavins, thearubigins, and theabrownins is a primary factor contributing to the alteration in color of rolled leaves. Herein, tea pigments are selected as the key quality indicators during rolling of black tea, aiming to establish rapid detection methods for them. A machine vision system is employed to extract nine color feature variables from the images of samples subjected to varying rolling times. Then, the tea pigment content in the corresponding samples is determined using a UV-visible spectrophotometer. In the meantime, the correlation between color variables and tea pigments is discussed. Additionally, Z-score and PCA are used to eliminate the magnitude difference and redundant information in original data. Finally, the quantitative prediction models of tea pigments based on the images' color features are established by using PLSR, SVR, and ELM. The data show that the Z-score–PCA–ELM model has the best prediction effect for tea pigments. The Rp values for the model prediction sets are all over 0.96, and the RPD values are all greater than 3.50. In this study, rapid determination methods for tea pigments during rolling of black tea are established. These methods offer significant technical support for the digital production of black tea.

Keywords: black tea rolling; color features; machine vision; machine learning

1. Introduction

As one of the most popular types of tea on the market, black tea is favored by consumers both domestically and internationally due to its distinctive flavor and functional components [1]. Compared to other types of tea, the formation of the flavor quality of black tea, as a fully fermented type of tea, is closely related to the processing technology used. The postharvest processing of black tea can be broadly classified into four stages: withering, rolling, fermentation, and drying [2]. Rolling not only significantly affects the appearance and texture of the leaves, but also lays a chemical substance basis for the subsequent fermentation stage. Simply speaking, rolling refers to the mechanical process of shaping the withered leaves into strips [3]. Under the action of external mechanical force, the cellular tissues of tea leaves are destroyed. Consequently, this damage also facilitates the release of intracellular substances, promoting polyphenols to interact with enzymes [4]. These polyphenols undergo oxidation reactions, resulting in the formation of pigment compounds such as theaflavins (TFs), thearubigins (TRs), and theabrownins (TBs) [5]. These pigment compounds are not only pivotal in the flavor of black tea, but also play a direct role in influencing the color changes observed in rolled leaves. Despite the fact that traditional stoichiometric methods can determine the content of these chemical components accurately, there is a lag in the collection, storage, transportation, and detection of the samples. This is not conducive to the efficient industrial production of black tea. At present, there are many non-destructive testing methods for tea [6–8], but there is still a lack of online testing technology for employment in the rolling process of black tea. Therefore, the real-time and rapid detection of tea pigments in rolled leaves is of great significance in black tea production and processing.

In recent research surrounding tea and other agricultural product testing, non-destructive testing technology has been widely used [9–11]. During tea processing, the use of sensors and other instruments to monitor the content of key substances online and effectively classify the processing degrees of samples are the goals of intelligent processing technology. As a kind of bionic vision technology, machine vision can obtain the color features and other information of samples through industrial cameras in real-time [12]. Unlike near-infrared spectroscopy, machine vision can indirectly associate certain chemical components with color changes in the appearance of samples. At present, it has been used in the detection of theaflavins and chlorophyll during the fermentation stage of black tea [13].

To summarize, machine vision technology is employed in this study to capture images of rolled leaves at various rolling times. Meanwhile, a rapid non-destructive detection method for tea pigments in the rolling process of black tea is established by combining image color characteristics and chemometrics methods. The color feature variables extracted from the images include RGB, HSV, and La*b*. Pearson correlation analysis is used to express the response intensity of color feature variables to tea pigments content. In addition, the application of Z-score and PCA effectively eliminates the information interference between color feature variables. Finally, a quantitative prediction model with strong generalization capabilities is established based on a linear model, nonlinear model, and neural network model. The analysis and comparison of various algorithm models is instrumental in developing effective methods for the rapid detection of tea pigments during the rolling of black tea.

2. Materials and Methods

2.1. Chemicals and Reagents

Ethyl acetate, n-butanol, oxalic acid, sodium bicarbonate, and ethanol were purchased from Aladdin Biochemical Technology Co., Ltd. (Shanghai, China). Pure water was prepared using a Pure Water System.

2.2. Sample Preparation

The tea variety selected in this study was Fuding Dabai, and the picking tenderness of fresh leaves was one bud and two leaves. The moisture content of the sample was determined by a moisture analyzer (MA35M-000230V1, Sartorious, Göttingen, Germany) during processing. The moisture content of fresh leaves was about 77.53%. First, the fresh leaves were evenly spread in a withering trough, maintained at a thickness of about 2–3 cm. At room temperature of about 24 °C and relative humidity of about 60%, the withering process ended after about 12 h. The moisture content of samples determined after withering was about 61.08%. Soon afterwards, the samples were evenly divided into nine groups of the same quality. They were then rolled in batches according to the set sampling time.

Rolling machines (6CR-45, Xiangfeng Machinery Co., Ltd., Tongxiang, China) were used, and the rolling parameters were set as follows: no-pressure for 20 min, light-pressure rolling for 20 min, heavy-pressure rolling for 10 min, light-pressure rolling for 10 min, heavy-pressure rolling for 10 min, and light-pressure rolling for a further 10 min. To collect samples of over-rolling, the final rolling time was extended by another 10 min, for a total of 90 min. From the end of withering to the end of rolling, sampling points were set every 10 min, for a total of 10 sampling points. After each batch of samples reached the corresponding rolling time, all samples were taken out. All samples taken each time were evenly divided into eight groups for data collection and preservation. The next day, the same operation was repeated. As a result, a total of 150 samples of rolled leaves were collected throughout the entire process.

2.3. Determination of Tea Pigments Content

The content of theaflavins (TFs), thearubigins (TRs), and theabrownins (TBs) was determined by referring to the system analysis method [14] and through quantitative analysis using a UV-visible spectrophotometer (P7, MAPADA, Shanghai, China). First, $NaHCO_3$ (2.5%) solution and saturated oxalic acid solution were prepared. A total of 3 g of tea sample was weighed out and transferred to a 250 mL conical bottle, and 125 mL boiling water was added. The tea sample was then soaked in a boiling water bath for 10 min and shaken intermittently. After the extraction, the solution was strained immediately with filter cloth, and the filtrate was allowed to cool. This filtrate was marked as the test solution.

The test solution was shaken well. Following this, 30 mL of it was removed with a pipette gun and placed into a separating funnel, to which 30 mL of ethyl acetate was added. Then, an oscillator (185/min) was used to shake the mixture for five minutes. Once standing and layering had occurred, the lower water layer was released, and the ethyl acetate layer above was poured out. A 2 mL extract of the ethyl acetate layer was diluted to a total volume of 25 mL using 95% ethanol. The absorbance values of the solution were recorded as EA.

A 2 mL extract of the water layer was diluted to a total volume of 25 mL using 2 mL saturated oxalic acid, 6 mL pure water, and 95% ethanol. The absorbance values of the solution were recorded as ED.

Then, 15 mL of the ethyl acetate layer was taken, transferred to the separation funnel, and 15 mL of $NaHCO_3$ (2.5%) solution was added. The mixture was shaken for 30 s with the oscillator. Following the completion of layering, the lower sodium bicarbonate layer was discarded, and the ethyl acetate layer above was saved. A total of 4 mL of the ethyl acetate layer was diluted to a total volume of 25 mL using 95% ethanol. The absorbance values of the solution were recorded as EC.

Finally, 15 mL of test solution was taken and transferred to the separation funnel. Then, 15 mL of n-butanol was added, and the mixture was shaken with the oscillator for 3 min. Once layering had occurred, the lower water layer was saved. A total of 2 mL of this water layer was diluted to a total volume of 25 mL using 2 mL saturated oxalic acid, 6 mL pure water, and 95% ethanol. The absorbance values of the solution were recorded as EB.

The values of EA, EB, EC, and ED were measured at 380 nm with 95% ethanol as the control [15]. The formulas for calculating the content of each component are as follows:

$$\text{TFs}\ (\%) = \text{EC} \times 2.25 \tag{1}$$

$$\text{TRs}\ (\%) = 7.06 \times (2 \times \text{EA} + 2 \times \text{ED} - \text{EC} - 2 \times \text{EB}) \tag{2}$$

$$\text{TBs}\ (\%) = 2 \times \text{EB} \times 7.06 \tag{3}$$

2.4. Image Information Acquisition

In this research, a custom-designed machine vision system was employed for data acquisition. The machine vision system is an industrial integrated machine composed of a high fidelity industrial camera (MV-CS020-10GMGC, Hikvision Digital Technology Co., Ltd., Hangzhou, China), a professional industrial lens, a dome shadowless light source (pure white, LED), a control panel, visual inspection software, and other core components. The camera has a resolution of 20 million pixels. The color array camera was used to shoot the tea samples under the dome light source, and the closed equipment box was used to ensure that the detection was not affected by the ambient light and maintain detection stability. The distance between the camera lens and the sample was about 6 cm, and the exposure time was 0.08 ms. During the experiment, the sample was evenly spread in the preset shooting area, and the thickness of it was controlled at about 2 cm. In order to make the data obtained more reliable, two to three different images were taken for each group of samples. These images were used to calculate the average values of color feature variables. To sum up, the experimental process is shown in Figure 1.

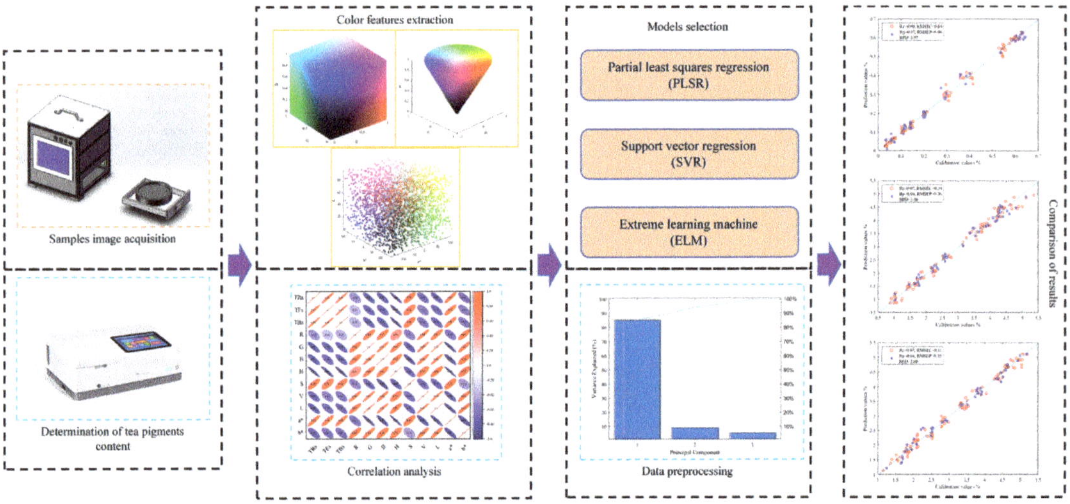

Figure 1. Flow chart of the experiment.

2.5. Color Features Information Extraction

Here, the machine vision system was used to automatically select a pixel region in the shooting area of rolled leaves, and this part of the region was regarded as the region of interest (ROI) [16]. Concurrently, the machine learning algorithms were used to extract color feature variables in the ROI. The average values of each color feature variable in the ROI were calculated, and they were used to represent the color feature information of each group of samples. As a result, each image was converted into nine color feature variables, including three color models, namely, RGB, HSV, and La*b*.

RGB is not only widely used in mobile phones, monitors, and other devices, but is also crucial in the research of graphics processing [17]. It consists of R, G, and B values (R: red, G: green, B: blue). Therefore, various other visible colors can be produced by changing the combination of R, G, and B values to different intensities. However, in some cases, it is found that this color model does not fully meet needs. For example, in terms of color perception, people's color cognition does not only rely on red, green, and blue. Therefore, in order to simulate human perception of color more accurately, HSV was introduced. It is composed of H, S, and V values (H: hue, S: saturation, V: brightness) [18]. Moreover, La*b* consists of L, a*, and b* values (L: brightness, a*: red-green degree, b*: yellow-blue degree), and is also widely used in color science [19]. In summary, the three color models of RGB, HSV, and La*b* have their own advantages and disadvantages, and are suitable for different scenes and needs. RGB plays a dominant role in digital image processing, while HSV is closer to human color perception.

2.6. Data Set Partitioning and Preprocessing

The random sampling method was used to divide the data set [20], and the ratio of training set to prediction set was set at 7:3. First, all the sample data were arranged in ascending order according to the corresponding tea pigments content. From top to bottom, each of the 10 sample data sets was divided into a group. In each group, random sampling was carried out, in which three groups of sample data were used as prediction set data, and the remaining seven groups of sample data were used as training set data. After partitioning, the obtained training set data contained 105 groups of sample data, while the prediction set data contained 45 groups of sample data. The color feature information in the images was derived from three distinct color models. The methods for calculating color features varied among these models, leading to differences in values. In order to eliminate the differences between them, the data standardization method (Z-score) was adopted as the pre-processing method of the original data before using the features information for modeling and analysis. Z-score is a widely used data processing technology that can ensure the comparability of data. In addition, to reduce the dimensionality of the data, principal component analysis (PCA) was used [21].

2.7. Establishment of Prediction Models

Partial least squares regression (PLSR), as a classical linear model, has been widely used in tea component content prediction research [22–24]. The PLSR model can not only avoid the collinearity problem, but also eliminate the noise effect on regression. Support vector regression (SVR), as a nonlinear modeling method, uses a kernel function to map problems in low-dimensional space to high-dimensional space for linear regression, so as to simplify complex problems [25,26]. In this study, the radial basis kernel function (RBF) was used as the kernel function of SVR for subsequent research and analysis [27]. Moreover, extreme learning machine (ELM), as a feedforward neural network analysis method, has also shown good performance in related research [28]. Some related studies have shown that the ELM model is superior to other classical learning algorithms in terms of learning efficiency and generalization ability. Accordingly, PLS, SVR, and ELM were used to establish prediction models of tea pigments.

2.8. Prediction Models Evaluation Indexes

The calibration correlation coefficient (Rc), prediction coefficient (Rp), calibration root-mean-square error (RMSEC), prediction root-mean-square error (RMSEP), and residual prediction deviation (RPD) were used as evaluation indexes of the models [29,30]. In general, higher Rc, Rp, and RPD values and lower RMSEC and RMSEP values represent better model performance. Here, the data analysis was carried out using MATLAB R2021b.

3. Results and Discussion

3.1. Results of Determination of Tea Pigments Content

The variations in content of TFs, TRs, and TBs in samples at various rolling time are presented in Table 1. TFs, TRs, and TBs are oxidative polymers formed from catechins through enzymatic and hydrothermal reactions [31]. They significantly influence the color and taste of black tea. The experimental data indicate that the content of TFs, TRs, and TBs in black tea increase with prolonged rolling time. This phenomenon also explains why the color of the leaves deepens as rolling time extends. However, the rates of alteration in the content of these three substances throughout this process is not uniform. This disparity might be closely associated with the degree of fragmentation of tea tissue cells and the process of oxidative polymerization.

Table 1. Changes in tea pigments content during rolling of black tea.

Rolling Time (min)	Number of Samples	TFs (%)	TRs (%)	TBs (%)
0	15	0.03 ± 0.00	0.99 ± 0.12	1.52 ± 0.19
10	15	0.06 ± 0.01	1.52 ± 0.11	1.83 ± 0.07
20	15	0.11 ± 0.01	1.76 ± 0.07	2.06 ± 0.04
30	15	0.14 ± 0.01	2.22 ± 0.06	2.44 ± 0.11
40	15	0.20 ± 0.01	2.57 ± 0.08	2.83 ± 0.06
50	15	0.29 ± 0.11	3.41 ± 0.19	3.40 ± 0.08
60	15	0.38 ± 0.02	3.79 ± 0.13	3.82 ± 0.10
70	15	0.54 ± 0.01	4.23 ± 0.08	4.40 ± 0.08
80	15	0.59 ± 0.01	4.32 ± 0.11	4.60 ± 0.03
90	15	0.63 ± 0.01	4.75 ± 0.31	4.94 ± 0.14

3.2. Response of Color Feature Variables During Rolling Process

The variations of each color feature variable in the image information in accordance the rolling time are shown in Figure 2. From a macro point of view, the color appearance of black underwent an obvious transformation process: from green to light reddish-brown, and the brightness gradually decreases, and loses luster. In HSV, both the H and V values exhibited a gradual decline as rolling time increased, whereas the S value tended to rise consistently. This showed the appearance characteristics of the sample: with the extension of rolling time, the hue and value of the rolled leaves gradually declined. In La*b*, the changes of L and a* values were more intuitive. With a continuous increase in rolling time, L gradually decreased, and a* gradually increased. This was due to the gradual deepening and reddening appearance of the rolled leaves. However, the regularity of b* was relatively poor, which may be related to the complex interaction of multiple facets of color information during the rolling. For the RGB color model, although the R, G, and B values all showed a downward trend, the R was relatively large and the rate of decline was slower as the rolling time increased. This is also consistent with the phenomenon of the deepened red appearance of the rolled leaves. These synergistic changes of color features from different color models confirm the overall rule of color change of rolled leaves from a microscopic point of view. Through the analysis of the aforementioned color feature variables extracted from the image, it is possible to quantitatively describe the color variations occurring during the rolling process.

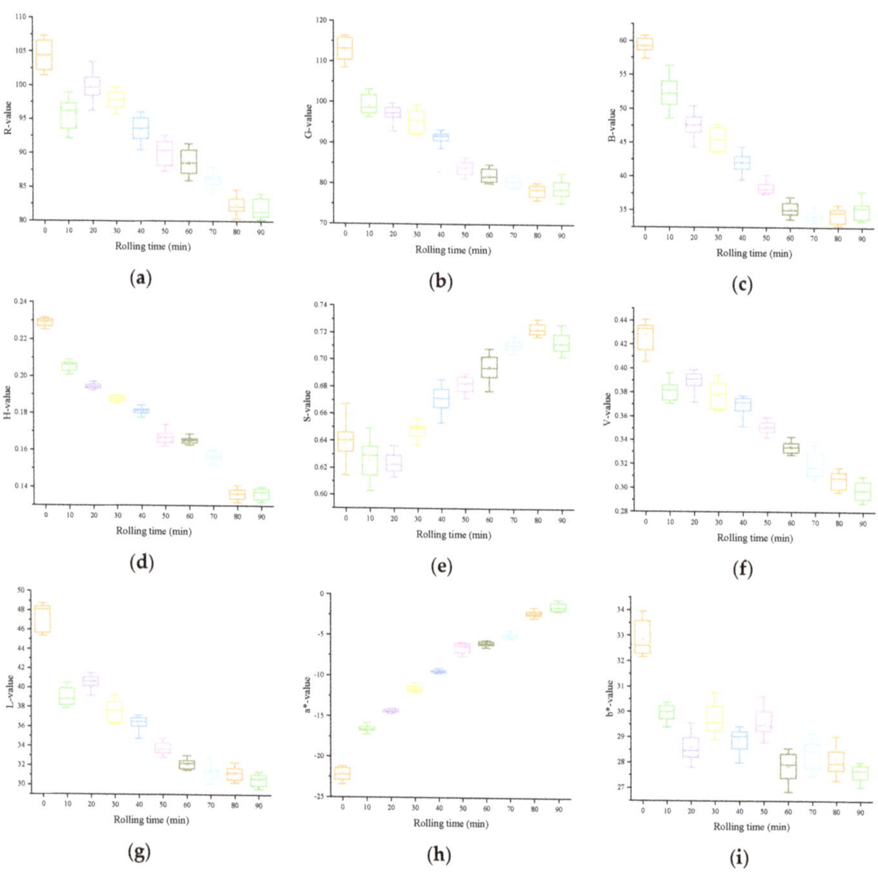

Figure 2. Image color feature variables: (**a**) R, (**b**) G, (**c**) B, (**d**) H, (**e**) S, (**f**) V, (**g**) L, (**h**) a*, (**i**) b*.

3.3. Correlation Analysis Between Color Features Variables and Tea Pigments

In order to further verify the relationship between tea pigments and color variables, Pearson correlation analysis method is used in this study. It is a commonly used method to measure the correlation between variables, and its correlation coefficient results are between −1 and 1, which can effectively reveal the relationship between quantitative data. The positive and negative values of the coefficient represent positive correlation and negative correlation, respectively. The closer the result value tends to 1 and −1, the stronger the correlation is; on the contrary, the closer it tends to 0, the weaker the correlation is. The results of the correlation analysis are shown in Figure 3. The results show that the G, B, H, L, and a* values were strongly correlated with TFs. The H and a* values also demonstrated a significant correlation with TRs and TBs, where the H exhibited a negative correlation and the a* showed a positive one. The absolute values of these correlation coefficients were all greater than 0.8, showing a significant linear relationship. By comparison, the R and V values were weakly correlated with TFs, TRs, and TBs. In addition, different color feature variables also showed different degrees of correlation strength in the analysis results. Therefore, the simultaneous use of these color characteristic variables may lead to information overlap in the data, and there may be information interference among the variables, which could have a negative impact on the model's predictive results. This further illustrates the rationality and effectiveness of selecting strongly correlated feature variables for model prediction analysis.

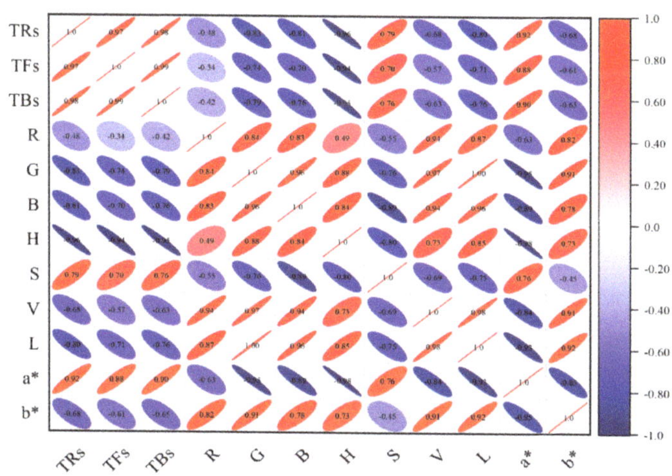

Figure 3. Correlation analysis diagram of tea pigments and image color feature variables.

3.4. Data Preprocessing and Principal Component Analysis

Since the color information variables in the image come from three different color models, the data information belongs to a high-dimensional array. Therefore, it is necessary to standardize the data of variables within different dimensions in order to enhance the comparability of data. Moreover, as mentioned above, some color variables are weakly correlated with tea pigments, and there are also correlations among different characteristic variables. The above situation may lead to information overlap and data redundancy. Hence, in the predictive analysis, the necessary dimensionality reduction of the data should be carried out. Based on this, Z-score is used in this study to normalize the magnitude differences among the data. Then, PCA is used to remove irrelevant variables, eliminate collinearity among the data, and reduce its dimension. The results of the PCA for color feature variables are shown in Figure 4. The explained variance for the first three principal components is 85.33%, 9.09%, and 5.29%, respectively. The cumulative contribution rate of the first three principal components is 99.71%, which reflects the effective information of the sample population.

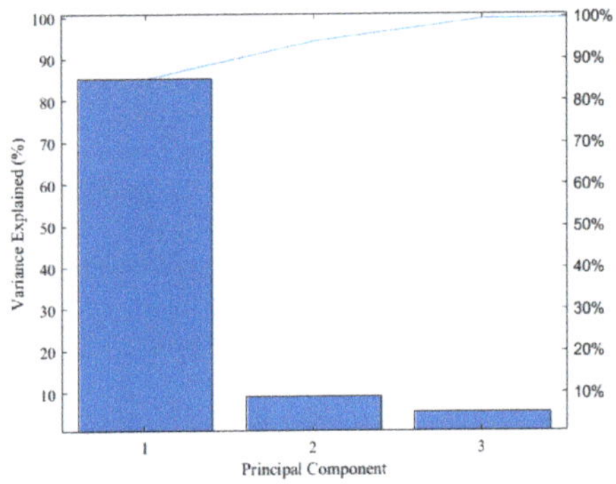

Figure 4. Explanatory variance in principal component analysis.

3.5. Prediction Models of Tea Pigments Based on Image Information

3.5.1. Prediction Results of TFs

The results of TFs prediction models based on image color feature information are shown in Table 2. Here, Z-score and Z-score–PCA were used as data preprocessing methods, respectively. These were used as model inputs, and corresponding TFs content values were used as model outputs. Meanwhile, PLSR, SVR, and ELM were used as modeling methods to establish the prediction models. The experimental data show that when the same modeling method was used, the prediction models based on Z-score–PCA data showed superior performance. This verifies that PCA can effectively eliminate redundant information in the data. Moreover, under the same data preprocessing conditions, the ELM model exhibited better performance in predicting TFs. This might be attributed to the fact that the training process of ELM is non-iterative. It initializes by randomly choosing the weight and bias of the hidden layer and subsequently directly resolves the weight of the output layer through an analytical approach. Owing to its non-iterative training mode, it can exhibit superior scalability when handling data. Consequently, the ELM will demonstrate better robustness against noise and abnormal data. In conclusion, the prediction model based on the Z-score–PCA–ELM model had the best prediction effect for TFs. Its Rc value was 0.99, RMSEC value was 0.04, Rp value was 0.97, and RMSEP value was 0.06. The RPD value reached 3.97.

Table 2. Results of theaflavins (TFs) prediction models based on machine vision technology.

Types of Models	Pretreatment Methods	Training Set		Prediction Set		
		Rc	RMSEC	Rp	RMSEP	RPD
PLSR	Z-score	0.93	0.08	0.91	0.12	2.01
	Z-score–PCA	0.94	0.07	0.94	0.10	2.19
SVR	Z-score	0.98	0.05	0.94	0.09	2.17
	Z-score–PCA	0.95	0.07	0.95	0.07	2.67
ELM	Z-score	0.99	0.04	0.96	0.07	3.65
	Z-score–PCA	0.99	0.04	0.97	0.06	3.97

The relationship between the prediction values and calibration values in the prediction set results of the ELM models are shown in Figure 5a,b. Among them, the horizontal and vertical coordinates corresponding to the red circles represent the prediction values and calibration values of the training set samples, respectively, and the horizontal and vertical coordinates corresponding to the blue asterisks represent the prediction values and calibration values of the prediction set samples, respectively. The closer the red scatter points to the blue solid line in Figure 5a,b, the better the training effect of the model, and the closer the blue scatter points to the blue solid line, the better the prediction effect of the model.

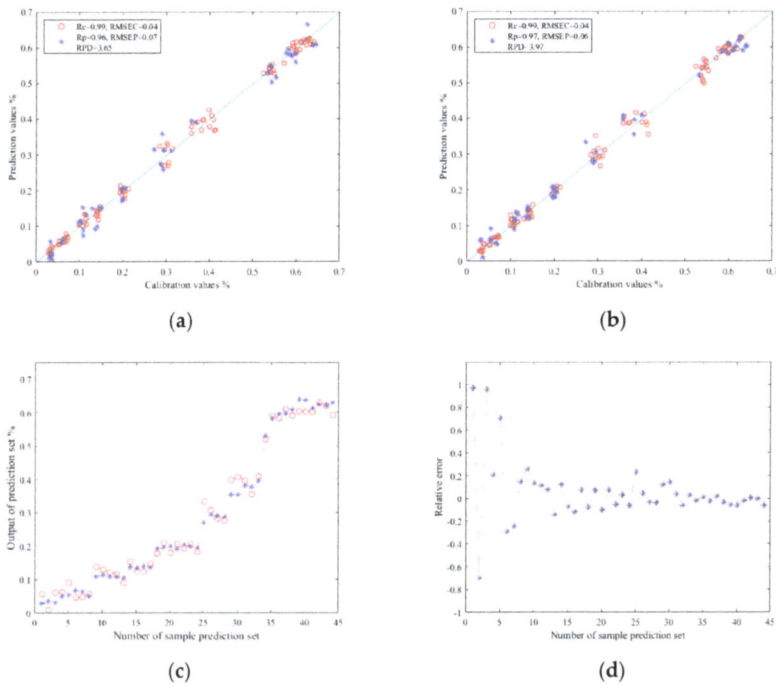

Figure 5. (**a**) Regression prediction scatter plot based on Z-score–ELM; (**b**) regression prediction scatter plot based on Z-score–PCA–ELM; (**c**) line chart of prediction results based on Z-score–ELM; (**d**) relative error chart of prediction results based on Z-score–PCA–ELM.

Figure 5c,d shows the relationship diagram and relative error diagram between each correction and prediction value in the prediction set of the Z-score–PCA–ELM model, respectively. The blue asterisks represents the calibration values of the model prediction set, and the red circles represent the prediction values of the model prediction set.

3.5.2. Prediction Results of TRs

The results of the TRs prediction models based on image color feature information are shown in Table 3. As mentioned above, Z-score and Z-score–PCA were used as data preprocessing methods. PLSR, SVR, and ELM were used to construct the prediction models of TRs. The results show that when Z-score–PCA was used as the data preprocessing method, the prediction model established by ELM had the best effect: the Rc value was 0.97, the RMSEC value was 0.34, the Rp value was 0.96, the RMSEP value was 0.36, and the RPD value was 3.56. To sum up, PCA can effectively eliminate irrelevant variables between data in the prediction study of TRs using machine vision technology, and the ELM model also showed better predictability. The relationship between the prediction values and calibration values in the prediction set results of the ELM models are shown in Figure 6a,b. In addition, Figure 6c,d shows the relationship diagram and relative error diagram between each correction and prediction value in the prediction set of the Z-score–PCA–ELM model, respectively.

Table 3. Results of thearubigins (TRs) prediction models based on machine vision technology.

Types of Models	Pretreatment Methods	Training Set		Prediction Set		
		Rc	RMSEC	Rp	RMSEP	RPD
PLSR	Z-score	0.96	0.35	0.93	0.51	2.35
	Z-score-PCA	0.95	0.41	0.94	0.44	2.71
SVR	Z-score	0.96	0.34	0.94	0.50	2.56
	Z-score-PCA	0.96	0.35	0.95	0.44	2.73
ELM	Z-score	0.96	0.36	0.94	0.44	3.01
	Z-score-PCA	0.97	0.34	0.96	0.36	3.56

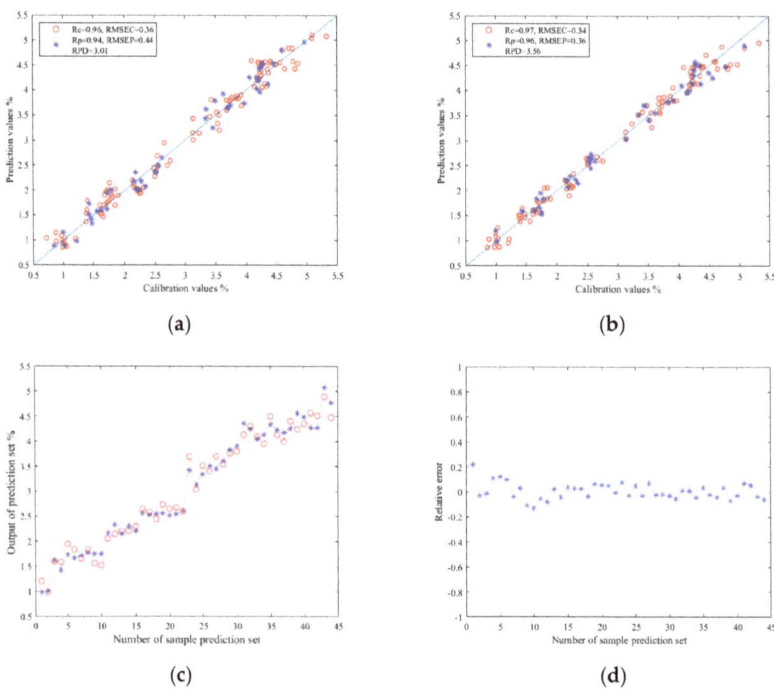

Figure 6. (**a**) Regression prediction scatter plot based on Z-score–ELM; (**b**) regression prediction scatter plot based on Z-score–PCA–ELM; (**c**) line chart of prediction results based on Z-score–ELM; (**d**) relative error chart of prediction results based on Z-score-PCA–ELM.

3.5.3. Prediction Results of TBs

The results of TBs prediction models based on image color feature information are shown in Table 4. By comparing the experimental data, it was found that the ELM was still better than PLSR and SVR in predicting TBs. The established Z-score–PCA–ELM model had the best prediction effect: the Rc value ws 0.97, the RMSEC value was 0.31, the Rp value was 0.96, the RMSEP value was 0.35, and the RPD value reached 3.60. The results show that this model can effectively and quickly predict TBs. The relationship between the prediction values and calibration values in the prediction set results of the ELM models are shown in Figure 7a,b. Moreover, Figure 7c,d shows the relationship diagram and relative error diagram between each correction and prediction value in the prediction set of the Z-score–PCA–ELM model, respectively. They provide an intuitive display of the results of the prediction model.

Table 4. Results of theabrownins (TBs) prediction models based on machine vision technology.

Types of Models	Pretreatment Methods	Training Set		Prediction Set		
		Rc	RMSEC	Rp	RMSEP	RPD
PLSR	Z-score	0.96	0.34	0.94	0.51	2.08
	Z-score–PCA	0.94	0.38	0.94	0.41	2.45
SVR	Z-score	0.97	0.32	0.94	0.40	2.82
	Z-score–PCA	0.95	0.37	0.95	0.37	3.02
ELM	Z-score	0.96	0.36	0.96	0.37	3.38
	Z-score–PCA	0.97	0.31	0.96	0.35	3.60

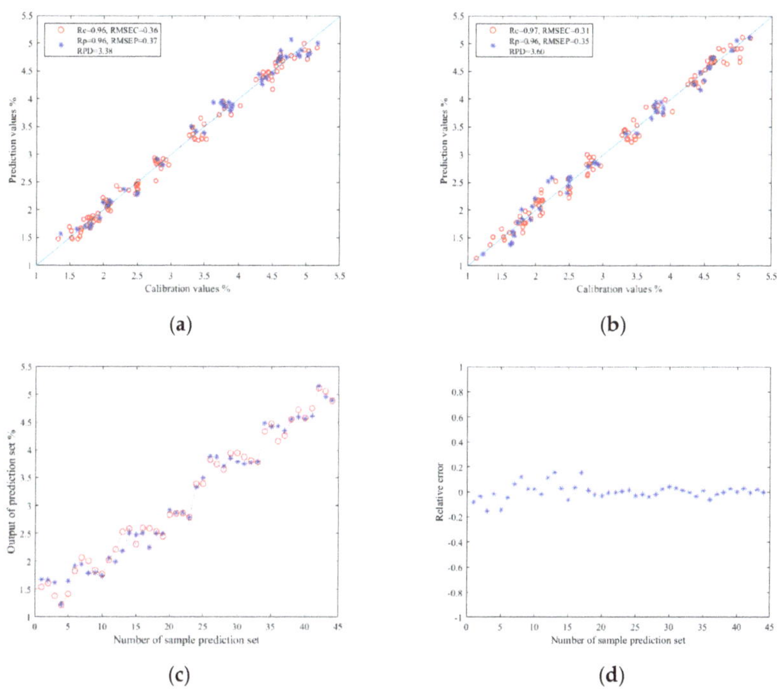

Figure 7. (**a**) Regression prediction scatter plot based on Z-score–ELM; (**b**) regression prediction scatter plot based on Z-score–PCA–ELM; (**c**) line chart of prediction results based on Z-score–ELM; (**d**) relative error chart of prediction results based on Z-score–PCA–ELM.

4. Conclusions

In this article, machine vision technology is employed to establish a rapid detection method for tea pigments during the rolling of black tea. The variation rules of color feature variables in the rolling process are discussed, and the correlation between color variables and tea pigments is characterized. The Z-score and Z-score–PCA are used as preprocessing techniques, which are subsequently combined with PLSR, SVR, and ELM for comparative analysis. The results show that Z-score–PCA is more advantageous for the application analysis of color feature variables. Meanwhile, ELM shows superior prediction accuracy. The Z-score–PCA–ELM model established in this study has the best prediction results for TFs, TRs, and TBs. This indicates that by combining machine vision technology with the machine learning algorithms, it is possible to quickly detect the content of tea pigments during rolling process. This study establishes innovative methods for the online evaluation of the rolling process in the high-efficiency industrial production of black tea.

Author Contributions: H.Z.: formal analysis, writing—original draft, writing—review and editing, methodology, and data curation. T.L.: data curation. Y.J.: investigation. X.-L.Y.: investigation and supervision. H.Y.: validation and supervision. All authors have read and agreed to the published version of the manuscript.

Funding: This work is supported by the Science and Technology Innovation Project of the Chinese Academy of Agricultural Sciences [grant number CAAS-ASTIP-TRICAAS].

Institutional Review Board Statement: Not applicable.

Informed Consent Statement: Informed consent was obtained from all subjects in this study.

Data Availability Statement: The original contributions presented in this study are included in the article. Further inquiries can be directed to the corresponding author.

Conflicts of Interest: The authors declare that they have no known competing financial interests or personal relationships that could have appeared to influence the work reported in this article.

References

1. Chen, Q.C.; Fu, Y.; Heng, W.T.; Yu, S.; Xie, F.; Dong, F.; Lin, Z.; Dai, W.D.; Fu, H.H. Re-rolling treatment in the fermentation process improves the taste and liquor color qualities of black tea. *Food Chem.-X* **2024**, *21*, 101143. [CrossRef] [PubMed]
2. Wang, A.L.; Lei, Q.Q.; Zhang, B.B.; Wu, J.H.; Fu, Z.Y.; He, J.F.; Wang, Y.B.; Wu, X.Y. Revealing novel insights into the enhancement of quality in black tea processing through microbial intervention. *Food Chem.-X* **2024**, *23*, 101743. [CrossRef] [PubMed]
3. Wu, S.M.; Yu, Q.Y.; Shen, S.; Shan, X.J.; Hua, J.J.; Zhu, J.Y.; Qiu, J.R.; Deng, Y.L.; Zhou, Q.H.; Jiang, Y.W.; et al. Non-targeted metabolomics and electronic tongue analysis reveal the effect of rolling time on the sensory quality and nonvolatile metabolites of congou black tea. *Lwt-Food Sci. Technol.* **2022**, *169*, 113971. [CrossRef]
4. Zhang, S.; Wu, S.M.; Yu, Q.Y.; Shan, X.J.; Chen, L.; Deng, Y.L.; Hua, J.J.; Zhu, J.Y.; Zhou, Q.H.; Jiang, Y.W.; et al. The influence of rolling pressure on the changes in non-volatile compounds and sensory quality of congou black tea: The combination of metabolomics, E-tongue, and chromatic differences analyses. *Food Chem.-X* **2023**, *20*, 100989. [CrossRef] [PubMed]
5. Hua, J.J.; Xu, Q.; Yuan, H.B.; Wang, J.J.; Wu, Z.Q.; Li, X.T.; Jiang, Y.W. Effects of novel fermentation method on the biochemical components change and quality formation of Congou black tea. *J. Food Compos. Anal.* **2021**, *96*, 103751. [CrossRef]
6. Liu, M.; Wang, R.X.; Shi, D.L.; Cao, R.Y. Non-destructive prediction of tea polyphenols during Pu-erh tea fermentation using NIR coupled with chemometrics methods. *J. Food Compos. Anal.* **2024**, *131*, 106247. [CrossRef]
7. Turgut, S.S.; Entrenas, J.A.; Taskin, E.; Garrido-Varo, A.; Pérez-Marín, D. Estimation of the sensory properties of black tea samples using non-destructive near-infrared spectroscopy sensors. *Food Control* **2022**, *142*, 109260. [CrossRef]
8. Luo, W.; Tian, P.; Fan, G.Z.; Dong, W.T.; Zhang, H.L.; Liu, X.M. Non-destructive determination of four tea polyphenols in fresh tea using visible and near-infrared spectroscopy. *Infrared Phys. Technol.* **2022**, *123*, 104037. [CrossRef]
9. Wen, J.; Xu, G.Q.; Zhang, A.; Ma, W.; Jin, G. Emerging technologies for rapid non-destructive testing of grape quality: A review. *J. Food Compos. Anal.* **2024**, *133*, 106446. [CrossRef]
10. Jiang, Y.Y.; Zhang, D.X.; Yang, L.; Cui, T.; He, X.T.; Wu, D.Y.; Dong, J.Q.; Li, C.; Xing, S.L. Design and experiment of non-destructive testing system for moisture content of in-situ maize ear kernels based on VIS-NIR. *J. Food Compos. Anal.* **2024**, *133*, 106369. [CrossRef]
11. Maniwara, P.; Nakano, K.; Ohashi, S.; Boonyakiat, D.; Seehanam, P.; Theanjumpo, P.; Poonlarp, P. Evaluation of NIRS as non-destructive test to evaluate quality traits of purple passion fruit. *Sci. Hortic.* **2019**, *257*, 108712. [CrossRef]
12. Wei, Y.; Wen, Y.Q.; Huang, X.L.; Ma, P.H.; Wang, L.; Pan, Y.; Lv, Y.J.; Wang, H.X.; Zhang, L.; Wang, K.B.; et al. The dawn of intelligent technologies in tea industry. *Trends Food Sci. Technol.* **2024**, *144*, 104337. [CrossRef]
13. Wang, Y.J.; Ren, Z.Y.; Chen, Y.Y.; Lu, C.Y.; Deng, W.W.; Zhang, Z.Z.; Ning, J.M. Visualizing chemical indicators: Spatial and temporal quality formation and distribution during black tea fermentation. *Food Chem.* **2023**, *401*, 134090. [CrossRef] [PubMed]
14. Zhu, H.K.; Liu, F.; Ye, Y.; Chen, L.; Li, J.Y.; Gu, A.H.; Zhang, J.Q.; Dong, C.W. Application of machine learning algorithms in quality assurance of fermentation process of black tea- based on electrical properties. *J. Food Eng.* **2019**, *263*, 165–172. [CrossRef]
15. Yao, L.H.; Jiang, Y.M.; Caffin, N.; D'Arcy, B.; Datta, N.; Liu, X.; Singanusong, R.; Xu, Y. Phenolic compounds in tea from Australian supermarkets. *Food Chem.* **2006**, *96*, 614–620. [CrossRef]
16. Ouyang, Q.; Wang, L.; Park, B.; Kang, R.; Chen, Q.S. Simultaneous quantification of chemical constituents in matcha with visible-near infrared hyperspectral imaging technology. *Food Chem.* **2021**, *350*, 129141. [CrossRef]
17. Chemov, V.; Alander, J.; Bochko, V. Integer-based accurate conversion between RGB and HSV color spaces. *Comput. Electr. Eng.* **2015**, *46*, 328–337. [CrossRef]
18. Paschos, G. Perceptually uniform color spaces for color texture analysis: An empirical evaluation. *IEEE Trans. Image Process.* **2001**, *10*, 932–937. [CrossRef]
19. Lan, T.M.; Shen, S.; Yuan, H.B.; Jiang, Y.W.; Tong, H.R.; Ye, Y. A Rapid Prediction Method of Moisture Content for Green Tea Fixation Based on WOA-Elman. *Foods* **2022**, *11*, 2928. [CrossRef]

20. Shen, S.; Hua, J.J.; Zhu, H.K.; Yang, Y.Q.; Deng, Y.L.; Li, J.; Yuan, H.B.; Wang, J.J.; Zhu, J.Y.; Jiang, Y.W. Rapid and real-time detection of moisture in black tea during withering using micro-near-infrared spectroscopy. *Lwt-Food Sci. Technol.* **2022**, *155*, 112970. [CrossRef]
21. Zheng, L.; Watson, D.G.; Johnston, B.F.; Clark, R.L.; Edrada-Ebel, R.; Elseheri, W. A chemometric study of chromatograms of tea extracts by correlation optimization warping in conjunction with PCA, support vector machines and random forest data modeling. *Anal. Chim. Acta* **2009**, *642*, 257–265. [CrossRef]
22. Asante, E.A.; Du, Z.; Lu, Y.Z.; Hu, Y.G. Detection and assessment of nitrogen effect on cold tolerance for tea by hyperspectral reflectance with PLSR, PCR, and LM models. *Inf. Process. Agric.* **2021**, *8*, 96–104. [CrossRef]
23. Amsaraj, R.; Mutturi, S. Real-coded GA coupled to PLS for rapid detection and quantification of tartrazine in tea using FT-IR spectroscopy. *Lwt-Food Sci. Technol.* **2021**, *139*, 110583. [CrossRef]
24. Sobhaninia, M.; Mani-Varnosfaderani, A.; Barzegar, M.; Sahari, M.A. Combining ion mobility spectrometry and chemometrics for detecting synthetic colorants in black tea: A reliable and fast method. *Food Chem.-X* **2024**, *21*, 101213. [CrossRef] [PubMed]
25. Xie, G.; Qian, Y.T.; Wang, S.Y. Forecasting Chinese cruise tourism demand with big data: An optimized machine learning approach. *Tour. Manag.* **2021**, *82*, 104208. [CrossRef]
26. Wei, X.; Li, S.; Zhu, S.P.; Zheng, W.Q.; Xie, Y.; Zhou, S.L.; Hu, M.D.; Miao, Y.J.; Ma, L.K.; Wu, W.J.; et al. Terahertz spectroscopy combined with data dimensionality reduction algorithms for quantitative analysis of protein content in soybeans. *Spectrochim. Acta Part A-Mol. Biomol. Spectrosc.* **2021**, *253*, 119571. [CrossRef]
27. Yang, Y.C.; Sun, D.W.; Wang, N.N. Rapid detection of browning levels of lychee pericarp as affected by moisture contents using hyperspectral imaging. *Comput. Electron. Agric.* **2015**, *113*, 203–212. [CrossRef]
28. Li, H.; Hu, Y.; Ma, S.; Haruna, S.A.; Chen, Q.; Zhu, W.; Xia, A. Porphyrin and pH sensitive dye-based colorimetric sensor array coupled chemometrics for dynamic monitoring of tea quality during ultrasound-assisted fermentation. *Microchem. J.* **2024**, *197*, 109813. [CrossRef]
29. Ren, G.X.; Yin, L.L.; Wu, R.; Ning, J.M. Rapid detection of ash content in black tea using a homemade miniature near-infrared spectroscopy. *Spectrochim. Acta Part A-Mol. Biomol. Spectrosc.* **2024**, *308*, 123740. [CrossRef]
30. Dai, F.S.; Shi, J.; Yang, C.S.; Li, Y.; Zhao, Y.; Liu, Z.Y.; An, T.; Li, X.L.; Yan, P.; Dong, C.W. Detection of anthocyanin content in fresh Zijuan tea leaves based on hyperspectral imaging. *Food Control* **2023**, *152*, 109839. [CrossRef]
31. Zhang, L.; Ho, C.T.; Zhou, J.; Santos, J.S.; Armstrong, L.; Granato, D. Chemistry and Biological Activities of Processed Camellia sinensis Teas: A Comprehensive Review. *Compr. Rev. Food Sci. Food Saf.* **2019**, *18*, 1474–1495. [CrossRef] [PubMed]

foods

MDPI

Article

Aerospace Mutagenesis Increases the Content of Characteristic Metabolites of Tea Tree Leaves and Enhances Taste Characteristics of Tea Leaves

Jiaming Chen [1,†], Yangxin Luo [2,†], Qi Zhang [1], Yulin Wang [2], Tingting Wang [2], Miao Jia [1], Yankun Liao [1], Xiaoli Jia [1], Haibin Wang [2,*] and Jianghua Ye [1,*]

[1] College of Tea and Food, Wuyi University, Wuyishan 354300, China; jiamingchen2009@outlook.com (J.C.)
[2] College of Life Science, Longyan University, Longyan 364012, China
[*] Correspondence: 82015018@lyun.edu.cn (H.W.); jhye1998@wuyiu.edu.cn (J.Y.)
[†] These authors contributed equally to this work.

Abstract: Aerospace mutagenesis can alter the physiological metabolism and growth of tea trees and affect tea leaf quality. In this study, the effect of aerospace mutagenesis on the metabolite content and taste characteristics of Dahongpao fresh leaves and raw tea were analyzed. The results showed that aerospace mutagenesis had little effect on the composition and total amount of metabolites in fresh leaves and raw tea, but significantly changed the content of different metabolites. Aerospace mutagenesis improved the content of lipids, lignans and coumarins, and amino acids and derivatives, which in turn enhanced the aroma and fresh and brisk taste of fresh leaves. Aerospace mutagenesis increased the content of alkaloids, tannins, flavonoids, lignans and coumarins, amino acids and derivatives, and organic acids in raw tea, and enhanced the bitterness, mellowness, and fresh and brisk taste of raw tea. This study provides a basis for the development of aerospace mutagenesis Dahongpao tea products and the establishment of processing techniques.

Keywords: aerospace mutagenesis; tea tree; metabolomics; characteristic metabolites; taste characteristics

1. Introduction

Aerospace mutagenesis is a special and novel technology for plant breeding. Plant seeds undergo significant changes in their heredity and gene regulation under the combined interference of multiple factors in unique environments, such as intense radiation, microgravity, high vacuum, and extreme temperatures, which do not occur under Earth's gravity [1], which in turn facilitates rapid and efficient screening of high-quality germplasm resources [2]. China has made significant achievements in aerospace mutation breeding, and through aerospace breeding, more than 200 new varieties of plants have been developed and utilized in social production practices, which have made significant contributions to the development of agriculture [3,4]. For example, the Chinese Academy of Sciences (CAS) utilized aerospace mutagenesis to breed tomatoes and produced two high-yielding, disease-resistant, and high-quality varieties [5]. The new varieties of *Salvia miltiorrhiza* bred by Tasly Group using aerospace mutagenesis have significantly higher active ingredient content and three times the quality of other *S. miltiorrhiza* [6]. It is evident that aerospace mutagenesis has achieved remarkable results in the screening and breeding of excellent plant varieties.

Wuyishan is a premium tea-producing area in China. Dahongpao (*Camellia sinensis*) is one of the most important cultivated varieties in Wuyishan [7]. In 2011, Dahongpao seeds were placed onboard the unmanned spacecraft Shenzhou VIII for the first time to carry out space mutation, which lasted 16 days, 13 h, and 34 min. This is the first and, so far, the only time that Wuyi Rock Tea seeds have been obtained through aerospace mutagenesis.

After the mutagenized Dahongpao tea tree seeds were germinated and planted, their tea tree leaf morphology already showed significant differences compared to the control at the seedling stage [8]. Jia et al. [9] analyzed the physiological characteristics and growth of tea tree, and found that, compared with the control, the surface of tea leaves after aerospace mutagenesis was smoother, the edge serrations were more compact and regular, and the photosynthetic capacity, resistance, and yield of tea trees were stronger. It is evident that the Dahongpao tea tree after aerospace mutagenesis had significant changes in its growth and yield. However, there are few reports on whether the quality of Dahongpao has also changed after aerospace mutagenesis in terms of quality.

Taste characteristics are one of the important indexes for evaluating tea quality, while the types of inclusions and their contents in tea leaves determine taste characteristics and affect tea quality [10,11]. Metabolomics technology can comprehensively and rapidly analyze metabolites in tea leaves and evaluate tea quality by metabolite content and taste characteristics [12,13]. Secondly, metabolomics technology can analyze and obtain key differential metabolites among different samples, which in turn facilitates rapid characterization of tea quality [14,15]. It was hypothesized that after aerospace mutagenesis, the type or content of metabolites in Dahongpao leaves changed significantly, which in turn affected the taste characteristics of fresh leaves or raw tea; especially, the type and content of characteristic metabolites in the leaves were closely related to taste characteristics. Based on this, Dahongpao fresh leaves were collected from aerospace-mutagenized and non-aerospace-mutagenized Dahongpao tea tree in this study. Metabolomics techniques were used to analyze the types of metabolites and their contents in tea tree fresh leaves and raw tea. We further screened and analyzed the characteristic metabolites of the two samples and analyzed their taste characteristics to clarify the effect of aerospace mutagenesis on the quality of Dahongpao tea, with a view to laying a foundation for the development of aerospace mutagenized Dahongpao tea products and the establishment of processing technology.

2. Materials and Methods

2.1. Materials

The experimental reagents were mainly methanol, acetonitrile, and formic acid, which were chromatographically pure and purchased from German Merck Shanghai branch (Shanghai, China).

Tea samples for this study were obtained from the Tea Tree Aerospace Breeding Experimental Base, Wuyishan, Fujian, China (117°59′47.7″ E, 27°44′8.4″ N). The research variety selected was Dahongpao (*C. sinensis*) tea tree, and the age of aerospace-mutagenized and non-aerospace-mutagenized tea trees at the time of sampling was 11 years. The main process of aerospace mutagenesis of tea tree seeds was, at 5:58 on 1 November 2011, the same batch of tea tree seeds mixed well, part of the seeds with the "Shenzhou VIII" unmanned spacecraft launched for aerospace mutagenesis, and part of the seeds remained on the ground as the control.

Two days after the lift-off of Shenzhou VIII, which carried tea tree seeds, it docked with the Tiangong I target vehicle and operated as a single unit for 12 days. And at 18:30 on 16 November 2011, the assemblage separated. The capsule successfully returned to Earth at 19:32 on 17 November 2011, with a total of 16 days, 13 h, and 34 min of space mutation time. In April 2012, the mutagenized seeds of the tea tree were germinated and then planted at the same time as the control, on the same plot, and the tea trees were uniformly managed in the same way.

In May 2023, Dahongpao fresh leaves without aerospace mutagenesis (CK) and with aerospace mutagenesis (TM) were picked in three replicates of 50 kg each. The picking standard for tea tree fresh leaves was one bud and three leaves. After each replicate sample was thoroughly mixed, 2 kg of each replicate were used for leaf metabolite extraction and identification. The remaining 48 kg of fresh leaves were taken for the preliminary processing to make raw tea. The method of preliminary processing of tea leaves was in accordance with the standard preliminary processing method of Wuyi rock tea, with specific reference to the method of Ye et al. [16]. Briefly, fresh tea leaves were withered for 30 min under 57,000 lux light, shaken for 450 min at room temperature of 27 °C and humidity of 70%, killed for 8 min at 260 °C, kneaded for 8 min, and dried for 50 min at 110 °C. Then, the raw tea could be obtained. Raw tea processed from CK (FCK) and raw tea processed from TM (FTM) were collected in 2 kg of three independent replicates per sample. The collected raw tea was used for metabolite extraction and identification.

2.2. Determination of Tea Metabolites

Extraction and determination of tea metabolites followed the method of Zhou et al. [17]. The tea samples were freeze-dried by vacuum freeze-dryer (Scientz-100F, Zhejiang, China) and then ground to powder. Then, 50 mg powder was added to 1.2 mL methanol solution (70%) before vortex oscillation for 30 s. The mixture was vortexed and shaken for 30 s at 30 min intervals 6 times. After the above mixture was then centrifuged at 12,000 rpm for 3 min, it was passed through a 0.22 μm filter membrane for ultra-performance liquid chromatography/tandem mass spectrometry (UPLC-MS/MS) analysis.

The above filtrates were determined by UPLC-MS/MS (UPLC: Shimadzu, Nexera X2, Kyoto, Japan; MS/MS: Applied Biosystems 4500 QTRAP, Framingham, MA, USA). The UPLC column utilized was Agilent SB-C18 (1.8 μm, 2.1 mm × 100 mm, Santa Clara. CA, USA). For the mobile phase, ultrapure water (containing 0.1% formic acid) served as phase A, and acetonitrile (also containing 0.1% formic acid) served as phase B. The UPLC assay was performed using a gradient elution mode as follows: the phase B ratio was 5% at 0.00 min, the phase B ratio increased linearly to 95% within 9.00 min and was maintained at 95% for 1 min (10.00–11.10 min), the phase B ratio was reduced to 5% and equilibrated at 5% up to the 14th min. The flow rate was set to 0.35 mL/min, and the column temperature was kept at 40 °C. Additionally, an injection volume of 2 μL was employed. MS/MS experiments were carried out under optimized conditions, specifically an electrospray ionization temperature of 500 °C. The positive ion mode was set to 5500 V and the negative mode was adjusted to −4500 V in the ion spray voltage settings. The ion source gas I was set to 50 psi, the ion source gas II was set to 60 psi, and the captain gas was set to 25 psi. The collision-activated dissociation was set to high. Triple-series quadrupole mass spectrometry was employed and operated in multiple reaction monitoring mode, with the collision gas set to medium. The declustering potential and collision energy were refined and optimized to obtain the appropriate declustering potential and collision energy of each multiple reaction monitoring ion. Then, based on the metabolites eluted within each period, each period was set to monitor a specific set of multiple reaction monitoring ion pairs.

The qualitative determination of metabolites was briefly described by first processing the mass spectrometry data using Analyst (1.6.3) software to obtain retention times and ion current intensities of metabolites, and then comparing them with the NIST20 mass spectrometry database to qualify metabolites. Quantitative determination of metabolites was briefly described as follows: The triple quadrupole was used to screen for characteristic ions of each substance. The detector was used to obtain the signal intensities of these characteristic ions. MultiQuant (3.0) software was used to integrate and correct the peaks, and the peak area of each peak represented the relative content of the corresponding substance [18].

2.3. Statistical Analysis

Excel 2020 was used for preliminary statistical analysis of raw metabolite data obtained, including metabolite classification, content calculation, etc. Analysis of paired Student's *t*-tests and variance (ANOVA) were used uniformly to analyze differences in data between different samples. Graphs were produced using Rstudio software (v 4.2.3), with box plots produced using the R package gghalves 0.1.4, principal component plots produced using the R package ggbiplot 0.55, bubble heat maps produced using the R package ggplot2 3.5.1, and volcano plots produced using the R package ggplot2 3.5.0. The R packages used for orthogonal partial least squares discrimination analysis (OPLS-DA) model construction for CK and TM, and FCK and FTM were all ropls and mixOmics. The R package used for bubble feature map production was ggplot2 3.4.4. The R packages used for taste wheel production were ComplexHeatmap 2.16.0, circlize v0.4.15, and RColorBrewer v1.1.3.

3. Results and Discussion

3.1. Quantity and Type Analysis of Metabolites in Dahongpao Fresh Leaves and Raw Tea

Aerospace mutation breeding is an efficient and fast breeding method [19]. Plants after aerospace mutagenesis experience significant changes in growth, physiology, and metabolism, which in turn lead to significant differences in metabolite composition and content [20]. Xia et al. [3] used GC-MS to find the changes in metabolite composition of andrographis paniculata after aerospace mutagenesis and indicated that aerospace mutagenesis effectively improved the content of different metabolites and the quality of andrographis paniculata. In the present study, metabolomics was utilized to analyze the effects of spaceflight mutagenesis on metabolites in fresh leaves and raw tea of Dahongpao (Tables S1 and S2). Among them, metabolite analysis of fresh leaves showed (Figure 1A) that a total of 2526 metabolites were detected, and non-aerospace-mutagenized Dahongpao fresh leaves (CK) and aerospace-mutagenized Dahongpao fresh leaves (TM) did not differ significantly in the total number of metabolites ($p > 0.05$). The obtained 2526 metabolites were further classified and analyzed for their contents, and the results showed (Table S1) that the detected metabolites could be classified into 12 groups according to the first-level classification method, and the top 5 metabolites with the highest number were flavonoids (23.40%), phenolic acids (17.14%), alkaloids (8.79%), amino acids and derivatives (7.76%), and lipids (7.60%). PCA analysis with metabolites detected in CK and TM and their contents showed (Figure 1B) that the two principal components could distinguish CK and TM effectively with an overall contribution of 84.01%. Further analysis showed that there were five categories of metabolites associated with CK, namely, quinones, phenolic acids, nucleotides and derivatives, phenolic acids, lipids, flavonoids, and seven categories associated with TM, namely, terpenoids, organic acids, lignans and coumarins, amino acids and derivatives, tannins, alkaloids, and others. It is evident that the composition and total number of metabolites in fresh leaves did not change significantly after aerospace mutagenesis, but different metabolites contained significant differences in their content.

Therefore, further analysis of metabolite content in this study revealed (Figure 1C) that 2526 metabolites could be categorized into 49 groups according to the secondary classification method, of which 10 groups were significantly greater in TM than CK, 16 groups were significantly less in TM than CK, and 23 groups were insignificantly different between TM and CK, whereas, among the 12 groups of metabolites obtained according to the first-level classification method, the content of 1 group (tannins) was significantly higher than that of CK, the content of 4 groups (quinones, phenolic acids, nucleotides and derivatives, lipids) was significantly lower than that of CK, and the content of 7 groups did not differ significantly between TM and CK. It is evident that the metabolite content of fresh leaves changed significantly after aerospace mutagenesis.

Regarding raw tea, the present study also analyzed the composition and content of metabolites, and the results revealed (Figure 2A) that a total of 2529 metabolites were obtained from raw tea. In terms of total number of metabolites, the difference between raw tea without aerospace mutagenesis (FCK) and raw tea after aerospace mutagenesis (FTM) was not significant ($p > 0.05$). The metabolites obtained were analyzed according to the first-level classification method and found (Table S2) to be classified into 12 groups, of which the top 5 groups with the greatest number were flavonoids (23.41%), phenolic acids (17.16%), alkaloids (8.82%), amino acids and derivatives (7.79%), and lipids (7.59%). PCA analysis showed (Figure 2B) that the two principal components could effectively differentiate FCK and FTM with a total contribution of 78.99%. Of these, the six groups associated with FCK were quinones, lipids, terpenoids, amino acids and derivatives, phenolic acids, and nucleotides and derivatives, while the six groups associated with FTM were tannins, organic acids, lignans and coumarins, flavonoids, alkaloids, and others. As can be seen, FCK and FTM remained similar in metabolite composition, but there were significant differences in the content of each metabolite. Accordingly, the classification of metabolites and the analysis of their content showed (Figure 2C) that 2529 metabolites could be classified into 49 groups according to the second-level classification method, in which the content of 10 groups was significantly higher in FTM than in FCK, the content of 8 groups was significantly lower in FTM than in FCK, and the content of 31 groups was insignificant between FTM and FCK. In contrast, metabolites were analyzed according to the first-level classification method into 12 groups, in which the content of tannins and others was significantly higher in FTM than in FCK, while the content of terpenoids, quinones, and lipids was significantly less in FTM than in FCK. It is evident that after aerospace mutagenesis, the metabolites in fresh leaves and raw tea are more similar in composition, and there is no significant difference in the total number of metabolites, but the content of each metabolite has changed significantly.

To sum up, aerospace mutagenesis can change the content of different metabolites in fresh leaves, which in turn leads to significant differences in the content of different metabolites in raw tea.

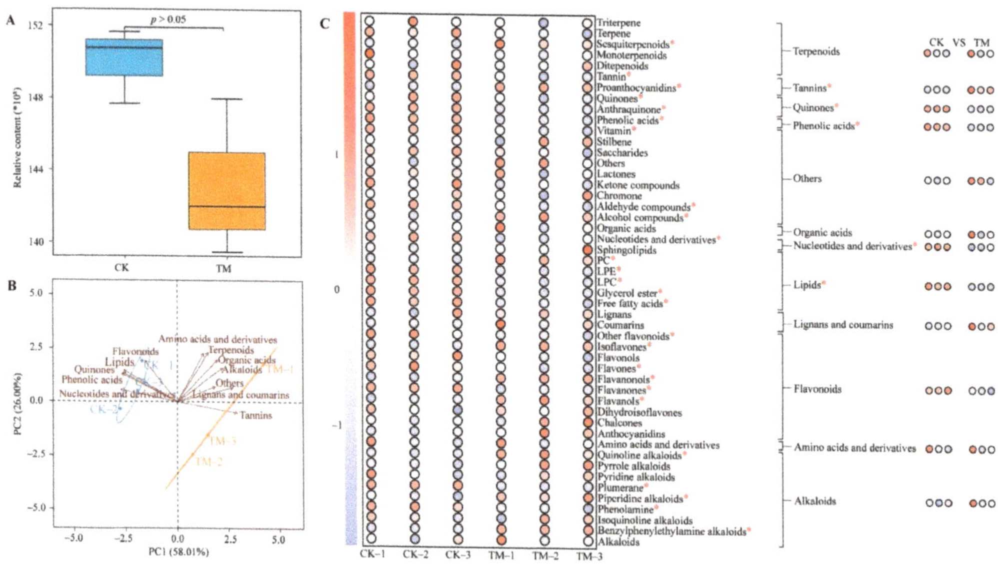

Figure 1. Analysis of metabolite content and types in fresh leaves of Dahongpao tea tree. Note: CK: Dahongpao fresh leaves without aerospace mutagenesis; TM: Dahongpao fresh leaves after aerospace mutagenesis; (**A**) analysis of total amount of metabolites in CK and TM; (**B**) PCA analysis after categorization of metabolites in CK and TM; (**C**) content analysis of different categories of metabolites in CK and TM; * denotes that the difference between two samples reaches the $p < 0.05$ level.

3.2. *Screening for Characteristic Metabolites in Dahongpao Fresh Leaves and Raw Tea*

After the above analysis, the present study further analyzed characteristic metabolites that significantly altered in fresh leaves and raw tea after aerospace mutagenesis. The OPLS-DA model is an effective way to obtain key metabolites that can distinguish different samples [21]. By constructing the model, inter- and intragroup differences of different samples can be analyzed, while variable importance projection (VIP) values of different metabolites in distinguishing different samples can be obtained, and then key metabolites with VIP greater than 1 can be obtained [22]. However, after the OPLS-DA model is constructed, the model needs to be evaluated, and the model can be used for analysis if the model's goodness-of-fit and predictability are both met [23]. Accordingly, in this study, the volcano diagram was first used for the preliminary screening of metabolites in fresh leaves, and then the OPLS-DA model was constructed to screen the key metabolites. The results of volcano diagram analysis showed (Figure 3A) that 148 metabolite contents in TM increased significantly, 230 metabolite contents decreased significantly, and 2148 metabolite contents were not significantly different from CK. The OPLS-DA model of CK and TM was further constructed based on 378 metabolites and their contents that were screened for significant differences in the volcano diagram, and it was found (Figure 3B) that the R^2Y value of the fit of the constructed model reached a significant level ($p < 0.005$), as well as the Q^2 value of predictability ($p < 0.005$). It can be seen that the constructed model meets the requirements and can be used for analysis. From the score plot of the model, the model can effectively differentiate between CK and TM, with a 93.20% between-group difference and less than 2.27% within-group difference. It is evident that the OPLS-DA model of CK and TM can effectively distinguish between the two samples. Based on this, VIP values of different metabolites distinguishing CK and TM were derived from the S-plot of the model, and 253 key metabolites with VIP > 1 were screened, of which the content of 105 metabolites was significantly larger in TM compared to CK and the content of 148 metabolites was significantly smaller in TM. A bubble feature map was further used to screen for characteristic metabolites distinguishing CK and TM, and it was found (Figure 3C) that 87 characteristic metabolites were obtained, of which the content of 35 metabolites was significantly larger in TM compared to CK, and the content of 52 metabolites was significantly smaller in TM. The results of characteristic metabolite classification and content analysis of each group showed (Figure 3D) that 87 metabolites could be classified into 23 groups according to the second-level classification method, of which 11 groups were significantly higher in TM compared to CK, 10 groups were significantly less in TM, and the difference of 2 groups between CK and TM was not significant. In addition, according to the first-level classification method, the 87 metabolites could be classified into 9 groups, in which the content of lipids, alkaloids, lignans and coumarins, amino acids and derivatives, and others was significantly higher in TM than in CK, and the content of tannins, flavonoids, phenolic acids and nucleotides, and derivatives was significantly smaller in TM.

It is evident that the metabolite content in fresh leaves has changed significantly after aerospace mutagenesis, especially characteristic metabolites, and this change may affect the content of each metabolite in their raw tea.

However, it is worth noting that the transformation of substances within tea leaves during processing ultimately leads to the formation of the distinctive taste that makes tea unique. This flavor development is driven by the content of different metabolites initially present in fresh leaves, which act as the substrate for the chemical reactions that occur during processing [24,25]. The higher the metabolite content and the richer the type of metabolites in fresh leaves, the more conducive they are to processing, resulting in the accumulation of different types of metabolites in raw tea and the formation of a unique taste of made tea [26,27]. Hereby, the present study further analyzed the characteristic metabolites in raw tea after aerospace mutagenesis, and the volcano diagram analysis showed (Figure 4A) that the content of 41 metabolites increased significantly, the content of 182 metabolites decreased significantly, and the difference in the content of 2206 metabolites was not significant in FTM compared to FCK. The OPLS-DA model based on FCK and FTM was further constructed, and showed (Figure 4B) that the R2Y value of the model fit and the Q2 value of predictability reached significant levels ($p < 0.005$). The model was effective in distinguishing FCK from FTM, with a difference of 93.50% between groups and less than 1.82% within groups. The OPLS-DA model of FCK and FTM meets the requirements and can effectively distinguish the two samples. Secondly, the VIP values of different metabolites were derived from the S-plot of the model, and 229 key metabolites with VIP > 1 were obtained, of which the content of 105 metabolites was significantly higher in FTM compared with FCK, and the content of 124 metabolites was significantly less in FTM. The bubble feature map was further used to screen for characteristic metabolites distinguishing FCK and FTM, and showed (Figure 4C) that 97 characteristic metabolites were obtained, of which 49 metabolites were significantly higher in FTM compared to FCK, and 48 metabolites were significantly lower in FTM. The classification of the characteristic metabolites and the content of each group classified showed (Figure 4D) that 97 metabolites could be classified into 24 groups according to the second-level classification, 16 groups of FTM had a significantly greater content than FCK, 7 groups of FTM had a significantly smaller content than FCK, and 1 group had a nonsignificant difference between FCK and FTM. However, according to the first-level classification, the 97 metabolites were categorized into 11 groups, in which the content of terpenoids, organic acids, lignans and coumarins, tannins, flavonoids, amino acids and derivatives, alkaloids, and others was significantly greater in FTM compared to FCK, while the content of phenolic acids, nucleotides and derivatives, and lipids was significantly less in FTM. It is evident that the metabolite content between FTM and FCK differed significantly. It was also found in this study that some metabolites with higher contents in TM compared to CK, e.g., alkaloids, amino acids and derivatives, and lignans and coumarins, remained at higher levels after processing, and FTM was significantly greater than FCK. Conversely, some metabolites with lower levels in TM, e.g., phenolic acids, nucleotides, and derivatives, remained low after processing, and FTM was significantly smaller than FCK. It is evident that there is a close relationship between the metabolite content of fresh leaves and the metabolite content of raw tea, especially characteristic metabolites.

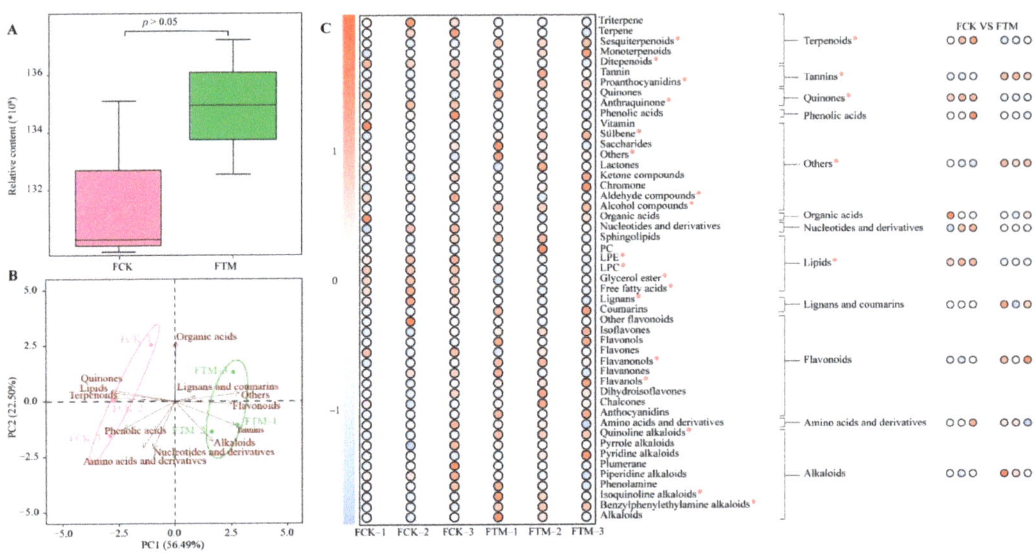

Figure 2. Analysis of metabolite content and types in raw tea of Dahongpao. Note: FCK: raw tea processed from Dahongpao fresh leaves without aerospace mutagenesis; FTM: raw tea processed from Dahongpao fresh leaves after aerospace mutagenesis; (**A**) analysis of total amount of metabolites in FCK and FTM; (**B**) PCA analysis after categorization of metabolites in FCK and FTM; (**C**) content analysis of different groups of metabolites in FCK and FTM; * denotes that the difference between two samples reaches the $p < 0.05$ level.

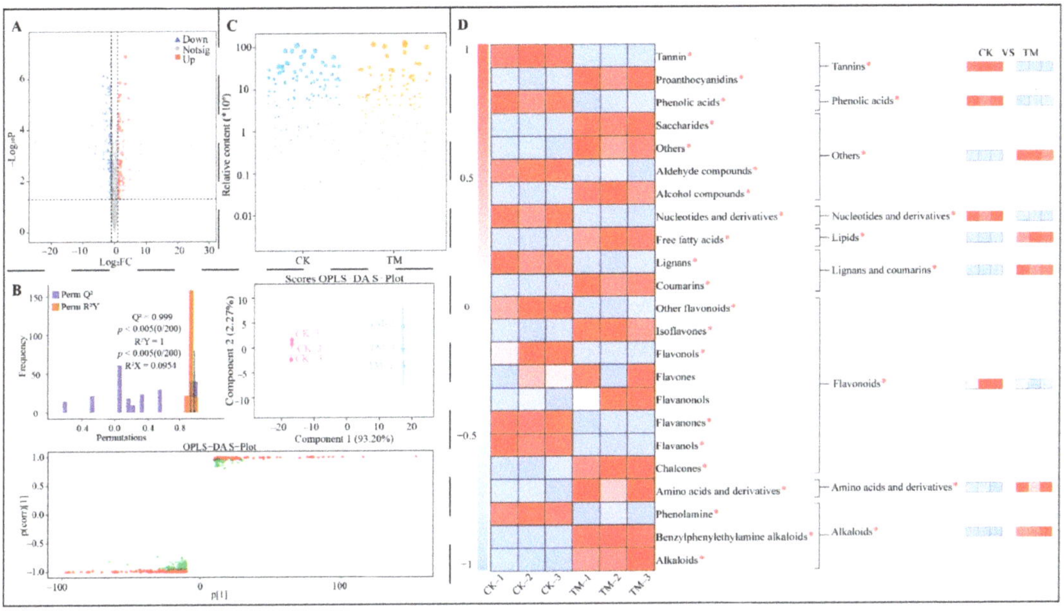

Figure 3. Screening and content analysis of characteristic metabolites in fresh leaves of Dahongpao tea tree. Note: CK: Dahongpao fresh leaves without aerospace mutagenesis; TM: Dahongpao fresh leaves after aerospace mutagenesis; (**A**) screening of differential metabolites in CK and TM by volcano map; (**B**) OPLS-DA model construction of CK and TM to screen key differential metabolites; (**C**) screening of characteristic metabolites in CK and TM by bubble feature map; (**D**) classification and content analysis of characteristic metabolites; * denotes that the difference between two samples reached the $p < 0.05$ level.

3.3. Taste Characteristics and Variation Analysis of Characteristic Metabolites in Dahongpao Fresh Leaves and Raw Tea

The evaluation of tea quality is significantly influenced by taste, and the formation of tea taste characteristics is highly correlated with the types and proportions of metabolites present in tea [28]. It has been reported that nucleotides and derivatives, lignans and coumarins, amino acids, and derivatives are found in tea and the main taste characteristic is fresh and brisk taste [29–31]. The taste characteristics of phenolic acids, flavonoids, tannins, and alkaloids mainly exhibit bitterness [32–34]. The taste characteristics of terpenoids and lipids mainly present aroma [35,36], whereas the taste characteristics of organic acids present a mellowness [37]. Accordingly, the present study analyzed taste characteristics based on the abovementioned characteristic metabolites obtained from fresh leaves and raw tea, and found that the characteristic metabolites of CK and TM exhibited mainly fresh and brisk taste, bitterness, and aroma (Figure 5A). However, the characteristic metabolites of FCK and FTM exhibited mainly mellowness, bitterness, fresh and brisk taste, and aroma (Figure 5B). Further analysis of the taste characteristics of fresh leaves and raw tea revealed (Figure 5C) that, from the taste characteristics of fresh leaves, aroma, and fresh and brisk taste of TM were higher than that of CK, while its bitterness was lower than that of CK; from the taste characteristics of raw tea, bitterness, mellowness, and fresh and brisk taste of FTM were greater than those of FCK, while its aroma was lower than that of FCK. Analysis from the point of view of taste characteristics revealed that the processing of fresh tea leaves into raw tea resulted in a significant enhancement of both aroma and mellowness, while bitterness and fresh and brisk taste were significantly reduced. It is evident that processing increases the intensity of aroma and mellowness of tea and reduces bitterness and fresh and brisk taste. Compared to CK, TM has a stronger aroma and fresh and brisk taste, while FTM has a stronger bitterness, fresh and brisk taste, and mellowness compared to FCK.

In addition, further analysis of metabolites with different taste characteristics in fresh leaves and raw tea in the present study revealed (Figure 5D) that from the perspective of fresh leaves, the content of lipids, alkaloids, lignans and coumarins, amino acids, and derivatives was greater in TM than in CK, whereas the content of tannins, flavonoids, phenolic acids, nucleotides, and derivatives was lower than CK. From the point of raw tea, FTM had a higher content of terpenoids, alkaloids, tannins, flavonoids, lignans and coumarins, organic acids, and amino acids and derivatives than FCK, while its content of lipids, nucleotides, and derivatives were lower than FCK. Analysis of metabolite content revealed that the content of terpenoids, lipids, nucleotides and derivatives, tannins, and organic acids increased significantly from fresh leaves to raw tea, while the content of alkaloids, flavonoids, lignans and coumarins, phenolic acids, amino acids, and derivatives decreased significantly. It can be seen that TM has a higher content of lipids, amino acids, lignans and coumarins, and derivatives and its aroma and fresh and brisk taste is stronger as compared to CK, whereas FTM has higher content of alkaloids, tannins, flavonoids, lignans and coumarins, organic acids, amino acids, and derivatives, and its bitterness, fresh and brisk taste, and mellowness were stronger than FCK.

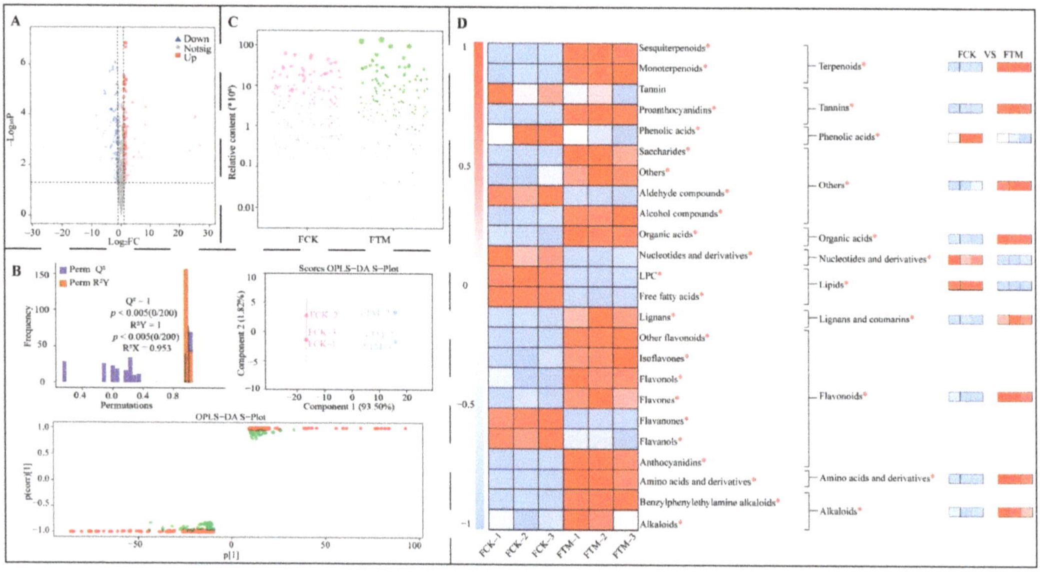

Figure 4. Screening and content analysis of characteristic metabolites in raw tea of Dahongpao tea tree. Note: FCK: raw tea processed from Dahongpao fresh leaves without aerospace mutagenesis; FTM: raw tea processed from Dahongpao fresh leaves after aerospace mutagenesis; (**A**) screening of differential metabolites in FCK and FTM by volcano diagram; (**B**) OPLS-DA model construction of FCK and FTM to screen key differential metabolites; (**C**) screening of characteristic metabolites in FCK and FTM by bubble feature plot; (**D**) classification and content analysis of characteristic metabolites; * indicates that the difference between two samples reached the $p < 0.05$ level.

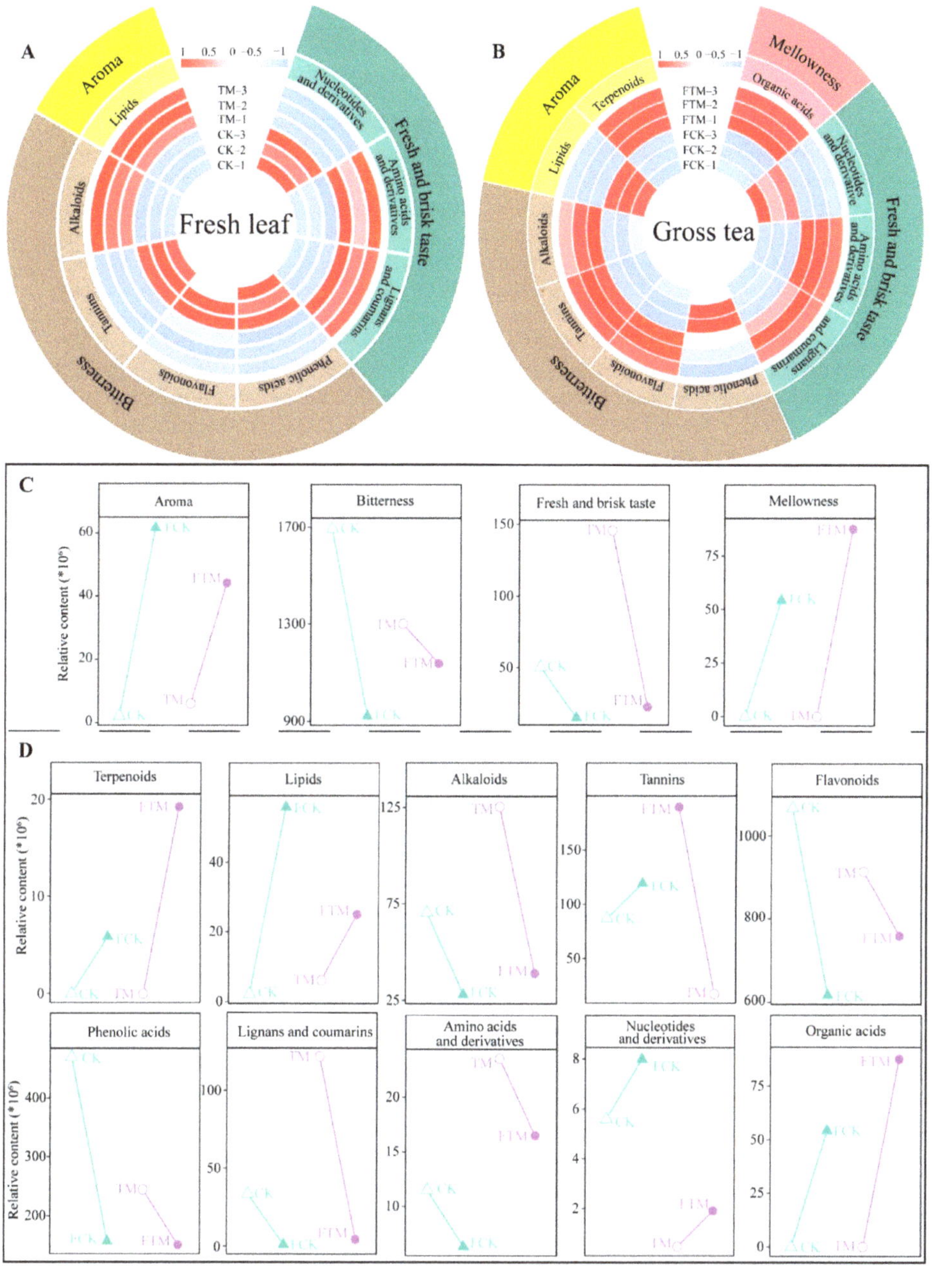

Figure 5. Taste and content analysis of characteristic metabolites in fresh leaves and raw tea of Dahongpao tea tree. Note: CK: Dahongpao fresh leaves without aerospace mutagenesis; TM: Dahongpao fresh leaves after aerospace mutagenesis; FCK: raw tea processed from Dahongpao fresh leaves without aerospace mutagenesis; FTM: raw tea processed from Dahongpao fresh leaves after aerospace mutagenesis; (**A**) taste and intensity analysis of the characteristic metabolites in CK and TM; (**B**) taste and intensity analysis of the characteristic metabolites in FCK and FTM; (**C**) changes in the content of different metabolites in fresh leaves and raw tea; (**D**) changes in the taste of the characteristic metabolites in fresh leaves and raw tea.

4. Conclusions

In the present study, the effect of aerospace mutagenesis on metabolite contents and taste characteristics in Dahongpao fresh leaves and raw tea was analyzed, and it was found (Figure 6) that the metabolites in fresh leaves and raw tea after aerospace mutagenesis were more similar in composition, and there was no significant difference in the total number of metabolites, but the content of different metabolites changed significantly. There were 87 characteristic metabolites distinguishing CK from TM and 97 characteristic metabolites distinguishing FCK from FTM. Taste characteristics analysis showed that taste characteristics of the characteristic metabolites in CK and TM were mainly fresh and brisk taste, bitterness, and aroma, whereas taste characteristics of characteristic metabolites in FCK and FTM were mainly fresh and brisk taste, bitterness, mellowness, and aroma. Aroma and fresh and brisk taste in TM were higher than CK, while bitterness was lower than CK. Fresh and brisk taste, bitterness, and mellowness were higher in FTM than in FCK, while aroma was lower than FCK. Processing increased the strength of aroma and mellowness and decreased bitterness and fresh and brisk taste of the tea. Aroma and fresh and brisk taste were stronger in TM compared to CK, while fresh and brisk taste, bitterness, and mellowness were stronger in FTM compared to FCK. In conclusion, TM had higher content of lipids, amino acids and derivatives, and lignans and coumarins, and its aroma and fresh and brisk taste was stronger compared to CK, while FTM had a higher content of alkaloids, tannins, amino acids and derivatives, flavonoids, lignans and coumarins, amino acids and derivatives, and organic acids, and its bitterness, fresh and brisk taste, and mellowness were stronger than FCK. It is evident that aerospace mutagenesis effectively altered the content of characteristic metabolites in tea leaves, which in turn altered their taste characteristics. This study provides a basis for the development of Dahongpao tea products by aerospace mutagenesis and the establishment of processing technology.

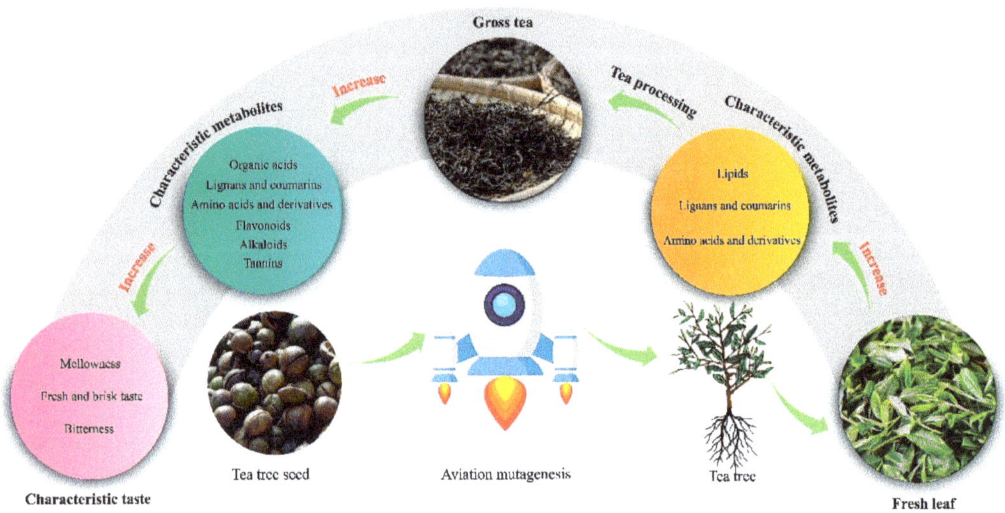

Figure 6. Mechanism analysis of the influence of aerospace mutagenesis on tea taste characteristics.

Supplementary Materials: The following supporting information can be downloaded at: https://www.mdpi.com/article/10.3390/foods13223538/s1, Table S1: Raw data of metabolites in fresh leaves of Dahongpao; Table S2: Raw data of metabolites in raw tea of Dahongpao.

Author Contributions: J.C. and Y.L. (Yangxin Luo): Methodology, conceptualization, formal analysis, data curation, writing—original draft. Q.Z., Y.W., T.W. and M.J.: Methodology, writing—original draft. Y.L. (Yankun Liao) and X.J.: Validation, formal analysis, data curation, writing—review and editing. H.W. and J.Y.: Methodology, conceptualization, formal analysis, supervision, project administration, writing—original draft, writing—review and editing. All authors have read and agreed to the published version of the manuscript.

Funding: This research was funded by Natural Science Foundation of Fujian Province (2024N0009, 2024J01861, 2024J01866, 2021J011137); Construction of first-class undergraduate specialty (tea science) in Fujian Province (SJZY2019004); Nanping City Science and Technology Plan Project (NP2021KTS06, NP2021KTS05, N2021Z012); Wuyi University Horizontal Research Program (2023-WHFW-017); Innovation and Entrepreneurship Training Program for College Students (202411312002, S202210397033, 202310397003, 202310397013); Faculty and students co-creation team of Wuyi University (2021-SSTD-01, 2021-SSTD-05).

Institutional Review Board Statement: Not applicable.

Informed Consent Statement: Not applicable.

Data Availability Statement: The original contributions presented in the study are included in the article/Supplementary Materials, further inquiries can be directed to the corresponding authors.

Conflicts of Interest: The authors declare that they have no conflicts of interest.

References

1. Prasad, B.; Richter, P.; Vadakedath, N.; Haag, F.W.; Strauch, S.M.; Mancinelli, R.; Schwarzwälder, A.; Etcheparre, E.; Gaume, N.; Lebert, M. How the space environment influences organisms: An astrobiological perspective and review. *Int. J. Astrobiol.* **2021**, *20*, 159–177. [CrossRef]
2. Furukawa, S.; Nagamatsu, A.; Nenoi, M.; Fujimori, A.; Kakinuma, S.; Katsube, T.; Wang, B.; Tsuruoka, C.; Shirai, T.; Nakamura, A.; et al. Space radiation biology for "Living in Space". *BioMed Res. Int.* **2020**, *2020*, 25. [CrossRef] [PubMed]
3. Xia, P.; Li, Q.; Liang, Z.; Zhang, X.; Yan, K. Spaceflight breeding could improve the volatile constituents of *Andrographis paniculata*. *Ind. Crop. Prod.* **2021**, *171*, 113967. [CrossRef]
4. Mohanta, T.K.; Mishra, A.K.; Mohanta, Y.K.; Al-Harrasi, A. Space breeding: The next-generation crops. *Front. Plant Sci.* **2021**, *12*, 2407. [CrossRef]
5. Li, J.; Liu, M.; Wang, P.; Zhang, J.; Xue, H.; Guo, Y. Effects of space conditions on mutation and inheritance of tomato. *Aerospace Med. Med. Eng.* **2000**, *2*, 114–118. [CrossRef]
6. Ma, L.; Kong, F.; Sun, K.; Wang, T.; Guo, T. From classical radiation to modern radiation: Past, present, and future of radiation mutation breeding. *Front. Public Health* **2021**, *9*, 768071. [CrossRef]
7. Li, L.; Hu, Y.; Wu, L.; Chen, R.; Luo, S. The complete chloroplast genome sequence of *Camellia sinensis* cv. Dahongpao: A most famous variety of Wuyi tea (Synonym: *Thea bohea* L.). *Mitochondrial DNA B* **2021**, *6*, 3–5. [CrossRef] [PubMed]
8. Liu, J.; Chen, Y.; Wang, W.; Wang, H.; Cai, J.; Wang, M.; Li, D.; Zhang, B.; Huang, K. Effects of space treatment on biological and growth characteristics of *Camellia sinensis*. *Chin. Bull. Bot.* **2020**, *55*, 564–572. [CrossRef]
9. Jia, M.; Chen, Y.; Zhang, Q.; Wang, Y.; Li, M.; Pang, X.; Hong, L.; Lin, S.; Jia, X.; Ye, J.; et al. Changes in the growth and physiological property of tea tree after aviation mutagenesis and screening and functional verification of its characteristic hormones. *Front. Plant Sci.* **2024**, *15*, 1402451. [CrossRef]
10. He, Y.; Zhang, Q.; Inostroza, A.C.; Kierszniowska, S.; Liu, L.; Li, Y.; Ruan, J. Application of metabolic fingerprinting in tea quality evaluation. *Food Control* **2024**, *160*, 110361. [CrossRef]
11. Yang, C.; Hu, Z.; Lu, M.; Li, P.; Tan, J.; Chen, M.; Lv, H.; Zhu, Y.; Zhang, Y.; Guo, L.; et al. Application of metabolomics profiling in the analysis of metabolites and taste quality in different subtypes of white tea. *Food Res. Int.* **2018**, *106*, 909–919. [CrossRef] [PubMed]
12. Gai, Z.; Wang, Y.; Jiang, J.; Xie, H.; Ding, Z.; Ding, S.; Wang, H. The quality evaluation of tea (*Camellia sinensis*) varieties based on the metabolomics. *HortScience* **2019**, *54*, 409–415. [CrossRef]
13. Farag, M.A.; Elmetwally, F.; Elghanam, R.; Kamal, N.; Hellal, K.; Hamezah, H.S.; Zhao, C.; Mediani, A. Metabolomics in tea products; a compile of applications for enhancing agricultural traits and quality control analysis of *Camellia sinensis*. *Food Chem.* **2023**, *404*, 134628. [CrossRef] [PubMed]
14. Wen, M.; Zhu, M.; Han, Z.; Ho, C.T.; Granato, D.; Zhang, L. Comprehensive applications of metabolomics on tea science and technology: Opportunities, hurdles, and perspectives. *Compr. Rev. Food Sci. F* **2023**, *22*, 4890–4924. [CrossRef] [PubMed]
15. Cai, H.; Zhong, Z.; Li, Z.; Zhang, X.; Fu, H.; Yang, B.; Zhang, L. Metabolomics in quality formation and characterisation of tea products: A review. *Int. J. Food Sci. Tech.* **2022**, *57*, 4001–4014. [CrossRef]
16. Ye, J.; Zhang, Q.; Cheng, P.; Wang, Y.; Zou, J.; Lin, S.; Li, M.; Jia, M.; Chen, Y.; Jia, X.; et al. Aviation mutagenesis alters the content of volatile compounds in Dahongpao (*Camellia sinensis*) leaves and improves tea quality. *Foods* **2024**, *13*, 946. [CrossRef]

17. Zhou, X.; Hu, L.; Hoang, N.H.; Thanh, T.L.; Zhou, C.; Mei, X.; Buensanteai, K. The changes in metabolites, quality components, and antioxidant activity of tea (*Camellia sinensis*) infected with exobasidium vexans by applying UPLC-MS/MS-based widely targeted metabolome and biochemical analysis. *Phytopathology* **2024**, *114*, 164–176. [CrossRef]
18. Wen, O.; Ning, J.; Zhu, X.; Jiang, Y.; Wang, J.; Yuan, H.; Hua, J. UPLC-ESI-MS/MS analysis revealed the dynamic changes and conversion mechanism of non-volatile metabolites during green tea fixation. *LWT* **2024**, *198*, 116010. [CrossRef]
19. Bado, S.; Forster, B.P.; Maghuly, F. Physical and chemicals mutagenesis in plant breeding. In *Mutation Breeding for Sustainable Food Production and Climate Resilience*; Penna, S., Jain, S.M., Eds.; Springer: Singapore, 2023. [CrossRef]
20. Zeng, D.; Cui, J.; Yin, Y.; Dai, C.; Zhao, H.; Song, C.; Guan, S.; Cheng, D.; Sun, Y.; Lu, W. Combining proteomics and metabolomics to analyze the effects of spaceflight on rice progeny. *Front. Plant Sci.* **2022**, *13*, 900143. [CrossRef]
21. Li, Z.; Sun, X.; Xu, T.; Dai, W.; Yan, Q.; Li, P.; Fang, Y.; Ding, J. Insight into the dynamic variation and retention of major aroma volatile compounds during the milling of Suxiang japonica rice. *Food Chem.* **2023**, *405*, 134468. [CrossRef]
22. Gong, R.; Huang, D.; Chen, Y.; Li, H.; Wang, Z.; Zhou, D.; Zhao, L.; Pan, Y.; Chang, Y.; Xiang, Y.; et al. Comparative metabolomics analysis reveals the variations of eating quality among three high-quality rice cultivars. *Mol. Breed.* **2020**, *40*, 112. [CrossRef]
23. Rivera-Pérez, A.; Romero-González, R.; Frenich, A.G. Fingerprinting based on gas chromatography-orbitrap high-resolution mass spectrometry and chemometrics to reveal geographical origin, processing, and volatile markers for thyme authentication. *Food Chem.* **2022**, *393*, 133377. [CrossRef] [PubMed]
24. Liao, Y.; Zhou, X.; Zeng, L. How does tea (*Camellia sinensis*) produce specialized metabolites which determine its unique quality and function: A review. *Crit. Rev. Food Sci.* **2022**, *62*, 3751–3767. [CrossRef]
25. Guo, L.; Chen, M.; Guo, Y.; Lin, Z. Variations in fatty acids affected their derivative volatiles during Tieguanyin tea processing. *Foods* **2022**, *11*, 1563. [CrossRef] [PubMed]
26. Wong, M.; Sirisena, S.; Ng, K. (2022). Phytochemical profile of differently processed tea: A review. *J. Food Sci.* **2022**, *87*, 1925–1942. [CrossRef]
27. Fang, X.; Xu, W.; Jiang, G.; Sui, M.; Xiao, J.; Ning, Y.; Niaz, R.; Wu, D.; Feng, X.; Chen, J.; et al. Monitoring the dynamic changes in aroma during the whole processing of Qingzhuan tea at an industrial scale: From fresh leaves to finished tea. *Food Chem.* **2024**, *439*, 137810. [CrossRef]
28. Zhang, L.; Cao, Q.Q.; Granato, D.; Xu, Y.Q.; Ho, C.T. Association between chemistry and taste of tea: A review. *Trends Food Sci. Tech.* **2020**, *101*, 139–149. [CrossRef]
29. Long, P.; Rakariyatham, K.; Ho, C.T.; Zhang, L. Thearubigins: Formation, structure, health benefit and sensory property. *Trends Food Sci. Tech.* **2023**, *133*, 37–48. [CrossRef]
30. Huang, D.; Wang, Y.; Chen, X.; Wu, J.; Wang, H.; Tan, R.; Jiao, L.; Mao, Y. Application of tea-specific fertilizer combined with organic fertilizer improves aroma of green tea. *Horticulturae* **2022**, *8*, 950. [CrossRef]
31. Hua, J.; Ouyang, W.; Zhu, X.; Wang, J.; Yu, Y.; Chen, M.; Yang, L.; Yuan, H.; Jiang, Y. Objective quantification technique and widely targeted metabolomic reveal the effect of drying temperature on sensory attributes and related non-volatile metabolites of black tea. *Food Chem.* **2024**, *439*, 138154. [CrossRef] [PubMed]
32. Wang, X.; Xiong, H.; Wang, S.; Zhang, Y.; Song, Z.; Zhang, X. Physicochemical analysis, sensorial evaluation, astringent component identification and aroma-active compounds of herbaceous Peony (*Paeonia lactiflora* Pall) black tea. *Ind. Crops Prod.* **2023**, *193*, 116159. [CrossRef]
33. Mostafa, H.S. Production of low-tannin *Hibiscus sabdariffa* tea through D-optimal design optimization of the preparation conditions and the catalytic action of new tannase. *Food Chem. X* **2023**, *17*, 100562. [CrossRef]
34. González-Muñoz, B.; Garrido-Vargas, F.; Pavez, C.; Osorio, F.; Chen, J.; Bordeu, E.; O'Brien, J.; Brossard, N. Wine astringency: More than just tannin–protein interactions. *J. Sci. Food Agr.* **2022**, *102*, 1771–1781. [CrossRef]
35. Ye, J.; Wang, Y.; Lin, S.; Hong, L.; Kang, J.; Chen, Y.; Li, M.; Jia, Y.; Jia, X.; Wu, Z.; et al. Effect of processing on aroma intensity and odor characteristic of Shuixian (*Camellia sinensis*) tea. *Food Chem. X* **2023**, *17*, 100616. [CrossRef]
36. Huang, F.F.; Yang, P.D.; Bai, S.L.; Liu, Z.H.; Li, J.; Huang, J.A.; Xiong, L.G. Lipids: A noteworthy role in better tea quality. *Food Chem.* **2024**, *431*, 137071. [CrossRef]
37. Zhang, X.; Du, X.; Li, Y.Z.; Nie, C.N.; Wang, C.M.; Bian, J.L.; Luo, F. Are organic acids really related to the sour taste difference between Chinese black tea and green tea? *Food Sci. Nutr.* **2022**, *10*, 2071–2081. [CrossRef]

foods

MDPI

Article

Profiling of Potential Anti-Diabetic Active Compounds in White Tea: An Integrated Study of Polyphenol-Targeted Metabolomics, Network Pharmacology, and Computer Simulation

Weiwei Wu, Zhiqiang Zheng, Zhihui Wang, Chenxi Gao, Yilin Liang, Wen Zeng and Weijiang Sun *

College of Horticulture, Fujian Agriculture and Forestry University, Fuzhou 350002, China; wweiwei0601@163.com (W.W.); zzq000422@163.com (Z.Z.); wzh1246900265@163.com (Z.W.); gaochenxii@163.com (C.G.); lyl908213@163.com (Y.L.); wenzeng_t@163.com (W.Z.)

* Correspondence: 000q020007@fafu.edu.cn

Abstract: Diabetes remains a critical global public health challenge, posing a growing threat to human health and well-being. White tea is a lightly fermented tea and one of the six traditional tea categories in China. Owing to its rich content of bioactive compounds such as catechins and alkaloids, it has demonstrated potential anti-diabetic properties. However, its precise bioactive components, mechanisms of action, and relevant molecular targets require further investigation. In this study, an integrated approach combining polyphenol-targeted metabolomics, in vitro antioxidant assays, α-glucosidase inhibition tests, network pharmacology analysis, GEO database exploration, molecular docking, and molecular dynamics simulations was employed to identify the potential anti-diabetic compounds, targets, and mechanisms of white tea. The findings revealed that white tea is particularly abundant in 10 bioactive compounds, including epigallocatechin gallate, epicatechin gallate, and catechin, all of which exhibit significant anti-diabetic potential. These compounds were found to exert their effects by interacting with core molecular targets, namely cathepsin V (CTSV) and nucleotide-binding oligomerization domain-containing protein 1 (NOD1), and engaging in pathways related to signal transduction, apoptosis, and immune responses. This study establishes a strong theoretical basis for advancing white tea research and underscores new opportunities for applying natural products in diabetes therapy.

Keywords: white tea; diabetes; polyphenol-targeted metabolomics; network pharmacology; molecular dynamic simulation

1. Introduction

White tea (WT), one of the six traditional tea types in China, is produced from the tender leaves of *Camellia sinensis* and is renowned for its distinctive flavor and significant health benefits [1]. Primarily cultivated in Fujian Province, WT undergoes minimal processing, remaining unfermented, which preserves a high concentration of natural active compounds [2]. Research has indicated that WT is notably rich in catechins and alkaloids—compounds recognized for reducing oxidative stress and modulating immune responses—thus aiding in the prevention and management of chronic conditions such as cardiovascular disease, cancer, and diabetes [3].

Diabetes mellitus (DM) is a significant global public health issue, currently affecting over 500 million people, with projections suggesting that this number could rise to nearly 700 million by 2045 [4]. The pathogenesis of DM is closely associated with insulin resistance or insufficient insulin secretion, leading to abnormal elevations in blood glucose levels [5]. If left uncontrolled, DM can result in severe complications, including kidney and cardiovascular diseases, significantly increasing healthcare costs [6]. Reports indicate that individuals with diabetes utilize healthcare resources more frequently and extensively, leading to significantly higher medical expenses compared to those without diabetes [7]. Thus, developing effective strategies for preventing and treating DM is crucial for improving global public health.

In recent years, increasing attention has been given to the potential role of WT in the prevention and treatment of DM. Studies have shown that the polyphenol compounds in WT can effectively inhibit the activity of α-glucosidase [8]. Additionally, catechins alleviate DM-induced tubular injury by inhibiting endoplasmic reticulum stress and NLRP3-mediated inflammatory responses [9]. Furthermore, catechins exert their protective effects against DM by regulating the epithelial–mesenchymal transition (EMT) process through multiple pathways [10]. Despite these promising findings, the specific bioactive components, molecular mechanisms, and relevant targets of WT's anti-DM effects require further elucidation.

Advances in metabolomics have significantly facilitated the identification and quantification of polyphenolic compounds, enabling deeper insights into their bioactivities [11]. Furthermore, developments in bioinformatics and related databases have promoted the application of network pharmacology in drug research, revealing the intricate relationships between bioactive compounds, targets, signaling pathways, and diseases [12]. Molecular docking analyses are frequently used to validate the interactions between bioactive compounds identified through network pharmacology and their target proteins, while molecular dynamics simulations further refine the binding modes of these compounds, offering a more accurate understanding of these interactions at the molecular level [13].

The present study aims to systematically investigate the anti-DM active components of WT through the integration of polyphenol-targeted metabolomics, in vitro antioxidant activity assays, α-glucosidase inhibition experiments, network pharmacology, GEO database analysis, molecular docking, and molecular dynamics simulations. By identifying the key molecular targets and elucidating the underlying mechanisms, this research provides a theoretical foundation for the comprehensive utilization of WT and offers new insights into the application of natural products in DM therapy.

2. Materials and Methods

2.1. Tea Samples and Chemicals

Tea leaf samples of the *Camellia sinensis* (L.) O. Kuntze cv. Fuding Dahao cultivar, specifically fresh leaves classified as one bud with one or two leaves, were sourced from Liumiao White Tea Co., Ltd., located in Fuding City, China. In the spring of 2024, fresh leaves at the one-bud-two-leaf stage were harvested and immediately transported to the withering facility. The leaves were evenly spread on bamboo trays, maintaining a thickness of 2–3 cm, and subjected to natural indoor withering for 48 h under controlled conditions (ambient temperature: 20–25 °C, humidity: 60–70%). Following the withering process, the leaves were transferred to a drying machine and dried at 60 °C for 3 h, producing the WT samples. After collection, all samples were stored in the dark at −20 °C until further analysis.

Vitamin C was obtained from Macklin Biochemical Technology Co., Ltd. (Shanghai, China), while acarbose was provided by Nanjing Dolay Biotechnology Co., Ltd. (Nanjing, China). Antioxidant assay kits for DPPH, ABTS, FRAP, and HRSA, as well as the α-glucosidase activity assay kits, were purchased from Suzhou Comin Biotechnology Co., Ltd. (Suzhou, China). Standards including gallic acid (GA, >99%), epigallocatechin gallate (EGCG, >99%), epicatechin (EC, >99%), catechin gallate (CG, >99%), epigallocatechin (EGC, >99%), catechin (C, >99%), gallocatechin gallate (GCG, >99%), epicatechin gallate (ECG, >99%), caffeine (>99%), theophylline (>99%), and theobromine (>99%) were purchased from Sigma-Aldrich (St. Louis, MO, USA). Methanol, acetonitrile, and acetic acid (HPLC grade) were sourced from Aladdin Biochemical Technology Co., Ltd. (Shanghai, China).

2.2. Determination of Antioxidant Activity and α-Glucosidase Inhibition of WT

The antioxidant activities of WT, including the 2,2-diphenyl-1-picrylhydrazyl (DPPH) radical scavenging assay, the 2,2′-azino-bis(3-ethylbenzothiazoline-6-sulfonic acid) (ABTS) radical scavenging assay, the ferric reducing antioxidant power (FRAP) assay, and the hydroxyl radical scavenging activity (HRSA) assay, as well as its inhibitory effect on α-glucosidase activity, were determined using respective assay kits. Vitamin C was used as the positive control in the antioxidant assays, while distilled water was the negative control. For the α-glucosidase inhibition assay, acarbose was used as the positive control and distilled water as the negative control.

2.2.1. DPPH Radical Scavenging Activity

To determine the DPPH radical scavenging activity, the method described by Peng et al. was followed [14]. A 0.1 g of tea powder or vitamin C was mixed with 1 mL of extraction solution 1, and the mixture was homogenized on ice. The homogenate was centrifuged at 10,000× g for 10 min at 4 °C. The supernatant was collected and stored on ice for analysis. A 20 μL aliquot of each sample (WT, vitamin C, or distilled water) was mixed with 380 μL of reaction solution 1 (A_{test1}) and incubated in the dark at room temperature for 20 min. After incubation, 200 μL of the mixture was transferred to a 96-well plate for absorbance measurement at 515 nm. A blank control (A_{blank1}) was prepared by adding 20 μL of extraction solution to 380 μL of reaction solution. The DPPH radical scavenging activity was calculated as follows:

$$\text{DPPH}(\%) = \frac{A_{blank1} - A_{test1}}{A_{blank1}} \times 100\%, \quad (1)$$

2.2.2. ABTS Radical Scavenging Activity

For the ABTS assay, the method described by Yan et al. was followed [15]. A mixture of 0.1 g of tea powder or vitamin C and 1 mL of extraction solution 2 was prepared and homogenized on ice. After centrifugation at 10,000× g for 10 min at 4 °C, the supernatant was collected and kept on ice. A 10 μL aliquot of each sample (WT, vitamin C, or distilled water) was mixed with 190 μL of reaction solution 2 (A_{test2}) and incubated for 10 min. The absorbance at 734 nm was then measured. A blank control (A_{blank2}) was prepared by mixing 10 μL of extraction solution with 190 μL of reaction solution. The ABTS radical scavenging activity was calculated as follows:

$$\text{ABTS}(\%) = \frac{A_{blank2} - A_{test2}}{A_{blank2}} \times 100\%, \quad (2)$$

2.2.3. HRSA Radical Scavenging Activity

To evaluate HRSA, the method described by Chen et al. was employed [16]. A mixture of 0.1 g of tea powder or vitamin C and 1 mL of extraction solution 3 was homogenized on ice, followed by centrifugation at 10,000× g for 10 min at 4 °C. The supernatant was placed on ice for subsequent analysis. A 200 μL aliquot of each sample (WT, vitamin C, or distilled water) was mixed with 100 μL of reaction solution 3, 100 μL of reaction solution 4, 200 μL of reaction solution 5, and 300 μL of distilled water (A_{test3}), and incubated at 37 °C for 20 min. A 200 μL aliquot of the reaction mixture was transferred to a 96-well plate, and the absorbance was measured at 510 nm. In the control group ($A_{control1}$), 100 μL of reaction solution 3, 100 μL of reaction solution 4, 200 μL of reaction solution 5, and 500 μL of distilled water were used, while in the blank group (A_{blank3}), 100 μL of reaction solution 3, 200 μL of reaction solution 5, and 600 μL of distilled water were used. The HRSA was calculated using the following formula:

$$\text{HRSA}(\%) = \frac{A_{blank2} - A_{test3}}{A_{blank3}} \times 100\%, \quad (3)$$

2.2.4. FRAP Radical Scavenging Activity

For the FRAP assay, the method described by Jiao et al. was followed [17]. A mixture of 0.1 g of tea powder or vitamin C and 1 mL of extraction solution 4 was homogenized on ice and then centrifuged at 10,000× *g* for 10 min at 4 °C. The supernatant was collected and placed on ice for further analysis. A 10 μL aliquot of each sample (WT, vitamin C, or distilled water) was mixed with 190 μL of reaction solution 6 (A_{test4}), and the mixture was incubated for 20 min. The absorbance at 593 nm was measured in a 96-well plate. A blank control (A_{blank4}) was prepared by adding 10 μL of extraction solution to 190 μL of reaction solution. The FRAP value (μmol Trolox/g) was calculated as follows:

$$\text{FRAP}(\mu\text{mol Trolox/g}) = \frac{0.8054 \times [(A_{test4} - A_{blank4}) - 0.0134]}{0.1}, \quad (4)$$

2.2.5. α-Glucosidase Inhibition Activity

To determine the inhibitory effect of WT on α-glucosidase activity, the method described by Yu et al. was followed [18]. A mixture of 0.1 g of tea powder or acarbose and 1 mL of extraction solution 5 was homogenized on ice, followed by centrifugation at 15,000× *g* for 10 min at 4 °C. The supernatant was collected and stored on ice for analysis. A 30 μL aliquot of each sample (WT, acarbose, or distilled water) was mixed with 120 μL of reaction solution 7 and 150 μL of reaction solution 8 (A_{test5}) and incubated at 37 °C for 30 min. The reaction mixture was then transferred to a 95 °C water bath for 5 min. After cooling, the mixture was centrifuged at 8000× *g* for 5 min at 4 °C. A 70 μL aliquot of the supernatant was mixed with 130 μL of reaction solution 9, and the mixture was incubated at room temperature for 2 min before measuring absorbance at 400 nm. In the control group ($A_{control2}$), 30 μL of sample (WT, acarbose, distilled water) was mixed with 120 μL of distilled water and 150 μL of reaction solution 8. The α-glucosidase inhibitory activity was calculated using the following formula:

$$\alpha - \text{Glucosidase inhibition}(\%) = \frac{85.47 \times [(A_{test5} - A_{control2})]}{0.1} \quad (5)$$

2.3. *Targeted Polyphenol Metabolomics Analysis*

Alkaloids (theophylline, theobromine, caffeine), gallic acid (GA), and catechins (GC, EGC, C, EC, EGCG, GCG, ECG) were quantified using a high-performance liquid chromatography (HPLC) system (E2695, Waters, Milford, MA, USA) equipped with a Phenyl-Hexyl 100A column (250 mm × 4.6 mm, 5 μm, Phenomenex, Los Angeles, CA, USA). Sample preparation was performed according to the Chinese National Standard GB/T 8313—2018, which specifies methods for the determination of polyphenols and catechins in tea [19]. The mobile phase consisted of solution A (2% acetic acid) and solution B (100% acetonitrile). The gradient conditions were set as follows: 0–25 min (91% A, 9% B); 25–35 min (68.3% A, 31.7% B); 35–40 min (100% B). The flow rate was 1 mL/min, with an injection volume of 10 μL. The column temperature was maintained at 35 °C, and the detection wavelength was set at 278 nm for ultraviolet (UV) detection.

2.4. *Screening of WT Active Components' Targets and Construction of WT-Component-Target Network*

The SMILES information for alkaloids (theophylline, theobromine, caffeine), gallic acid (GA), and catechins (GC, EGC, C, EGCG, GCG, and ECG) was retrieved from the PubChem database (https://pubchem.ncbi.nlm.nih.gov/, accessed on 10 July 2024). The potential targets of these compounds were predicted using the SwissTargetPrediction (http://www.swisstargetprediction.ch/, accessed on 10 July 2024) and SEA (https://sea.bkslab.org/, accessed on 10 July 2024) databases. Subsequently, a network of active components and their target genes was constructed using Cytoscape (version 3.9.1), revealing the core relationships between WT components and their targets.

2.5. Identification of Differentially Expressed Genes (DEGs) in DM

Gene expression profiles related to DM were obtained from human samples in the GEO database (http://www.ncbi.nlm.nih.gov/geo/, accessed on 10 July 2024). Differential gene analysis was conducted using the R package "limma", and results were displayed in boxplots based on the criteria of $p < 0.05$ and $|\log_2 FC| > 0.5$ (Supplementary Figure S1A,B). A volcano plot was generated using the "ggplot2" package in R (version 4.3).

2.6. Weighted Gene Co-Expression Network Analysis (WGCNA)

Gene modules were defined using the WGCNA algorithm, which identifies groups of genes with similar expression patterns. The weighted gene co-expression network was constructed using the "WGCNA" package in R. Soft threshold power was chosen based on the criterion of scale-free topology ($R^2 > 0.85$). The adjacency matrix was transformed into a topological overlap matrix (TOM) to quantify the degree of overlap between genes. Modules were identified using hierarchical clustering and dynamic tree-cut methods, with a minimum module size of 30 genes. Modules showing a high degree of similarity (>0.75) were merged to reduce redundancy, and module-trait relationships were calculated using Pearson correlation coefficients. Different modules were represented by distinct colors, and modules with similar expression patterns were merged. The modules with the highest correlations to healthy and DM samples were selected, and genes closely associated with these samples were extracted for further analysis.

2.7. Screening of Key Targets and Functional Enrichment Analysis

Potential anti-DM therapeutic targets were identified by intersecting the predicted targets of WT active components, DEGs, and genes from the pink and turquoise modules using the "UpSetR" package in R. To explore the biological functions of the identified key target genes, Gene Ontology (GO) analysis and KEGG pathway enrichment were performed using the "clusterProfiler" package, and the results were ranked by *p*-value. GO analysis included terms from three categories: molecular function (MF), cellular component (CC), and biological process (BP). The top 10 most significant terms ($p < 0.05$) from each category were selected and visualized using the "GOplot" package. Similarly, the top 10 most significant KEGG pathways ($p < 0.05$) were visualized with the "GOplot" package

2.8. Molecular Docking Verification

The diagnostic performance of key targets in AD was evaluated using receiver operating characteristic (ROC) curves, and the area under the curve (AUC) was calculated using the "pROC" package. Targets with an AUC greater than 0.7 were identified as core targets. The expression differences in core targets between healthy and AD samples were analyzed using boxplots and visualized with the "ggplot2" package.

Ligand files in SDF format for 10 WT active components, including EGCG and ECG, as well as the positive drug acarbose, were obtained from the PubChem database. The 3D structure models of the CTSV (PDB ID: 7PK4) and NOD1 (PDB ID: 4JQW) proteins were retrieved from the UniProt database (https://www.uniprot.org/, accessed on 10 July 2024). Molecular docking was performed using AutoDock Vina 1.2.5 with an exhaustiveness value of 8. The docking results were evaluated based on the binding affinity (expressed in kcal/mol), and poses with root mean square deviation (RMSD) values of less than 2 Å were considered valid. The docking results were further analyzed and visualized with Discovery Studio 2019.

2.9. Molecular Dynamics Simulation

To further investigate the stability of protein-ligand interactions, molecular dynamics (MD) simulations of the EGCG-CTSV, ECG-CTSV, and C-NOD1 complexes were performed using GROMACS 2020.6 software. The AMBER99SB force field and SPC water model were applied, with a time step of 2 fs, and the particle mesh Ewald (PME) method was used for long-range electrostatic interactions. The system was equilibrated for 100 ps under NVT conditions and 100 ps under NPT conditions. The production MD run was carried out for 50 ns at 300 K with a cut-off distance of 1.2 nm. The root mean square deviation (RMSD), root mean square fluctuation (RMSF), and potential energy were monitored to confirm system stability and convergence. Energy minimization was performed first, followed by equilibration to stabilize the system, after which MD simulations were conducted. Free energy calculations were carried out using MM/GBSA and MM/PBSA methods, and the results were visualized using Xmgrace 5.1.25 software.

2.10. Statistical Analysis

Statistical analysis was conducted using SPSS 19.0, and all data were presented as mean ± standard deviation (SD). One-way analysis of variance (ANOVA) was used, followed by post hoc analysis with Duncan's multiple range test. A value of $p < 0.05$ was considered statistically significant.

3. Results and Discussion

3.1. Antioxidant Activity and α-Glucosidase Inhibition of WT

Antioxidants play a critical role in protecting pancreatic β-cell function by alleviating oxidative stress. This protection contributes to the management of DM and its complications [20]. α-Glucosidase inhibitors are crucial in controlling postprandial blood glucose levels by delaying carbohydrate absorption [21]. Therefore, the antioxidant activity and α-glucosidase inhibitory activity of WT were evaluated in this study. The results demonstrated that the DPPH, ABTS, HRSA, and FRAP values in the Control group were 1.31%, 1.27%, 0.64%, and 0.13 µmol Trolox/g, respectively, while in the WT group, these values were 71.50%, 85.43%, 6.57%, and 9.46 µmol Trolox/g, respectively. In the Vitamin C group, the DPPH, ABTS, HRSA, and FRAP values were 74.50%, 94.92%, 15.99%, and 28.49 µmol Trolox/g, respectively (Table 1). Statistical analysis indicated that both the WT and Vitamin C groups had significantly higher DPPH values compared to the Control group, and the Vitamin C group exhibited significantly higher ABTS, HRSA, and FRAP values than both the WT and Control groups. Additionally, the WT group showed significantly higher ABTS, HRSA, and FRAP values than the Control group. In the α-glucosidase inhibition assay, the α-glucosidase activity in the acarbose, WT, and Control groups was 8.74 nmol/min/g, 14.09 nmol/min/g, and 163.13 nmol/min/g, respectively. As shown in Table 1, α-glucosidase activity was significantly higher in the control group compared to both the acarbose and WT groups. The results of this study align with those reported by Zhou et al. and Esposito et al., confirming that WT demonstrates pronounced antioxidant activity and significant inhibition of α-glucosidase [8,22]. Collectively, these findings suggest that WT may offer promising antidiabetic properties.

Table 1. Results of antioxidant activity and α-glucosidase inhibition assays.

Samples	DPPH	ABTS	FRAP	HRSA	α-Glucosidase Activity
Control	1.31 ± 0.19^{b}	1.27 ± 0.31^{c}	0.13 ± 0.03^{c}	0.63 ± 0.08^{c}	163.13 ± 4.82^{a}
acarbose	74.50 ± 0.93^{a}	94.92 ± 0.44^{a}	28.49 ± 0.15^{a}	15.99 ± 0.02^{a}	8.74 ± 0.12^{b}
WT	71.50 ± 0.39^{a}	85.43 ± 0.10^{b}	9.46 ± 0.14^{b}	6.57 ± 0.45^{b}	14.09 ± 1.48^{b}

Different letters indicate a significant difference at the $p < 0.05$ level.

3.2. Screening of Key Active Components and Relevant Targets of WT

Targeted polyphenol metabolomics analysis (Figure 1, Supplementary Table S1) revealed that the contents of theophylline, theobromine, and caffeine were 1.24 mg/g, 1.52 mg/g, and 39.87 mg/g, respectively, while the content of GA was 0.73 mg/g. The contents of catechins, including GC, EGC, C, EC, EGCG, GCG, and ECG, were 5.97 mg/g, 7.07 mg/g, 0.75 mg/g, 5.24 mg/g, 102.63 mg/g, 7.25 mg/g, and 23.7 mg/g, respectively. To further identify potential anti-DM active components of WT, a correlation analysis was conducted between these components and antioxidant activity as well as α-glucosidase inhibitory activity. The results (Figure 1A, Supplementary Table S2) indicated significant positive correlations ($r > 0.8$) between DPPH and theophylline, theobromine, GC, EGC, EGCG, ECG, and GA. ABTS was significantly positively correlated ($r > 0.8$) with theophylline, theobromine, caffeine, GC, GA, EGCG, GCG, and ECG. HRSA showed significant positive correlations ($r > 0.8$) with theophylline, caffeine, EGCG, GCG, and ECG. Additionally, FRAP exhibited a significant positive correlation ($r > 0.8$) with C, while α-glucosidase inhibitory activity was significantly negatively correlated ($r < -0.8$) with theophylline, theobromine, caffeine, GC, EGCG, ECG, and GA. Based on these findings, theophylline, theobromine, caffeine, GA, GC, EGC, C, EGCG, GCG, and ECG were identified as potential anti-DM active components of WT, and these 10 components were subjected to further analysis. The SMILES information for these compounds was obtained from the PubChem database, and their potential targets were predicted using the SwissTargetPrediction and SEA databases. After removing duplicates, a total of 175 potential targets and their corresponding gene symbols were identified (Figure 1B, Supplementary Table S3). Theophylline was associated with 40 potential targets, theobromine with 42, caffeine with 48, GA with 72, GC with 13, EGC with 13, C with 24, EGCG with 56, GCG with 56, and ECG with 57 potential targets.

3.3. Screening of Key Genes in DM

In this study, the GSE76896 dataset was selected for analysis, which contains 171 pancreatic islet samples, including 116 from healthy individuals and 55 from DM patients (Supplementary Table S4). A total of 14,435 genes were identified from the GSE76896 dataset. DEGs were screened based on the criteria of $p < 0.05$ and $|\log2FC| > 0.5$. By comparing healthy and DM samples, 1195 DEGs were identified, with 645 genes being upregulated and 550 downregulated (Figure 1C, Supplementary Table S5). WGCNA was used to construct a scale-free gene co-expression network, correlating gene expression patterns with clinical data. By correlating gene expression patterns with clinical data, this systems biology approach overcomes the limitations of traditional methods that focus solely on gene expression without considering gene-gene relationships [23]. WGCNA of the 14,435 genes resulted in the division of these genes into nine co-expression modules, and the correlation between these modules and both healthy and DM samples was displayed (Figure 1D). Among the identified modules, the pink and turquoise modules exhibited significant correlations with DM samples. The pink module (125 genes) was positively correlated with DM samples ($r = -0.26$, $p = 7 \times 10^{-4}$) and negatively correlated with healthy samples ($r = -0.26$, $p = 7 \times 10^{-4}$). Conversely, the turquoise module (6231 genes) was negatively correlated with DM samples ($r = -0.35$, $p = 2 \times 10^{-6}$) and positively correlated with healthy samples ($r = 0.35$, $p = 2 \times 10^{-6}$). Based on these findings, the clinical significance of the pink and turquoise modules was further explored.

To identify key anti-DM genes affected by WT components, an intersection analysis was conducted between the predicted targets of WT active components, DEGs, and genes from the pink and turquoise modules. Figure 1E shows the overlap between predicted targets of WT active components and DEGs. This analysis identified 19 intersecting genes. Additionally, two intersecting genes were found between WT targets and the pink module, while 38 were found between WT targets and the turquoise module. Additionally, one gene overlapped between the WT targets, DEGs, and the pink module, and three genes overlapped between the WT targets, DEGs, and the turquoise module (Supplementary Table S6). After deduplication, a total of 63 intersecting genes were identified, suggesting that WT active components may exert their anti-DM effects by targeting these 63 key genes.

3.4. Functional Enrichment Analysis of Key Targets

To further elucidate the biological functions of the key targets, GO and KEGG enrichment analyses were performed on the 63 key genes. GO enrichment analysis showed that these genes were enriched in 548 BP terms, 23 CC terms, and 94 MF terms, accounting for 82.40%, 3.46%, and 14.14% of the total, respectively. In the BP category, the key targets were primarily enriched in processes related to signal transduction and apoptosis (Regulation of apoptotic signaling pathway, Positive regulation of MAPK cascade, Regulation of ERK1 and ERK2 cascade and Regulation of protein tyrosine kinase activity), immune and inflammatory responses (Positive regulation of cytokine production, Regulation of leukocyte migration, and Regulation of inflammatory response), responses to external stimuli and cellular stress (Response to oxidative stress and Response to reactive oxygen species), and Hormone metabolic process (Figure 2A). In the CC category, the key targets were mainly enriched in membrane structures and associated regions (Membrane raft, Membrane microdomain, Mitochondrial outer membrane, Basolateral plasma membrane, and Organelle outer membrane), endosome and lysosome-associated structures (Endosome lumen, Lysosomal lumen, Endolysosome, and Recycling endosome), and Vacuolar lumen (Figure 2B). In the MF category, the key targets were enriched in protease activity and regulation (Cysteine-type endopeptidase activator activity involved in apoptotic process, Cysteine-type endopeptidase regulator activity involved in apoptotic process and Serine-type endopeptidase activity), signal transduction and receptor binding (Tumor necrosis factor receptor superfamily binding and Transmembrane receptor protein tyrosine kinase activity), enzymatic activity and redox reactions (Oxidoreductase activity, acting on paired donors, with incorporation or reduction in molecular oxygen and Monooxygenase activity), and interactions with proteins and transcriptional regulators (Nuclear receptor binding, RNA polymerase II-specific DNA-binding transcription factor binding and p53 binding) (Figure 2C). Additionally, KEGG pathway analysis revealed that the key targets were primarily involved in signaling and disease-related pathways (AGE-RAGE signaling pathway in diabetic complications, p53 signaling pathway, Toll-like receptor signaling pathway, and MAPK signaling pathway), apoptosis and cancer-related pathways (Apoptosis, Hepatocellular carcinoma, Central carbon metabolism in cancer, and Lipid and atherosclerosis), and metabolic and biosynthesis pathways (Tyrosine metabolism, Steroid hormone biosynthesis, and Arginine and proline metabolism) (Figure 2D). These results from GO and KEGG pathway enrichment analyses suggest that the key targets play significant roles in signal transduction, apoptosis, and immune responses.

3.5. Screening of Core Targets

The ROC curve is commonly used to evaluate the diagnostic efficiency of core genes, with an AUC greater than 0.7 typically indicating high diagnostic value [24]. In this study, 60 key genes were evaluated using ROC curves, with AUC > 0.75 set as the screening threshold. The results indicated that two targets, CTSV and NOD1, had AUC values exceeding 0.75 (Figure 3A). To verify the reliability of the ROC curve results, box plots were employed to analyze the differential expression of these core targets between healthy and DM samples. It was observed that NOD1 was significantly upregulated in DM samples, whereas CTSV was significantly downregulated ($p < 0.05$) (Figure 3B). NOD1, a receptor involved in immune responses, can trigger inflammation by activating the NF-κB and MAPK signaling pathways, which interfere with insulin signaling and lead to insulin resistance [25]. In adipose tissue, the activation of NOD1 increases lipolysis and reduces insulin-stimulated glucose uptake, further exacerbating glucose metabolism disorders [26]. The aberrant activation of NOD1 has been strongly linked to chronic low-grade inflammation, a critical pathological hallmark of type 2 diabetes [27]. Consequently, NOD1 is believed to play a pivotal role in the development and progression of DM. Therefore, therapeutic strategies targeting NOD1 could hold significant potential for the treatment of DM by mitigating inflammation-associated disease mechanisms. CTSV, a cysteine protease mainly active in lysosomes, is involved in protein degradation [28]. CTSV plays a critical role in apoptosis, with its dysregulated expression accelerating the apoptosis of pancreatic β-cells, a key contributor to type 2 diabetes pathogenesis [29]. Thus, targeting CTSV may protect pancreatic islet function by reducing β-cell apoptosis, offering a promising therapeutic approach for DM management. Understanding the precise function of these core targets in DM pathology holds significant clinical relevance. Firstly, these core targets could serve as diagnostic biomarkers, facilitating early detection and risk assessment of DM.

Secondly, targeting these core genes and their related pathways may offer new avenues for the development of DM therapeutic strategies. Modulating their expression or activity could potentially aid in the regulation of immune responses and slow the progression of DM.

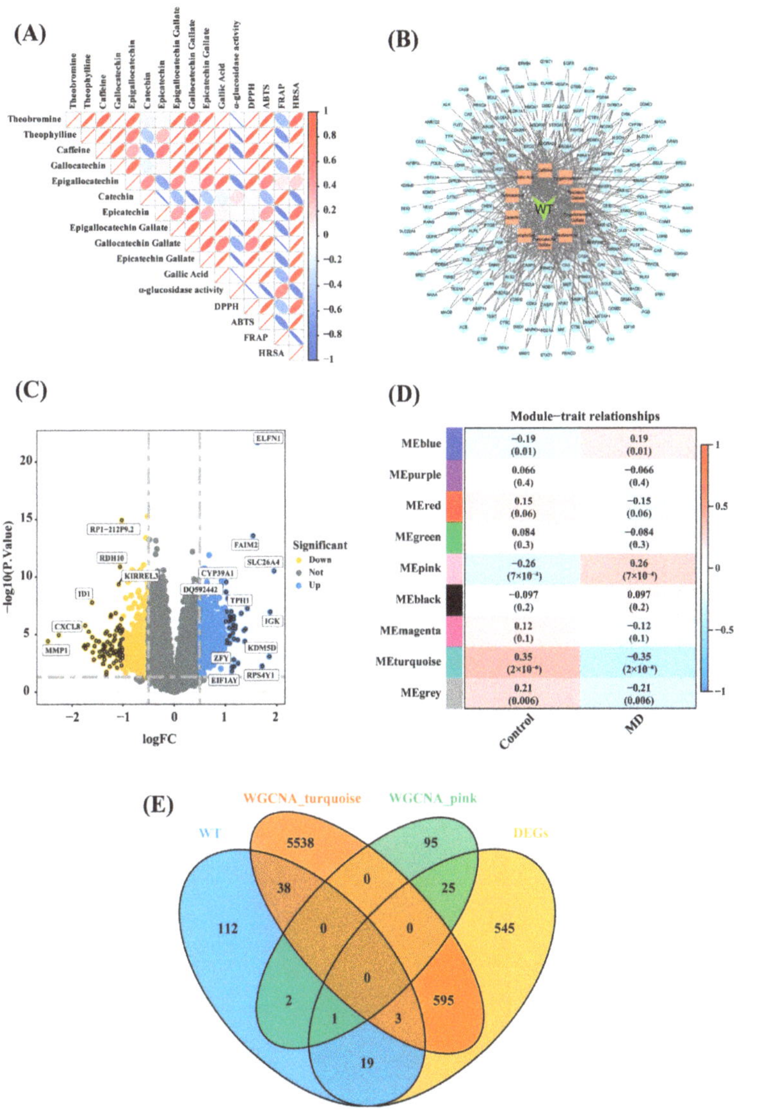

Figure 1. (**A**) The correlation between WT bioactive compounds and antioxidant activity and α-glucosidase inhibition indices. Correlation coefficients are displayed in a heatmap, where blue indicates a negative correlation, and red signifies a positive correlation; (**B**) A WT-bioactive compound-target network diagram. In the diagram, WT is shown in Yellow Green, bioactive compounds in Light Salmon, and their targets in Pale Turquoise. Each line connecting the compounds and targets represents potential interactions; (**C**) Volcano plot of differentially expressed genes (DEGs) from GSE76896. Blue dots represent upregulated genes, and yellow dots represent downregulated genes; (**D**) Heatmap showing correlations between gene co-expression modules and phenotypic traits. Red represents positive correlations, while blue represents negative correlations; (**E**) Venn diagram.

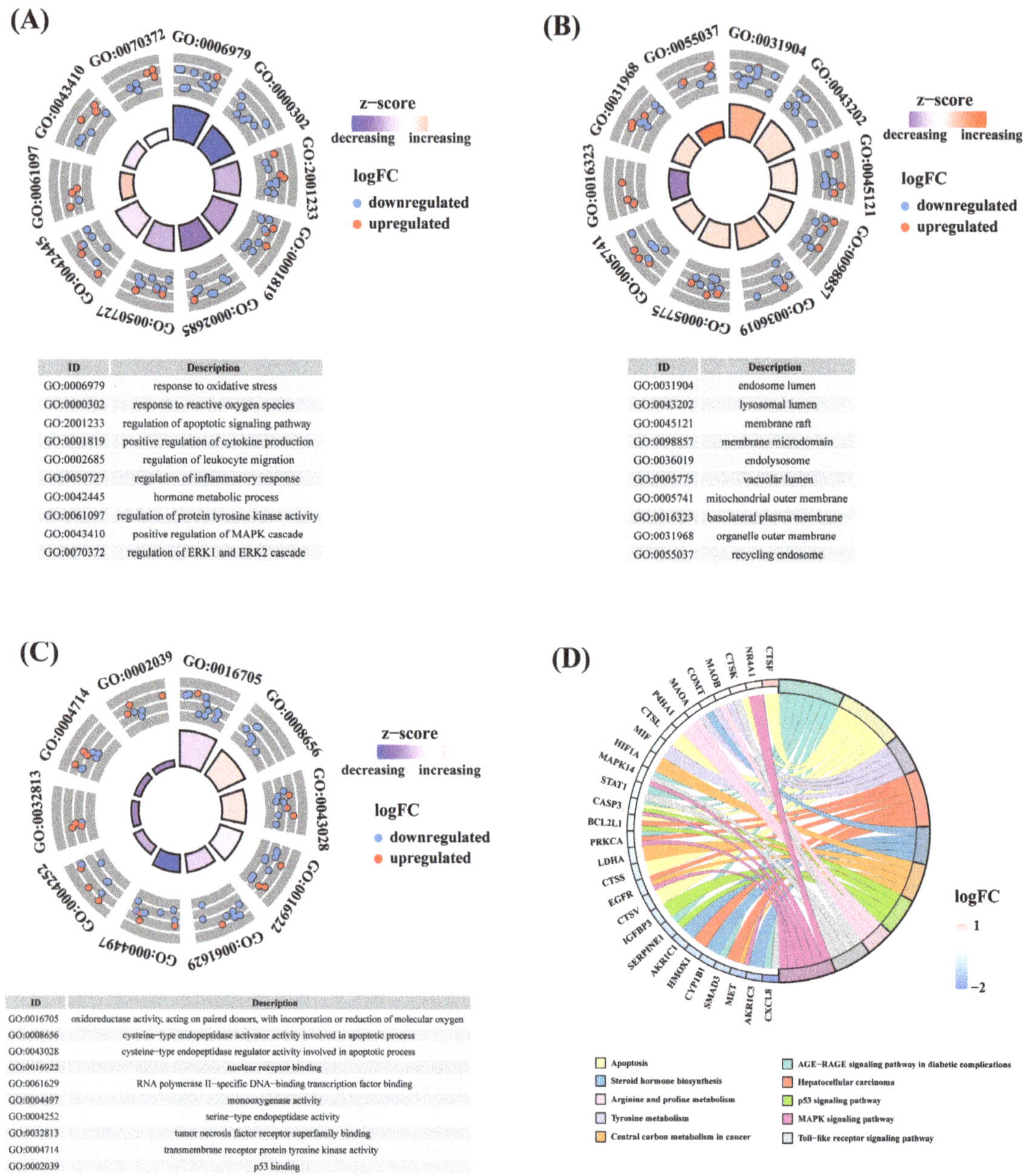

Figure 2. Functional enrichment analysis of key targets was performed as follows. (**A**) Biological process (BP) enrichment analysis of key targets; (**B**) Cellular component (CC) enrichment analysis of key targets; (**C**) Molecular function (MF) enrichment analysis of key targets; (**D**) KEGG pathway enrichment analysis of key targets.

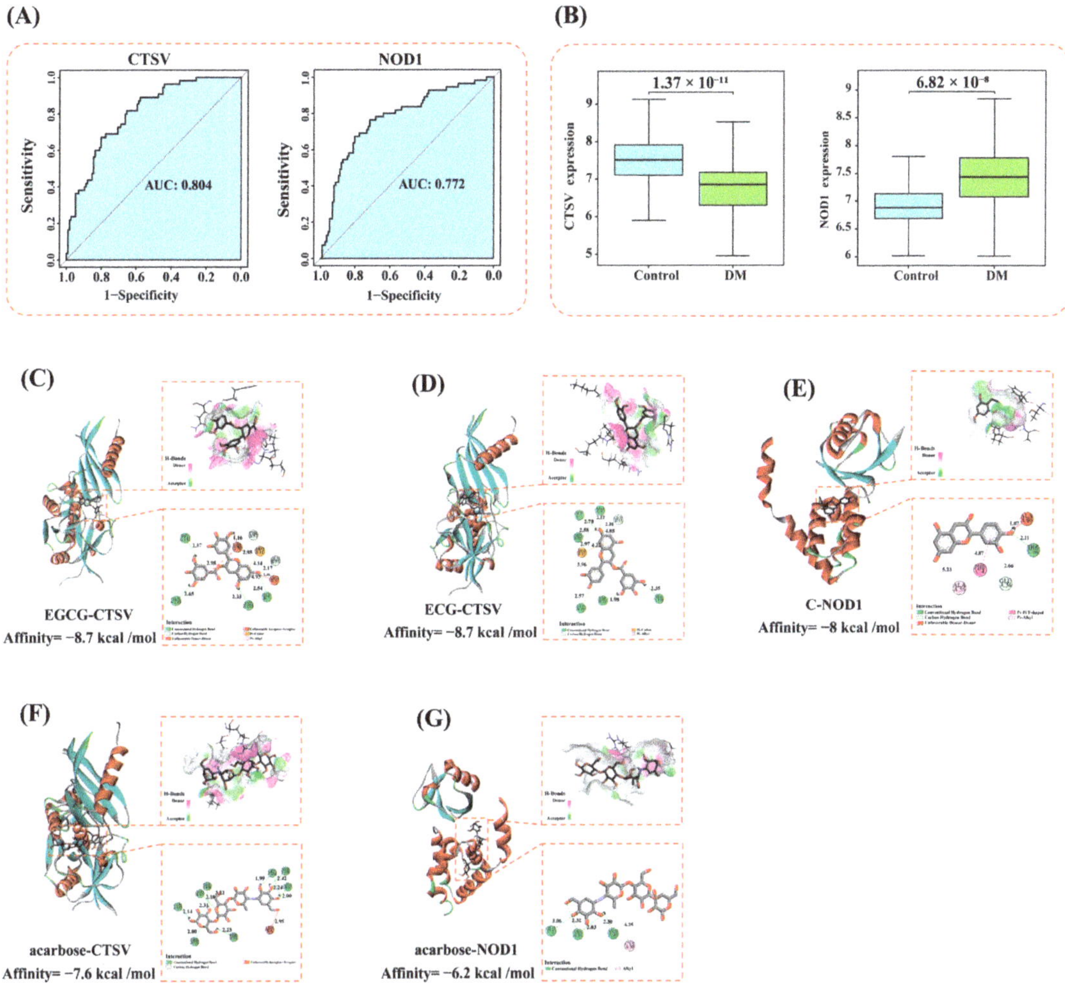

Figure 3. (**A**) Receiver Operating Characteristic (ROC) curve analysis of CTSV and NOD1; (**B**) Box plot comparing CTSV and NOD1 gene expression levels between the Control group (light blue) and DM group (light green), with error bars representing variability; (**C**) Interaction diagram of the EGCG-CTSV complex; (**D**) Interaction diagram of the ECG-CTSV complex; and (**E**) Interaction diagram of the C-NOD1 complex; (**F**) Interaction diagram of the acarbose-CTSV complex; (**G**) Interaction diagram of the acarbose—NOD1 complex.

3.6. Molecular Docking

Molecular docking is a computational method used to predict the binding modes and interaction strengths between proteins and small molecules. In this study, molecular docking was performed to investigate the interactions between ten active compounds of WT and two core targets, and their docking scores and binding modes were evaluated. It is generally accepted that a docking score of less than 0 kcal/mol indicates that the active compound can spontaneously bind to the target, a docking score below −5 kcal/mol suggests good binding activity, and a docking score below −7 kcal/mol indicates strong binding activity [30,31]. The results showed that the docking scores of the 10 active compounds of WT with the 2 core targets ranged from −5.3 kcal/mol to −8.7 kcal/mol (Supplementary Table S7). Among these, 10 combinations had docking scores between −5 kcal/mol and

−7 kcal/mol, and 10 combinations had docking scores below −7 kcal/mol, indicating strong binding activity between the WT active compounds and the 2 core targets. The docking scores of the reference drug acarbose with CTSV and NOD1 were −7.6 kcal/mol and −6.2 kcal/mol, respectively. Four active compounds demonstrated higher docking scores with CTSV than acarbose, while 10 active compounds exhibited higher docking scores with NOD1. This suggests that these 10 compounds are key contributors to the anti-DM effect of WT.

Among them, theobromine improves non-alcoholic fatty liver in obese mice by regulating the mTOR signaling pathway, lowering blood glucose and lipid levels, and additionally reducing blood glucose by inhibiting α-glucosidase activity [31,32]. Theophylline enhances insulin secretion and improves hypoglycemia awareness [33]. It has been reported that theophylline protects pancreatic β-cells through its immunosuppressive effects [34]. Caffeine lowers blood glucose and improves insulin sensitivity by reducing the expression of inflammatory factors and improving fatty liver [35]. When used in combination with anti-DM drugs, caffeine enhances drug bioavailability and significantly lowers blood glucose, demonstrating a synergistic anti-DM effect [36]. In addition, GC helps reduce postprandial hyperglycemia by inhibiting the activities of α-amylase and α-glucosidase [37]. C lowers blood glucose and alleviates DM symptoms by improving insulin sensitivity, inhibiting α-glucosidase activity, and enhancing antioxidant capacity. Furthermore, it reduces oxidative stress, preventing DM complications [38–40]. EGCG exhibits significant anti-DM effects, including lowering blood glucose levels by improving insulin sensitivity, inhibiting hepatic gluconeogenesis, and inhibiting α-glucosidase [41–43]. GCG suppresses carbohydrate absorption, enhances insulin sensitivity, and possesses anti-glycation properties [37,44]. ECG reduces the risk of DM complications and postprandial hyperglycemia by inhibiting protein glycation and the activities of α-amylase and α-glucosidase [45]. GA protects pancreatic cells and reduces DM-induced oxidative stress and complications by improving glucose metabolism and enhancing antioxidant activity [46,47]. GA, when combined with conventional drugs such as metformin, enhances the hypoglycemic effect [48].

Upon ranking the docking scores, it was found that EGCG and ECG were the best ligands for CTSV, with docking scores of −8.7 kcal/mol. C was the best ligand for NOD1, with a docking score of −8 kcal/mol. EGCG formed six hydrogen bonds with CTSV residues ASP163 (2.33 Å), GLY7 (2.54 Å), HIS141 (2.37 Å), GLN145 (2.65 Å), THR52 (2.95 Å), and GLN53 (2.17 Å), a hydrophobic interaction with ARG9 (4.93 Å), an electrostatic interaction with ARG11 (4.14 Å), and repulsive forces with THR58 (2.98 Å and 1.16 Å) and ARG9 (1.46 Å) (Figure 3C). ECG formed eight hydrogen bonds with CTSV residues GLY7 (2.70 Å), THR52 (2.58 Å and 2.97 Å), HIS141 (2.35 Å), THR58 (1.98 Å), ARG9 (2.12 Å), LYS19 (2.57 Å), and GLN53 (2.28 Å), a hydrophobic interaction with ARG9 (4.85 Å), and electrostatic interactions with ARG11 (3.96 Å and 4.22 Å) (Figure 3D). C formed two hydrogen bonds with NOD1 residues THR66 (2.11 Å) and GLU56 (2.66 Å), a hydrophobic interaction with ALA55 (5.23 Å) and PHE4 (4.87 Å), and repulsive forces with ALA60 (1.07 Å) (Figure 3E). In CTSV, acarbose formed ten hydrogen bonds with the residues ASN161 (2.42 Å), LEU162 (2.24 Å and 1.99 Å), GLN145 (2.14 Å), GLY140 (2.10 Å), THR54 (2.25 Å), THR58 (2.00 Å), ARG11 (2.31 Å), ARG9 (2.00 Å), and GLY140 (3.22 Å), as well as a hydrophobic interaction with LEU4 (2.95 Å) (Figure 3F). In NOD1, acarbose established four hydrogen bonds with ARG35 (2.30 Å), ASN36 (2.32 Å and 2.03 Å), and GLN2 (3.06 Å), along with a hydrophobic interaction with CYS39 (4.25 Å) (Figure 3G). In comparison, the binding of EGCG and ECG to CTSV relied on multiple hydrogen bonds, hydrophobic interactions, and electrostatic forces, demonstrating superior binding affinity compared to acarbose. C exhibited the strongest binding affinity to NOD1, primarily through hydrogen bonds and hydrophobic interactions. Although acarbose formed several hydrogen bonds, its binding affinity to CTSV and NOD1 was relatively weaker due to insufficient hydrophobic and electrostatic interactions.

EGCG and ECG have demonstrated significant binding affinity to CTSV, highlighting their potential as lead compounds for the development of anti-diabetic therapies targeting this protease. Given that CTSV plays a crucial role in regulating pancreatic β-cell apoptosis, inhibiting CTSV may help preserve β-cell function and maintain islet integrity. Both EGCG and ECG interact with multiple CTSV residues, reinforcing their therapeutic potential. However, the catechol groups in EGCG and ECG are part of pan-assay interference compounds (PAINS), which are prone to generating false-positive results in high-throughput screenings, as noted by Baell and Holloway [49]. This raises concerns about the reliability of initial pharmacological assessments.

To ensure the validity of their therapeutic effects, future research should focus on improving the bioavailability and absorption of EGCG and ECG while rigorously validating their anti-diabetic activity to eliminate the risk of PAINS-related artifacts. Further experimental studies are essential to accurately evaluate the therapeutic potential of these compounds. Additionally, strategies such as chemical modification or advanced drug delivery systems, including nanoparticles, could be explored to enhance the bioavailability and efficacy of EGCG and ECG in diabetes treatment. A deeper understanding of their molecular interactions will provide a critical foundation for the development of diabetes treatments based on these natural compounds.

3.7. Molecular Simulation

To further investigate the stability of protein-ligand interactions, molecular dynamics simulations were conducted on the EGCG-CTSV, ECG-CTSV, and C-NOD1 protein-ligand complexes. The root mean square deviation (RMSD) curve was used to reveal fluctuations in protein conformation, where RMSD values below 1 nm generally indicate the relative stability of protein-ligand interactions under physiological conditions [50]. The results showed that the RMSD of EGCG-CTSV, ECG-CTSV, and C-NOD1 fluctuated within 0.27 ± 0.03 Å, 0.22 ± 0.03 Å, and 0.28 ± 0.05 Å, respectively, suggesting that these complexes formed relatively tight interactions during binding (Figure 4A). The radius of gyration (Rg) was used to assess the structural stability of the protein-ligand complexes [51]. The Rg values for the three complexes were 2.05 ± 0.02 nm, 2.06 ± 0.01 nm, and 1.80 ± 0.02 nm, respectively, indicating that the complexes adopted a more compact and stable structure (Figure 4B). Solvent-accessible surface area (SASA) was employed to provide insights into protein folding and hydrophobicity [52]. The SASA values of the three complexes remained relatively stable at 146.85 ± 2.07 nm^2, 147.56 ± 2.04 nm^2, and 100.81 ± 1.97 nm^2, respectively, indicating that no significant structural changes occurred during the simulations (Figure 4C). The number of hydrogen bonds reflected the strength of protein-ligand binding [53], and as shown in Figure 4D, the hydrogen bond density and strength for EGCG-CTSV, ECG-CTSV, and C-NOD1 were consistently high.

To further explore the stability of protein-ligand interactions, MM/GBSA and MM/PBSA methods were used to evaluate the binding free energy of the three complexes. Both methods are widely applied to assess the binding potential of protein complexes. The MM/GBSA results (Table 2) indicated that the binding free energy of EGCG-CTSV, ECG-CTSV, and C-NOD1 was −25.48 kcal/mol, −27.52 kcal/mol, and −27.52 kcal/mol, respectively. The binding free energy is composed of van der Waals energy, electrostatic energy, solvation energy, and solvent-accessible surface area energy. The van der Waals energy for the three complexes was −44.67 kcal/mol, −40.04 kcal/mol, and −36.84 kcal/mol, respectively; the electrostatic energy was −4.80 kcal/mol, −7.5 kcal/mol, and −7.13 kcal/mol, respectively; the solvation energy was 28.78 kcal/mol, 24.78 kcal/mol, and 20.62 kcal/mol, respectively; and the solvent-accessible surface area energy was −4.78 kcal/mol, −4.71 kcal/mol, and −4.71 kcal/mol, respectively. These results suggest that van der Waals energy, electrostatic energy, and solvent-accessible surface area energy contributed to complex binding, while solvation energy hindered complex binding. The MM/PBSA results (Table 3) were similar to those of MM/GBSA, with ECG-CTSV showing the highest binding free energy, followed by C-NOD1 and EGCG-CTSV. The binding free energy for the three complexes was −26.93 kcal/mol, −25.39 kcal/mol, and −23.29 kcal/mol, respectively.

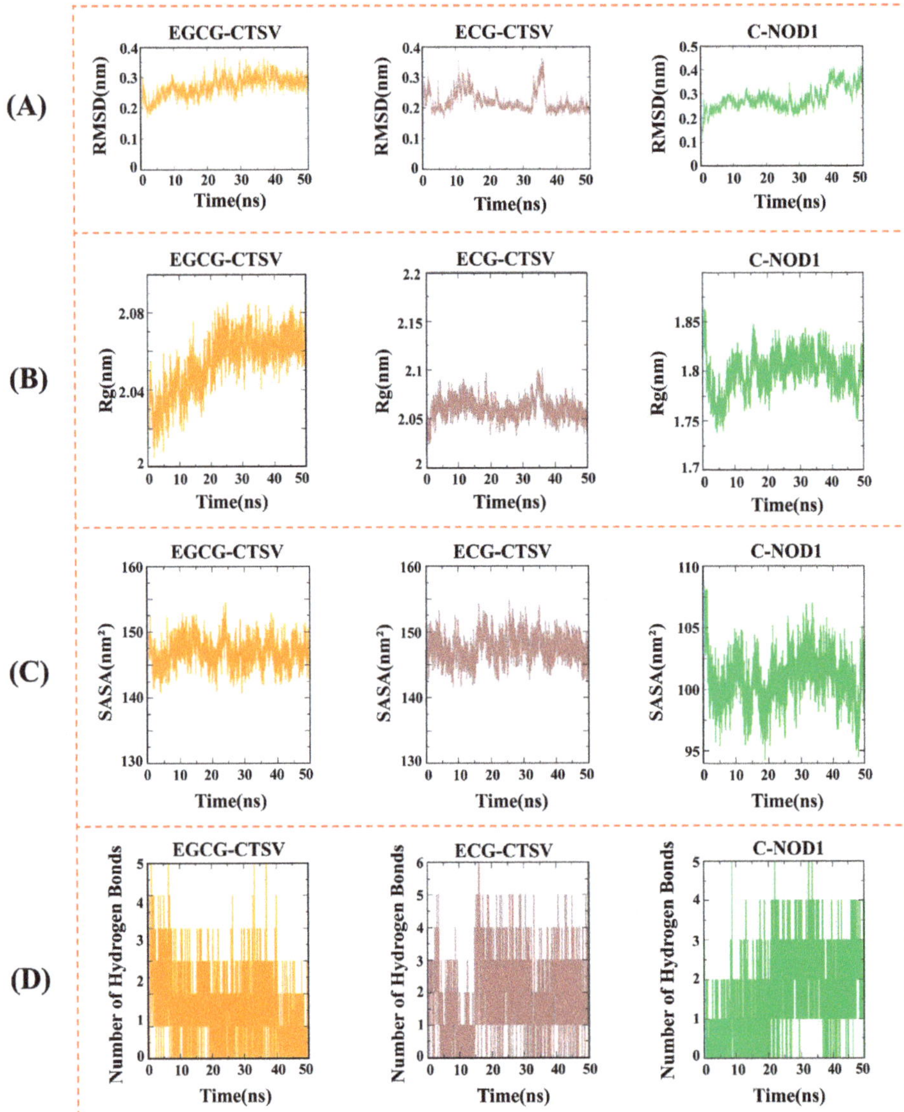

Figure 4. Molecular dynamics simulations of the three complexes were performed as follows: (**A**) Root Mean Square Deviation (RMSD) of the three complexes; (**B**) Radius of gyration (Rg) of the three complexes; (**C**) Solvent Accessible Surface Area (SASA) of the three complexes; and (**D**) Number of hydrogen bonds in the three complexes.

In conclusion, molecular dynamics simulations revealed that the EGCG-CTSV, ECG-CTSV, and C-NOD1 complexes exhibited stable binding, characterized by favorable contributions from van der Waals, electrostatic, solvation, and solvent-accessible surface area energies.

Table 2. The binding energy of the ligand-receptor complex under the MM/GBSA method is expressed in kcal/mol.

Parameters	EGCG-CTSV	ECG-CTSV	C-NOD1
Delta Van Der Waals Energy	−44.67 ± 0.66	−40.04 ± 2.58	−36.84 ± 0.14
Delta Electrostatic Energy	−4.80 ± 2.41	−7.56 ± 1.55	−7.13 ± 2.04
Delta Generalized Born Energy	28.78 ± 0.21	24.78 ± 0.30	20.62 ± 0.53
Delta Solvent Accessible Surface Area Energy	−4.78 ± 0.02	−4.71 ± 0.23	−4.17 ± 0.15
Total Binding Energy	−25.48 ± 2.51	−27.52 ± 3.03	−27.52 ± 2.11

Table 3. The binding energy of the ligand-receptor complex under the MM/PBSA method is expressed in kcal/mol.

Parameters	EGCG-CTSV	ECG-CTSV	C-NOD1
Delta Van Der Waals Energy	−44.67 ± 0.66	−40.04 ± 2.58	−36.84 ± 0.14
Delta Electrostatic Energy	−4.80 ± 2.41	−7.56 ± 1.55	−7.13 ± 2.04
Delta Polar Solvation Energy	29.67 ± 0.72	24.08 ± 1.35	21.34 ± 0.63
Delta Non-Polar Solvation Energy	−3.51 ± 0.02	−3.42 ± 0.16	−2.76 ± 0.03
Total Binding Energy	−23.29 ± 2.60	−26.93 ± 3.30	−25.39 ± 2.14

4. Conclusions

This study employed a multi-faceted approach to systematically investigate the anti-DM active compounds in WT and their potential mechanisms. This approach included polyphenol-targeted metabolomics, in vitro antioxidant and α-glucosidase activity assays, network pharmacology, GEO data analysis, molecular docking, and molecular dynamics simulations. These findings, combined with antioxidant and α-glucosidase activity assays, identified 10 key active compounds in WT that exhibit significant anti-DM activity. Additionally, network pharmacology, GEO data analysis, and ROC analysis highlighted CTSV and NOD1 as critical targets of these compounds, implicating pathways related to signal transduction, apoptosis, and immune response in their anti-DM effects. Crucially, molecular docking revealed that these active compounds bind to the core targets via hydrogen bonding and hydrophobic interactions. Molecular dynamics simulations further supported the stability of these protein–ligand interactions.

The previous research has demonstrated that WT can regulate blood glucose levels and improve insulin sensitivity. This study systematically evaluated the antidiabetic potential of WT, identifying its key bioactive compounds, target proteins, and related mechanisms of action. These findings provide valuable insights for developing diabetes treatments based on natural products. Despite the significant discoveries revealed in this study, the limitations of computational simulations should be acknowledged. The complexity of biological systems makes it difficult for models to fully replicate all variables and interactions. Therefore, future research should aim to further validate and explore the mechanisms by which these 10 active compounds regulate diabetes-related pathways through in vitro and in vivo studies, as well as molecular biology approaches.

Supplementary Materials: The following supporting information can be downloaded at: https://www.mdpi.com/article/10.3390/foods13213354/s1, Supplementary Figure S1. Targeted Metabolomics Profiling of Polyphenols in White Tea; Supplementary Figure S2. Boxplot of Standardized GSE76896 Dataset; Supplementary Table S1. Targeted Metabolomics Analysis of Polyphenol Content in White Tea; Supplementary Table S2. Correlation between Active Components of White Tea, Antioxidant Activity, and α-Glucosidase Activity; Supplementary Table S3. Target Proteins of 10 Active Components in White Tea; Supplementary Table S4. Sample Information for GSE76896; Supplementary Table S5. Differentially Expressed Genes Information; Supplementary Table S6. Intersection Analysis; Supplementary Table S7. Docking Scores of 10 Active Compounds from White Tea and Positive Control Drugs with Two Core Targets.

Author Contributions: Conceptualization, Writing—original draft, Visualization, W.W.; Formal analysis, Investigation, Z.Z.; Software, Visualization, Z.W.; Data curation, Visualization, C.G.; Data curation, Formal analysis, Y.L.; Visualization, Investigation, W.Z.; Conceptualization, Writing—review & editing, Funding acquisition, Project administration, W.S. All authors have read and agreed to the published version of the manuscript.

Funding: This work was supported by the National Natural Science Foundation of China (Grant No. 32372769) and the Technology Innovation of Fujian Zhang Tianfu Tea Development Foundation (FJZTF01). The funders had no role in the design, determination, and interpretation of data, or in writing the manuscript.

Institutional Review Board Statement: Not Applicable.

Informed Consent Statement: Not Applicable.

Data Availability Statement: The original contributions presented in the study are included in the article/Supplementary Material, further inquiries can be directed to the corresponding author.

Conflicts of Interest: The authors declare no conflicts of interest.

References

1. Zhang, C.; Zhou, C.; Xu, K.; Tian, C.; Zhang, M.; Lu, L.; Zhu, C.; Lai, Z.; Guo, Y. A Comprehensive Investigation of Macro-Composition and Volatile Compounds in Spring-Picked and Autumn-Picked White Tea. *Foods* **2022**, *11*, 3628. [CrossRef] [PubMed]
2. Zhang, L.; Dai, H.; Zhang, J.; Zheng, Z.; Song, B.; Chen, J.; Lin, G.; Chen, L.; Sun, W.; Huang, Y. A Study on Origin Traceability of White Tea (White Peony) Based on Near-Infrared Spectroscopy and Machine Learning Algorithms. *Foods* **2023**, *12*, 499. [CrossRef]
3. Zhou, S.; Zhang, J.; Ma, S.; Ou, C.; Feng, X.; Pan, Y.; Gong, S.; Fan, F.; Chen, P.; Chu, Q. Recent Advances on White Tea: Manufacturing, Compositions, Aging Characteristics and Bioactivities. *Trends Food Sci. Technol.* **2023**, *134*, 41–55. [CrossRef]
4. Essaouiba, A.; Jellali, R.; Gilard, F.; Gakière, B.; Okitsu, T.; Legallais, C.; Sakai, Y.; Leclerc, E. Investigation of the Exometabolomic Profiles of Rat Islets of Langerhans Cultured in Microfluidic Biochip. *Metabolites* **2022**, *12*, 1270. [CrossRef]
5. Teja, C.; Babu, S.N.; Noor, A.; Daniel, J.A.; Devi, S.A.; Khan, F.R. Cu/TEMPO Catalyzed Dehydrogenative 1,3-Dipolar Cycloaddition in the Synthesis of Spirooxindoles as Potential Antidiabetic Agents. *RSC Adv.* **2020**, *10*, 12262–12271. [CrossRef]
6. Hogan, J.P.; Peercy, B.E. Flipping the Switch on the Hub Cell: Islet Desynchronization through Cell Silencing. *PLoS ONE* **2021**, *16*, e0248974. [CrossRef] [PubMed]
7. Jacobs, E.; Hoyer, A.; Brinks, R.; Icks, A.; Kuß, O.; Rathmann, W. Healthcare Costs of Type 2 Diabetes in Germany. *Diabet. Med.* **2017**, *34*, 855–861. [CrossRef]
8. Esposito, F.; Pala, N.; Carcelli, M.; Boateng, S.T.; D'Aquila, P.S.; Mariani, A.; Satta, S.; Chamcheu, J.C.; Sechi, M.; Sanna, V. α-Glucosidase Inhibition by Green, White and Oolong Teas: In Vitro Activity and Computational Studies. *J. Enzyme Inhib. Med. Chem.* **2023**, *38*, 2236802. [CrossRef]
9. Zhang, X.; Huo, Z.; Jia, X.; Xiong, Y.; Li, B.; Zhang, L.; Li, X.; Li, X.; Fang, Y.; Dong, X.; et al. (+)-Catechin Ameliorates Diabetic Nephropathy Injury by Inhibiting Endoplasmic Reticulum Stress-Related NLRP3-Mediated Inflammation. *Food Funct.* **2024**, *15*, 5450–5465. [CrossRef]
10. Ni, Y.; Zhu, D.; Chen, C.; Wang, F.; Miu, Y.; Zhang, W. (+)-Catechins Play a Protective Role in Diabetic Kidney Disease by Alleviating EMT through Multiple Pathways. *Mol. Nutr. Food Res.* **2024**, *68*, e2400387. [CrossRef]
11. Cai, J.; Qiu, Z.; Liao, J.; Li, A.; Chen, J.; Wu, Z.; Khan, W.; Sun, B.; Liu, S.; Zheng, P. Comprehensive Analysis of the Yield and Leaf Quality of Fresh Tea (*Camellia sinensis* cv. Jin Xuan) under Different Nitrogen Fertilization Levels. *Foods* **2024**, *13*, 2091. [CrossRef] [PubMed]
12. Wen, X.; Lv, C.; Zhou, R.; Wang, Y.; Zhou, X.; Qin, S. The Molecular Mechanism Underlying the Therapeutic Effect of Dihydromyricetin on Type 2 Diabetes Mellitus Based on Network Pharmacology, Molecular Docking, and Transcriptomics. *Foods* **2024**, *13*, 344. [CrossRef] [PubMed]
13. Tuo, Y.; Lu, X.; Tao, F.; Tukhvatshin, M.; Xiang, F.; Wang, X.; Shi, Y.; Lin, J.; Hu, Y. The Potential Mechanisms of Catechins in Tea for Anti-Hypertension: An Integration of Network Pharmacology, Molecular Docking, and Molecular Dynamics Simulation. *Foods* **2024**, *13*, 2685. [CrossRef] [PubMed]
14. Peng, M.; Gao, Z.; Liao, Y.; Guo, J.; Shan, Y. Development of Citrus-Based Functional Jelly and an Investigation of Its Anti-Obesity and Antioxidant Properties. *Antioxidants* **2022**, *11*, 2418. [CrossRef]
15. Yan, Z.; Zhou, Z.; Jiao, Y.; Huang, J.; Yu, Z.; Zhang, D.; Chen, Y.; Ni, D. Hot-Air Drying Significantly Improves the Quality and Functional Activity of Orange Black Tea Compared with Traditional Sunlight Drying. *Foods* **2023**, *12*, 1913. [CrossRef]
16. Chen, B.; Ma, Y.; Li, H.; Chen, X.; Zhang, C.; Wang, H.; Deng, Z. The Antioxidant Activity and Active Sites of Delphinidin and Petunidin Measured by DFT, in Vitro Chemical-Based and Cell-Based Assays. *J. Food Biochem.* **2019**, *43*, e12968. [CrossRef]
17. Jiao, Y.; Song, Y.; Yan, Z.; Wu, Z.; Yu, Z.; Zhang, D.; Ni, D.; Chen, Y. The New Insight into the Effects of Different Fixing Technology on Flavor and Bioactivities of Orange Dark Tea. *Molecules* **2023**, *28*, 1079. [CrossRef]

18. Yu, L.; Bian, C.; Zhu, N.; Shen, Y.; Yuan, H. Enhancement of Methane Production from Anaerobic Digestion of Waste Activated Sludge with Choline Supplement. *Energy* **2019**, *173*, 1021–1029. [CrossRef]
19. Chen, Y.-H.; Zhang, Y.-H.; Chen, G.-S.; Yin, J.-F.; Chen, J.-X.; Wang, F.; Xu, Y.-Q. Effects of Phenolic Acids and Quercetin-3-O-Rutinoside on the Bitterness and Astringency of Green Tea Infusion. *NPJ Sci. Food* **2022**, *6*, 8. [CrossRef]
20. Okubo, H.; Syddall, H.E.; Phillips, D.I.W.; Sayer, A.A.; Dennison, E.M.; Cooper, C.; Robinson, S.M. Dietary Total Antioxidant Capacity Is Related to Glucose Tolerance in Older People: The Hertfordshire Cohort Study. *Nutr. Metab. Cardiovasc. Dis.* **2013**, *24*, 301–308. [CrossRef]
21. Sun, H.; Wang, D.; Song, X.; Zhang, Y.; Ding, W.; Peng, X.; Zhang, X.; Li, Y.; Ma, Y.; Wang, R.; et al. Natural Prenylchalconaringenins and Prenylnaringenins as Antidiabetic Agents: α-Glucosidase and α-Amylase Inhibition and in Vivo Antihyperglycemic and Antihyperlipidemic Effects. *J. Agric. Food Chem.* **2017**, *65*, 1574–1581. [CrossRef]
22. Zhou, B.; Wang, Z.; Yin, P.; Ma, B.; Ma, C.; Xu, C.; Wang, J.; Wang, Z.; Yin, D.; Xia, T. Impact of Prolonged Withering on Phenolic Compounds and Antioxidant Capability in White Tea Using LC-MS-Based Metabolomics and HPLC Analysis: Comparison with Green Tea. *Food Chem.* **2021**, *368*, 130855. [CrossRef] [PubMed]
23. Cheng, J.; Zhang, S.; Fan, A.; Li, Y.; Xu, P.; Huang, J.; He, M.; Wang, H. An Immune-related Gene Signature for the Prognosis of Human Bladder Cancer Based on WGCNA. *Comput. Biol. Med.* **2022**, *151*, 106186. [CrossRef] [PubMed]
24. Prakash, O.; Bettadiah, B.K.; Kudachikar, V.B. Chemical Composition and in Vitro Antihyperglycemic Potential of Kainth Fruit (*Pyrus pashia* Buch.-Ham Ex D. Don). *Food Biosci.* **2021**, *42*, 101119. [CrossRef]
25. Rivers, S.L.; Klip, A.; Giacca, A. NOD1: An Interface Between Innate Immunity and Insulin Resistance. *Endocrinology* **2019**, *160*, 1021–1030. [CrossRef]
26. Gulzar, F.; Ahmad, S.; Singh, S.; Kumar, P.; Sharma, A.; Tamrakar, A.K. NOD1 Activation in 3T3-L1 Adipocytes Confers Lipid Accumulation in HepG2 Cells. *Life Sci.* **2023**, *316*, 121400. [CrossRef]
27. Qiu, J.; Wu, J.; Chen, W.; Ruan, Y.; Mao, J.; Li, S.; Tang, X.; Zhao, L.; Li, S.; Li, K.; et al. NOD1 Deficiency Ameliorates the Progression of Diabetic Retinopathy by Modulating Bone Marrow–Retina Crosstalk. *Stem Cell Res. Ther.* **2024**, *15*, 1–17. [CrossRef]
28. Gureeva, T.A.; Timoshenko, O.S.; Kugaevskaya, E.V.; Solovyova, N.I. Cysteine Cathepsins: Structure, Physiological Functions, and the Role in Carcinogenesis. *Biochem. Mosc. Suppl. B Biomed. Chem.* **2022**, *16*, 91–103. [CrossRef]
29. Yang, L.; Zeng, Q.; Deng, Y.; Qiu, Y.; Yao, W.; Liao, Y. Glycosylated Cathepsin V Serves as a Prognostic Marker in Lung Cancer. *Front. Oncol.* **2022**, *12*, 876245. [CrossRef]
30. Zhou, X.; Ni, J.; Ge, W.; Wang, X.; Li, Y.; Wang, H.; Ma, C. Machine Learning-Based Virtual Screening of Multi-Target Anti-Obesity Compounds from Medicinal and Edible Plants: A Combined in Silico and in Vitro Study. *Food Biosci.* **2024**, *59*, 104077. [CrossRef]
31. Wei, D.; Wu, S.; Liu, J.; Zhang, X.; Guan, X.; Gao, L.; Xu, Z. Theobromine Ameliorates Nonalcoholic Fatty Liver Disease by Regulating Hepatic Lipid Metabolism via MTOR Signaling Pathway in Vivo and in Vitro. *Can. J. Physiol. Pharmacol.* **2020**, *99*, 775–785. [CrossRef] [PubMed]
32. Indrianingsih, A.W.; Wulanjati, M.P.; Windarsih, A.; Bhattacharjya, D.K.; Suzuki, T.; Katayama, T. In Vitro Studies of Antioxidant, Antidiabetic, and Antibacterial Activities of *Theobroma cacao*, *Anonna muricata* and *Clitoria ternatea*. *Biocatal. Agric. Biotechnol.* **2021**, *33*, 101995. [CrossRef]
33. de Galan, B.E.; Tack, C.J.; Lenders, J.W.; Pasman, J.W.; Elving, L.D.; Russel, F.G.; Lutterman, J.A.; Smits, P. Theophylline Improves Hypoglycemia Unawareness in Type 1 Diabetes. *Diabetes* **2002**, *51*, 790–796. [CrossRef]
34. Rabinovitch, A.; Sumoski, W.L. Theophylline Protects against Diabetes in BB Rats and Potentiates Cyclosporine Protection. *Diabetologia* **1990**, *33*, 506–508. [CrossRef]
35. Yamauchi, R.; Kobayashi, M.; Matsuda, Y.; Ojika, M.; Shigeoka, S.; Yamamoto, Y.; Tou, Y.; Inoue, T.; Katagiri, T.; Murai, A.; et al. Coffee and Caffeine Ameliorate Hyperglycemia, Fatty Liver, and Inflammatory Adipocytokine Expression in Spontaneously Diabetic KK-Ay Mice. *J. Agric. Food Chem.* **2010**, *58*, 5597–5603. [CrossRef]
36. Alshabi, A.M.; Alkahtani, S.A.; Shaikh, I.A.; Habeeb, M.S. Caffeine Modulates Pharmacokinetic and Pharmacodynamic Profiles of Pioglitazone in Diabetic Rats. *Saudi Med. J.* **2021**, *42*, 151–160. [CrossRef]
37. Wu, X.; Ding, H.; Hu, X.; Pan, J.; Liao, Y.; Gong, D.; Zhang, G. Exploring Inhibitory Mechanism of Gallocatechin Gallate on A-Amylase and a-Glucosidase Relevant to Postprandial Hyperglycemia. *J. Funct. Foods* **2018**, *48*, 200–209. [CrossRef]
38. Nazir, N.; Zahoor, M.; Ullah, R.; Ezzeldin, E.; Mostafa, G.A.E. Curative Effect of Catechin Isolated from *Elaeagnus Umbellata* Thunb. Berries for Diabetes and Related Complications in Streptozotocin-Induced Diabetic Rats Model. *Molecules* **2020**, *26*, 137. [CrossRef]
39. Mrabti, H.N.; Jaradat, N.; Fichtali, I.; Ouedrhiri, W.; Jodeh, S.; Ayesh, S.; Cherrah, Y.; Faouzi, M.E.A. Separation, Identification, and Antidiabetic Activity of Catechin Isolated from *Arbutus unedo* L. Plant Roots. *Plants* **2018**, *7*, 31. [CrossRef]
40. Ihm, S.-H.; Lee, J.-O.; Kim, S.-J.; Seung, K.-B.; Schini-Kerth, V.B.; Chang, K.; Oak, M.-H. Catechin Prevents Endothelial Dysfunction in the Prediabetic Stage of OLETF Rats by Reducing Vascular NADPH Oxidase Activity and Expression. *Atherosclerosis* **2009**, *206*, 47–53. [CrossRef]
41. Zhang, Z.F.; Li, Q.; Liang, J.; Dai, X.Q.; Ding, Y.; Wang, J.B.; Li, Y. Epigallocatechin-3-O-Gallate (EGCG) Protects the Insulin Sensitivity in Rat L6 Muscle Cells Exposed to Dexamethasone Condition. *Phytomedicine* **2009**, *17*, 14–18. [CrossRef]
42. Waltner-Law, M.E.; Wang, X.L.; Law, B.K.; Hall, R.K.; Nawano, M.; Granner, D.K. Epigallocatechin Gallate, a Constituent of Green Tea, Represses Hepatic Glucose Production. *J. Biol. Chem.* **2002**, *277*, 34933–34940. [CrossRef] [PubMed]

43. Li, T.; Liu, J.; Zhang, X.; Ji, G. Antidiabetic Activity of Lipophilic (-)-Epigallocatechin-3-Gallate Derivative under Its Role of Alpha-Glucosidase Inhibition. *Biomed. Pharmacother.* **2007**, *61*, 91–96. [CrossRef]
44. Ma, H.; Liu, W.; Frost, L.; Kirschenbaum, L.J.; Dain, J.A.; Seeram, N.P. Glucitol-Core Containing Gallotannins Inhibit the Formation of Advanced Glycation End-Products Mediated by Their Antioxidant Potential. *Food Funct.* **2016**, *7*, 2213–2222. [CrossRef]
45. Wu, X.; Zhang, G.; Hu, X.; Pan, J.; Liao, Y.; Ding, H. Inhibitory Effect of Epicatechin Gallate on Protein Glycation. *Food Res. Int.* **2019**, *122*, 230–240. [CrossRef]
46. Huang, D.-W.; Chang, W.-C.; Wu, J.S.-B.; Shih, R.-W.; Shen, S.-C. Gallic Acid Ameliorates Hyperglycemia and Improves Hepatic Carbohydrate Metabolism in Rats Fed a High-Fructose Diet. *Nutr. Res.* **2016**, *36*, 150–160. [CrossRef]
47. Punithavathi, V.R.; Prince, P.S.M.; Kumar, R.; Selvakumari, J. Antihyperglycaemic, Antilipid Peroxidative and Antioxidant Effects of Gallic Acid on Streptozotocin Induced Diabetic Wistar Rats. *Eur. J. Pharmacol.* **2010**, *650*, 465–471. [CrossRef]
48. Oboh, G.; Ogunbadejo, M.D.; Ogunsuyi, O.B.; Oyeleye, S.I. Can Gallic Acid Potentiate the Antihyperglycemic Effect of Acarbose and Metformin? Evidence from Streptozotocin-Induced Diabetic Rat Model. *Arch. Physiol. Biochem.* **2020**, *128*, 619–627. [CrossRef] [PubMed]
49. Baell, J.B.; Holloway, G.A. New Substructure Filters for Removal of Pan Assay Interference Compounds (PAINS) from Screening Libraries and for Their Exclusion in Bioassays. *J. Med. Chem.* **2010**, *53*, 2719–2740. [CrossRef]
50. Wu, Q.; Xu, C.; Tian, Y.; Han, A.; Liu, Y. Simulation Screening of Umami Peptides in the Microbiota of Soy Sauce Based on Molecular Docking and Molecular Dynamics. *Food Biosci.* **2024**, *61*, 104722. [CrossRef]
51. Jin, M.; Sun, K.; Guo, S.; Wang, Y.; Jin, J.; Qi, H. Novel Stabilization of Chlorophyll by Sea Cucumber (*Apostichopus japonicas*) Protein: Multi-Spectral and Molecular Dynamics Analysis. *Food Biosci.* **2024**, *61*, 104861. [CrossRef]
52. Li, B.; Jiang, G.; Li, S.; Zhu, J.; Ding, X.; Song, Y.; Ding, Z. Integrating Label-Free Proteomics and Molecular Dynamics to Investigate the Interactions between Proteins and Bioactive Compounds in Rosa Roxburghii Tratt Fermented by *Lactobacillus acidophilus* GIM1.208. *Food Biosci.* **2024**, *58*, 103748. [CrossRef]
53. Dong, Y.; Chang, R.; Ji, Z.; Xu, Y.; Ren, Q.; Zhou, Z.; Mao, J. Unraveling Umami Complexity: From Exploring Umami Peptides in Fermented Soybean Curd to Molecular Elucidation of Taste Mechanisms. *Food Biosci.* **2024**, *59*, 103951. [CrossRef]

foods

MDPI

Article

Dynamic Changes in Aroma Compounds during Processing of Flat Black Tea: Combined GC-MS with Proteomic Analysis

Cun Ao [1], Xiaojun Niu [1], Daliang Shi [1], Xuxia Zheng [1], Jizhong Yu [1,*] and Yingbin Zhang [2,*]

[1] Tea Research Institute, Hangzhou Academy of Agricultural Science, Hangzhou 310024, China; aocun123@163.com (C.A.); xiaojunwords@126.com (X.N.); sdl7698@126.com (D.S.); xxzheng43@163.com (X.Z.)
[2] Tea Research Institute, Chinese Academy of Agricultural Sciences, Hangzhou 310008, China
* Correspondence: hchyu@126.com (J.Y.); zhyb@tricaas.com (Y.Z.)

Abstract: Flat black tea (FBT) has been innovatively developed to alleviate homogenisation competition, but the dynamic changes in aroma components during the process remain unclear. This study employed HS-SPME-GC-MS to analyse the aroma components of tea samples from various processing stages of FBT, and to make a comparative assessment with conventional strip-like Congou black tea (SBT). Additionally, a proteomic analysis was conducted on fresh leaves, withered leaves, and frozen–thawed leaves. Significant changes were observed in the aroma components and proteins during the processing. The results of the multivariate and odour activity value analysis demonstrated that the principal aroma components present during the processing of FBT were linalool, (E)-2-hexen-1-al, methyl salicylate, geraniol, hexanal, benzeneacetaldehyde, (Z)-3-hexenyl butyrate, dimethyl sulphide, 2-methylbutanal, 2-ethylfuran, nonanal, nonanol, 3-methylbutanal, (Z)-3-hexen-1-ol, 2-pentylfuran, linalool oxide I, and β-myrcene. Freezing–thawing and final roasting are the key processing steps for forming the aroma quality of FBT. The final roasting yielded a considerable quantity of pyrazines and pyrroles, resulting in a high-fried aroma, but caused a significant reduction in linalool, geraniol, β-myrcene, and esters, which led to a loss of floral and fruity aromas. The freezing–thawing treatment resulted in an accelerated loss of aroma substances, accompanied by a decrease in the expression level of lipoxygenase and 2-C-methyl-D-erythritol 2,4-cyclodiphosphate synthase. The formation of aroma substances in the linoleic acid metabolic pathway and terpenoid metabolic process was hindered, which had a negative impact on tea aroma. This study elucidates the causes of unsatisfactory aroma quality in tea products made from frozen tea leaves, providing theoretical support for the utilisation of frostbitten tea leaves, and helps us to understand the mechanism of aroma formation in black tea.

Keywords: black tea; aroma compound; dynamic change; processing technology

1. Introduction

Black tea is the most widely produced tea in the world, ranking second in production in China. The China Tea Marketing Association has reported that the production of Chinese black tea in 2023 reached 491,200 tons, with an output value of CNY 51.97 billion. Of this total, 379,000 tons was sold domestically [1]. Black tea is classified into three main categories: broken black tea, Congou black tea, and Souchong black tea. Congou black tea represents the most consumed category of black tea in China, characterised by a sweet and mellow taste, as well as intact leaves [2,3]. The primary processing techniques employed in the production of Congou black tea include withering, rolling, fermentation, and drying [4]. The tea cells are crushed through rolling, which gives rise to the quality characteristics of red soup and red leaves in fermentation [5]. Notable Congou black teas produced in China include Keemun black tea, Dianhong black tea, Jinjunmei black tea, Yingde black tea, Jiuquhongmei tea, and others. The quality and flavour of these teas vary according to the variety and origin [6].

Aroma is the second most important factor in the evaluation of black tea quality after taste [7]. It is a significant determinant of consumer preferences and product choices.

The principal aroma types of black tea are sweet, floral, fruity, honey, fired, and smoky pine [8]. The formation of different aroma types is dependent on the specific types and proportions of the aroma components. Over 400 varieties of volatile compounds have been identified in black tea [9]. However, only a small number of these compounds have been found to exhibit aroma characteristics and are, therefore, considered to be odour-active substances [10]. Keemun black tea is considered to be among the world's four most highly regarded black teas due to its distinctive aroma, which is known as the "Keemun aroma". The primary aroma-active compounds responsible for this characteristic aroma are geraniol, linalool, benzeneaceetalhydrate, 2-pentylfuran, and methyl salicylate [11]. The key volatile substances responsible for the floral and caramel aromas in Dianhong black tea are identified as linalool, geraniol, phenylethanol, and 3-ethyl-2,5-dimethylpyrazine [12]. The floral and fruity aroma components of Jinmudan black tea were identified as linalool, phenylacetaldehyde, and δ-decalactone [13]. The principal components of the floral and honey-like fragrance in Hunan black tea are nonanal, nerolidol, benzyl alcohol, and phenylethanol [14].

The aroma of tea is subject to a number of influences, including the variety of tea trees, the ecological environment, and the processing technology employed [15]. However, the application of tea production technology, along with the use of different tea varieties, cultivation measures, and processing techniques, has led to a significant increase in product homogeneity. Consequently, some product developers have sought inspiration in the shaping process of green tea to develop distinctive black tea products with the aim of stimulating market consumption. They have successfully developed black tea products with a flat, needle-like, whorl-like, or granular appearance [16,17]. Researchers had previously attempted to card tea leaves into a straight shape after rolling and then flattening them to process FBT. However, the rolling process resulted in a curved and wrinkled shape, which was not sufficiently smooth and straight. FBT with a flat and straight shape has been achieved through technological innovation. The carding treatment was conducted with pressure, rather than rolling, to crush the tea cells, in conjunction with the flattening process observed in Longjing tea [18]. The additional findings indicated that freezing and thawing can effectively disrupt the structure of tea cells, accelerate the fermentation process, and enhance the colour and flavour of black tea [19]. Furthermore, this process preserves the natural shape of the buds and leaves, facilitating the formation of a flat, straight, and smooth tea shape.

In recent years, the mechanisms of aroma formation during the entire tea processing have become a subject of considerable research interest, with studies reported in six major tea categories: green tea [20–22], black tea [23–25], oolong tea [26,27], dark tea [28,29], yellow tea [30], and white tea [31]. The majority of research on the dynamic changes in black tea aroma during processing has focused on the SBT. The shaping process has been identified as a significant factor influencing the quality of black tea [32]. It may, therefore, be inferred that there are differences in aroma between FBT and SBT. Nevertheless, there is a paucity of literature examining the changes in aroma during the processing of FBT. In order to emphasise the advantages of Zhejiang Province in flat tea production, we successfully obtained FBT with a shape similar to Longjing tea and a black colour through process optimisation. The tea cells were broken by a combination of freezing and thawing with subsequent carding treatment, and the aroma was enhanced by high-temperature roasting. In this study, SBT was employed as a control, with proteomics and HS-SPME-GC-MS technology used to gain insight into the dynamic changes in aroma during the processing of FBT. This research aims to provide theoretical guidance on the formation of black tea aroma and the innovation of black tea products.

2. Materials and Methods

2.1. Chemicals and Reagents

Ethyl decanoate (99%, analytical reagent) was procured from Aladdin Biochemical Technology Co., Ltd. (Shanghai, China). n-Alkanes (C7-C30) and solid-phase microextraction (SPME) fibre (Supelco, CAR/DVB/PDMS, 50/30 μm, 2 cm) were obtained from Merck KGaA (Darmstadt, Germany).

2.2. Preparation of Tea Samples

Fresh tea leaves of the Yingshuang variety (Chinese National Approved Tea Variety, Number: GS13041-1987, one of the main tea varieties grown in Zhejiang Province) were harvested from Huanghuyunding Farm in Yuhang District, Hangzhou City, with one bud and one leaf. The tea processing flow is illustrated in Figure 1, and the specific parameters are presented in Table S1. A total of 12 tea samples were collected from each processing step, encompassing fresh leaves (FL), withered leaves (SWL and FWL), rolled leaves (SRL), fermented leaves (SFL and FFL), initial dried leaves (SIL), final dried leaves (SBT), frozen–thawed leaves (FTL), carded leaves (FCL), the leaves shaped into flat (FSL), and final roasted leaves (FBT). A total of 250 g of tea samples from each processing step (including fresh leaves) were rapidly frozen with liquid nitrogen and stored at −80 °C for the subsequent analysis. With the exception of the dried samples, all the samples were subjected to freeze-drying (at −55 °C for 48 h) and subsequently pulverised to a particle size of less than 1 mm prior to GC-MS analysis.

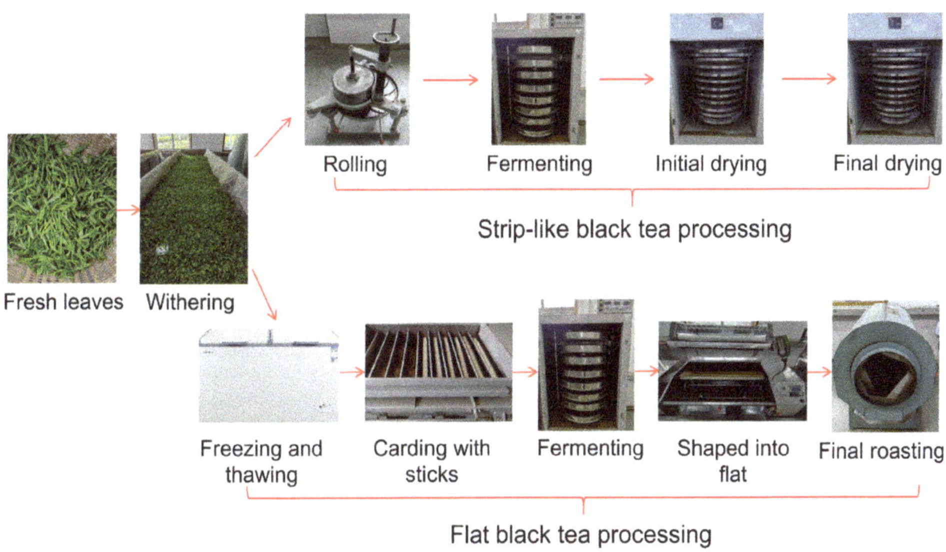

Figure 1. Processing flowchart of black tea.

2.3. Sensory Evaluation Method

Tea infusions were prepared and subjected to sensory evaluation in accordance with the Chinese national standard (GB/T 23776-2018) [7] and the reference [33]. In summary, 3.00 g of the tea samples were brewed with 150 mL of boiling water for 5 min. The sensory evaluation of the aroma qualities of the tea soups was conducted by seven evaluators, comprising three men and four women. All the participants had undergone sensory evaluation training for a period exceeding two years. Initially, a preliminary evaluation was conducted, after which the sensory evaluation group deliberated and determined that the aroma should be evaluated according to six dimensions: overall intensity, floral, fruity, high-fired, sweet, and dull.

2.4. GC-MS Analysis Conditions

In total, 1.60 g of tea sample was transferred into a 100 mL sample bottle containing 80 mL of purified water and 10 μL of ethyl decanoate (0.863 g/L). Following a thorough shaking, the sample was subjected to a 10 min water bath at 50 °C to allow for equilibrium. The SPME fibre (50/30 μm DVB/CAR/PDMS) was introduced into the headspace (HS) for a period of 50 min at 50 °C. Subsequently, the fibre was maintained within the gas chromatograph injector at 250 °C for 5 min, during which time volatile desorption occurred.

An analysis of the aroma compounds was conducted using an Agilent (Santa Clara, CA, USA) 8890 gas chromatograph (GC), which was coupled to an Agilent 5977C mass spectrometer (MS). The gas chromatography column was of the HP-INNOWAX capillary type (60 m × 0.25 mm × 2.5 μm). The carrier gas was helium (99.999%), with a constant flow rate of 1.6 mL/min and a solvent delay of four minutes. The injector temperature was 250 °C. The oven temperature was regulated as follows: maintained at 50 °C for 5 min, increased at a rate of 3 °C/min to 180 °C, maintained for 1 min, increased at a rate of 10 °C/min to 240 °C, and maintained for 8 min. The ion source temperature was 230 °C under electron ionisation mode, with an electron energy of 70 eV. The scanning mode was set to full scan, with a scanning range of m/z 50–550.

2.5. *Qualitative and Semi-Quantitative Analysis of Aroma Components*

The Agilent MassHunter Workstation Software Unknowns Analysis 12.0 was used for qualitative analysis. The aroma compounds were identified by comparing the spectra with the NIST 20 mass spectral database (https://webbook.nist.gov/chemistry/ (accessed on 23 August 2024)). The retention indices were calculated according to the linear formula of the n-alkanes (C7–C30). Those with a matching rate exceeding 85% were identified as volatile substances. The concentration of each aroma substance was calculated in relation to the internal standard using the following equation: $C_{(i)} = A_{(i)} \times C_{(s)}/A_{(s)}$.

$C_{(i)}$: concentration of the aroma substances;

$A_{(i)}$: area under the curve for the aroma substances;

$C_{(s)}$: concentration of the internal standard;

$A_{(s)}$: area under the curve for the internal standard.

The relative odour-active values (rOAVs) were calculated using the following equation: $rOAV_{(i)} = C_{(i)}/OT_{(i)}$. $OT_{(i)}$ denotes the odour threshold of aroma components in water. Aroma compounds with rOAV > 1 were deemed to exert a notable influence on the aroma.

2.6. *Proteomic Analysis Method*

A 4D label-free proteomic analysis was performed on the fresh leaves, withered leaves and frozen–thawed leaves. The proteomic analysis was entrusted to Meiji Biotechnology Co., Ltd. (Shanghai, China) for testing. Protein extraction, digestion, quantification, and LC-MS/MS analysis were performed according to the literature method [34]. MS/MS spectra were searched using the MaxQuant version 2.0.3.1 software against the Camellia database. Data analysis was performed using the Majorbio Cloud platform (https://cloud.majorbio.com (accessed on 7 December 2023)). The thresholds of fold change (FC > 2 or FC < 0.5) and p-value < 0.05 were employed to identify differentially expressed proteins (DEPs). The DEPs were subsequently employed in GO and KEGG enrichment analyses. The functional annotation of all the identified proteins was performed using the Gene Ontology (GO) database (http://geneontology.org/ (accessed on 22 December 2023)) and the Kyoto Encyclopedia of Genes and Genomes (KEGG) pathway database (http://www.genome.jp/kegg/ (accessed on 22 December 2023)).

2.7. *Data Analysis*

All the data were expressed as mean ± SD of three replicates and significant differences were analysed by Duncan's multiple range test using the SAS 9.3 software with a significance threshold of 0.05. The PCA and OPLS-DA of the aroma compound content were performed using SIMCA 14.1. Heat maps of aroma compounds and bubble plots of protein functional enrichment were performed using Metware Cloud, a free online data analysis platform (https://cloud.metware.cn (accessed on 13 September 2024)). Line graphs and column plots were generated using Origin 2021b.

3. Results and Discussion

3.1. Sensory Evaluation Results

The tea samples and infusions are shown in Figure 2. The FBT was flat, straight, and smooth in shape, similar to Longjing tea, while the SBT was curved, tightly bound, and had obvious golden hairs. The difference in appearance between the two is obvious. The infusion colour of the FBT was significantly redder than that of the other one. It can be seen that the final roasting process promoted the dissolution of internal substances, resulting in a deeper colour, which is consistent with previous research [35]. The brewed leaves of the flat black tea were intact and uniform.

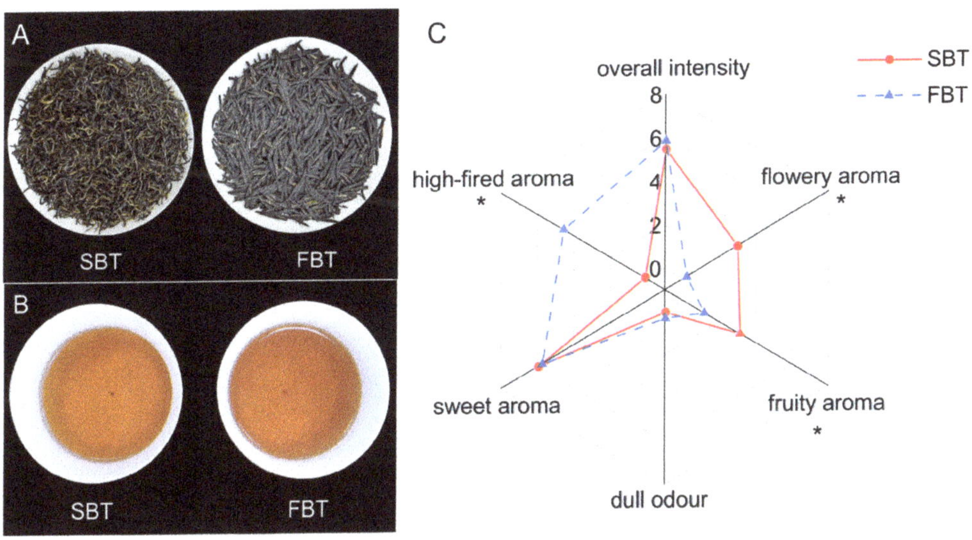

Figure 2. The sensory qualities of black tea. (**A**) Dried tea samples; (**B**) tea infusions; (**C**) the aroma profiles of the tea infusions. (*: $p < 0.05$).

The quantitative description method was used to evaluate the aroma of the tea samples. The results showed that the floral and fruity aroma intensities of the SBT were significantly higher than those of the FBT, while the high-fried aroma intensity of the FBT was significantly higher than that of the SBT. The overall intensity and sweet intensity of both were similar. It was shown that different processing methods had a significant effect on the aroma of the black tea. In order to reduce the dull odour caused by the freezing–thawing process, the FBT was finally roasted at 120–130 °C, resulting in a high-fried aroma. Long-time final drying at high temperatures is not conducive to the retention of floral aroma [36]. Meanwhile, the SBT was dried at a lower temperature, retaining more floral and fruity aroma intensity. Although we obtained an FBT which is similar in shape to Longjing tea, its aroma is not as good as that of SBT.

3.2. Screening of Differential Aroma Compounds

By searching the MS database and comparing retention indices, a total of 92 aroma components were identified, including 25 alcohols, 18 aldehydes, 10 esters, 6 ketones, 16 heterocyclic compounds, 15 hydrocarbons, and 2 sulphur compounds (Table S2). The PCA result showed that there was a significant difference between the samples (Figure 3A). The contents of alcohols, aldehydes, esters, terpenes, aromatic hydrocarbons, and sulphur substances, and the total amount of aroma components in SBT were significantly higher than those in FBT, while the contents of pyrazines, pyrroles, and furans were in contrast. Many of the pyrazines and pyrroles exhibited a roasted aroma linked to the Maillard reaction [36,37]. On the other hand, terpenoids and esters play an important role in floral or fruity aromas [38].

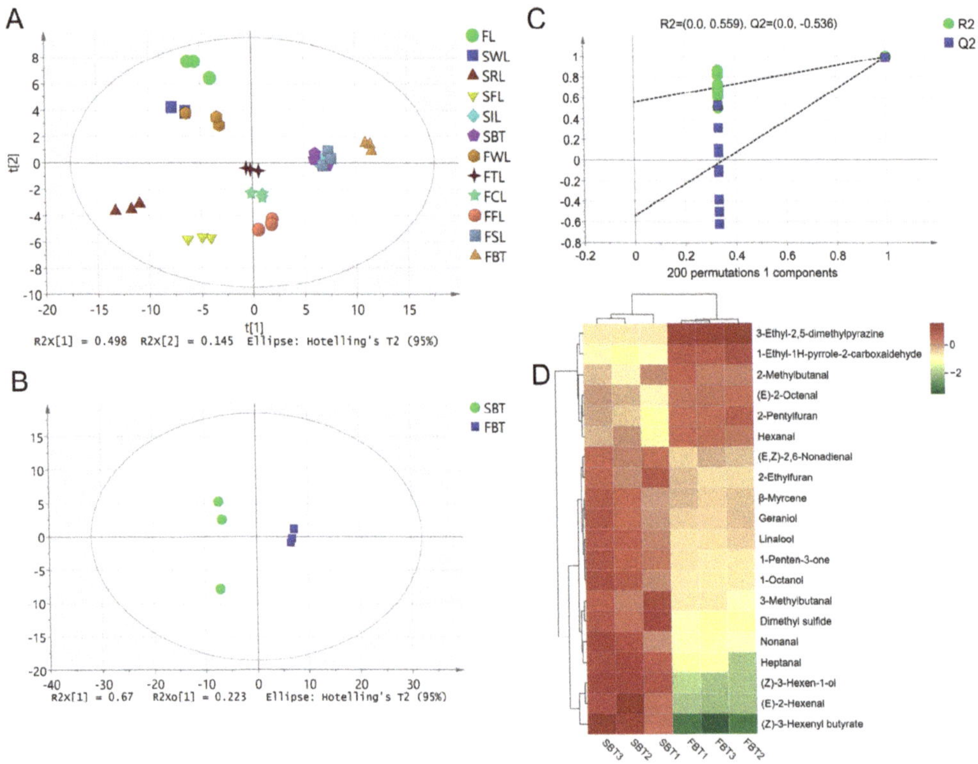

Figure 3. Screening of differential aroma compounds. (**A**) PCA chart of all the samples during the black tea processing; (**B**) scores of OPLS-DA (R2Y = 0.999, Q2 = 0.990); (**C**) cross-validation by 200-times permutation test (R2 = 0.599, Q2 = −0.536); (**D**) heatmap visualisation constructed with the key differential components (VIP > 1.0, OAV > 1.0, and $p < 0.05$).

An OPLS-DA analysis was performed on the aroma components between the SBT and FBT. The model achieved good discrimination results with the R2X, R2Y, and Q2 values of 0.893, 0.999, and 0.99, respectively (Figure 3B). The cross-validation results of 200 permutation tests showed that the OPLS-DA model was reliable and there was no overfitting phenomenon (R2 = 0.599, Q2 = −0.536) (Figure 3C). Among them, there were 55 aroma components with VIP values > 1. To further screen for the most important differential components, the rOAVs of the aroma components were calculated (Table S3). There were 18 components with VIP and rOAV > 1, and *p*-value < 0.05, including dimethyl sulphide, heptanal, (Z)-3-hexen-1-ol, 3-ethyl-2,5-dimethylpyrazine, 2-methylbutanal, β-myrcene, linalool, 3-methylbutanal, 1-penten-3-one, 2-pentylfuran, (E)-2-hexen-1-al, octanol, nonanal, 2-ethylfuran, geraniol, (E)-2-octen-1-al, (E,Z)-2,6-nonadien-1-al, and hexanal.

Linalool and geraniol are clearly floral components [39–41], with the contents in the SBT being 1.90 and 1.68 times those in the FBT, respectively. 2-Methylbutanal [42,43], 2-pentylfuran [43], 3-methylbutanal [44], octanol [45], β-myrcene [45], heptanal [46], and nonanal [46] were identified as fruity aroma compounds by GC-O. 3-Ethyl-2,5-dimethylpyrazine were identified as roasted aroma compounds [12,46]. In addition, the rOAV of 1-ethyl-1H-pyrrole-2-carboxaldehyde in the FBT was 0.75 and 2.79 times that of the SBT. It has been reported as a roasting aroma [12,47] and may contribute to the high-fired aroma. The OTs of most of the esters were high and the rOAVs were low. However, the OT of (Z)-3-hexenyl butyrate with a fruity odour was not found [48]. Its content was as high as 13.76 μg/mL and its fold change between the SBT and FBT was 12.94. These components may be the key components leading to the difference in aroma between the SBT and FBT.

Components with similar structures or aroma characters have synergistic and additive effects [49], and at low or moderate concentrations, aroma components are more likely to have additive or synergistic effects [50]. The greater the difference in chain length between similar substances, the greater the synergistic effect [51]. Esters, especially acetate esters, can enhance floral and fruity flavours through synergistic effects [52–54]. The ester components in this paper are mostly derivatives of hexanol, 3-hexen-1-ol, and 2-hexen-1-ol, which have similar chemical structures. It can be inferred that these substances have a synergistic effect to present fruity aroma. The aroma characteristics, aroma thresholds, and interactions of these unique ester compounds in tea are not yet very clear and are worth further research. Additionally, research has shown that (Z)-3-hexen-1-ol, linalool, and β-myrcene have a synergistic effect with fruit aroma [51]. Therefore, the high levels of these substances in the SBT may further enhance the intensity of fruit aroma.

3.3. *Dynamic Changes in Aroma Components*

The manufacturing process of the FBT was significantly different from that of the SBT. The former employed a combination of freezing–thawing and carding treatment instead of rolling to break down tea cells for the subsequent fermentation. In the drying process of the FBT, stir-frying was used, whereas in the drying process of the SBT, baking was employed. The changes in the content of different types of aroma compounds during processing are shown in Figure S1. During the processing of the SBT, the content of alcohols, aldehydes, esters, and terpenes, and the total amount of aroma substances showed a trend of first increasing and then decreasing, with the highest content observed in the SRL. The increase in terpenoids during withering and rolling is beneficial for aroma formation [25]. Furans, pyrroles, and sulphur compounds showed a gradually increasing trend. Alkanes showed a first decreasing and then increasing trend. During the processing of FBT, the content of aldehydes and esters, and the total amount of aroma compounds showed a trend of first increasing and then decreasing, with the highest content in the FWL. However, alcohols, terpenes, benzenes, and ketones showed a gradually decreasing trend. Pyrazines and pyrroles showed a gradually increasing trend. During the heat treatment process, they increased rapidly due to the Maillard reaction [55]. However, the total amount of aroma compounds rapidly decreased, which was consistent with previous studies [25].

The proportions of different types of aroma compounds are shown in Figure 4. The proportions of alcohols and terpenes gradually decrease during the processing of both SBT and FBT, while the proportions of furans, pyrroles, and pyrazines gradually increase during the processing. The proportion of benzenes, alkanes, and ketones first decreases and then increases, while the proportions of aldehydes and sulphur compounds first increase and then decrease. In the SBT, the proportion of esters first increases and then decreases, with the highest proportion in the FRL. In the FBT, however, it showed a decreasing trend. Compared with the changes in aroma components between the SBT and FBT, the contents of alcohols, aldehydes, esters, ketones, and terpenes, and the total amount of aroma components decreased significantly after freezing–thawing during the FBT process. And in the later stages, aroma compounds such as alcohols, aldehydes, esters, and ketones did not increase significantly, while the rolling stage in SBT caused a significant increase in these substances. It can be seen that freezing and thawing is a key step in changing the aroma.

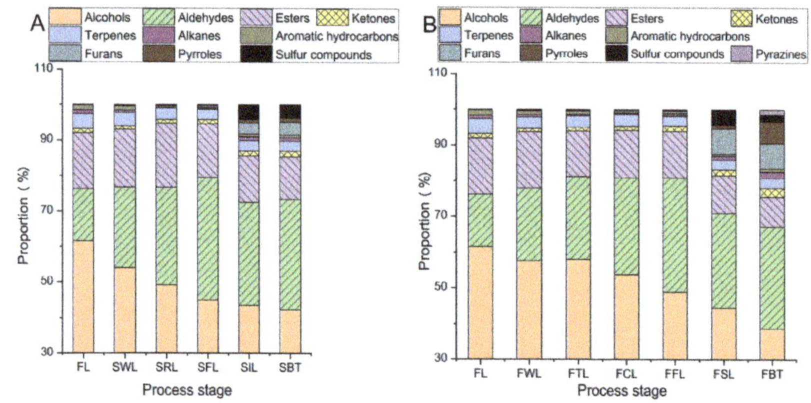

Figure 4. The proportion of different types of aroma substances during processing. (**A**) The manufacturing process of SBT; (**B**) the manufacturing process of FBT.

The OPLS-DA analysis was used to establish a discriminant model for samples at different stages of FBT processing. The model worked well and the model was reliable with no overfitting phenomenon (Figure 5A,B). There were 23 components with VIP > 1 (Figure 5C). A total of 17 aroma components were identified as the key aroma-active compounds, excluding those with rOAV < 1, including linalool, (E)-2-hexen-1-al, methyl salicylate, geraniol, hexanal, (Z)-3-hexenyl butyrate, dimethyl sulphide, 2-methylbutanal, 2-ethylfuran, nonanal, nonanol, benzene acetaldehyde, 3-methylbutanal, (Z)-3-hexen-1-ol, 2-pentylfuran, linalool oxide I, and β-myrcene. Notably, these compounds largely corresponded to the differential components between FBT and SBT.

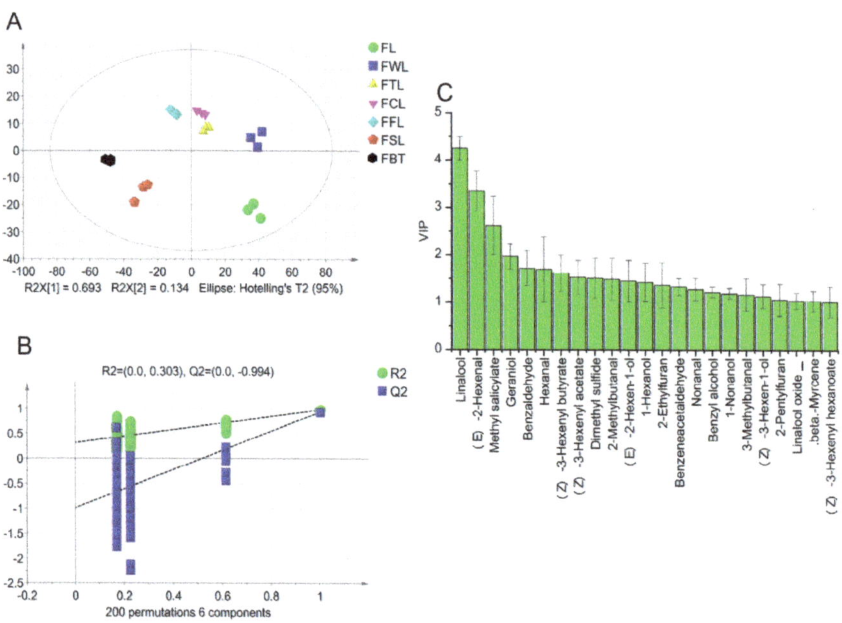

Figure 5. Key aroma components in FBT manufacturing process. (**A**) Scores of OPLS-DA (R2Y = 0.968, Q2 = 0.900); (**B**) cross-validation by 200-times permutation test (R2 = 0.303, Q2 = −0.994); (**C**) aroma components of variables important in projecting (VIP > 1).

3.4. Proteomic Analysis Results

The proteomic analysis was conducted on FL, FWL, and FTL. A total of 20,092 proteins were identified, comprising 19,938 common proteins and 58 unique proteins (Figure 6A). The PCA demonstrated that the first two principal components explained 52.1% of the original variable information, and a clear distinction was evident between the three (Figure 6B). DEPs were identified through the screening of the samples subjected to different treatments, with FC > 2 or FC < 0.5, and $p < 0.05$. The proteomic analysis revealed that the withering and freezing–thawing processes significantly impact the protein profile of the tea leaves. Specifically, FWL exhibited the upregulation of 218 proteins and downregulation of 173 proteins in comparison to FL (Figure 6C). In contrast, FTL displayed the upregulation of 236 proteins and downregulation of 323 proteins in comparison to FWL (Figure 6D).

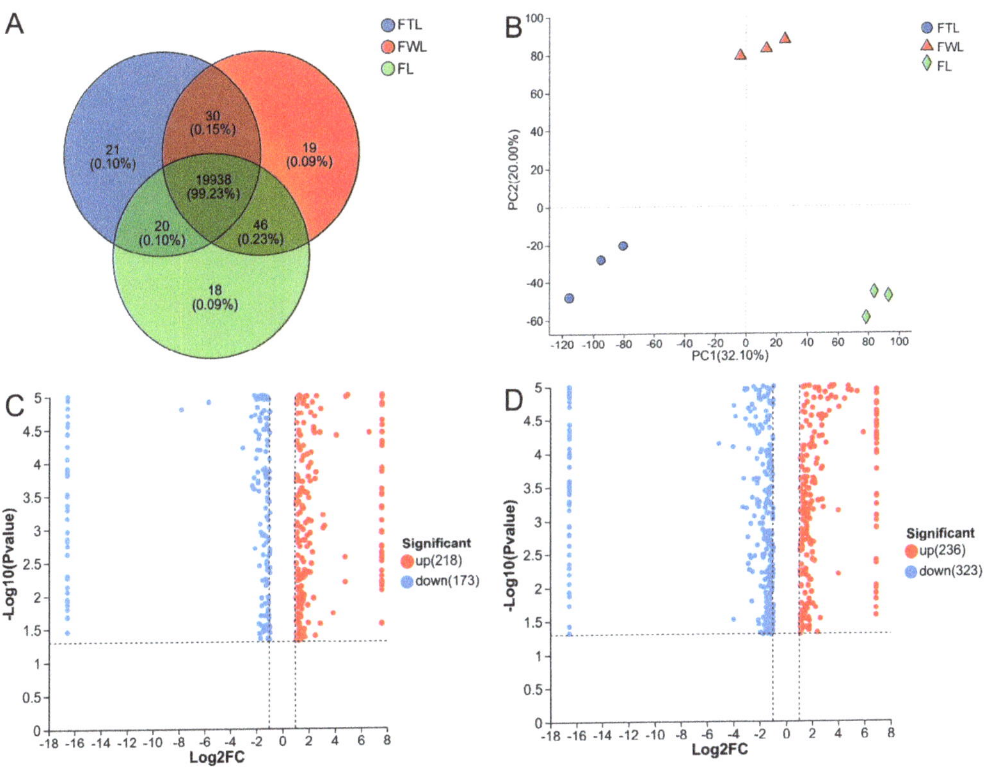

Figure 6. The number of proteins identified in the study. (**A**) The number of proteins identified in different sample; (**B**) PCA chat; (**C**) differential proteins as a volcano plot of FWL vs. FL; (**D**) differential proteins as a volcano plot of FTL vs. FWL.

The functional enrichment analysis was conducted on the differential proteins with a particular emphasis on the pathways associated with the metabolisms of aroma components. The GO functional enrichment analysis between the FWL and FL revealed significant differences in the biological processes, including the L-phenylalanine metabolic process, aromatic compound catabolic process, benzene-containing compound metabolic process, olefinic compound metabolic process, carboxylic acid metabolic process, carotene metabolic process, and terpene metabolic process. In the case of the upregulated proteins, significant differences were observed with regard to sesquiterpenoid biosynthetic processes, hydrolase activity (acting on glycosyl bonds), and β-glucosidase activity. The KEGG functional enrichment analysis revealed that the proteins involved in the metabolism of phenylalanine exhibited significant differences. The proteins involved in carotenoid biosynthesis, terpenoid backbone biosynthesis, fatty acid metabolism, pyruvate metabolism, glycolysis/gluconogenesis, and α-linolenic acid metabolism were found to be enriched, although no significant difference was observed (Figure S2).

The GO functional enrichment analysis between the FTL and FWL revealed significant differences in cellular component proteins, including the integral and intrinsic components of the membrane, as well as cellular anatomical entities. Additionally, there were notable distinctions in biological process proteins, particularly in the positive regulation of hydrolase activity. The changes in cellular component proteins indicated that freezing and thawing had caused significant damage to cellular structures. In the case of the upregulated proteins, a significant difference was observed in the acyl-CoA hydrolase activity proteins. The KEGG functional enrichment analysis revealed that the proteins involved in ether lipid metabolism exhibited notable differences. Diterpenoid biosynthesis, linoleic acid metabolism, terpenoid backbone biosynthesis, butanoate metabolism, carotenoid biosynthesis, citrate cycle (TCA cycle), fatty acid metabolism, and glycolysis/gluconeogenesis were found to be enriched, yet no significant difference was observed (Figure S3). The number of proteins associated with aroma metabolism during withering is significantly greater than that observed during freezing and thawing. The primary pathways for the production of volatile substances in tea are lipid degradation, terpenoid synthesis, amino acid metabolism, carotenoid degradation, and glycoside hydrolysis [3,39]. Terpenoids, phenylpropanoids/benzenoids, and fatty acid derivatives were identified as the primary volatile substances [56,57]. The aforementioned enriched pathways directly or indirectly influence the formation of aroma components.

3.5. Metabolic Pathway Analysis

The analysis of specific DEPs revealed that the proteins responsible for β-glucosidase activity, α-galactosidase, and a protein involved in the synthesis of methyl jasmonate in the α-linolenic acid metabolism pathway exhibited a notable increase in FWL relative to FL (Table S4). It has been demonstrated that a range of aroma precursors exist in glycoside form, and that glycosidases are capable of breaking down these precursors in order to release aromas [55,58]. The upregulation of these proteins is beneficial for enhancing the aroma during the withering process. The lipoxygenase (K00454 and K15718) and 2-C-methyl-D-erythritol 2,4-cyclodiphosphate synthase (K01770) were found to be significantly downregulated in FTL in comparison to FWL. Lipoxygenase is a pivotal enzyme in α-linolenic acid metabolism, directly influencing the synthesis of aromatic compounds such as aliphatic alcohols, aldehydes, and esters [59]. Geranyl diphosphate is a crucial component in the synthesis of terpenes, including geraniol, linalool, and myrcene, in the monoterpenoid biosynthesis pathway [3]. The downregulation of these proteins is unfavourable for the formation of aroma.

The metabolic pathways of the key aroma components in this study are the terpene metabolism pathway and the linoleic acid metabolism pathway. The alterations in the key aroma components associated with the regulation of these pathways during FBT processing are illustrated in Figure 7. During the withering process, the contents of (Z)-3-hexen-1-ol, (E)-2-hexen-1-al, (E)-2-hexen-1-ol and related esters, (Z)-3-nonen-1-ol, 2-nonen-1-ol, (E,Z)-3,6-nonadien-1-ol, and geraniol exhibited a marked increase. The rise in glycosidase activity is advantageous for the breakdown of glycoside precursors, resulting in an elevated geraniol content. Following the freezing and thawing process, a notable decline was observed in the contents of (E)-2-hexen-1-al, (Z)-3-hexen-1-ol and related esters, (E,Z)-3,6-nonadien-1-ol, β-myrcene, limonene, β-ocimene, α-terpinene, α-terpineol, linalool, geraniol, heptanal, nonanal, (Z)-3-nonen-1-ol, and 2-nonen-1-ol.

Among them, the concentrations of (Z)-3-hexen-1-ol, linalool, and geraniol exhibited a reduction of 27.1% to 32.2%. The concentrations of (Z)-3-hexenyl butyrate, methyl salicylate, (Z)-3-hexen-1-ol acetate, and (Z)-3-hexenyl hexanoate exhibited a notable decline, ranging from 23.0% to 86.4%. The significant decrease in content resulted in a notable weakening of the floral and fruity aroma of the FBT. It can be observed that the process of freezing and thawing not only caused cell damage and accelerated the loss of aroma compounds, but also led to a reduction in the expression of lipoxygenase and 2-C-methyl-D-erythritol 2,4-cyclodiphosphate synthase, impeding the synthesis of aroma compounds in the linoleic acid metabolism pathway and terpene synthesis pathway, particularly esters and terpenes with a floral or fruity aroma. This also elucidates the reason for the poor aroma quality of tea products manufactured from fresh tea leaves that have been frostbitten in late spring.

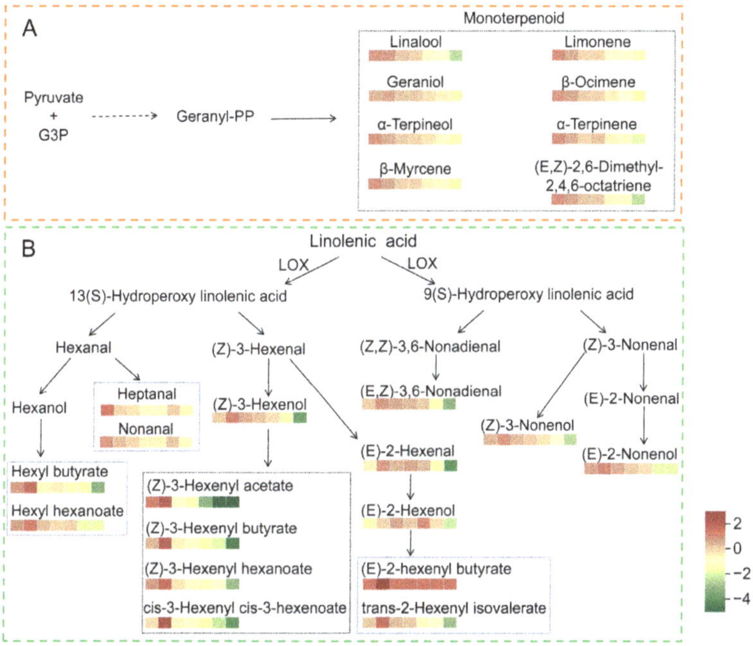

Figure 7. Metabolic pathway of key aroma compounds. (**A**) The synthesis pathway of terpenes; (**B**) the metabolic pathway of fatty acid-derived aroma compounds. The boxes of different colours from left to right are FL, FWL, FTL, FCL, FFL, FSL, and FBT; red and green colours represent above-mean level and below-mean level, respectively. G3P: D-glyceradehyde-3-phosphate; Geranyl-PP: geranyl diphosphate; LOX: lipoxygenase.

4. Conclusions

This study combined freezing–thawing and carding with sticks instead of rolling to crush tea cells to produce FBT with a shape similar to Longjing tea. The differences in aroma quality between SBT and FBT, as well as the changes in aroma components during processing were investigated. Additionally, proteomic analysis was conducted on the fresh leaves, withered leaves, and frozen–thawed leaves. The results showed that the combination of freezing–thawing and carding can achieve a good cell disruption effect, and the FBT produced has a flat and straight appearance, but it has a negative impact on the formation of floral and fruity aromas. The floral and fruity aroma of the SBT was absent in the FBT. Instead, the FBT had a pronounced high-fired aroma. The principal aroma components during the FBT process were linalool, (E)-2-hexen-1-al, methyl salicylate, geraniol, hexanal, (Z)-3-hexenyl butyrate, dimethyl sulphide, 2-methylbutanal, 2-ethylfuran, benzeneacetaldehyde, nonanal, nonanol, 3-methylbutanal, (Z)-3-hexen-1-ol, 2-pentylfuran, linalool oxide I, and β-myrcene. The key steps in the aroma formation of the FBT were freezing–thawing and final roasting. The freezing–thawing processing resulted in a reduction in the activity of lipoxygenase and 2-C-methyl-D-erythritol 2,4-

cyclodiphosphate synthase, which impeded the synthesis of aroma compounds in the linoleic acid metabolism pathway and the terpenoid synthesis pathway. Additionally, it accelerated the conversion or loss of aroma substances, which is detrimental to the formation of floral and fruity aromas. The high-temperature final roasting produces pyrazines and pyrroles while causing a significant reduction in linalool, geraniol, myrcene, and esters. Therefore, the freezing–thawing and high-temperature final drying processes should be avoided when processing floral and fruity flavoured black teas.

Supplementary Materials: The following supporting information can be downloaded at https://www.mdpi.com/article/10.3390/foods13203243/s1, Figure S1: The changes in the content of different types of aroma compounds during processing; Figure S2: Bubble charts of GO and KEGG functional enrichment analysis between FWL and FL; Figure S3: Bubble charts of GO and KEGG functional enrichment analysis between FTL and FWL; Table S1: The specific parameters of black tea processing; Table S2: The content of aroma components in different tea samples; Table S3: The rOAVs of aroma compounds in SBT and FBT [60–64]; Table S4: Detail table of differentially expressed proteins.

Author Contributions: Conceptualization, C.A. and Y.Z.; methodology, C.A. and Y.Z.; software, C.A. and X.N.; validation, X.N., J.Y. and Y.Z.; investigation, C.A., D.S. and X.Z.; resources, D.S. and X.Z.; data curation, C.A. and X.N.; writing—original draft preparation, C.A.; writing—review and editing, J.Y. and Y.Z.; project administration, J.Y.; funding acquisition, J.Y. and C.A. All authors have read and agreed to the published version of the manuscript.

Funding: This research was funded by the National Tea Industry Technology System (CARS-19-02A), the Zhejiang Province Basic Public Welfare Research Program (LGN21C160009), and the Zhejiang Province Agricultural Major Technology Collaborative Promotion Project (2024ZDXT05).

Institutional Review Board Statement: The national laws do not require ethical approval for sensory evaluation. There are no human ethics committees or formal documentation procedures available for sensory evaluation.

Informed Consent Statement: Verbal informed consent was obtained from the participants. Verbal consent was obtained rather than written because in China, the sensory evaluation of tea is a widely used method for evaluating tea quality. The Chinese National Standard "Methodology for sensory evaluation of tea" (GB/T 23776-2018) specifies sensory evaluation methods and personnel requirements.

Data Availability Statement: The original contributions presented in the study are included in the article/Supplementary Materials, further inquiries can be directed to the corresponding authors.

Conflicts of Interest: The authors declare no conflicts of interest.

References

1. Mei, Y.; Liang, X. Analysis of China's tea production, sales, import and export situation in 2023. *China Tea* **2024**, *46*, 18–26.
2. Hua, J.J.; Yuan, H.B.; Jiang, Y.W. Status of industry, progress of processing, and prospect of black tea in China. *J. Huazhong Agric. Univ.* **2022**, *41*, 16–23.
3. Liu, Y.; Chen, Q.C.; Liu, D.C.; Yang, L.; Hu, W.; Kuang, L.Q.; Huang, Y.J.; Teng, J.; Liu, Y. Multi-omics and enzyme activity analysis of flavour substances formation: Major metabolic pathways alteration during Congou black tea processing. *Food Chem.* **2023**, *403*, 134263. [CrossRef] [PubMed]
4. Aaqil, M.; Peng, C.X.; Kamal, A.; Nawaz, T.; Zhang, F.; Gong, J.S. Tea harvesting and processing techniques and its effect on phytochemical profile and final quality of black tea: A review. *Foods* **2023**, *12*, 4467. [CrossRef] [PubMed]
5. Chang, X.J.; Zhou, Z.W.; Zhu, C.; Wang, Z.; Zhao, S.S.; Guo, Y.Q. Research advances of processing technology of Congou black tea. *J. Anhui Agric. Sci.* **2016**, *44*, 66–68+94.
6. Li, C.; Yue, C.N.; Yang, P.X.; Cao, H.H.; Zhu, Y.H.; Lin, S.H.; Jiang, X.F. Research progress on characteristic aroma of Congou black tea. *J. Food Saf. Qual.* **2021**, *12*, 8834–8842.
7. *GB/T 23776-2018*; Methodology for Sensory Evaluation of Tea. Chinese National Standard: Beijing, China, 2018.
8. Dai, Q.L.; Liu, G.Q.; Zhang, R.; Niu, M.S.; Tu, T.Y.; Wang, S.T.; Shen, C.H.; Zheng, L. Quantitative description analysis of black tea aroma characteristics and construction of black tea aroma flavor wheel. *Food Ferment. Sci. Technol.* **2023**, *59*, 103–108+175.
9. Shi, Y.F.; Di, T.M.; Yang, S.L.; Wu, L.Y.; Chen, Y.Q.; Xia, T.; Zhang, X.F. Changes in aroma components in the processing of flowery black tea. *Food Sci.* **2018**, *39*, 167–175.

10. Wang, Q.W.; Xie, J.L.; Wang, L.L.; Jiang, Y.W.; Deng, Y.L.; Zhu, J.Y.; Yuan, H.B.; Yang, Y.Q. Comprehensive investigation on the dynamic changes of volatile metabolites in fresh scent green tea during processing by GC-E-Nose, GC–MS, and GC × GC-TOFMS. *Food Res. Int.* **2024**, *187*, 114330. [CrossRef]
11. Su, D.; He, J.J.; Zhou, Y.Z.; Li, Y.L.; Zhou, H.J. Aroma effects of key volatile compounds in Keemun black tea at different grades: HS-SPME-GC-MS, sensory evaluation, and chemometrics. *Food Chem.* **2022**, *373*, 131587. [CrossRef]
12. Ma, L.J.; Gao, M.M.; Zhang, L.Q.; Qiao, Y.; Li, J.X.; Du, L.P.; Zhang, H.L.; Wang, H. Characterization of the key aroma-active compounds in high-grade Dianhong tea using GC-MS and GC-O combined with sensory-directed flavor analysis. *Food Chem.* **2022**, *378*, 132058. [CrossRef] [PubMed]
13. Wu, Q.Y.; Zhou, Z.W.; Zhang, Y.N.; Huang, H.Q.; Ou, X.X.; Sun, Y. Identification of key components responsible for the aromatic quality of Jinmudan black tea by means of molecular sensory science. *Foods* **2023**, *12*, 1794. [CrossRef] [PubMed]
14. Yin, X.; Huang, J.A.; Huang, J.; Wu, W.L.; Tong, T.; Liu, S.J.; Zhou, L.Y.; Liu, Z.H.; Zhang, S.G. Identification of volatile and odor-active compounds in Hunan black tea by SPME/GC-MS and multivariate analysis. *LWT-Food Sci. Technol.* **2022**, *164*, 113656. [CrossRef]
15. Liao, X.L.; Yan, J.N.; Wang, B.; Meng, Q.; Zhang, L.Y.; Tong, H.R. Identification of key odorants responsible for cooked corn-like aroma of green teas made by tea cultivar 'Zhonghuang 1'. *Food Res. Int.* **2020**, *136*, 109355. [CrossRef]
16. Gao, J.; Jiang, L.K.; Dong, W.H.; Zhang, Y.P.; Zhou, B.X.; Cai, X. Technology of mechanically cleaner production of high-quality black tea with different shapes. *J. Tea* **2010**, *36*, 111–114.
17. Tao, M.; Xiao, Z.P.; Huang, A.; Chen, J.Y.; Yin, T.J.; Liu, Z.Q. Effect of 1–20 years storage on volatiles and aroma of Keemun congou black tea by solvent extraction-solid phase extraction-gas chromatography-mass spectrometry. *LWT-Food Sci. Technol.* **2021**, *136*, 110278.
18. He, W.Z.; He, W.D.; Zheng, S.H.; Miao, Y.M.Z.; Yan, F. Processing technology of flat-shaped black tea. *J. Anhui Agric. Sci.* **2016**, *44*, 114–115.
19. Zhang, Y.F.; Li, L.X.; Zhang, X.F.; Lu, Z.; Yu, P.L.; Xue, C.; Hua, Z.X.; Mei, Y. Effects of freezing on quality of black tea. *J. Tea Sci.* **2013**, *33*, 370–376.
20. Wang, H.J.; Hua, J.J.; Yu, Q.Y.; Li, J.; Wang, J.J.; Deng, Y.L.; Yuan, H.B.; Jiang, Y.W. Widely targeted metabolomic analysis reveals dynamic changes in non-volatile and volatile metabolites during green tea processing. *Food Chem.* **2021**, *363*, 130131. [CrossRef] [PubMed]
21. Yang, Y.Q.; Qian, M.C.; Deng, Y.L.; Yuan, H.B.; Jiang, Y.W. Insight into aroma dynamic changes during the whole manufacturing process of chestnut-like aroma green tea by combining GC-E-Nose, GC-IMS, and GC × GC-TOFMS. *Food Chem.* **2022**, *387*, 132813. [CrossRef]
22. Jiang, G.X.; Xue, R.; Xiang, J.; Wang, Y.F.; Liu, B.; Yuan, Y.; Pu, Q.; Fang, X.; Hu, X.M.; Liu, X.Y.; et al. Dynamic changes in the aroma profiles and volatiles of Enshi Yulu tea throughout its industrial processing. *Food Chem.* **2024**, *458*, 140145. [CrossRef] [PubMed]
23. Yang, Y.Q.; Xie, J.L.; Wang, Q.W.; Deng, Y.L.; Zhu, L.; Zhu, J.Y.; Yuan, H.B.; Jiang, Y.W. Understanding the dynamic changes of volatile and non-volatile metabolites in black tea during processing by integrated volatolomics and UHPLC-HRMS analysis. *Food Chem.* **2024**, *432*, 137124. [CrossRef] [PubMed]
24. Wu, H.L.; Huang, W.J.; Chen, Z.J.; Chen, Z.; Shi, J.F.; Kong, Q.; Sun, S.L.; Jiang, X.H.; Chen, D.; Yan, S.J. GC–MS-based metabolomic study reveals dynamic changes of chemical compositions during black tea processing. *Food Res. Int.* **2019**, *120*, 330–338. [CrossRef]
25. Jiang, B.; Yang, L.R.; Luo, X.P.; Huang, R.Y.; Jiao, W.W.; Zhong, X.X.; Li, L.X.; Wang, Q.; Liu, M.L.; Liu, K.Y. Aroma formation and dynamic changes during Sichuan black tea processing by GC–MS-Based metabolomics. *Fermentation* **2023**, *9*, 686. [CrossRef]
26. Chen, S.; Liu, H.H.; Zhao, X.M.; Li, X.L.; Shan, W.N.; Wang, X.X.; Wang, S.S.; Yu, W.Q.; Yang, Z.B.; Yu, X.M. Non-targeted metabolomics analysis reveals dynamic changes of volatile and non-volatile metabolites during oolong tea manufacture. *Food Res. Int.* **2020**, *128*, 108778. [CrossRef]
27. Liu, Z.B.; Chen, F.C.; Sun, J.Y.; Ni, L. Dynamic changes of volatile and phenolic components during the whole manufacturing process of Wuyi Rock tea (Rougui). *Food Chem.* **2022**, *367*, 130624. [CrossRef]
28. Liu, P.P.; Zheng, P.C.; Feng, L.; Gong, Z.M.; Zheng, L.; Gao, S.W.; Wang, X.P.; Ye, F.; Huang, J.A.; Liu, Z.H. Dynamic changes in the aroma profile of Qingzhuan tea during its manufacture. *Food Chem.* **2022**, *375*, 131847. [CrossRef]
29. Fang, X.; Xu, W.C.; Jiang, G.X.; Sui, M.Y.; Xiao, J.Y.; Ning, Y.Y.; Niaz, R.; Wu, D.W.; Feng, X.G.; Chen, J.H.; et al. Monitoring the dynamic changes in aroma during the whole processing of Qingzhuan tea at an industrial scale: From fresh leaves to finished tea. *Food Chem.* **2024**, *439*, 137810. [CrossRef] [PubMed]
30. Wei, Y.M.; Li, T.H.; Xu, S.S.; Ni, T.C.; Deng, W.W.; Ning, J.M. The profile of dynamic changes in yellow tea quality and chemical composition during yellowing process. *LWT-Food Sci. Technol.* **2021**, *139*, 110792. [CrossRef]
31. Chen, Q.C.; Zhu, Y.; Dai, W.D.; Lv, H.P.; Mu, B.; Li, P.L.; Tan, J.F.; Ni, D.J.; Lin, Z. Aroma formation and dynamic changes during white tea processing. *Food Chem.* **2019**, *274*, 915–924. [CrossRef]
32. Zan, M.L.; Du, M.J.; Yang, W.Q. Effect of shaping technology on main aroma components of black tea in Yunnan Fengqing. *J. Kunming Univ.* **2017**, *39*, 46–49.
33. Teng, R.M.; Ao, C.; Huang, H.H.; Shi, D.L.; Mao, Y.X.; Zheng, X.X.; Zhao, Y. Research of processing technology of Longjing tea with 'Baiye 1' based on non-targeted aroma metabolomics. *Foods* **2024**, *13*, 1338. [CrossRef] [PubMed]

34. Wu, S.N.; Liu, X.Z.; Cheng, L.; Wang, D.J.; Qin, G.X.; Zhang, X.F.; Zhen, Y.G.; Wang, T.; Sun, Z. Protective Mechanism of Leucine and Isoleucine against H_2O_2-Induced Oxidative Damage in Bovine Mammary Epithelial Cells. *Oxidative Med. Cell. Longev.* **2022**, *2022*, 4013575. [CrossRef]
35. Ye, F.; Qiao, X.Y.; Gui, A.H.; Wang, S.P.; Liu, P.P.; Wang, X.P.; Teng, J.; Zheng, L.; Feng, L.; Han, H.S.; et al. Metabolomics provides a novel interpretation of the changes in main compounds during black tea processing through different drying methods. *Molecules* **2021**, *26*, 6739. [CrossRef]
36. Yang, Y.Q.; Hua, J.J.; Deng, Y.L.; Jiang, Y.W.; Qian, M.C.; Wang, J.J.; Li, J.; Zhang, M.M.; Dong, C.W.; Yuan, H.B. Aroma dynamic characteristics during the process of variable-temperature final firing of Congou black tea by electronic nose and comprehensive two-dimensional gas chromatography coupled to time-of-flight mass spectrometry. *Food Res. Int.* **2020**, *137*, 109656. [CrossRef]
37. Ma, J.Q.; Wang, Y.J.; Li, J.Y.; Zhang, J.X.; Wei, Y.M.; Yan, Y.X.; Wang, H.P.; Pan, Y.; Xiong, Z.C.; Wang, R.J.; et al. Aroma formation mechanism by the drying step during Congou black tea processing: Analyses by HP-SPME and SAFE with GC-MS. *LWT-Food Sci. Technol.* **2024**, *198*, 116019. [CrossRef]
38. Kang, S.Y.; Yan, H.; Zhu, Y.; Liu, X.; Lv, H.P.; Zhang, Y.; Dai, W.D.; Guo, L.; Tan, J.F.; Peng, Q.H.; et al. Identification and quantification of key odorants in the world's four most famous black teas. *Food Res. Int.* **2019**, *121*, 73–83. [CrossRef] [PubMed]
39. Yin, P.; Kong, Y.S.; Liu, P.P.; Wang, J.J.; Zhu, Y.; Wang, G.M.; Sun, M.F.; Chen, Y.; Guo, G.Y.; Liu, Z.H. A critical review of key odorants in green tea: Identification and biochemical formation pathway. *Trends Food Sci. Technol.* **2022**, *129*, 221–232. [CrossRef]
40. Chen, X.H.; Sun, H.Y.; Qu, D.; Yan, F.; Jin, W.G.; Jiang, H.; Chen, C.; Zhang, Y.F.; Li, C.Y.; Xu, Z.M. Identification and characterization of key aroma compounds in Chinese high altitude and northernmost black tea (*Camellia sinensis*) using distillation extraction and sensory analysis methods. *Flavour Fragr. J.* **2020**, *35*, 666–673. [CrossRef]
41. Zhu, J.C.; Zhu, Y.; Wang, K.; Niu, Y.W.; Xiao, Z.B. Characterization of key aroma compounds and enantiomer distribution in Longjing tea. *Food Chem.* **2021**, *361*, 130096. [CrossRef]
42. Sarhir, S.T.; Amanpour, A.; Selli, S. Characterization of ayran aroma active compounds by solvent-assisted flavor evaporation (SAFE) with gas chromatography–mass spectrometry–olfactometry (GC–MS–O) and aroma extract dilution analysis (AEDA). *Anal. Lett.* **2019**, *52*, 2077–2091. [CrossRef]
43. Chen, X.H.; Chen, D.J.; Jiang, H.; Sun, H.Y.; Zhang, C.; Zhao, H.; Li, X.S.; Yan, F.; Chen, C.; Xu, Z.M. Aroma characterization of Hanzhong black tea (*Camellia sinensis*) using solid phase extraction coupled with gas chromatography–mass spectrometry and olfactometry and sensory analysis. *Food Chem.* **2019**, *274*, 130–136. [CrossRef] [PubMed]
44. Niu, Y.W.; Deng, J.M.; Xiao, Z.B.; Zhu, J.C. Characterization of the major aroma-active compounds in peach (*Prunus persica* L. Batsch) by gas chromatography–olfactometry, flame photometric detection and molecular sensory science approaches. *Food Res. Int.* **2021**, *147*, 110457. [CrossRef]
45. Gong, X.W.; Han, Y.; Zhu, J.C.; Hong, L.; Zhu, D.L.; Liu, J.H.; Zhang, X.; Niu, Y.W.; Xiao, Z.B. Identification of the aroma-active compounds in Longjing tea characterized by odor activity value, gas chromatography- olfactometry, and aroma recombination. *Int. J. Food Prop.* **2017**, *20*, S1107–S1121. [CrossRef]
46. Jia, X.; Deng, Q.C.; Yang, Y.N.; Xiang, X.; Zhou, X.P.; Tan, C.B.; Zhou, Q.; Huang, F.H. Unraveling of the aroma-active compounds in virgin camellia oil (*Camellia oleifera* Abel) using gas chromatography–mass spectrometry–olfactometry, aroma recombination, and omission studies. *J. Agric. Food Chem.* **2021**, *69*, 9043–9055. [CrossRef]
47. Wang, H.J.; Shen, S.; Wang, J.J.; Jiang, Y.W.; Li, J.; Yang, Y.Q.; Hua, J.J.; Yuan, H.B. Novel insight into the effect of fermentation time on quality of Yunnan Congou black tea. *LWT-Food Sci. Technol.* **2022**, *155*, 112939. [CrossRef]
48. Li, H.H.; Luo, L.Y.; Ma, M.J.; Zeng, L. Characterization of volatile compounds and sensory analysis of jasmine scented black tea produced by different scenting processes. *J. Food Sci.* **2018**, *83*, 2718–2732. [CrossRef] [PubMed]
49. Zhu, J.C.; Chen, F.; Wang, L.Y.; Niu, Y.W.; Xiao, Z.B. Evaluation of the synergism among volatile compounds in Oolong tea infusion by odour threshold with sensory analysis and E-nose. *Food Chem.* **2017**, *221*, 1484–1490. [CrossRef] [PubMed]
50. Xiao, Z.B.; Zhu, J.C.; Niu, Y.W.; Wang, P.P.; Sun, X.X.; Wang, R.L. Synergistic effect of aromatic compounds. *J. Food Sci. Technol.* **2018**, *36*, 1–7.
51. Xiao, Z.B.; Liu, L.X.; He, J.B.; Niu, Y.W.; Zhu, J.C.; Zhang, J. The perception interaction of complex fruits aroma system: Including binary, ternary, quaternary, and with different chain length, functional groups and substituent positions compounds. *LWT-Food Sci. Technol.* **2024**, *205*, 116458. [CrossRef]
52. Mao, D.S.; Liu, H.; Li, Z.F.; Niu, Y.W.; Xiao, Z.B.; Zhang, F.M.; Zhu, J.C. Impact of sensory interactions among volatile compounds of juice of Red Delicious apples. *Hortic. Environ. Biotechnol.* **2020**, *61*, 197–206. [CrossRef]
53. Na Li, N.; Guanyu Li, G.Y.; Aihua Li, A.H.; Yongsheng Tao, Y.S. Synergy effect between fruity esters and potential odorants on the aroma of Hutai-8 rose wine revealed by threshold, s-curve, and σ–τ plot methods. *J. Agric. Food Chem.* **2023**, *71*, 13869–13879.
54. Lytra, G.; Tempere, S.; Floch, A.L.; Revel, G.D.; Barbe, J.C. Study of sensory interactions among red wine fruity esters in a model solution. *J. Agric. Food Chem.* **2013**, *61*, 8504–8513. [CrossRef]
55. Liu, F.; Wang, Y.; Zhang, T.; Tang, X.B.; Wang, X.P.; Li, C.H. Review on aroma change during black tea processing. *J. Tea Sci.* **2018**, *38*, 9–19.
56. Shi, J.; Wang, J.T.; Lv, H.P.; Peng, Q.H.; Schreiner, M.; Baldermann, S.; Lin, Z. Integrated proteomic and metabolomic analyses reveal the importance of aroma precursor accumulation and storage in methyl jasmonate-primed tea leaves. *Hortic. Res.* **2021**, *8*, 95. [CrossRef]

57. Niu, X.J.; Ao, C.; Yu, J.Z.; Zhao, Y.; Huang, H.T. GC-MS combined with Proteomic analysis of volatile compounds and formation mechanisms in green teas with different aroma types. *Foods* **2024**, *13*, 1848. [CrossRef]
58. Huang, W.J.; Fang, S.M.; Wang, J.; Zhuo, C.; Luo, Y.H.; Yu, Y.L.; Li, L.Q.; Wang, Y.J.; Deng, W.W.; Ning, J.M. Sensomics analysis of the effect of the withering method on the aroma components of Keemun black tea. *Food Chem.* **2022**, *395*, 133549. [CrossRef] [PubMed]
59. Dudareva, N.; Klempien, A.; Muhlemann, J.K.; Kaplan, I. Biosynthesis, function and metabolic engineering of plant volatile organic compounds. *New Phytol.* **2013**, *198*, 16–32. [CrossRef]
60. Zhai, X.T.; Zhang, L.; Granvogl, M.; Ho, C.T.; Wan, X.C. Flavor of tea (*Camellia sinensis*): A review on odorants and analytical techniques. *Compr. Rev. Food Sci. Food Saf.* **2022**, *21*, 3867–3909. [CrossRef]
61. Xie, J.L.; Wang, L.L.; Deng, Y.L.; Yuan, H.B.; Zhu, J.Y.; Jiang, Y.W.; Yang, Y.Q. Characterization of the key odorants in floral aroma green tea based on GC-E-Nose, GC-IMS, GC-MS and aroma recombination and investigation of the dynamic changes and aroma formation during processing. *Food Chem.* **2023**, *427*, 136641. [CrossRef]
62. Joshi, R.; Gulati, A. Fractionation and identification of minor and aroma-active constituents in Kangra orthodox black tea. *Food Chem.* **2015**, *167*, 290–298. [CrossRef] [PubMed]
63. Odor & Flavor Detection Thresholds in Water. Available online: http://www.leffingwell.com/odorthre.htm (accessed on 23 August 2024).
64. Guo, X.Y.; Ho, C.T.; Wan, X.C.; Zhu, H.; Liu, Q.; Wen, Z. Changes of volatile compounds and odor profiles in Wuyi rock tea during processing. *Food Chem.* **2021**, *341*, 128230. [CrossRef] [PubMed]

foods

MDPI

Article

Metabolite Profiling and Identification of Sweet/Bitter Taste Compounds in the Growth of *Cyclocarya Paliurus* Leaves Using Multiplatform Metabolomics

Liang Chen [1,2], Dai Lu [1], Yuxi Wan [1], Yaqian Zou [1,2], Ruiyi Zhang [1,2], Tao Zhou [1], Bin Long [1], Kangming Zhu [3], Wei Wang [1] and Xing Tian [1,2,4,*]

1 TCM and Ethnomedicine Innovation & Development International Laboratory, Innovative Material Medical Research Institute, School of Pharmacy, Hunan University of Chinese Medicine, Changsha 410208, China; chenliang_201230@163.com (L.C.); x944084134ld@163.com (D.L.); wanyuxi2005@163.com (Y.W.); zouyaqian2003@163.com (Y.Z.); zhangry2003812@163.com (R.Z.); zhoutao20021108@163.com (T.Z.); 19873178048@163.com (B.L.); wangwei402@hotmail.com (W.W.)

2 Department of Food and Drug Engineering, School of Pharmacy, Hunan University of Chinese Medicine, Changsha 410208, China

3 School of Informatics, Hunan University of Chinese Medicine, Changsha 410208, China; 13526009623@163.com

4 Hunan Engineering and Technology Research Center for Health Products and Life Science, Changsha 410208, China

* Correspondence: acctianxing@hotmail.com

Abstract: *Cyclocarya paliurus* tea, also known as "sweet tea", an herbal tea with *Cyclocarya paliurus* leaves as raw material, is famous for its unique nutritional benefits and flavor. However, due to the unique "bittersweet" of *Cyclocarya paliurus* tea, it is still unable to fully satisfy consumers' high-quality taste experience and satisfaction. Therefore, this study aimed to explore metabolites in *Cyclocarya paliurus* leaves during their growth period, particularly composition and variation of sweet and bitter taste compounds, by combining multi-platform metabolomics analysis with an electronic tongue system and molecular docking simulation technology. The results indicated that there were significant differences in the contents of total phenols, flavonoids, polysaccharides, and saponins in *C. paliurus* leaves in different growing months. A total of 575 secondary metabolites were identified as potential active metabolites related to sweet/bitter taste using nontargeted metabolomics based on UHPLC-MS/MS analysis. Moreover, molecular docking technology was utilized to study interactions between the candidate metabolites and the sweet receptors T1R2/T1R3 and the bitter receptors T2R4/T2R14. Six key compounds with high sweetness and low bitterness were successfully identified by using computational simulation analysis, including cis-anethole, gluconic acid, beta-D-Sedoheptulose, asparagine, proline, and citrulline, which may serve as candidates for taste modification in *Cyclocarya paliurus* leaves. These findings provide a new perspective for understanding the sweet and bitter taste characteristics that contribute to the distinctive sensory quality of *Cyclocarya paliurus* leaves.

Keywords: *Cyclocarya paliurus* leaves; sweet/bitter/bittersweet; nontargeted metabolomics; electronic tongue; molecular docking

1. Introduction

Cyclocarya paliurus (Batalin) Iljinskaja (*C. paliurus*), also known as "sweet tea tree" and "the third tree", is a deciduous tree belonging to the genus *Cyclocarya Iljinskaja (Juglangdaceae)* and is widely distributed in south China, with the southwestern area of Hunan province being one of its important native habitats [1]. As a traditional edible and medicinal plant, *C. paliurus* leaves taste sweet, often as a "sweet tea" traditionally consumed daily by the local Chinese populace for more than 1000 years, and its safety and flavor have long been recognized [2]. As a traditional Chinese medicine, the leaves of *C. paliurus* contain rich biological active substances, such as flavonoids, terpenoids, sterols, acids, glycosides, and trace elements [3], for the treatment of diabetes, hypertension, and hyperlipidemia, as well as for reducing cholesterol and modulating functions of the immune system [4]. However, limited information is available about acute toxicity of extracts from *C. paliurus* leaves [5]. In 2013, *C. paliurus* leaves were approved by the National Health and Family Planning Commission as a new food ingredient, which can be eaten or drunk as ordinary food [6]. Recently, tea processed from *C. paliurus* leaves has become the first health tea approved by the US Food and Drug Administration in China and has been industrialized, which has broad market prospects [7]. Of note, despite the many reputations and valuable effects of *C. paliurus* leaves, due to the widespread coexistence of "sweetness and bitterness" (sweetness and bitterness are often perceived at the same time during drinking *C. paliurus* tea), they are still unable to deliver an exceptional taste experience that meets consumers' expectations for quality and satisfaction, resulting in limited product market recognition [8]. In fact, research on processing and utilization of *C. paliurus* leaves mainly focuses on volatile compounds and biological activities, while research on sweet and bitter substances in *C. paliurus* leaves has not been reported [9]. Therefore, exploring the potential "sweet and bitter" substances in the leaves of *C. paliurus*, and further screening the key taste substances, will definitely contribute to expansion of the application field of the whole industrial chain of *C. paliurus*, which has important social value and economic benefits.

"Bitterness" and "sweetness" are often functionally and metaphorically opposed, and this antagonism is complicated by the fact that some compounds exhibit both sweet and bitter tastes. In fact, in previous reports the sweet/bitter substances in plants were mainly composed of terpenoids (diterpenoids or triterpenes), flavonoids, glycosides, sweet/bitter amino acids, and phenolic acid [10,11]. Detection of sweet/bitter compounds in foods or medicinal and edible plants has traditionally relied on sensory-guided fractionation via preparative chromatography; however, use of this method has been limited because it is time consuming and laborious [12]. Thus, there is an urgent need to use faster and more efficient methods to screen sweet/bitter/bittersweet substances in plants. Moreover, few studies [3,13] have been conducted on the core flavor metabolites that influence the sensory qualities of *C. paliurus* leaves.

As a rapid and repeatable high-throughput detection technology, metabolomics analysis can identify a wide variety of metabolites in biological samples and has been widely used to investigate formation and transformation of various intracellular metabolisms and compounds in plants [14]. The electronic tongue system, as an intelligent instrument that simulates human taste, has been reported to be able to quantitatively and qualitatively analyze the taste of different medicinal and edible plants [15,16]. Generally, nontargeted metabolomics has been used to assess the relationship between metabolites and sensory quality, such as in identifying key taste compounds in food, combined with sensory evaluation and intelligent taste analysis [12]. Moreover, databases related to food flavor have also been established in recent years, such as VirtualTaste and BitterDB [17,18]. Thus, LC-MS-based nontargeted metabolomics combined with electronic tongue analysis is an excellent method to establish the relationship between *C. paliurus* leaves' core metabolites and sweetness/bitterness/bittersweetness.

A number of studies have focused on binding interactions between key taste compounds and taste receptors [19]. Sweet/bitter receptors play an irreplaceable role in the process of taste perception, and their structure and function has become a new research focus [20]. The perception of bitterness and sweetness stems from complex interactions between compounds and receptors. Sweetness begins with interaction between the sweet-active substance and a pair of special receptors (T1R2/T1R3) located on taste buds. Therefore, the perception of sweetness is extremely low when T1R2 or T1R3 is present alone, but once the two monomers are combined, taste receptor cells can directly convert information content of chemical stimulation into neural signals and transmit them to the brain, which can perceive almost all sweet tastes [21,22]. Bitter taste receptors (hT2R4 and hT2R14) are known to be activated by a large number of compounds originating from different chemical categories [23,24]. In fact, in silico molecular modeling techniques have been successfully used to screen and evaluate the binding mechanisms of interesting compounds to the taste receptors, including sweet and bitter receptors [19,22].

The chemical composition, metabolites, and taste of *C. paliurus* leaves may vary somewhat depending on its growth period. Thus, the present study aimed to investigate the flavor-related metabolites that contributed to the sweet/bitter/bittersweet taste of *C. paliurus* leaves harvested in different growing months by using multi-platform metabolomics analysis, including the electronic tongue system, ultra-performance liquid chromatography/tandem mass spectrometry (UHPLC-MS/MS), and molecular docking for metabolite profiling. This approach is expected to provide new insights into the key taste components of *C. paliurus* leaves and contribute to development of strategies for enhancing the sensory quality of their products.

2. Materials and Methods

2.1. Reagents and Chemicals

The samples of *C. paliurus* leaves from various growth months were procured from Hunan Yueling Junshan Agroforestry Technology Co., Ltd. (Hunan, China). Folin–Ciocalteu's phenol reagent was obtained from Hefei Bomei Biotechnology Co., Ltd. (Hefei, China). Anhydrous ethanol was acquired from Anhui Ante Food Co., Ltd. (Anhui, China). Phenol was purchased from Guangdong Guanghua Technology Co., Ltd. (Guangdong, China). Sulfuric acid and perchloric acid were obtained from Hunan Huihong Reagent Co., Ltd. (Hunan, China). Gallic acid, rutin solution, glucose, and oleanolic acid were guaranteed reagents provided by Chengdu Lemeitian Pharmaceutical Technology Co., Ltd. (Chengdu, China). Anhydrous sodium carbonate, sodium nitrate (III), aluminum nitrate(V), vanillin, and glacial acetic acid were sourced from Sinopharm Chemical Reagent Co., Ltd. (Shanghai, China).

2.2. *Cyclocarya Paliurus Samples*

All samples of *C. paliurus* leaves were collected from the Huping Mountain in Shimen County, Hunan, P.R. China (110.773665° N, 29.94036696° E). To investigate the cumulative dynamics of substances in the leaves of *C. paliurus* throughout its growing period, approximately 200 $\pm$ 0.50 g of fresh, fully mature leaves was collected at four-week intervals, starting from May and ending in September 2023 (1 May, 1 June, 1 July, 1 August, and 1 September). These samples were randomly gathered from three trees with similar canopy characteristics and were mixed to create a pool representing each month. All samples were dried in an oven at 60 °C until a constant weight was achieved. Subsequently, the *C. paliurus* leaf samples were ground into powder and refrigerated at -20 °C for further analyses.

2.3. *Electronic Tongue Analysis for C. Paliurus Samples*

Taste attributes (saltiness, bitterness, sourness, umami, umami aftertaste, astringency, bitter aftertaste, and astringency aftertaste) of the five different growth months of *C. paliurus* leaf samples were determined using a TS-sa402b electronic tongue system (Intelligent Sensor Technology Co., Ltd., Atsugi, Japan) with wide area selection specific artificial lipid membrane sensors. The taste attribute indices for the tested samples were calculated based on the absolute value of the lipid membrane potential for each artificial sensor, referenced against the potential of the solution. The output value of the reference solution was considered the tasteless point, set at 0 for different taste components, and values bigger than these tasteless points were considered meaningful. A precisely weighed (2.00 $\pm$ 0.05) g *C. paliurus* leaf sample was steeped in 150 mL of boiling water for 1 h. Subsequently, the aqueous extract was filtered through three layers of gauze and allowed to cool to 25 °C before undergoing analysis using the TS-sa402b electronic tongue system. Each sample was cycled 4 times, and the mean values of the last three cycles were used for statistical analysis using TS-sa402b Library search software (Intelligent Sensor Technology Co., Ltd., Atsugi, Japan).

2.4. Determination of Total Phenolic Content (TPC), Total Flavonoid Content (TFC), Total Polysaccharide Content (TP), and Total Saponin Content (TSC)

To determine total polyphenols, total flavonoids, total polysaccharides, and total saponins, 6.0 ± 0.01 g of each *C. paliurus* leaf-powder sample was mixed with 120 mL of distilled water in a 250 mL round-bottomed flask and extracted at 90 °C for 40 min in a water bath. After extraction, each extracted solution was centrifuged at 4,500 rpm for 5 min at 4 °C. Then, all the resulting water extract was filtered through three layers of gauze three times to obtain the filtrate. The filtrate was concentrated under reduced pressure (75 kPa) and a temperature of 65 °C, followed by freeze-drying using a freeze dryer (Xinzhi freeze-drying equipment Co., LTD, Ningbo, China). After drying, the lyophilized *C. paliurus* powder was obtained, sealed, and refrigerated at −20 °C for further analyses. The TPC, TFC, TP, and TSC were determined using lyophilized *C. paliurus* powder. The extraction yield was calculated as the percentage of the initial fresh weight to the final dry weight after lyophilization, and all samples were analyzed in parallel three times.

The TPC in the *C. paliurus* leaf samples was determined by using the Folin–Ciocalteu colorimetric method, as described by Musci et al [25]. Specifically, an aliquot of 250 μL of sample infusion (solid–liquid ratio: 1:100) or gallic acid standard solution were pipetted into a cuvette containing 1 mL of Folin–Ciocalteu reagent. After 5 min at room temperature, 1 mL of 10% Na_2CO_3 was added. The absorbance was measured at 765 nm against a blank sample. The TPC was calculated based on the standard curve using gallic acid and represented as gallic acid equivalents (GAE) mg/g of leaves on dry weight.

The TFC was assayed using the aluminum chloride colorimetric technique in accordance with Yang et al. [26]. Specifically, 2 mL of tea extract was mixed with 6.4 mL of a 1.5% $AlCl_3$ solution and 3.2 mL of sodium acetate buffer solution (0.1 M, pH 5.5). The combined volume was brought up to 20 mL with 50% ethanol, incubated in the dark at 25 °C for 30 min, and the absorbance was measured at 415 nm. A standard curve was prepared using different concentrations of rutin solution, and the results were expressed as rutin equivalents (RE) mg/g of leaves on dry weight.

The TP was analyzed using the phenol–sulfuric acid method [27]. A total of 1.0 g lyophilized powder was mixed with 20 mL of distilled water and extracted twice at 90 °C for 2 h. After extraction, the filtrate was adjusted to constant volume 50 mL. After diluting 50 times, 1 mL of diluted test samples was taken out, and 1 mL of 5% phenol and 5 mL of concentrated sulfuric acid were mixed. The mixture was incubated in a water bath at 40 °C for 30 min. After cooling, the absorbance of the sample was measured at 490 nm. The results were expressed as glucose equivalents (GE) mg/g of leaves on dry weight.

The detection method for the TSC was as referred to by de Aguiar NS [28]. Using the spectrophotometric method, it is possible to quantify TSC relative to the triterpene nucleus released after a hydrolysis reaction, with subsequent reaction with vanillin. Additionally, a standard curve was prepared using different concentrations of oleanolic acid, and the results were expressed as oleanolic acid equivalents (OAE) mg/g of leaves on dry weight.

2.5. Nontargeted Metabolomics Analysis by UHPLC-MS/MS

2.5.1. Metabolite Extraction

Fifty milligrams of each *C. paliurus* leaf sample was accurately weighed. Metabolites were extracted using 1 mL precooled mixtures of methanol, acetonitrile, and water (v/v/v, 2:2:1) and then placed for 1 h of ultrasonic shaking in ice baths. Following sonication, the mixture was placed at −20 °C for 1 h and centrifuged at 14,000 rpm for 20 min at 4 °C. The supernatants were carefully recovered and concentrated to dryness under vacuum conditions. To prepare the samples for mass spectrometry, 150 μL of precooled methanol–water (4:1, v/v) solution was added to redissolve the dried sample. The resulting mixture was centrifuged at 20,000 rpm for 20 min at 4 °C, and the supernatant was collected for further analysis.

2.5.2. UHPLC-MS/MS Analysis

The metabolomics profiling was conducted using a UPLC-ESI-Q-Orbitrap-MS system (UHPLC, Shimadzu Nexera X2 LC-30AD, Shimadzu, Japan) coupled with a Q-Exactive Plus instrument (Thermo Scientific, San Jose, CA, United States of America).

For liquid chromatography (LC) separations, samples were analyzed using an ACQUITY UPLC® HSS T3 column (2.1 × 100 mm, 1.8 µm). The flow rate was 0.3 mL/min, and the mobile phase consisted of: A: 0.1% formic acid (FA) in water and B: 100% acetonitrile (ACN). The gradient was 0% buffer B for 2 min and was linearly increased to 48% in 4 min and then up to 100% in 4 min and maintained for 2 min before decreasing to 0% mobile phase B in 0.1 min.

For mass spectrometry (MS) data acquisition, electrospray ionization (ESI) was performed with positive and negative modes applied separately. The following HESI source conditions were set: spray voltage: 3.8kV (positive) and 3.2kV (negative); capillary temperature: 320 °C; sheath gas (nitrogen) flow: 30 arb (arbitrary units); aux gas flow: 5 arb; probe heater temp: 350 °C; S-Lens RF Level: 50. The instrument was set to acquire data over the *m/z* range of 70-1050 Da for full MS scans. Full MS scans were acquired at a resolution of 70,000 at *m/z* 200, with a maximum acquisition time of 100 ms for MS scans and 50 ms for MS/MS scans. The isolation window for MS2 was set to 2 *m/z*, and the normalized collision energy (stepped) was set as 20, 30, and 40 for fragmentation.

2.5.3. Metabolome Data Preprocessing

The raw mass spectrometry (MS) data were processed using data-independent MS/MS deconvolution for comprehensive metabolome analysis; (MS-DIAL), a software tool for data-independent analysis (DIA) in mass spectrometry, was utilized for the analysis of the samples [29], for peak alignment, retention time correction, and peak area extraction. The identification of metabolites was based on accurate mass (tolerance < 10 ppm) and MS/MS data (tolerance < 0.02 Da) matched against databases such as HMDB (https://hmdb.ca/ (accessed on 9 March 2024)) and MassBank (https://massbank.eu/MassBank/ (accessed on 13 March 2024)). Only features with more than 50% nonzero measurement values in at least one group were retained for further analyses. In addition, the sweet–bitterness-related metabolites were identified through the screening process conducted by BitterDB (https://bitterdb.agri.huji.ac.il/dbbitter.php (accessed on 9 March 2024)) and VirtualTaste (https://insilico-cyp.charite.de/VirtualTaste (accessed on 13 March 2024)). Based on the predicted results, core taste metabolites were selected for molecular docking analysis.

2.6. Computational Study of Sweet/Bitter Compound Binding to T1R2/T1R3 and T2R4/T2R14 Receptors

The core sweet–bitterness-related differentially expressed metabolites were selected. Molecular docking analysis was used to clarify key sweet/bitter/bittersweet substances using a T1R2/T1R3 model and a T2R4/T2R14 model. Homology modeling was employed to generate the three-dimensional (3D) structures of human T1R2/T1R3 and T2R4/T2R14 receptors, as their structures were not available. The amino acid sequences of the sweet taste receptors hT1R2 and hT1R3 and the bitter taste receptors hT2R4 and hT2R14 were retrieved from the protein database (NCBI, http://www.ncbi.nlm.nih.gov/protein/ (accessed on 13 June 2024)) with accession numbers NP_689418.2, NP_689414.1, NP_058640.1, and NP_076411 [28]. These sequences were then submitted to the I-TASSER online modeling server, which utilized the LOMETS multi-threaded method to identify structural templates in the PDB database. Using protein sequences, 3D structures were constructed and evaluated for quality using ERRAT, Ramachandran plots, and C-scores. The best model was selected for molecular docking, with the DoGSiteScorer website (https://proteins.plus/ (accessed on 13 March 2024)) predicting docking sites for the receptor models [30]. Sweet /bitter/bittersweet compounds were docked into the predicted sites using AutoDock Vina, generating 20 poses for each ligand to interact with the protein binding site [31]. The binding energy (Affinity) of the ligand conformation was a key parameter, with lower binding energy indicating a more stable conformation and stronger binding affinity.

2.7. Statistical Analysis

The taste attributes obtained via electronic tongue analyses are expressed as means ± standard deviation and were subjected to one-way analysis of variance (ANOVA) with Duncan's test using SPSS (25.0, IBM) software. Hierarchical clustering analysis was performed using TBtools software (TBtools, Guangzhou, China), while SIMCA-P 14.1 multivariate statistical software (Umetrics, Umea, Sweden) was used for generating radar plots, principal component analysis (PCA), and orthogonal partial least squares discriminant analysis (OPLS-DA) along with other visual representations. The molecular docking results were visualized using PyMOL 2.2.0 software (PyMOL, DeLano Scientific LLC).

3. Results and Discussion

3.1. Electronic Tongue Analysis and Correlation Analysis

The taste attributes of the *C. paliurus* leaves in the five different months were measured by the electronic tongue system, as shown in Table 1. The tasteless points for all indices were set at 0, and values exceeding these tasteless points were deemed meaningful [32,33]. The taste values of sweetness, bitterness, and umami all had values higher than the tasteless points (p <0.05), signifying that these were the effective sensory indices for *C. paliurus* leaves in different months. As an herbal tea, *C. paliurus* tea has gradually entered the field of view of more consumers, but the widespread coexistence of sweetness and bitterness observed in the taste attributes, which cannot meet the high-quality taste expectations of consumers, has resulted in low market recognition of *C. paliurus* tea products. Therefore, this study primarily focused on the sweetness and bitterness during the five different growth months of *C. paliurus* leaves. As illustrated in Figure 1, the sweetness of *C. paliurus* leaves harvested in July was slightly lower than that in other months, meanwhile, the bitterness was significantly lower than that in other months ($p < 0.5$). According to previous reports [34], sweetness and bitterness of herbal teas were mostly determined by chemical composition, such as flavonoids, polyphenols, polysaccharides, and saponins. Hence, further main active compounds analysis and nontargeted metabolomics analysis of the five different growth months of *C. paliurus* leaves were carried out.

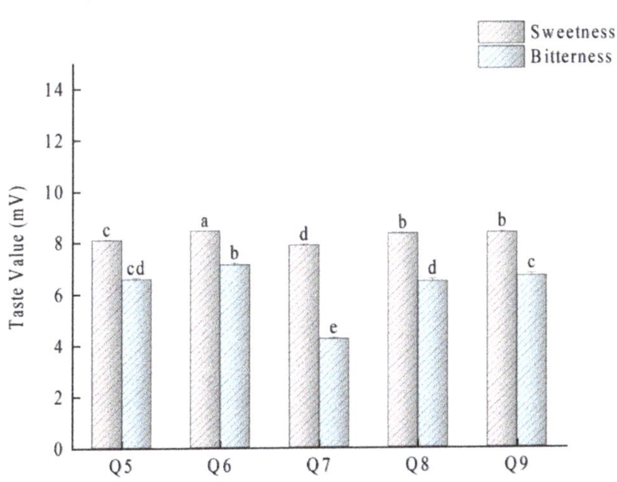

Figure 1. Bar chart of sweetness and bitterness values of *C. paliurus* leaves in different growth months. Notes: Q5–Q9 refer to the samples of the leaves of *C. paliurus* collected from May to September. [a–e] Means within the same taste value with different superscripts differ significantly ($p < 0.05$).

Table 1. Determination of taste characteristics by electronic tongue of *C. paliurus* harvested in different months.

Taste Characteristic	Samples of *C. paliurus*				
	Q5	Q6	Q7	Q8	Q9
Sweetness	8.09 ± 0.02 [c]	8.45 ± 0.00 [a]	7.86 ± 0.04 [d]	8.33 ± 0.01 [b]	8.36 ± 0.04 [b]
Sourness	−24.20 ± 0.15 [c]	−26.25 ± 0.06 [e]	−22.77 ± 0.23 [a]	−24.55 ± 0.11 [d]	−23.12 ± 0.13 [b]
Bitterness	6.57 ± 0.05 [cd]	7.12 ± 0.06 [b]	4.23 ± 0.04 [e]	6.46 ± 0.1 [d]	6.67 ± 0.10 [c]
Astringency	3.10 ± 0.23 [a]	2.21 ± 0.24 [c]	2.46 ± 0.16 [bc]	0.67 ± 0.31 [e]	1.24 ± 0.24 [d]
Bitter Aftertaste	2.26 ± 0.06 [c]	2.09 ± 0.05 [d]	1.59 ± 0.04 [f]	1.85 ± 0.05 [e]	2.40 ± 0.08 [b]
Astringency Aftertaste	2.76 ± 0.15 [a]	2.13 ± 0.11 [b]	2.74 ± 0.16 [a]	1.39 ± 0.09 [d]	1.70 ± 0.11 [c]
Umami	6.35 ± 0.19 [b]	7.90 ± 0.17 [a]	4.65 ± 0.21 [d]	5.55 ± 0.19 [c]	3.98 ± 0.20 [e]
Richness	2.48 ± 0.23 [b]	2.08 ± 0.14 [c]	2.33 ± 0.12 [bc]	1.50 ± 0.07 [d]	1.56 ± 0.19 [d]

Notes: Standard error of means (n = 3). [a–f] Means within the same row with different superscripts differ significantly ($p < 0.05$). Q5–Q9 refer to the samples of the leaves of *C. paliurus* collected from May to September, 2023.

3.2. Changes in TPC, TFC, TP and TSC of C. Paliurus Leaves in Different Growth Periods

Flavonoids, polyphenols, polysaccharides, and saponins were widely present in the *C. paliurus* leaves, which were harvested in the subtropical monsoon climate of Huping Mountain, Changde, Hunan Province [35,36]. The TPC, TFC, TP, and TSC of the *C. paliurus* leaves in the different growth months are presented in Figure 2. There was no significant difference in the TPC of the *C. paliurus* leaves in the different growth months ($p < 0.05$), suggesting that the phenolic levels may not have been significantly influenced by the progression of the growth period, or that other factors, such as regional climate conditions, might have played a more dominant role in modulating the TPC. The TFC could have been linked to increased solar radiation and temperatures in the early growth months, promoting synthesis of flavonoids, followed by decline as the season progressed and temperatures began to moderate. In terms of the TP and TSC, a reversal was observed in the TP of the *C. paliurus* leaves in the different growth periods, where an initial decline was succeeded by an upward movement, indicating a potential turning point or recovery. This pattern might have been related to the plants' adaptation to climatic conditions, such as increased rainfall in the early growth months providing sufficient hydration for polysaccharide synthesis, followed by a period of adjustment as the climate became hotter and drier. Moreover, the TSC of the *C. paliurus* leaves displayed periodic fluctuations, which could have been associated with seasonal factors. The periodic nature of these fluctuations might have been a response to changing environmental conditions, including variations in temperature, rainfall, and humidity throughout the growth months. In fact, phenolic acids are responsible for *C. paliurus* tea's distinctive color and taste, and bioactive components contribute to its antibacterial, antiviral, antioxidation, antihypertension, and hypolipidemic activities [3,37,38]. Previous studies on chemical composition showed that *C. paliurus* leaves contain rich phenolic acid compounds, especially flavonoids [4,39,40]. As shown in Figure 2, the active components of the *C. paliurus* exhibited dynamic seasonal variations, and the highest TFC, TP, and TSC were found in the *C. paliurus* leaves harvested in July and could be attributed to the climatic conditions of July. In fact, July is in the middle of summer compared to other months, and generally Huping Mountain experiences higher temperatures, higher humidity, and more variable rainfall patterns in July compared to other months. These conditions could favor biosynthesis and accumulation of secondary metabolites such as flavonoids, polysaccharides, and saponins in *C. paliurus* leaves. However, there was no significant correlation between the sweetness and bitterness of the *C. paliurus* leaves and the TPC, TFC, TP and TSC (Figure 3). Nevertheless, previous research has suggested that a higher polyphenol content contributes to increased bitterness [34]. The bitterness induced by polyphenols is influenced not only by their overall content but also by the specific composition of the polyphenols, their bitterness thresholds, and the effective concentrations at which they are present. Furthermore, empirical evidence has suggested that perceived intensity of bitterness does not exhibit a direct linear correlation with concentration of these compounds [41]. Consequently, nontargeted metabolomics and molecular docking technology were employed to further analyze the intrinsic relationship between the sweet /bitter/bittersweet substances.

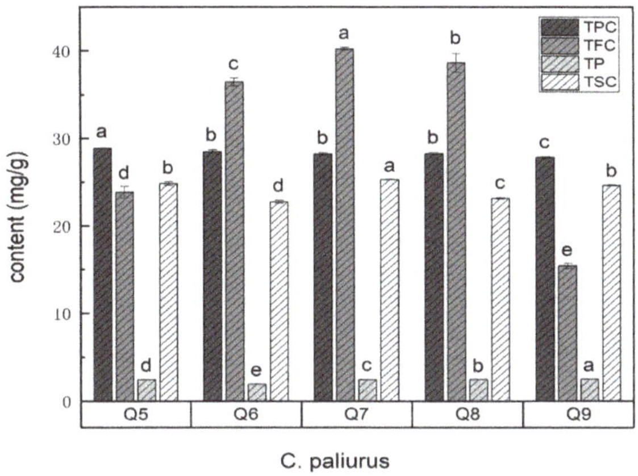

Figure 2. TPC, TFC, TP, and TSC from *C. paliurus* leaves. Notes: TPC (total phenolic content), TFC (total flavonoid content), TP (total polysaccharide content), and TSC (total saponin content). Q5–Q9 refer to the samples of the leaves of *C. paliurus* collected from May to September. [a–e] Means within the same indicator with different superscripts differ significantly ($p < 0.05$).

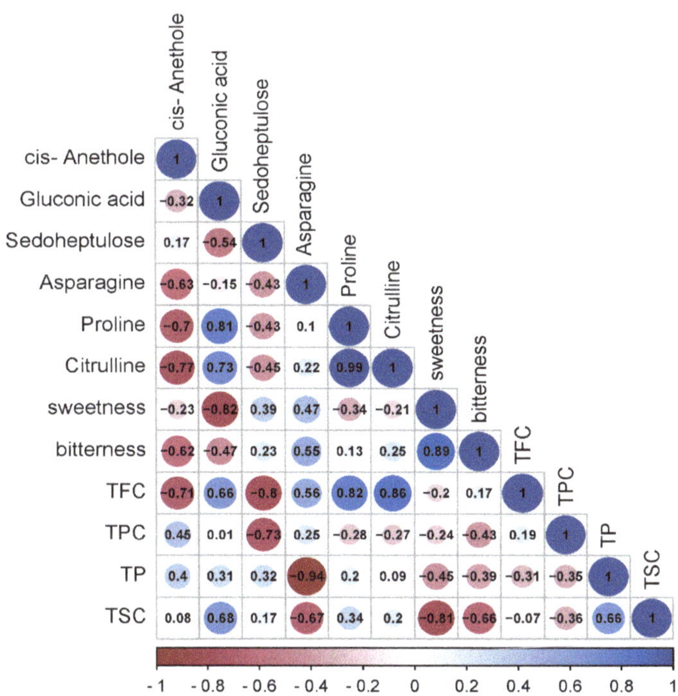

Figure 3. Associations between six key metabolites and the rest of the indices. Notes: The color gradient from red to blue on the chart illustrates the intensity of correlation coefficients, with red indicating the weakest association and blue representing the strongest.

3.3. Nontargeted Metabolomics Analysis of C. Paliurus Leaves

Initial investigation revealed that *C. paliurus* leaves possess notable bitterness in addition to their sweetness, factors that significantly affect their sensory quality. In this study, nontargeted metabolomics based on UHPLC-MS/MS analysis was undertaken to identify the potential sweet and bitter components within the *C. paliurus* leaves. Utilizing the PRIMe database (PRIMe: Platform for RIKEN Metabolomics) for search and metabolite analysis, a total of 1571 secondary metabolites were successfully characterized from the *C. paliurus* leaves, with both positive and negative patterns detected. A circus plot was used to analyze the associations between the differential metabolites and for classification. Metabolite name, HMDB classification of metabolite, *p*-value, VIP from OPLS-DA analysis, and correlation line were listed from outside to inside. As shown in Figure 4A, these different metabolites included 603 positive correlations (38.4%) and 441 negative correlations (28.1%). Of these, 575 differential metabolites were identified and systematically categorized into 15 distinct classes (Figure 4B). These categories included 80 prenol lipids (13.91%), 62 organooxygen compounds (10.78%), 56 flavonoids (9.74%), 48 fatty acyls (8.35%), 33 carboxylic acids and derivatives (5.74%), 32 steroids and steroid derivatives (5.57%), 31 benzene and substituted derivatives (5.39%), 21 coumarins and derivatives (3.65%), 20 phenols (3.48%), 15 cinnamic acids and derivatives (2.61%), 11 indoles and derivatives (1.91%), 8 organonitrogen compounds (1.39%), 8 purine nucleosides (1.39%), 6 glycerophospholipids (1.04%), and 144 undefined compounds (25.04%).

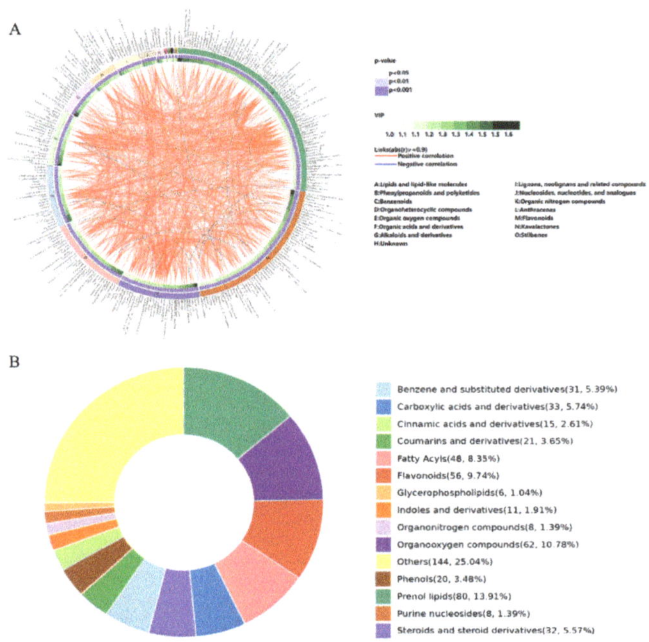

Figure 4. Nontargeted metabolomics analysis. (**A**) The circus plot of differential metabolites. Notes: Circus circle plot is arranged from the outside inwards in the following order: metabolite names, HMDB classifications of metabolites, p-values, VIP from PLS-DA analysis, and correlation lines. Arcs connecting any two points on the circle are called chords, indicating a correlation between the two points. (**B**) The proportion of various metabolites in *C. paliurus* leaf samples. Notes: Various color-coded sections denote items categorized under distinct chemical groups, with the corresponding percentages indicating the proportion of items within each chemical category. The proportion of metabolites is calculated as a fraction of the total metabolites detected. Those metabolites lacking a defined chemical classification are categorized as "others".

The principal component analysis (PCA), partial least squares discriminant analysis (PLS-DA), and orthogonal projections to latent structures discriminant analysis (OPLS-DA) have been used to pinpoint the metabolite combinations that explain the highest variance and to illustrate the clustering patterns of the tea samples [42]. All the metabolites were subjected to multivariate analysis using SIMCA-P 14.1 multivariate statistical software. The results of the PCA (Figure 5A) illustrated that PC1 and PC2 accounted for 34.49% and 17.51% of the total variation, respectively. In addition, five samples were completely separated in the PCA map, which indicated that there were significant differences in the metabolite content of samples from different growth months. The OPLS-DA utilizes a predictive principal component (t1) for group difference detection and multiple orthogonal components for intra-group variability. Therefore, to obtain a higher level of population separation and better understand the differences between the *C. paliurus* leaves in different months, the OPLS-DA was used for classification (Figure 5B). Permutation test cross-validation confirmed the robustness of the OPLS-DA model, exhibiting R2 and Q2 intercepts of 0.999 and 0.9975, which proved that the model had good prediction ability (Table S2).

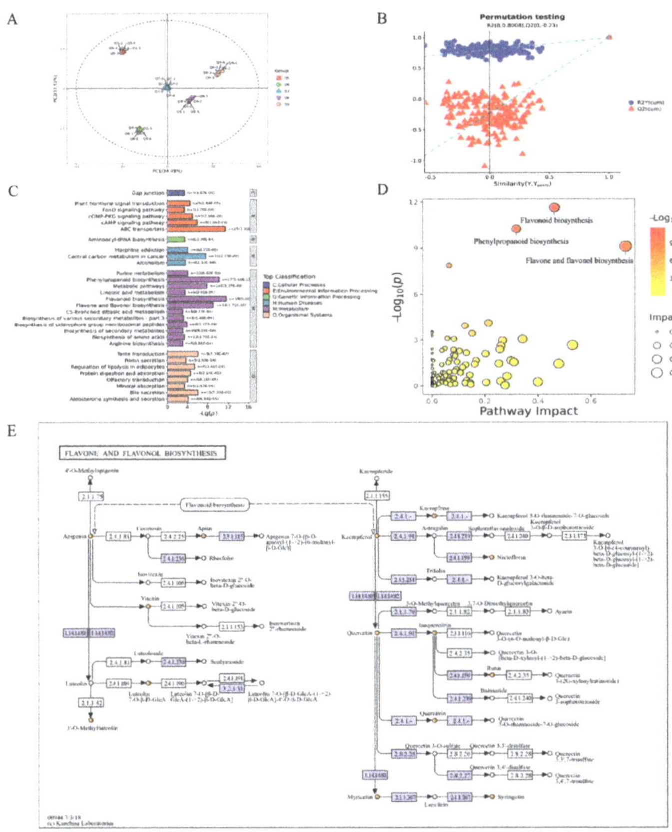

Figure 5. Multivariate analyses of *C. paliurus* leaf samples. (**A**) Principal component analysis (PCA) of metabolite profiles across different groups. (**B**) OPLS-DA permutation test plot of the comparison groups. (**C**) The bar chart of significant KEGG enrichment pathways. (**D**) The diagram of pathway impact. (**E**) The flavone and flavanol biosynthesis pathways.

Since different metabolites coordinate their biological functions, the Kyoto Encyclopedia of Genes and Genomes (KEGG) pathway-based analysis is instrumental in further elucidating their roles [43]. Fisher's exact test was used to analyze and calculate the significance level of the metabolite enrichment in each pathway to identify the metabolic and signal transduction pathways that were significantly affected. The KEGG enrichment pathway map is presented in Figure 5C. The x-axis represents the negative logarithm of the *p*-values, while the y-axis indicates the specific pathways. Colors of the bars were assigned based on the KEGG pathway level 1 classifications.

As shown in Figure 5D, the majority of the differentially expressed metabolites were primarily associated with biosynthetic pathways, such as flavonoid and flavanol biosynthesis, flavonoid biosynthesis, phenylpropanoid biosynthesis, and other pathways highlighted by the KEGG enrichment analysis, all of which exhibited statistical significance($p < 0.05$). As the most significant pathways, the flavone and flavanol biosynthesis pathways have been extensively studied in *C. paliurus* [44]. Given the visual representation of the pathways in Figure 5E, the complex network of synthesis and transformation of key flavonoid components such as apigenin, luteolin, and their glycosides and derivatives is revealed. The diagram captures the interplay between these compounds and their conversions into various forms. Additionally, it highlights synthesis and transformation of three flavanol components: kaempferol, quercetin, and myricetin, as well as their derivatives. Flavanols belong to the class of flavonoids, which mainly exist as glycosides in *C. paliurus* leaves, and have a certain contribution to the taste quality of tea, especially bitterness, that has been noted in other herbal teas [45]. In this study, the data of the nontargeted metabolomics analysis showed a correlation with the taste indices of the electronic tongue analysis, in which luteolin (VIP = 1.22) and apigenin (VIP = 1.14) were the core contributions to the difference between these *C. paliurus* leaf samples as indicated by high VIP values, which were responsible for the tea infusion's bitterness [45]. During the growth of the *C. paliurus* leaves in the different months, the concentration of luteolin exhibited a pattern of initial increase peaking in June, followed by a decrease. In contrast, the concentration of apigenin followed an inverse trend, initially declining and then rising, with the lowest concentration observed in June. Overall, the core bitterness-related metabolic (luteolin and apigenin) variation between the *C. paliurus* leaves' growth stages was revealed via nontargeted metabolomics analysis. However, *C. paliurus* leaves have an observed widespread coexistence of sweetness and bitterness in the taste attributes, and key metabolites of the potential "sweet-bitterness" in *C. paliurus* leaves have not been reported. Therefore, a deeper investigation into the differential metabolites that contribute to the characteristic sweet-bitterness of *C. paliurus* leaves was carried out.

To identify the differential metabolites that may be related to the sweet/bitter taste of *C. paliurus*, the 575 different metabolites were further screened by BitterDB (https://bitterdb.agri.huji.ac.il/dbbitter.php (accessed on 9 March 2024)) and the related literature and using VirtualTaste (https://insilico-cyp.charite.de/VirtualTaste (accessed on 13 June 2024)) to predict the sweet and bitter taste. A total of 68 metabolites were predicted to be active in sweetness. The Variable Importance for the Projection (VIP) score measures metabolite expression patterns' impact on sample classification, identifying the key discriminative. Therefore, the core metabolites with VIP ≥ 1 were selected from VirtualTaste and were used to predict important explanatory variables for sweet and bitter substances. Detailed information about these metabolites is provided in Table S1. These metabolites included 20 phenylpropanoids and polyketides (29.0%), 17 lipids and lipid-like molecules (24.6%), 12 organic oxygen compounds (17.4%), 8 organic acids and derivatives (11.6%), 6 benzenoids (8.7%), 2 lignans, neolignanes, or related compounds (2.9%), 1 kavalactone (1.4%), 1 organic nitrogen compound (1.4%), 1 organoheterocyclic compound (1.4%), and 1 superclass (1.4%).

3.4. Molecular Docking of Candidate Sweet/Bitter Substances with the T1R1/T1R3 and T2R4/T2R14 Receptors

To investigate the potential sweet/bitter/bittersweet substances, molecular docking analysis was conducted to evaluate the stability of and the interactions between these substances and the human sweet receptors (T1R1/TIR3) and bitter receptors (T2R4/T2R14).

The 68 core sweetness metabolites predicted as active in sweetness were further screened (Sweet Prediction > 0.7, Bitter Prediction > 0.7) and categorized into four groups for detailed analysis: Category 1 included the bitter substances identified in BitterDB with active Bitter Predictions in VirtualTaste; Category 2 comprised the bitter substances listed in BitterDB but predicted as inactive in VirtualTaste; Category 3 encompassed the non-bitter substances according to BitterDB with active Bitter Predictions in VirtualTaste; Category 4 consisted of the bitter substances found in BitterDB but rated as inactive in VirtualTaste (Table S2).

The amino acid sequences of the sweet taste receptors hT1R2 and hT1R3 and the bitter taste receptors hT2R4 and hT2R14 (IDs NP_689418.2, NP_689414.1, NP_058640.1, NP_076411.1, respectively) were downloaded from the NCBI database (Figure S1). Each sequence was submitted individually to the I-TASSER online modeling server, which employs the LOMETS multi-threaded method to identify structural templates in the PDB database. The server selected the top-scoring template from each of the 10 threaded programs based on Z-score, a measure of template alignment quality. Following template selection, I-TASSER constructed a full-length protein model by assembling contiguous segments derived from the threaded alignment through a Monte Carlo simulation that replicated an exchange process. This computational approach generated comprehensive 3D protein models suitable for further analysis and molecular docking studies [46]. The quality of the resulting models was assessed using ERRAT, Ramachandran plots, and C-score to select the optimal model for the subsequent molecular docking (Figure 6A).

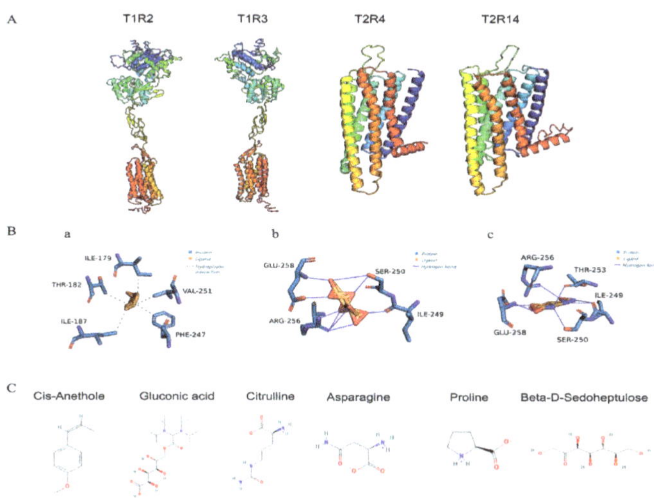

Figure 6. Homologous modeling and screening process of molecular docking. (**A**) Sweet receptors (T1R1/TIR3) and bitter receptors (T2R4/T2R14). (**B**) Interaction patterns of hT2R14 with cis-anethole, gluconic acid and citrulline; (**a**) hT2R14–cis-anethole, (**b**) hT2R14–gluconic acid, (**c**) hT2R14–citrulline. (**C**) Six core sweet- and bitter-related metabolites.

The DoGSiteScorer (https://proteins.plus/ (accessed on 13 June 2024)) website was utilized to predict docking sites for the hT1R2, hT1R3, hT2R4, and hT2R14 proteins. Sweet taste receptor binding sites (for hT1R2 and hT1R3) were identified in the Venus Flytrap Module (VFTM), situated in the central cleft between the two larger lobes. In contrast, bitter taste receptor binding sites (for hT2R4 and hT2R14) were located in the central cavity formed by seven transmembrane helices, consistent with the prior research on docking sites of the sweet and bitter taste receptors with various ligands [47,48].

The small molecules, after undergoing pre-treatment, were subjected to docking simulations with the four pre-processed receptor models using the Vina software 1.2.0. The binding energy (Affinity) of the ligand conformation is defined as a key parameter in the process of molecular docking, where a lower binding energy indicates a more stable ligand conformation. This stability correlates with a stronger binding affinity to the protein receptor, suggesting a potentially higher inhibitory effect on the receptor's activity [49]. Given the direct correlation between the sweetness of a ligand and its interaction energy with sweetness receptors, hT1R2 and hT1R3 are capable of independently yet synergistically generating sweetness signals. This dual capacity facilitates a cumulative effect, wherein the combined action of these receptors can significantly amplify the perceived sweetness [50]. To elucidate the differential effects on taste perception, an analysis comparing the binding affinities of the sweet and bitter taste receptors was conducted. Specifically, this study calculated the sum of the lowest binding energies (denoted as As) for the binding sites of the sweet taste receptors hT1R2 and hT1R3. At the same time, the sum of the lowest binding energies of the bitter receptors hT2R4 and hT2R14 (expressed as Ab) was also predicted. This comparative approach allowed us to assess the relative inhibitory potential of the ligands on the receptors responsible for sweet and bitter taste sensations. The relevant docking results are shown in Table 2. The data of correlation analysis pointed out a positive correlation between the sum of the lowest binding energies of the sweet receptor (As) and the Sweet Predict scores, indicating that the likelihood of the ligand being perceived as sweet increased with an increase in As. Conversely, a negative correlation was observed between the sum of the lowest binding energies for the bitter taste receptors (Ab) and the Bitter Predict scores, indicating that higher Ab values were linked to a reduced likelihood of bitterness perception. Thus, As ≥ -15 and Ab ≥ -15 were identified as thresholds for the identification of high sweetness and low bitterness metabolites. Finally, six compounds (cis-anethole, gluconic acid, beta-D-Sedoheptulose, asparagine, proline, citrulline) were selected to match the conditions (As ≥ -15 and Ab ≥ -15) (Figure 6C). Among them, cis-anethole belongs to benzenoids; gluconic acid and beta-D-Sedoheptulose belong to organic oxygen compounds. Asparagine, proline, and citrulline belong to organic acids and derivatives. Based on the average standardized peak area data from Table 3, the levels of the six metabolites (cis-anethole, gluconic acid, beta-D-Sedoheptulose, asparagine, proline, and citrulline) in the *C. paliurus* leaves remained high from May to September. Sedoheptulose exhibited the highest content, with significant fluctuations across different months. In contrast, the levels of the other five metabolites were relatively consistent during June, July, and August. Therefore, these six core metabolites with high sweetness and low bitterness could be potential candidates for taste modulation in *C. paliurus* leaves. These findings suggest that by selectively enriching these compounds or by employing targeted breeding and genetic modification strategies, it may be possible to enhance the palatability of *C. paliurus* products. Moreover, this method enabled the targeted identification of compounds that were likely contributors to the sweetness and bitterness, thereby deepening our comprehension of the molecular underpinnings of taste perception within this specific context.

Currently, the hT2R14 receptor has been recognized as one of the most broadly responsive bitterness receptors, capable of detecting a wide array of bitter compounds [51,52]. Potency of these bitter substances can be significantly mitigated through inhibition of the hT2R14 receptor, offering a potential strategy for modulating bitterness intensity in various applications. Thr (threonine), Asp (aspartic acid), and Phe (phenylalanine) have been identified as the key amino acids responsible for docking of bitter taste blockers with hT2R14 in molecular docking calculations [53]. The interactions between the hT2R14 receptor and cis-anethole, gluconic acid, and citrulline were assessed using the PLIP web service (Protein-Ligand Interaction Profiler, https://projects.biotec.tu-dresden.de/plip-web/plip/index (accessed on 11 June 2024)), which provides an online platform for predicting and analyzing protein–ligand interactions. The results of this prediction were then imported into PyMOL, a molecular visualization tool, to render three-dimensional representations of the complexes. According to Figure 6B, the analyses of the ligands' interactions with the hT2R14 model at the docking site revealed distinct binding characteristics for the core metabolites from the *C. paliurus* leaves. Cis-anethole demonstrated no interaction with Asp but formed hydrogen bonds with Thr-182 and Phe-247. Citrulline was found to engage in hydrophobic interactions with Thr-253. In contrast, gluconic acid did not create any significant bonds with amino acids Thr, Asp, or Phe. The differential binding profiles of these metabolites to the hT2R14 receptor were reflected in their respective binding energies (Ab), with cis-anethole showing the lowest, followed by citrulline, with gluconic acid exhibiting the highest. This order of binding energies (cis-anethole<citrulline<gluconic acid) suggests a potential inverse relationship between the binding strength to the receptor and the perceived bitterness of the metabolites. This methodological approach, therefore, offers a promising strategy for predicting bitterness intensity of differential metabolites in *C. paliurus* leaves. By identifying metabolites that exhibit weaker binding to the hT2R14 receptor, it may be possible to select or modify compounds that contribute favorably to the taste profile of *C. paliurus* tea, enhancing palatability and potential health benefits. Furthermore, the identification of key amino acids involved in the binding of bitter compounds to the hT2R14 receptor opens up avenues for developing taste blockers or modifiers that can reduce the bitterness of *C. paliurus* leaves. This could involve the design of small molecules that can competitively inhibit binding of bitter compounds to the receptor, thereby reducing perceived bitterness.

Table 2. Relevant metabolites screened for molecular docking.

Metabolite Name	hT1R2	hT1R3	hT2R4	hT2R14	As	Ab
		Category 1				
Ganoderic Acid B	−13.0	−11.9	−12.3	−11.7	−24.9	−24.0
Deoxycholic acid	−10.9	−10.8	−11.7	−10.7	−21.7	−22.4
Indole-3-acetyl-valine	−9.2	−8.9	−8.6	−7.9	−18.1	−16.5
		Category 2				
Ginsenoside Rf	−10.1	−12.7	−10.6	−10.4	−22.8	−21.0
FA 18:3;O	−6.8	−8.7	−7.9	−7.0	−15.5	−14.9
cis-Anethole	−5.8	−5.6	−5.8	−5.6	−11.4	−11.4
Icariside F2	−10.1	−9.7	−8.4	−9.0	−19.8	−17.4
Forsythoside E	−10.3	−10.7	−8.0	−8.7	−21.0	−16.7
Gluconic acid	−5.7	−5.6	−5.0	−5.1	−11.3	−10.1
Olivil 4′-O-Glucoside	−9.2	−10.6	−9.2	−8.7	−19.8	−17.9
		Category 3				
Centaurein	−10.5	−10.0	−8.4	−9.1	−20.5	−17.5
Iridin	−9.2	−10.0	−8.8	−8.0	−19.2	−16.8
Syringetin	−9.9	−9.6	−8.3	−8.3	−19.5	−16.6
Dihydromelilotoside	−8.6	−8.5	−7.8	−7.1	−17.1	−14.9
Lamiide	−10.3	−10.1	−8.6	−8.0	−20.4	−16.6
		Category 4				
Sedoheptulose	−7.0	−6.8	−5.8	−6.6	−13.8	−12.4
Hematoxylin	−10.5	−9.3	−9.4	−8.9	−19.8	−18.3
Asparagine	−5.1	−5.4	−4.6	−5.2	−10.5	−9.8
Proline	−5.2	−5.3	−5.0	−5.4	−10.5	−10.4
Citrulline	−5.7	−5.8	−5.1	−5.7	−11.5	−10.8
Gaultherin	−9.6	−9.0	−7.8	−8.1	−18.6	−15.9
Apiopaeonoside	−9.7	−9.3	−8.5	−8.2	−19.0	−16.7

Notes: the sum of the lowest binding energies for the sweet taste receptors (As); the sum of the lowest binding energies for the bitter taste receptors (Ab).

Table 3. The mean values of the normalized peak areas of core compounds of *C. paliurus* leaves.

Metabolite Name	Mean Value of Normalized Peak Area				
	Q5	Q6	Q7	Q8	Q9
cis-Anethole	$2.19 \times 10^{8\ \mathrm{d}}$	$2.63 \times 10^{8\ \mathrm{c}}$	$3.68 \times 10^{8\ \mathrm{a}}$	$3.57 \times 10^{8\ \mathrm{a}}$	$3.15 \times 10^{8\ \mathrm{b}}$
Gluconic acid	$9.52 \times 10^{10\ \mathrm{a}}$	$1.11 \times 10^{10\ \mathrm{c}}$	$6.85 \times 10^{10\ \mathrm{b}}$	$1.86 \times 10^{10\ \mathrm{c}}$	$9.58 \times 10^{9\ \mathrm{c}}$
Sedoheptulose	$1.08 \times 10^{9\ \mathrm{b}}$	$2.15 \times 10^{9\ \mathrm{b}}$	$1.85 \times 10^{9\ \mathrm{b}}$	$2.26 \times 10^{9\ \mathrm{b}}$	$1.04 \times 10^{10\ \mathrm{a}}$
Asparagine	$1.90 \times 10^{8\ \mathrm{b}}$	$4.33 \times 10^{8\ \mathrm{a}}$	$3.22 \times 10^{7\ \mathrm{c}}$	$9.19 \times 10^{7\ \mathrm{c}}$	$1.51 \times 10^{7\ \mathrm{d}}$
Proline	$2.59 \times 10^{10\ \mathrm{a}}$	$3.70 \times 10^{9\ \mathrm{c}}$	$5.04 \times 10^{9\ \mathrm{b}}$	$5.75 \times 10^{9\ \mathrm{b}}$	$2.97 \times 10^{9\ \mathrm{c}}$
Citrulline	$3.59 \times 10^{8\ \mathrm{a}}$	$1.00 \times 10^{8\ \mathrm{b}}$	$6.53 \times 10^{7\ \mathrm{c}}$	$1.09 \times 10^{8\ \mathrm{b}}$	$5.82 \times 10^{7\ \mathrm{c}}$

Notes: Q5–Q9 refer to the samples of the leaves of *C. paliurus* collected from May to September, 2023. $^{\mathrm{a–d}}$ Means within the same row with different superscripts differ significantly ($p < 0.05$).

To statistically calculate the relationships between the core six metabolites and the taste intensity, Spearman's correlation analysis coefficient was utilized in Figure 6. There was a significant correlation between the sensory characteristics (sweetness and bitterness) and the six core metabolites. The absolute value of a correlation coefficient in statistics indicates the strength of the correlation: the closer the coefficient is to 1 or −1, where −1 means completely negative correlation, 1 means completely positive correlation, and 0 means completely irrelevant, the stronger the correlation. [54]. According to Figure 6, asparagine exhibited a very strong negative correlation with the TP (−0.94), sedoheptulose showed a very strong negative correlation with the TFC (−0.8), and cis-anethole had a strong negative correlation with the TFC (−0.71). Gluconic acid demonstrated a strong positive correlation with both the TSC (0.68) and TFC (0.66), while asparagine had a strong negative correlation with the TSC (−0.67). These correlations hinted at a fascinating interaction between the metabolite profiles and the taste profiles of the *C. paliurus* leaves. It was reasonable to deduce that the concentrations of cis-anethole and sedoheptulose could significantly influence the bitterness intensity of the *C. paliurus* leaf samples, considering their direct associations with the TFC and TPC, respectively. In fact, bitterness was generally considered undesirable; still, they were important for providing the complex sensory perceptions of *C. paliurus* teas. Therefore, it is possible to design small molecule compounds that can competitively inhibit the binding of bitter compounds such as cis-anethole and sedoheptulose to receptors, thereby reducing the perceived bitterness. Meanwhile, asparagine had high Pearson correlation coefficients with bitterness and sweetness and negative correlation with the TSC and TP; the concentration of asparagine in the *C. paliurus* leaves was likely a significant determinant of the sweetness intensity. However, as one of the components of the synthetic sweetener aspartame, L-asparagine was without taste, while D-asparagine was intensely sweet. Details and mechanisms of interactions between chiral sweet molecules and sweet receptors remain to be elucidated [55]. Therefore, this study advanced our understanding of the metabolic changes and the sweetness and bitterness taste in the different growth months of the *C. paliurus* leaves and provided a new method for detailing the interactions between sweet molecules and sweet receptors.

Moreover, growing demand for tasty and healthy food has driven the development of low-calorie sweeteners originated from natural sources [56]. Natural sweeteners generally refer to a kind of sweet chemical component directly extracted from nature or properly modified, mainly secondary metabolites of plants, which include steviol glycosides, rebaudiosides, Luo Han Guo (monk fruit), etc. [57]. Natural sweeteners such as stevia have been widely used in high-end beverages, food, and health products and have been recognized by consumers, with an increasing position in the international context of sugar reduction [58]. Of note, leaves of the stevia plant have been widely used to make a sugar substitute called stevia. However, stevia has a bitter aftertaste, which makes it similar to the leaves of *C. paliurus*. Therefore, the leaves of *C. paliurus* have potential application value for developing natural sweeteners. This study provided a theoretical basis and method for design and development of a *C. paliurus* natural sweetener and related products.

4. Conclusions

This study explored the differential metabolites contributing to the sweet and bitter taste in *C. paliurus* leaves from different growth months using a multiplatform metabolomics approach, including UHPLC-MS/MS, electronic tongue assay, and molecular docking approaches. The results indicated significant variations in the taste characteristics and the content of bioactive substances, with leaves harvested in July exhibiting relatively high sweetness and the lowest bitterness. Then nontargeted metabolomics revealed 1571 differential metabolites, with 575 identified as key metabolites based on VIP values. Molecular docking analysis elucidated the interactions between these metabolites and the sweetness (hT1R2/hT1R3) and bitterness (hT2R4/hT2R14) receptors, revealing a positive correlation between the binding affinities and perceived sweetness, and a negative correlation for bitterness. Notably, six core metabolites (cis-anethole, gluconic acid, beta-D-Sedoheptulose, asparagine, proline, and citrulline) were identified as potentially modulating the sweet–bitter taste of the *C. paliurus* leaves. This discovery highlighted the effectiveness of the integrated analytical and in-computer simulation screening method in uncovering and characterizing compounds critical to taste profiles of food products. Moreover, this study provided a foundation for enhancing the taste attributes of *C. paliurus* leaves, which could increase their marketability and meet consumer demands for high-quality taste experiences. Further studies are ongoing in the author's lab, focusing on the formation mechanisms of the key sweet–bitterness related components in *C. paliurus* leaves. We hope to report more about these advancements in the future.

Supplementary Materials: The following supporting information can be downloaded at: https://www.mdpi.com/article/10.3390/foods13193089/s1, Figure S1: Template used for each receptor modeling; Table S1. The compound information for different growth months (Q5–Q9) detected in UHPLC-MS/MS analysis; Table S2. Molecular docking results of ligands with T1R2\T1R3 and T2R4\T2R14.

Author Contributions: Conceptualization, L.C., B.L.; methodology, L.C., D.L., Y.W., T.Z. and B.L.; formal analysis, L.C., D.L., Y.W., Y.Z., R.Z. and X.T.; investigation, L.C., Y.W., Y.Z. and R.Z.; writing—original draft, L.C. and Y.Z.; writing—review and editing. L.C., W.W. and X.T.; data curation, D.L., T.Z., B.L. and K.Z.; visualization, K.Z.; supervision, W.W. and X.T.; funding acquisition, W.W. and X.T. All authors have read and agreed to the published version of the manuscript.

Funding: This research was funded by "Hunan Natural Science Foundation Program" (Project No. 2023JJ30445, 2024JJ6337), "Hunan Provincial Department of Education Outstanding Youth Project" (Project No.22B0378), "Hunan Province Traditional Chinese Medicine Research Project" (Project No. B2024012), "Hunan University of Chinese Medicine "Double First-class" construction discipline project" (Project No.2022ZYT029), "The State-sponsored Postdoctoral Researcher Program of China" (Project No.GZB20230206).

Data Availability Statement: The original contributions presented in the study are included in the article/Supplementary Material; further inquiries can be directed to the corresponding author.

Acknowledgments: The authors thank Hunan Yueling Junshan Agriculture and Forestry Technology Co., Ltd. for providing the *Cyclocarya paliurus* leaves and support.

Conflicts of Interest: The authors declare that they have no known competing financial interests or personal relationships that could have appeared to influence the work reported in this paper.

References

1. Yang, Y.; Guo, T.; Huang, F.; Zheng, H.; Li, W.; Yuan, H.; Xie, Q.; Hussain, N.; Wang, W.; Jian, Y. α-Glucosidase inhibitory flavonol glycosides from *Cyclocarya paliurus* (Batalin) Iljinskaja and their kinetics characteristics. *Phytochemistry* **2024**, *225*, 114–195. [CrossRef]
2. Fang, S.; Sun, D.; Shang, X.; Fu, X.; Yang, W. Variation in radial growth and wood density of *Cyclocarya paliurus* across its natural distribution. *New For.* **2020**, *51*, 453–467. [CrossRef]
3. Zhao, J.; Wang, Z.; Xu, D.; Sun, X. Advances on *Cyclocarya paliurus* polyphenols: Extraction, structures, bioactivities and future perspectives. *Food Chem.* **2022**, *396*, 133667. [CrossRef] [PubMed]
4. Chen, Z.; Jian, Y.; Wu, Q.; Wu, J.; Sheng, W.; Jiang, S.; Shehla, N.; Aman, S.; Wang, W. *Cyclocarya paliurus* (Batalin) Iljinskaja: Botany, Ethnopharmacology, phytochemistry and pharmacology. *J. Ethnopharmacol.* **2022**, *285*, 114912. [CrossRef] [PubMed]
5. Liu, C.; Zhao, M.; Wen, L.; Zhao, H. The chemical composition and toxic effects of aqueous extracts of *Cyclocarya paliurus* leaves. *Front. Nutr.* **2022**, *9*, 994055. [CrossRef] [PubMed]
6. Khan, I.; Hou, F.; Le, H.P. The impact of natural resources, energy consumption, and population growth on environmental quality: Fresh evidence from the United States of America. *Sci. Total Environ.* **2021**, *754*, 142–222. [CrossRef]
7. Li, Y.; Quan, H.; Liang, L.; Yang, T.; Feng, L.; Mao, X.; Wang, Y. Nontargeted metabolomics reveals the discrimination of *Cyclocarya paliurus* leaves brewed by different methods. *Food Res. Int.* **2021**, *142*, 110221. [CrossRef]
8. Liu, W.; Wu, Y.; Hu, Y.; Qin, S.; Guo, X.; Wang, M.; Wu, L.; Liu, T. Effects of *Cyclocarya paliurus* Aqueous and Ethanol Extracts on Glucolipid Metabolism and the Underlying Mechanisms: A Meta-Analysis and Systematic Review. *Front. Nutr.* **2020**, 7. [CrossRef]
9. Shen, Y.; Peng, Y.; Zhu, X.; Li, H.; Zhang, L.; Kong, F.; Wang, J.; Yu, D. The phytochemicals and health benefits of *Cyclocarya paliurus* (Batalin) Iljinskaja. *Front Nutr.* **2023**, *10*, 1158158. [CrossRef] [PubMed]
10. Zhang, S.; Nie, L.; Zhao, W.; Cui, Q.; Wang, J.; Duan, Y.; Ge, C. Metabolomic analysis of the occurrence of bitter fruits on grafted oriental melon plants. *PLoS ONE* **2019**, *14*, 0223707. [CrossRef]

11. Kim, S.S.; Kwak, H.S.; Kim, M.J. The effect of various salinity levels on metabolomic profiles, antioxidant capacities and sensory attributes of doenjang, a fermented soybean paste. *Food Chem.* **2020**, *328*, 127176. [CrossRef] [PubMed]
12. Liu, S.; Shao, L.; Gong, J.; Sheng, J.; Ning, Z.; Xu, X.; Wang, H. Discovery of a temperature-dependent protease spoiling meat from *Pseudomonas fragi*: Target to myofibrillar and sarcoplasmic proteins rather than collagen. *Food Chem.* **2024**, *457*, 140155. [CrossRef] [PubMed]
13. Kakar, M.U.; Naveed, M.; Saeed, M.; Zhao, S.; Rasheed, M.; Firdoos, S.; Manzoor, R.; Deng, Y.; Dai, R. A review on structure, extraction, and biological activities of polysaccharides isolated from *Cyclocarya paliurus* (Batalin) Iljinskaja. *Int. J. Biol. Macromol.* **2020**, *156*, 420–429. [CrossRef] [PubMed]
14. Yu, L.; Hua, K.-D.; Chen, C.-Q.; Yue, J.-J.; Yuan, J.-L. Metabolomic and transcriptomic analysis of bitter compounds in Dendrocalamopsis oldhamii shoots. *J. Food Compos. Anal.* **2024**, *130*. [CrossRef]
15. Zhang, L.; Cao, Q.-Q.; Granato, D.; Xu, Y.-Q.; Ho, C.-T. Association between chemistry and taste of tea: A review. *Trends Food Sci. Technol.* **2020**, *101*, 139–149. [CrossRef]
16. Tian, X.; Wang, H.; Chen, L.; Yuan, H.; Peng, C.; Wang, W. Distinct Changes in Metabolic Profile and Sensory Quality with Different Varieties of *Chrysanthemum* (Juhua) Tea Measured by LC-MS-Based Untargeted Metabolomics and Electronic Tongue. *Foods* **2024**, *13*, 1080. [CrossRef]
17. Fritz, F.; Preissner, R.; Banerjee, P. VirtualTaste: A web server for the prediction of organoleptic properties of chemical com-pounds. *Nucleic Acids Res.* **2021**, *49*, 679–684. [CrossRef] [PubMed]
18. Dagan-Wiener, A.; Di Pizio, A.; Nissim, I.; Bahia, M.S.; Dubovski, N.; Margulis, E.; Niv, M.Y. BitterDB: Taste ligands and receptors database in 2019. *Nucleic Acids Res.* **2018**, *47*, 1179–1185. [CrossRef]
19. Yu, Z.; Kang, L.; Zhao, W.; Wu, S.; Ding, L.; Zheng, F.; Liu, J.; Li, J. Identification of novel umami peptides from myosin via homology modeling and molecular docking. *Food Chem.* **2020**, *344*, 128728. [CrossRef]
20. Acevedo, W.; Ramírez-Sarmiento, C.A.; Agosin, E. Identifying the interactions between natural, non-caloric sweeteners and the human sweet receptor by molecular docking. *Food Chem.* **2018**, *264*, 164–171. [CrossRef] [PubMed]
21. Hutasingh, N.; Chuntakaruk, H.; Tubtimrattana, A.; Ketngamkum, Y.; Pewlong, P.; Phaonakrop, N.; Roytrakul, S.; Rungrotmongkol, T.; Paemanee, A.; Tansrisawad, N.; et al. Metabolite profiling and identification of novel umami compounds in the chaya leaves of two species using multiplatform metabolomics. *Food Chem.* **2023**, *404*, 134564. [CrossRef] [PubMed]
22. Spaggiari, G.; Di Pizio, A.; Cozzini, P. Sweet, umami and bitter taste receptors: State of the art of in silico molecular modeling approaches. *Trends Food Sci. Technol.* **2020**, *96*, 21–29. [CrossRef]
23. Ahmad, R.; Dalziel, J.E. G Protein-Coupled Receptors in Taste Physiology and Pharmacology. *Front. Pharmacol.* **2020**, *11*, 587664. [CrossRef] [PubMed]
24. Jaggupilli, A.; Howard, R.; Upadhyaya, J.D.; Bhullar, R.P.; Chelikani, P. Bitter taste receptors: Novel insights into the bio-chemistry and pharmacology. *Int. J. Biochem. Cell Biol.* **2016**, *77*, 184–196. [CrossRef]
25. Musci, M.; Yao, S. Optimization and validation of Folin–Ciocalteu method for the determination of total polyphenol content of Pu-erh tea. *Int. J. Food Sci. Nutr.* **2017**, *68*, 913–918. [CrossRef] [PubMed]
26. Xiao, Y.; He, C.; Chen, Y.; Ho, C.-T.; Wu, X.; Huang, Y.; Gao, Y.; Hou, A.; Li, Z.; Wang, Y.; et al. UPLC–QQQ–MS/MS-based widely targeted metabolomic analysis reveals the effect of solid-state fermentation with Eurotium cristatum on the dynamic changes in the metabolite profile of dark tea. *Food Chem.* **2022**, *378*, 131999. [CrossRef] [PubMed]
27. Wu, L.; Gao, Y.; Ren, W.-C.; Su, Y.; Li, J.; Du, Y.-Q.; Wang, Q.-H.; Kuang, H.-X. Rapid determination and origin identification of total polysaccharides contents in *Schisandra chinensis* by near-infrared spectroscopy. *Spectrochim. Acta Part A Mol. Biomol. Spectrosc.* **2021**, *264*, 120327. [CrossRef]
28. de Aguiar, N.S.; Hansel, F.A.; Reis, C.A.F.; Lazzarotto, M.; Wendling, I. Optimizing the Vanillin-Acid Sulfuric Method to Total Saponin Content in Leaves of Yerba Mate Clones. *Chem Biodivers.* **2024**, *21*, 202301883. [CrossRef]
29. Tsugawa, H.; Cajka, T.; Kind, T.; Ma, Y.; Higgins, B.; Ikeda, K.; Kanazawa, M.; VanderGheynst, J.; Fiehn, O.; Arita, M. MS-DIAL: Data-independent MS/MS deconvolution for comprehensive metabolome analysis. *Nat. Methods* **2015**, *12*, 523–526. [CrossRef]
30. Volkamer, A.; Kuhn, D.; Grombacher, T.; Rippmann, F.; Rarey, M. Combining Global and Local Measures for Structure-Based Druggability Predictions. *J. Chem. Inf. Model.* **2012**, *52*, 360–372. [CrossRef]
31. Eberhardt, J.; Santos-Martins, D.; Tillack, A.F.; Forli, S. AutoDock Vina 1.2.0: New Docking Methods, Expanded Force Field, and Python Bindings. *J. Chem. Inf. Model.* **2021**, *61*, 3891–3898. [CrossRef] [PubMed]
32. Gharibzahedi, S.M.T.; Barba, F.J.; Zhou, J.; Wang, M.; Altintas, Z. Electronic Sensor Technologies in Monitoring Quality of Tea: A Review. *Biosensors* **2022**, *12*, 356. [CrossRef] [PubMed]
33. Liu, Y.; Meng, F.; Tang, P.; Huang, D.; Li, Q.; Lin, M. Widely Targeted Metabolomics Analysis of the Changes to Key Non-volatile Taste Components in Stropharia rugosoannulata Under Different Drying Methods. *Front. Nutr.* **2022**, *9*, 884400. [CrossRef] [PubMed]
34. Osakabe, N.; Shimizu, T.; Fujii, Y.; Fushimi, T.; Calabrese, V. Sensory Nutrition and Bitterness and Astringency of Polyphenols. *Biomolecules* **2024**, *14*, 234. [CrossRef]
35. Sun, C.; Shang, X.; Ding, H.; Cao, Y.; Fang, S. Natural variations in flavonoids and triterpenoids of *Cyclocarya paliurus* leaves. *J. For. Res.* **2020**, *32*, 805–814. [CrossRef]
36. Bai, M.; Wang, L.L.; Wang, Y.B.; Liu, B.T.; Xu, G.F.; Qin, Y.C. Dynamic Change of Active Components in Leaves of *Cyclocarya paliurus*. *J. Zhejiang For. Sci. Technol.* **2018**, *38*, 21–26. [CrossRef]

37. Wei, L. The effect of cyclocarya flavonoids extrat on the liver insulin resistance of db/db mices and the Pi3K/AKT /FOXO1 pathway. *Beijing Univ. Chin. Med.* **2021**. [CrossRef]
38. Gong, X.; Ji, M.; Xu, J.; Zhang, C.; Li, M. Hypoglycemic effects of bioactive ingredients from medicine food homology and medicinal health food species used in China. *Crit. Rev. Food Sci. Nutr.* **2019**, *60*, 2303–2326. [CrossRef]
39. Xie, J.H.; Dong, C.J.; Nie, S.P.; Li, F.; Wang, Z.J.; Shen, M.Y.; Xie, M.Y. Extraction, chemical composition and antioxidant activity of flavonoids from *Cyclocarya paliurus* (Batal.) Iljinskaja leaves. *Food Chem.* **2015**, *186*, 97–105. [CrossRef]
40. Zhang, Y.; Zhao, M.; He, J.; Chen, L.; Wang, W. In vitro and in vivo immunomodulatory activity of acetylated polysaccharides from *Cyclocarya paliurus* leaves. *Int. J. Biol. Macromol.* **2024**, *259*, 129174. [CrossRef]
41. Chen, L.; Cao, H.; Xiao, J. Polyphenols: Absorption, bioavailability, and metabolomics. In *Polyphenols: Properties, Recovery, and Applications*; Galanakis, C.M., Ed.; Woodhead Publishing: Sawston, Cambridge, 2018; pp. 45–67.
42. Fraser, K.; Lane, G.A.; Otter, D.E.; Hemar, Y.; Quek, S.-Y.; Harrison, S.J.; Rasmussen, S. Analysis of metabolic markers of tea origin by UHPLC and high resolution mass spectrometry. *Food Res. Int.* **2013**, *53*, 827–835. [CrossRef]
43. Nakaya, A.; Katayama, T.; Itoh, M.; Hiranuka, K.; Kawashima, S.; Moriya, Y.; Okuda, S.; Tanaka, M.; Tokimatsu, T.; Yamanishi, Y. KEGG OC: A large-scale automatic construction of taxonomy-based ortholog clusters. *Nucleic Acids Res.* **2012**, *41*, 353–357. [CrossRef] [PubMed]
44. Xi, H.; Xu, W.; He, F.; Liu, Z.; Wang, Y.; Xie, J. Spatial metabolome of biosynthesis and metabolism in *Cyclocarya paliurus* leaves. *Food Chem.* **2024**, *443*, 138519. [CrossRef] [PubMed]
45. Li, Q.; Jin, Y.; Jiang, R.; Xu, Y.; Zhang, Y.; Luo, Y.; Huang, J.; Wang, K.; Liu, Z. Dynamic changes in the metabolite profile and taste characteristics of Fu brick tea during the manufacturing process. *Food Chem.* **2020**, *344*, 128576. [CrossRef] [PubMed]
46. Zheng, W.; Zhang, C.; Li, Y.; Pearce, R.; Bell, E.W.; Zhang, Y. Folding non-homologous proteins by coupling deep-learning contact maps with I-TASSER assembly simulations. *Cell Rep. Methods* **2021**, *1*, 100014. [CrossRef]
47. Singla, R.; Jaitak, V. Synthesis of rebaudioside A from stevioside and their interaction model with hTAS2R4 bitter taste receptor. *Phytochemistry* **2016**, *125*, 106–111. [CrossRef] [PubMed]
48. Servant, G.; Kenakin, T.; Zhang, L.; Williams, M.; Servant, N. Chapter Three—The function and allosteric control of the human sweet taste receptor. *Adv. Pharmacol.* **2020**, *3*, 59–82. [CrossRef]
49. Naguib, M.J.; Kamel, A.M.; Negmeldin, A.T.; Elshafeey, A.H.; Elsayed, I. Molecular docking and statistical optimization of taurocholate-stabilized galactose anchored bilosomes for the enhancement of sofosbuvir absorption and hepatic relative targeting efficiency. *Drug Deliv.* **2020**, *27*, 996–1009. [CrossRef]
50. Han, P.; Keast, R.S.J.; Roura, E. Salivary leptin and TAS1R2/TAS1R3 polymorphisms are related to sweet taste sensitivity and carbohydrate intake from a buffet meal in healthy young adults. *Br. J. Nutr.* **2017**, *118*, 763–770. [CrossRef]
51. Singh, N.; Shaik, F.A.; Myal, Y.; Chelikani, P. Chemosensory bitter taste receptors T2R4 and T2R14 activation attenuates proliferation and migration of breast cancer cells. *Mol. Cell. Biochem.* **2020**, *465*, 199–214. [CrossRef]
52. Xu, Q.; Singh, N.; Hong, H.; Yan, X.; Yu, W.; Jiang, X.; Chelikani, P.; Wu, J. Hen protein-derived peptides as the blockers of human bitter taste receptors T2R4, T2R7 and T2R14. *Food Chem.* **2019**, *283*, 621–627. [CrossRef] [PubMed]
53. Yu, Z.; Wang, Y.; Zhao, W.; Li, J.; Shuian, D.; Liu, J. Identification of Oncorhynchus mykiss nebulin-derived peptides as bitter taste receptor TAS2R14 blockers by in silico screening and molecular docking. *Food Chem.* **2022**, *368*, 130839. [CrossRef] [PubMed]
54. Schober, P.; Boer, C.; Schwarte, L.A. Correlation Coefficients: Appropriate Use and Interpretation. *Anesth. Analg.* **2018**, *126*, 1763–1768. [CrossRef] [PubMed]
55. Gal, J. The Discovery of Stereoselectivity at Biological Receptors: Arnaldo Piutti and the Taste of the Asparagine Enantiomers—History and Analysis on the 125th Anniversary. *Chirality* **2012**, *24*, 959–976. [CrossRef]
56. An, J.P.; Wang, Y.; Munger, S.D.; Tang, X. A review on natural sweeteners, sweet taste modulators and bitter masking com-pounds: Structure-activity strategies for the discovery of novel taste molecules. *Crit Rev Food Sci Nutr.* **2024**, *17*, 1–24. [CrossRef] [PubMed]
57. Mora, M.R.; Dando, R. The sensory properties and metabolic impact of natural and synthetic sweeteners. *Compr. Rev. Food Sci. Food Saf.* **2021**, *20*, 1554–1583. [CrossRef]
58. Sun, Y.; Liang, J.; Zhang, Z.; Sun, D.; Li, H.; Chen, L. Extraction, physicochemical properties, bioactivities and application of natural sweeteners: A review. *Food Chem.* **2024**, *457*, 140103. [CrossRef]

foods

MDPI

Article

Analysis of Metabolite Differences in Different Tea Liquors Based on Broadly Targeted Metabolomics

Xiongyu Li [1], Miao Niu [1], Hongyan Yang [1], Xianxiu Zhou [1], Jianliang Ding [1], Yawen Xu [2], Caiyou Lv [1] and Jiahua Li [1,*]

[1] College of Tea Science, Yunnan Agricultural University, Kunming 650201, China; lxy19912992159@163.com (X.L.); 2022210231@stu.ynau.edu.cn (M.N.); 13529725893@163.com (H.Y.); 2023210231@stu.ynau.edu.cn (X.Z.); 13937016027@163.com (J.D.); 15108721378@163.com (C.L.)

[2] College of Pu-Erh Tea, West Yunnan University of Applied Sciences, Puer 665000, China; 1575859889@163.com

* Correspondence: 15087172337@163.com; Tel.: +86-150-8717-2337

Abstract: To expand the development of characteristic extension products of Yunnan tea and improve the utilization rate of Yunnan tea resources, in this study, we compared the metabolite composition among raw Pu-erh tea, ripe Pu-erh tea prepared with glutinous rice (according to tea to glutinous rice ratio of 1:3), and ripe Pu-erh tea prepared with a mixture of sorghum, rice, glutinous rice, wheat, and corn as raw materials (according to a tea to glutinous rice ratio of 1:3). Rice flavor liquor prepared with 100% glutinous rice served as a control. The raw Pu-erh tea liquor (RAWJ), ripe Pu-erh tea liquor (RIPEJ), ripe Pu-erh tea mixed grain liquor (HHLSJ), and rice-flavor liquor (MJ) were all brewed by semi-solid fermentation. The non-volatile components of the liquor samples were analyzed by ultra-high-performance liquid chromatography-tandem mass spectrometry as a broadly targeted metabolomics technique. A total of 691 metabolites were identified from the four samples. Among them, 674, 671, 633, and 667 species were detected in RAWJ, RIPEJ, HHLSJ, and MJ samples, respectively. Venn diagram analysis demonstrated 19, 21, and 14 unique metabolites in RAWJ, RIPEJ, and HHLSJ, respectively, compared with the metabolite composition of MJ. Flavonoids are the most important differential metabolite between tea liquor and rice-flavor liquor. This study provides a theoretical basis for the development of tea liquor products and offers insight into the difference in non-volatile components between tea liquor and rice-flavor liquor.

Keywords: Pu-erh tea; tea liquor; metabolomics; non-volatile compound

1. Introduction

Tea is one of the most popular and economically non-alcoholic beverages in the world, which can be divided into six categories according to the degree of fermentation: green tea (unfermented), white tea (slightly fermented), yellow tea (slightly fermented), oolong tea (semi-fermented), black tea (fully fermented), and dark tea (post-fermented) [1]. As a unique distilled spirit in China, Chinese liquor, also known as baijiu, exhibits a diverse range of flavors, which are categorized into 12 distinct types such as sauce-flavor, strong-flavor, mild-flavor, and rice-flavor liquors. These variations are influenced by diverse factors, including the specific regions where the raw materials are sourced, which contribute distinct characteristics to the liquor, as well as the fermentation processes [2]. Both tea and liquor have a deep historical heritage and remain highly desired consumer items in China and internationally.

Tea liquor (characterized by an alcohol content greater than 40%) is an extended product of tea that is brewed with grain water and tea as the main raw materials, which combines the advantages of tea and liquor, including the beneficial health effects of tea and the desired quality and flavor characteristics of liquor [3–5]. The brewing process of tea liquor is generally divided into liquid fermentation [6], solid fermentation [7], and semi-solid fermentation [8,9]. In recent years, there has been increased research interest in improving the fermentation technology of tea liquor along with attempts to uncover the functional properties of the tea liquor. Qin et al. [10] optimized the production technology of tea liquor fermentation by inoculating bacteria in black tea and rice wine; Xu et al. [11] developed a new type of fermented tea wine that is rich in anthocyanins using Ziyan tea as the main raw material. Wang et al. [12] optimized the production process of glutinous rice tea wine (alcohol content below 15%) based on a response surface method. Moreover, Li et al. [13] and He et al. [14] showed that tea wine is rich in catechins, gallic acid, and other active antioxidant substances, resulting in various health effects such as lowering blood lipids, preventing cardiovascular and cerebrovascular diseases, and improving immunity, suggesting that moderate drinking of tea wine could be beneficial to health.

Yunnan is the only producing area of Pu-erh tea, which can be divided into raw and ripe Pu-erh tea. Raw Pu-erh tea is prepared using the fresh leaves of the Yunnan large-leafed species of tea as the raw material, followed by fixation, rolling, and sun-drying, whereas ripe Pu-erh tea is prepared from the sun-dried green tea and is then processed according to a specific fermentation process [15]. According to the statistics of the Department of Agriculture and Rural Affairs of Yunnan Province and the China Tea Distribution Association, in 2023, the tea-planting area of Yunnan Province was 5.36 billion m^2 and the total output of processed tea was 418,000 tons, representing an increase of 5.1% compared with the production levels of the previous year. However, the average price of Pu-erh tea is 168.3 CNY/kg, which is notably lower than that of other teas such as Longjing tea at 983.26 CNY/kg, resulting in a lack of motivation for tea farmers to invest in Pu-erh tea; consequently, it has become increasingly common for farmers to abandon the production of summer and autumn tea, seriously restricting the development of the Yunnan tea industry. The development of tea liquor and other processed tea products such as black tea bread [16] and Pu-erh tea paste [17] offers a conducive strategy to improve the utilization rate of Yunnan's tea resources, increasing the scale and efficiency of the tea industry, and ultimately promoting the transformation and advancement of Yunnan's tea industry by broadening consumption channels. In recent years, plant bioactive phenolic compounds have garnered widespread attention due to their myriad health benefits [18], and tea is abundant in phenolic compounds. Farag et al. [19]. utilized UPLC-MS and GC-MS for the analysis of metabolite fingerprints from four types of cinnamon.

To encourage such development, in this study, we used Pu-erh tea, glutinous rice, sorghum, rice, wheat, and corn as the main raw materials to prepare raw, ripe, and mixed-grain Pu-erh tea liquor, along with rice-flavor liquor, using semi-solid fermentation technology. The main metabolites of these samples were systematically analyzed and compared using ultra-high-performance liquid chromatography-tandem mass spectrometry (UPLC-MS/MS) combined with multivariate statistical analysis to comprehensively determine the main metabolic changes among the samples. The results of this study can therefore provide some theoretical support for the in-depth development of Yunnan Pu-erh tea liquor and other tea liquor products.

2. Materials and Methods

2.1. Raw Materials

Raw and ripe Pu-erh tea was obtained from Yunnan Yicang Trading Co., Ltd (Kunming, China). Sorghum was purchased from the Luancheng district Hebei Province Amoy grain shop (Luancheng, China). Rice and corn were obtained from Qiqihar North Kitchen Trading Co., Ltd. (Qiqihar, China). Glutinous rice was purchased from Wuchang Xingwang Rice Industry Co., Ltd. (Harbin, China). Wheat was obtained from Jincheng Pingle Flour Co. and Ltd. (Jincheng, China). Koji (small koji) was obtained from Angel Yeast Co., Ltd. (Yichang, China).

A refractometer was purchased from Hengshui Jinjun Trading Co., Ltd. (Hengshui, China). Alcohol meter Hengshui Green Trading Co., Ltd. (Hengshui, China). Electronic balance ME204/02 METTLER-TOLEDO Instrument Co., Ltd. (Zurich, Switzerland). Pure water (making tea liquor) was purchased from Kunming Zhen Ming Food Co., Ltd. (Kunming, China).

2.2. Preparation of Tea Liquor

The preparation process of tea liquor was based on the methodology described by Zou et al. [20] and the semi-solid fermentation method of the rice-flavor liquor followed the standard GB/T 10781.3-2006 [21]. We used the following process to slightly improve the production of raw Pu-erh tea liquor (RAWJ) and ripe Pu-erh tea liquor (RIPEJ). In brief, Angel liquor koji (0.6%) was added to cooked and cooled glutinous rice (1 kg); the mixture was then subjected to saccharification (30 °C, for 3 days). The tea soup (1.5 L) and tea leaves (330 g) were added to the glutinous rice-liquor mixture for secondary fermentation (28 °C for 7 days), followed by distillation (79–95 °C for 20 min at 20–25 mL/min) for liquor reduction. The ripe Pu-erh tea mixed-grain liquor (HHLSJ) was prepared with the same production process except for the use of 1 kg mixed grains (sorghum, rice, glutinous rice, wheat, and corn) rather than glutinous rice alone. The rice-flavor liquor sample, used as a control, was prepared in the same manner using glutinous rice as the raw material, although secondary fermentation was performed with pure water (Kunming Zhen Ming Food Co., Ltd.).

Fermentation was carried out in a stainless-steel distiller and fermentation barrel from Nanfeng County Daliang Brewing Equipment Factory using a constant-temperature fermentation box (LY small second layer; Dongguan Fenggang Lai Electrical Appliance Company, Dongguan, China). The fermentation conditions of all tea liquors were as follows: 0.6% koji, 1:3 tea-to-rice ratio, and 1:50 tea-to-water ratio of the secondary fermented tea soup (according to the national standard GB/T 23776-2018 [22] "Tea Sensory Evaluation Method"). The tea soup added amount was 1.5 times the mass of glutinous rice. The grain in the mixed-grain liquor was a crushed powder of 36% sorghum, 22% rice, 18% glutinous rice, 16% wheat, and 8% corn according to the standard GB/T 22211-2008 [23] of Wuliangye liquor, using Pu-erh ripe tea. The distillation method was based on the segmental liquor-receiving approach, with the first liquor receiving 60% alcohol by volume (ABV), and the last liquor having a volume below 45% alcohol by volume (ABV). The above processes were repeated three times and the alcohol content was standardized to 50% vol, with the volume of each bottle maintained at 500 mL. The samples were stored at −80 °C until analysis.

2.3. Sample Extraction

The metabolomics method was used to analyze the metabolite components of different tea liquors based on UPLC-MS/MS. The frozen and stored sample was thawed, 9 mL of this distillate sample was placed in a 50 mL centrifuge tube labeled with a unique code, and 300 μL of 70% methanol (chromatographically pure, Merck, Darmstadt, Germany) containing internal standard extraction solution was added to the tube facilitate dissolution. The mixture was vortexed for 15 min, followed by ultrasound treatment in an ice-water bath (0 °C, KQ5200E) for 10 min (ensure that the compounds in the sample are fully solubilized or released). The sample was then centrifuged at 12,000 rpm at 4 °C (5424R, Eppendorf, Hamburg, Germany) for 3 min, filtered through a microporous membrane with a pore size of 0.22 μm, and transferred to a sample bottle for storage until subsequent detection.

2.4. UPLC-MS/MS

The ExionLC™ AD (AB SCIEX Analytical Instrument Trading Co., Ltd., Shanghai, China.) and QTRAP 6500 MS/MS (Applied Biosystems, AB SCIEX, Shanghai, China) systems were used for UPLC-electrospray ionization (ESI)-MS/MS.

UPLC was performed with an Agilent SB-C18 column (1.8 μm, 2.1 mm × 100 mm). Mobile phase A was ultra-pure water (with 0.1% formic acid chromatographically pure, Aladdin Reagents Ltd., Shanghai, China) and mobile phase B was acetonitrile (with 0.1% formic acid added). The sample was measured using the following gradient procedure:

5%B at 0 min, which was increased linearly to 95% within 9 min, maintained at 95% for 1 min, decreased to 5% at 10–11.10 min, and then balanced at 5% to 14 min. The flow rate was 0.35 mL/min, the column temperature was 40 °C, and the sample volume size was 2 μL.

The operating parameters of the ESI source were as follows: source temperature of 550 °C, ion spray voltage of 5500 V (positive-ion mode) or −4500 V (negative-ion mode) and ion source gas I (GSI), gas II (GSII), and gas curtain gas (CUR) set to 50, 60, and 25 psi, respectively. The collision-induced ionization parameter was set to "high". The QQQ scan was run in multiple reaction monitoring (MRM) modes with the collision gas (nitrogen) set to "medium". The declustering potential (DP) and collision energy (CE) of each MRM ion pair were further optimized. A specific set of MRM (Figure S1) ion pairs was monitored at each period based on the eluted metabolites during each period according to the 3000 MRM product standard.

2.5. Qualitative Analysis of Metabolites

Qualitative analysis of metabolites was carried out according to secondary spectrum information (Figure S2), excluding isotopic signals, repetitive signals containing K^+, Na^+, NH_4^+ ions, and repetitive signals from fragment ions of other larger molecular weight substances. Detection of the relative content of species in different samples was performed using five parameters of substance detection (DP, CE, retention time, Q1, and Q3) to obtain qualitative data on the detected substances [24]. There are 3 levels for metabolite characterization: Level: Substance identification level, 1: sample substance secondary mass spectra (all fragmentation sub-ions of the substance), RT, and database substance matching score of 0.7 or more; 2: sample substance secondary mass spectra (all fragmentation sub-ions of the substance), RT, and database substance matching score of 0.5–0.7; 3: sample substance Q1, Q3, RT, DP, and CE checking consistency with the database substances.

2.6. Statistical Analysis

Analyst 1.6.3 software was used to process the mass spectrum data. Microsoft Office Excel 2019 was used for data processing with an online data analysis platform (https://cloud.metware.cn/). The statistical function in R (www.r-project.org) was used for unsupervised principal component analysis (PCA). The cor function of the R software package (version 3.5.1) was used to calculate Pearson correlation coefficients and determine their significance, with the data visualized in heat map form. The differences in metabolites between the two groups were analyzed using the projected variable importance (VIP) score according to a threshold of VIP ≥ 1 ($|\log2$ fold change (FC)$|$) ≥ 2.0 or <0.05 to indicate a significant difference. VIP values (including score plots) were extracted from orthogonal partial least squares discriminant analysis (OPLS-DA) and permutation plots were generated using the R software package MetaboAnalystR (ggplot2 3.3.0).

3. Results and Discussion

3.1. Overview of Non-Volatile Metabolites in the Four Samples

The metabolomics method was used to analyze the metabolite components of different tea liquor samples. Based on the UPLC-MS/MS detection platform, a total of 691 metabolites were identified in the RAWJ, RIPEJ, HHLSJ, and MJ samples. The proportions of each component in the four sample liquor are shown in Figure 1, including others (15.77%), flavonoids (14.91%), lipids (13.02%), alkaloids (11.14%), phenolic acids (10.71%), organic acids (10.13%), amino acids and derivatives (9.84%), terpenoids (7.38%), lignans and coumarins (3.18%), nucleotides and derivatives (2.75%), quinones (0.87%), and tannins (0.29%).

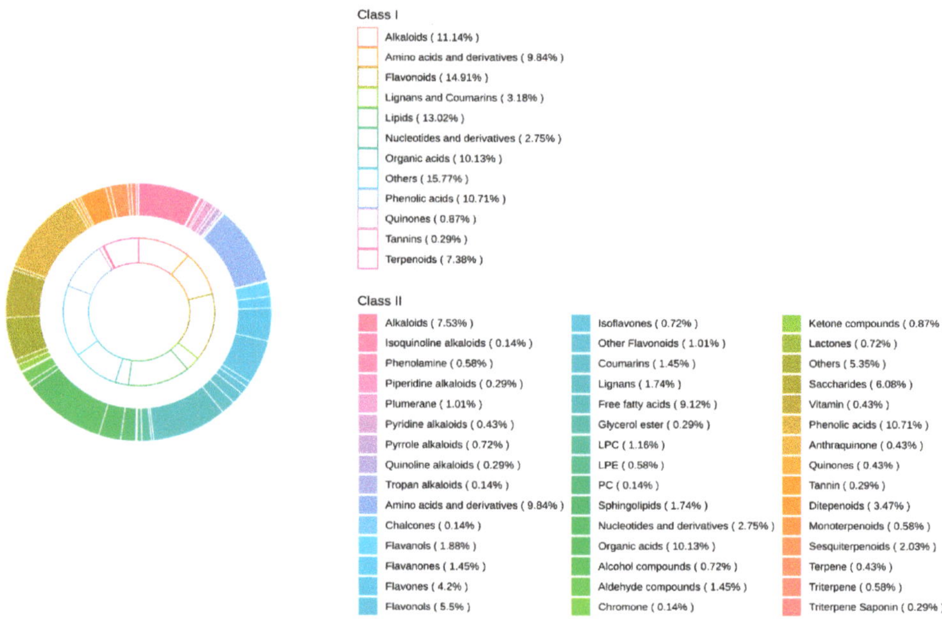

Figure 1. Ring chart of metabolite component proportions in the four liquor samples.

To clarify the differences between metabolites of the tea liquor and rice-flavor liquor metabolites, a Venn diagram was constructed. As shown in Figure 2, 605 metabolites were identified in the four samples, among which 674, 671, 633, and 667 metabolites were identified in RAWJ, RIPEJ, HHLSJ, and MJ, respectively. Compared with the MJ group, 19, 21, and 14 unique metabolites were identified in the RAWJ, RIPEJ, and HHLSJ groups, respectively. Compared with the other three groups, the unique metabolites in RIPEJ were quercetagetin, apigenin-7-O-glucoside, and apigenin-4′-O-glucoside, and the unique metabolite in MJ was 2′-deoxyuridine 5′-monophosphate. The results of this study are consistent with those reported previously, further supporting that the addition of tea is beneficial to enrich the quality components of liquor [25].

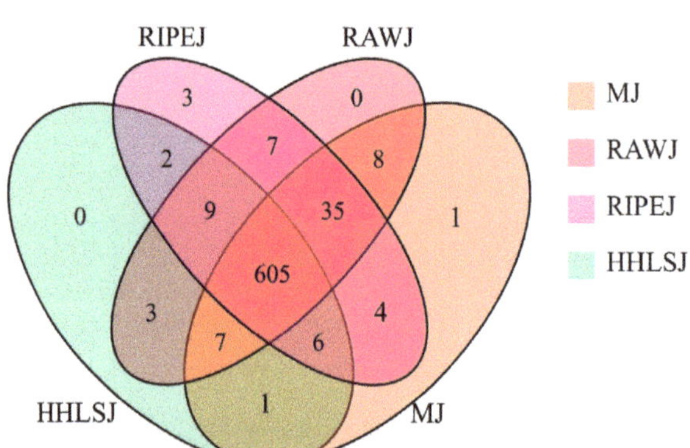

Figure 2. Venn diagram of the metabolites identified in the four liquor samples.

A total of 109 other compounds were detected, which could be divided into eight subclasses, including alcohol compounds (0.72%), aldehyde compounds (1.45%), chromone (0.14%), ketone compounds (0.87%), lactones (0.72%), saccharides (6.08%), vitamins (0.43%), and others (5.35%). Among them, MJ had the highest number of species (109), followed by RAWJ (108), RIPEJ (108), and HHLSJ (102).

A total of 103 flavonoids were detected, which could be divided into seven subclasses, including chalcones (0.14%), flavanols (1.88%), flavanones (1.45%), flavones (4.2%), flavonols (5.5%), isoflavones (0.72%), and other flavonoids (1.01%). Among them, RIPEJ had the highest number of species (101), followed by RAWJ (99), MJ (94), and HHLSJ (89). Flavonoid compounds are bioactive components that are widely present in plants and fruits. The RAWJ and RIPEJ samples had more types of flavonoids compared to MJ. This is attributed to the fact that tea is rich in polyphenols [26]; therefore, the flavonoid types in tea liquor are higher than those in rice-flavor liquor.

A total of 90 lipid compounds was detected in the four samples, which could be divided into six subclasses: free fatty acids (9.12%), glycerol ester (0.29%), lysophosphatidylcholines (1.16%), lysophosphatidylethanolamines (0.58%), phosphatidylcholines (0.14%), and sphingolipids (1.74%). Among them, RIPEJ and RAWJ had the highest number of lipid species (88), followed by MJ (86) and HHLSJ (85).

A total of 77 alkaloids were detected, which could be divided into the following eight subclasses: alkaloids (7.53%), isoquinoline alkaloids (0.14%), phenolamine (0.58%), piperidine alkaloids (0.29%), plumerane (1.01%), pyridine alkaloids (0.43%), pyrrole alkaloids (0.72%), quinoline alkaloids (0.29%), and tropane alkaloids (0.14%). Among them, RAWJ had the highest number of species (76), followed by MJ (75), RIPEJ (72), and HHLSJ (71). Alkaloids are a class of natural nitrogen-containing alkaline organic compounds. Some alkaloids have a bitter taste [27], which will make the flavor of the liquor more rich and complex.

A total of 74 phenolic acids were detected. RAWJ had the largest number of species in this class (72), followed by RIPEJ (68), MJ (67), and HHLSJ (61). Phenolic acid is a plant metabolic product [28] and tea is known to be rich in phenolic acid compounds, such as gallic acid and salicylic acid. These compounds have well-established health functions such as anti-oxidation and free radical-scavenging activity [29]. As the main bioactive compounds in the brewing process, the primary and secondary materials under the action of various enzyme systems and microorganisms will also generate some phenolic acids and phenolic acid esters; these biologically active substances enter the liquor body through the unique distillation process, resulting in an elevated concentration in the final liquor products.

A total of 70 organic acids were detected, with the highest numbers found in MJ and RIPEJ (70), followed by RAWJ (68) and HHLSJ (65). Tea is rich in organic acids, including monocarboxylic acid and polycarboxylic acid [30]. Organic acids are precursors of ester synthesis, which are of great significance in regulating the wine body's taste and also display health functions; in particular, esters can effectively reduce the bitterness of the liquor while increasing the sweetness. According to their chain length, organic acids can be divided into short-chain and long-chain acids. Short-chain acids have been associated with beneficial effects in treating alcoholism [31], whereas long-chain acids have been shown to reduce the risk of cardiovascular diseases while other acid compounds can protect biological systems from the influence of hydroxyl and peroxygen free radicals [32].

A total of 68 amino acids and their derivatives were detected with the highest numbers found in RIPEJ and HHLSJ (68), followed by MJ (67) and RAWJ (66). Amino acids and their derivatives are mainly generated from the fermentation process of grains, which contribute to the "sweet" and "fresh" tastes of liquor, along with the ability to moderate the sour and spicy taste of liquor [33]. Tea has a low content of free amino acids and rice flavor liquor has a high protein and amino acid content [34]. Given that the number of amino acids and their derivatives was highly similar among the four samples, the advantages of tea liquor with respect to the amino acid content are unclear.

A total of 51 kinds of terpenoids were detected, which were the most abundant in RAWJ, RIPEJ, and MJ (51), followed by HHLSJ (49). Terpenoids are the general term for all isoprene polymers and their derivatives, which can be divided into monoterpenoids, sesquiterpenoids, diterpenoids, triterpenoids, tetraterpenoids, or carotenes, which are characterized by a series of antibacterial and antioxidant properties [35]. In recent years, the research on terpenoids in liquor has become more extensive, and some studies have shown that terpenoids have important contributions to the aroma of liquor [36].

Compared to the other metabolites identified, the proportions of lignans and coumarins, nucleotides and their derivatives, quinones, and tannins in the four samples of liquor were lower, with only 22, 19, 6, and 2 of these metabolite types identified, respectively.

In summary, our results suggest that the mixed fermentation of tea and grain will not change the key flavor substances in liquor in general, and there is no obvious quantitative difference in substance types. Therefore, the appropriate use of tea and grain fermentation may impart unique flavor and aroma components to liquor.

3.2. Screening of Differential Metabolites

3.2.1. Data Quality Assessment

To further understand the overall metabolic differences among the four samples and the differences among samples in the same group, PCA was conducted on all metabolites in the four samples. As shown in Figure 3, RAWJ, RIPEJ, HHLSJ, and MJ were significantly separated along PC1 (41.85%) and PC2 (30.03%). The RAWJ and RIPEJ samples were close to each other and the total contribution rate of cumulative variance in PC1 and PC2 was 71.88%. The results showed significant differences in metabolites between different tea liquors and rice-flavor liquor.

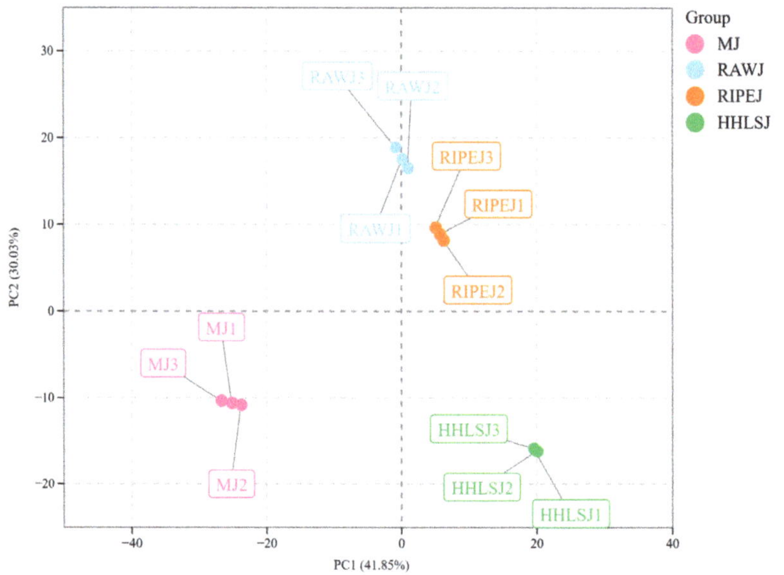

Figure 3. Principal component analysis of metabolites identified in four kinds of liquor samples.

Correlation analysis was used as an evaluation index of bioreplicated correlation where different colors represent different correlation coefficients; where red indicates a stronger positive correlation (cor value closer to 1), white indicates a weaker correlation, and blue indicates a stronger negative correlation (cor value closer to 1). As shown in Figure 4, all samples had cor values close to 1, indicating credible biological duplication in the same group of samples.

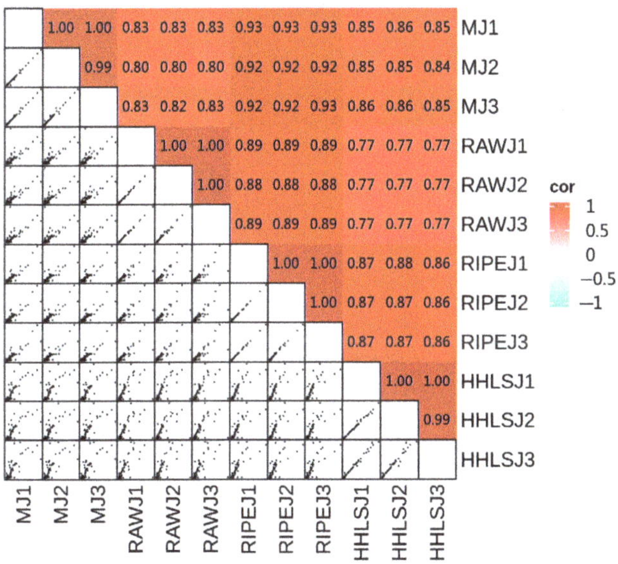

Figure 4. Correlation analysis between duplicated samples of the four kinds of liquor.

3.2.2. OPLS-DA Analysis

To further explore the differences in metabolite types of the four samples, the OPLS-DA method was used, which is a supervised statistical method of discriminant analysis. The key feature of OPLS-DA is that it can remove data changes unrelated to a categorical variable Y in an independent variable X so that the classification information is mainly concentrated along on one principal component [37]. R2X (cum) and R2Y (cum) are two parameters calculated by cross-validation procedures to evaluate the goodness of fit, while Q2 (cum) is a parameter used to describe the validity of the model [38]. In this study, Q2 = 0.997, R2X = 0.855, and R2Y = 1 in RAWJ and MJ; Q2 = 0.997, R2X = 0.856, R2Y = 1 in RIPEJ and MJ; and Q2 = 0.997, R2X = 0.896, and R2Y = 0.999 in HHLSJ and MJ, indicating that MJ is separated from the RAWJ, RIPEJ, and HHLSJ samples.

The top 20 metabolites with high VIP values in each group are shown in Figure 5. For example, 3-hydroxylup-20(29)-en-28-al, nicotinamide, 4-O-acetyl-3-O-caffeoyl-2-C-methyl-D-erythronate six amino acids and derivatives, six other compounds, four flavonoids, and one terpenoid, lipid, phenolic acid, and alkaloid were each identified in the RAWJ and MJ groups; 2,5-dihydroxybenzoic acid, quercetagetin, 2-amino-4-hydroxy-3-methylpentanoic acid, seven amino acids and derivatives, four organic acids, three phenolic acids, three alkaloids, and one flavonoid, other compounds, and lipid were each identified in the RIPEJ and MJ groups; and sanleng acid, 2-hydroxy-4-methyl-3-undecanoyloxypentanoic acid methyl ester, N-acetyl-D-galactosamine, five lipids, four other compounds, three organic acids, two phenolic acids, and one tannin, nucleotide, and derivative, lignan and coumarin, flavonoid, alkaloid, and amino acid and derivatives were each identified in the HHLSJ and MJ groups.

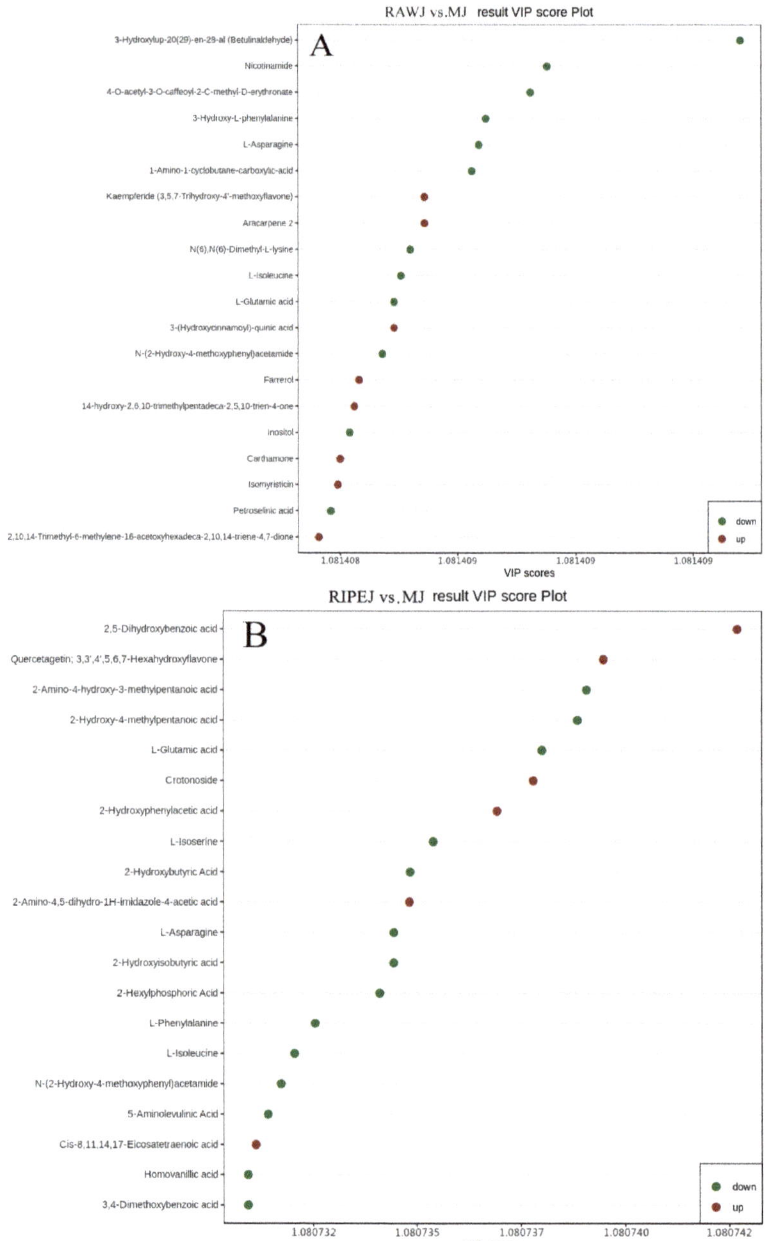

Figure 5. *Cont.*

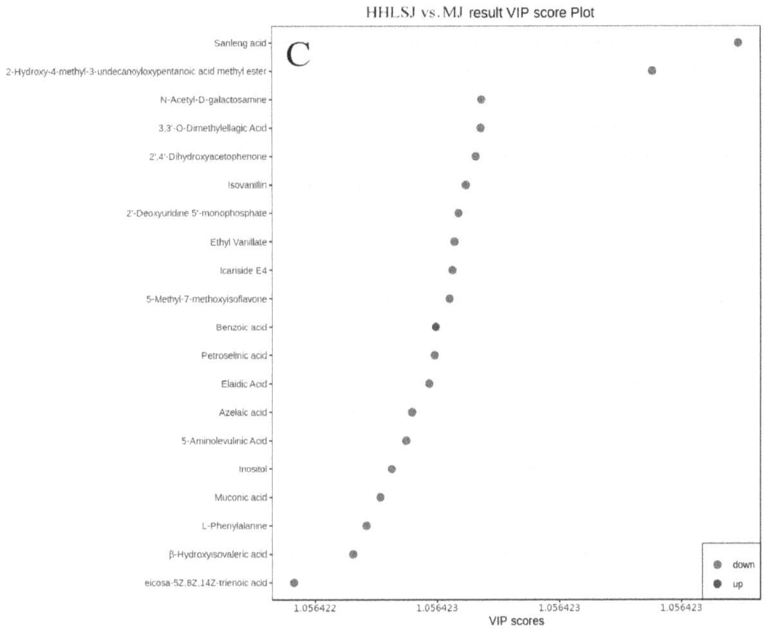

Figure 5. VIP values of metabolites compared between RAWJ and MJ (**A**), RIPEJ and MJ (**B**), and HHLSJ and MJ (**C**).

The FC in the metabolite contents of tea liquor and rice-flavor liquor was calculated to visually demonstrate their overall differential metabolite compositions. The metabolites were then arranged according to the FC values and a dynamic distribution diagram of metabolite content difference was generated (Figure 6). A total of 272, 246, and 125 metabolites with FC ≥ 1 or ≤ 0.5 were identified in RAWJ, RIPEJ, and HHLSJ samples compared with MJ, respectively. Figure 6 displays the top 10 up-regulated and down-regulated differential metabolites. For instance, sanshodiol showed the highest degree of up-regulation in RAWJ compared to MJ, while icariside E4 exhibited the lowest level of down-regulation. Similarly, sanshodiol had the highest up-regulation, while D-pantothenol had the lowest down-regulation in RIPEJ compared to MJ. In HHLSJ, epicatechin gallate was up-regulated with the highest FC, while rhamnetin was down-regulated with the lowest FC value compared to the corresponding levels in MJ.

To further comprehend the disparities in metabolites among RAWJ, RIPEJ, HHLSJ, and MJ, the FC values and VIP scores were used to identify significant differential metabolites among a total of 691 identified metabolites. According to the screening criteria (VIP ≥ 1; FC ≥ 2 or ≤ 0.5; $p \leq 0.05$) [37], 655 significantly distinct metabolites were identified (Table S1). The volcano plots for each group are depicted in Figure 7, with red and green dots denoting up-regulated and down-regulated differentiated metabolites, respectively, while gray dots represent non-differentiated metabolites among groups. In the RAWJ and MJ groups (315 species), 26 species were up-regulated and 40 species were down-regulated (Figure 7A). In the comparison of the RIPEJ and MJ groups (358 species), 26 were up-regulated and 41 were down-regulated (Figure 7B). In the comparison of the HHLSJ and MJ groups (306 species), 14 were up-regulated and 78 were down-regulated(Figure 7C). Statistically significant differences were found in the types of substances in each group, totaling 60 flavonoids, 34 phenolic acids, 34 others, 22 lipids, 21 alkaloids, 10 amino acids and their derivatives, 10 nucleotides and their derivatives, 9 terpenoids, 5 quinones, 4 lignans, and coumarins compounds. Among these, flavonoids showed the most important difference between tea wine and rice wine. The chemical structural formulas of the metabolites with significant differences in content in each group are presented in Figure 8.

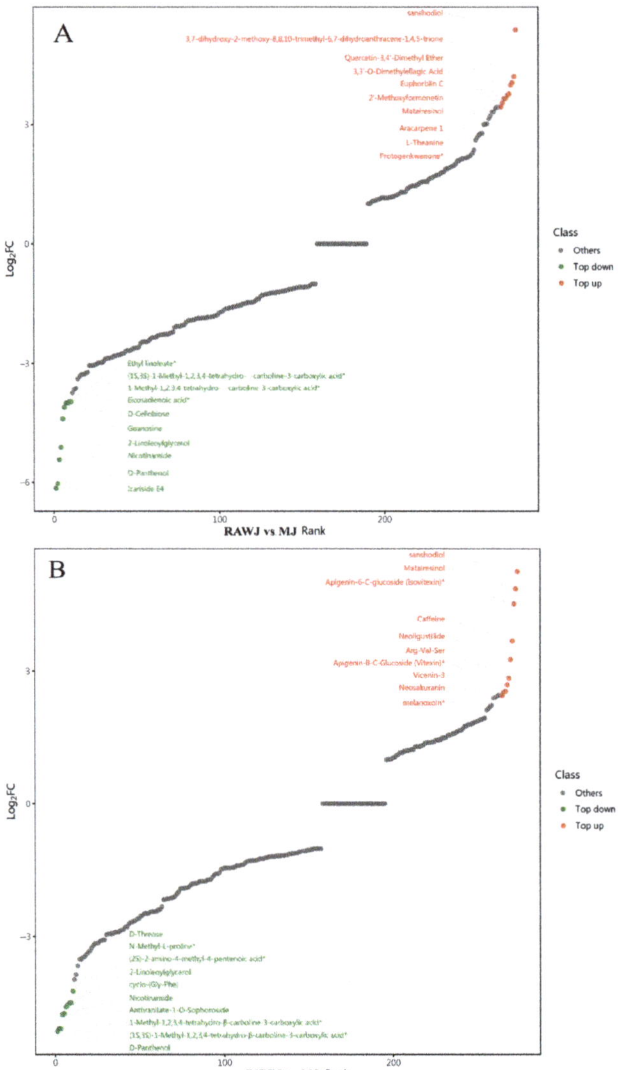

Figure 6. *Cont.*

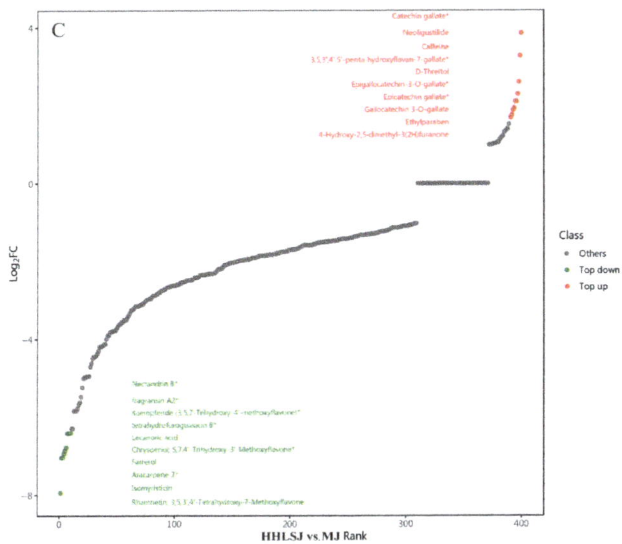

Figure 6. Dynamic distribution of metabolite content differences between RAWJ and MJ (**A**), RIPEJ and MJ (**B**), and HHLSJ and MJ (**C**). Note: The asterisk (*) in the diagram denotes the presence of isomers for the substance.

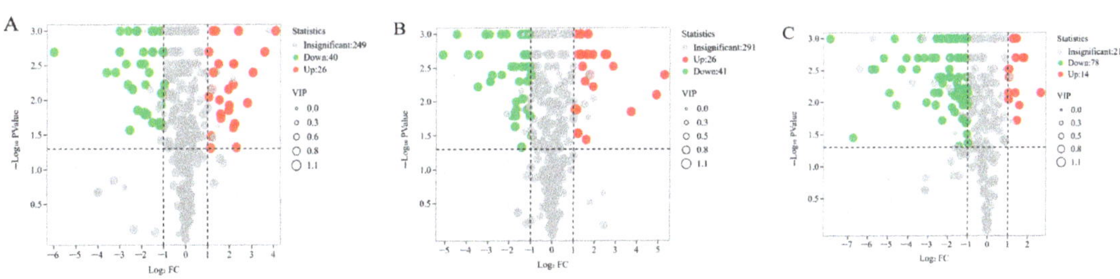

Figure 7. Volcano plots of differential nonvolatile metabolites for the comparisons of RAWJ vs. MJ (**A**), RIPEJ vs. MJ (**B**), and HHLSJ vs. MJ (**C**).

Finally, we evaluated the differences in the relative contents of non-volatile metabolites in different groups according to the unit variance scale. According to the K-means clustering method, 529 different metabolites were divided into seven subclasses (Table S2). There were 82 metabolites in RAWJ with higher contents than those found in MJ, RIPEJ, and HHLSJ. In this subgroup, the most common metabolites were flavonoids (29 species), including kaempferide and quercetin-3-O-glucoside. In subclass 2, the contents of 53 metabolites in MJ and RAWJ were higher than those in RIPEJ and HHLSJ, of which 12 were lipids, hydroperoxylinoleic acid, 9-hydroxy-12-oxo-15(Z)-octadecenoic acid, and others. In subclass 3 and subclass 6, the contents of 140 and 88 metabolites in MJ were higher than those in RAWJ, RPIEJ, and HHLSJ, including L-asparagine, DL-threonine, L-glutamine, and 46 other amino acids and derivatives. In subclass 4, the contents of 28 metabolites in HHLSJ were higher than those in MJ, RAWJ, and RIPEJ, with the majority represented by flavonoids (eight kinds), including catechin (+C), epicatechin gallate, and epigallocatechin-3-O-gallate. In subclass 5, 98 metabolites were found at higher abundance in RIPEJ than in MJ, RAWJ, and HHLSJ, which were also dominated by flavonoids (34 species), such as kaempferol-3-O-robinobioside(biorobin), kaempferol-3-O-glucoside-7-O-rhamnoside, and 4′,6-dihydroxyflavone. In subclass 7, a total of 40 metabolites were found at lower levels in RAWJ than in MJ, RAWJ, and HHLSJ, with alkaloids (eight species) being the most abundant, including betaine and acetaminophen. Among these, acetaminophen is mainly derived from chemical synthesis and is not found in natural products. Its detection in the sample may be due to contamination of the experimental materials or labeling errors.

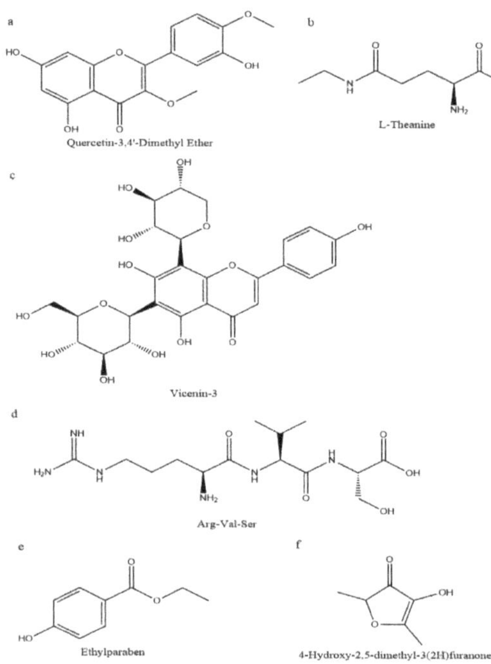

Figure 8. Simplified chemical structures of key differentiated substances in the three groups based on screening according to the fold change value, variable importance in projection (VIP) value, and *p*-value. (**a**,**b**) RAWJ vs. MJ; (**c**,**d**) RIPEJ vs. MJ; (**e**,**f**) HHLSJ vs. MJ.

4. Conclusions

In this study, based on the broadly targeted metabolomics method of UPLC-MS/MS, we identified differences in the metabolite composition between three types of tea liquor and rice-flavor liquor. A total of 691 metabolites were identified in the four liquor samples (RAWJ, RIPEJ, HHLSJ, and MJ), including 109 "other" metabolite types, 103 flavonoids, 90 lipids, 77 alkaloids, 74 phenolic acids, 70 organic acids, 68 amino acids and derivatives, 51 terpenoids, 22 lignans and coumarins, 19 nucleotides and derivatives, 6 quinones, and 2 tannins. The tea liquors RAWJ, RIPEJ, and HHLSJ had an additional 19, 21, and 14 unique metabolites compared with those found in the MJ sample, respectively. Flavonoids are the most important differential metabolites between tea liquor and rice-flavor liquor. This study systematically revealed the composition of non-volatile substances of tea liquor, which can provide a certain theoretical basis for further in-depth research and development of tea liquor.

Supplementary Materials: The following supporting information can be downloaded at https://www.mdpi.com/article/10.3390/foods13172800/s1: Figure S1 (a): MRM detection of multimodal maps positive Ion Mode; (b): MRM detection of multimodal maps negative ion mode; Figure S2 (a): QC_MS_tic_overlap-N; (b): QC_MS_tic_overlap-P; (c): QC_MS_TIC-N; (d): QC_MS_TIC-P; (e): Catechin second-order spectrum; (f): L-Theanine second-order spectrum; (g): Epicatechin gallate second-order spectrum; (h): Gallocatechin 3-O-gallate second-order spectrum; (i): Vicenin-3 second-order spectrum; (j): Epigallocatechin-3-O-gallate second-order spectrum; (k): Epicatechin second-order spectrum; Table S1: Metabolite Profiles Identified by FC, VIP, and P-Value Screening; Table S2: Metabolite Clustering Results from K-means Analysis.

Author Contributions: Sample Preparation, X.L. and M.N.; methodology, H.Y.; software, X.Z.; validation, J.D. and X.L.; data curation, X.L.; writing—original draft preparation, X.L.; formal analysis Y.X.; writing—review and editing, J.L.; project administration, C.L.; funding acquisition, C.L.; funding acquisition. All authors have read and agreed to the published version of the manuscript.

Funding: This research was funded by the National Modern Agricultural Tea Industry System Special Funds Program, grant number CARS-19. The funder: Caiyou Lv.

Institutional Review Board Statement: Not applicable.

Informed Consent Statement: Not applicable.

Data Availability Statement: The original contributions presented in the study are included in the article and Supplementary Materials, further inquiries can be directed to the corresponding author.

Conflicts of Interest: The authors declare no conflicts of interest.

References

1. Lin, F.J.; Wei, X.L.; Liu, H.Y.; Li, H.; Xia, Y.; Wu, D.T.; Zhang, P.Z.; Gandhi, G.R.; Li, H.-B.; Gan, R.Y. State-of-the-art review of dark tea: From chemistry to health benefits. *Trends Food Sci. Technol.* **2021**, *109*, 126–138. [CrossRef]
2. Jia, W.; Fan, Z.; Du, A.; Li, Y.; Zhang, R.; Shi, Q.; Shi, L.; Chu, X. Recent advances in baijiu analysis by chromatography based technology–A review. *Food Chem.* **2020**, *324*, 126899. [CrossRef]
3. Yue, C.; Zhao, Q.C. Research on Cordyceps militaris tea wine. *China Brew.* **2011**, *30*, 179–182.
4. Zou, C.; Xu, Y. Fermentation process optimization and chemical composition analysis on black tea wine. In *E3S Web of Conferences*; EDP Sciences: Les Ulis, France, 2021.
5. He, Y.; Liu, Z.; Qian, M.; Yu, X.; Xu, Y.; Chen, S. Unraveling the chemosensory characteristics of strong-aroma type baijiu from different regions using comprehensive two-dimensional gas chromatography–time-of-flight mass spectrometry and descriptive sensory analysis. *Food Chem.* **2020**, *331*, 127335. [CrossRef]
6. Song, S.; Pan, L.; Wu, X. Production of Dangshan Pear Wine by Liquid Fermentation Based on Orthogonal Design Optimal Experiments. *J. Food Eng. Technol.* **2015**, *4*, 17–23.
7. Pandey, A. Solid-state fermentation. *Biochem. Eng. J.* **2003**, *13*, 81–84. [CrossRef]
8. Ma, R.; Sui, L.; Zhang, J.; Hu, J.; Liu, P. Polyphasic characterization of yeasts and lactic acid bacteria metabolic contribution in semi-solid fermentation of Chinese baijiu (traditional fermented alcoholic drink): Towards the design of a tailored starter culture. *Microorganisms* **2019**, *7*, 147. [CrossRef]
9. Liu, X.; Ma, R.; Hu, J.; Zhou, X.; Wang, M.; Zhang, J.; Liu, P. Microbiota Dynamics in Semisolid Fermentation of Light-Flavor Liquor with Corn as Main Material by Illumina MiSeq Sequencing. *Res. Sq.* 2020, *preprint*. [CrossRef]
10. Qin, Y.; Yuan, Z.J.; Yang, F.Y.; Yu, Y.G. Development of a new type of Anhua black tea and its application: Black tea wine. *J. Food Process. Preserv.* **2021**, *46*, e15862. [CrossRef]
11. Xu, W.; Wang, X.; Jia, W.; Wen, B.; Liao, S.; Zhao, Y.; Tang, Q.; Li, K.; Hua, Y.; Yang, Y.; et al. Dynamic changes in the major chemical and volatile components during the "Ziyan" tea wine processing. *LWT Food Sci. Technol.* **2023**, *186*, 115273. [CrossRef]
12. Qi, W.; Wu, L.; Zhao, W.W.; Zhou, S.L. Response surface optimization of glutinous rice tea wine production process. *China Brew.* **2020**, *39*, 204–210.
13. Li, Y.; Zhang, S.; Sun, Y. Measurement of catechin and gallic acid in tea wine with HPLC. *Saudi J. Biol. Sci.* **2020**, *27*, 214–221. [CrossRef] [PubMed]
14. He, T.; Liu, F.; Du, L.; Xiao, D. Development of Pu'er tea fermented wine. *Liquor-Mak. Sci. Technol.* **2015**, *10*, 103–106.
15. *GB/T 22111-2008.2008-06-17*; Geographical Indication Products Pu'er Tea. Standards Press of China: Beijing, China, 2008.
16. Zehra, K.; Yaren, A. Assessment of instrumental and sensory quality characteristics of the bread products enriched with Kombucha tea. *Int. J. Gastron. Food Sci.* **2022**, *29*, 100562.
17. Yang, R.; Lin, W.; Liu, J.; Liu, H.; Fu, X.; Liu, H.; Han, Z.; Wang, L.; Wang, Y.; Ba, G. Formation mechanism and solution of Pu-erh tea cream based on non-targeted metabonomics. *LWT* **2023**, *173*, 114331. [CrossRef]
18. Ali, A.; Cottrell, J.J.; Dunshea, F.R. Characterization, Antioxidant Potential, and harmacokinetics Properties of Henolic Compounds from Native Astralian Herbs and Fruits. *Plants* **2023**, *12*, 993. [CrossRef]
19. Farag, M.A.; Kabbash, E.M.; Mediani, A.; Döll, S.; Esatbeyoglu, T.; Afifi, S.M. Comparative Metabolite Fingerprinting of Four Different Cinnamon Species Analyzed via UPLC–MS and GC–MS and Chemometric Tools. *Molecules* **2022**, *27*, 2935. [CrossRef]
20. Chen, W.; Gong, L.; Guo, Z.; Wang, W.; Zhang, H.; Liu, X.; Yu, S.; Xiong, L.; Luo, J. A novel integrated method for large-scale detection, identification, and quantification of widely targeted metabolites: Application in the study of rice metabolomics. *Mol. Plant* **2013**, *6*, 1769–1780. [CrossRef]
21. *GB/T 10781.3-2006*; Rice Aromatic Chinese Spirits. Standards Press of China: Beijing, China, 2006.
22. *GB/T 23776-2018*; Sensory Evaluation Method for Tea. Standards Press of China: Beijing, China, 2018.

23. *GB/T 22211-2008*; Fragrance Rice Baijiu. Standards Press of China: Beijing, China, 2008.
24. You, J.; Li, H.; Wang, Q.; Xu, F.; Lin, S.; Wang, X.; Huang, S.; Sheng, Y.; Zhu, B.; Zhang, Q.; et al. Establishment of male and female Eucommia fingerprints by UPLC combined with OPLS-DA model and its application. *Chem. Biodivers.* **2023**, *20*, e202201054. [CrossRef]
25. Şöhretoğlu, D.; Arroo, R.; Sari, S.; Huang, S. Flavonoids as Inducers of Apoptosis and Autophagy in Breast Cancer. In *Discovery and Development of Anti-Breast Cancer Agents from Natural Products*; Elsevier: Amsterdam, The Netherlands, 2021.
26. Ding, L.L.; Lin, X.Y. Status quo and research and development of bitter phytonutrients. *Food Res. Dev.* **2003**, *1*, 57–59.
27. Chen, C.H.; Wu, M.C.; Hou, C.Y.; Jiang, C.M.; Huang, C.M.; Wang, Y.T. Effect of phenolic acid on antioxidant activity of wine and inhibition of pectin methyl esterase. *J. Inst. Brew.* **2009**, *115*, 328–333. [CrossRef]
28. Zuo, Y.; Chen, H.; Deng, Y. Simultaneous determination of catechins, caffeine and Gallic acids in green, Oolong, black and pu-erh teas using HPLC with a photodiode array detector. *Talanta* **2002**, *57*, 307–316. [CrossRef] [PubMed]
29. Qu, Y.Q.; Tukhvatshin, M.; Zheng, D.; Zhang, Q.; Yang, J.; Ye, N. Determination of phenolic acid compounds in tea by high performance liquid chromatography. *Chin. J. Tea* **2019**, *6*, 60.
30. Wang, S.; Kai, L.; Yang, J.; Ji, X.; Qian, H.; Xu, Y.; Du, H. Effects of short-chain fatty acids and their ethyl esters in strong-flavored liquor on the human body. *Food Ferment. Ind.* **2020**, *46*, 7.
31. Chen, H.; Wu, Y.; Wang, J.; Hong, J.; Tian, W.; Zhao, D.; Sun, J.; Huang, M.; Li, H.; Zheng, F.; et al. Uncover the flavor code of roasted sesame for sesame flavor baijiu: Advance on the revelation of aroma compounds in sesame flavor baijiu by means of modern separation technology and molecular sensory evaluation. *Foods* **2022**, *11*, 11. [CrossRef]
32. Procopio, S.; Sprung, P.; Becker, T. Effect of amino acid supply on the transcription of flavour-related genes and aroma compound production during lager yeast fermentation. *LWT Food Sci. Technol.* **2015**, *63*, 289–297. [CrossRef]
33. Wen, C.K.; Chen, X.; Wang, Y.F.; Fang, S.L.; Xia, Y.F.; Li, Q.; Li, X.Q.; Chen, M.B. Research Progress on functional components of rice wine. *China Brew.* **2019**, *38*, 5–8.
34. Yasuda, S.; Horinaka, M.; Iizumi, Y.; Goi, W.; Sukeno, M.; Sakai, T. Oridonin inhibits SASP by blocking p38 and NF-κB pathways in senescent cells. *Biochem. Biophys. Res. Commun.* **2022**, *590*, 55–62. [CrossRef]
35. Wang, C.; Li, C.; Zhang, X.; Rao, J.; Xin, P.; Yin, J.; Song, Q. Determination of terpenoids in yam wine by liquid–liquid extraction coupled with gas chromatography-mass spectrometry. *Food Ferment. Ind.* **2019**, *45*, 213–217.
36. Worley, B.; Powers, R. PCA as a practical indicator of OPLS-DA model reliability. *Curr. Metab.* **2016**, *4*, 97–103. [CrossRef]
37. Yuan, Z.; Qin, Y.; Yang, F.; Deng, Z. Ultrasound-assisted technology to develop a new type of dark tea wine. *Food Sci. Technol.* **2021**, *11*, 46.
38. Eriksson, L.; Byrne, T.; Johansson, E.; Trygg, J.; Vikström, C. *Multi- and Megavariate Data Analysis: Part I: Basic Principles and Applications*; Umetrics Inc.: Wilmington, MA, USA, 2006.

foods

MDPI

Article

Rapid Color Quality Evaluation of Needle-Shaped Green Tea Using Computer Vision System and Machine Learning Models

Jinsong Li [1,2,3,†], Qijun Li [4,†], Wei Luo [1,2,3], Liang Zeng [1,2,3,*,‡] and Liyong Luo [1,2,3,‡]

1 Integrative Science Center of Germplasm Creation in Western China (CHONGQING) Science City, College of Food Science, Southwest University, Chongqing 400715, China; lly1979@swu.edu.cn (L.L.)

2 Chongqing Key Laboratory of Speciality Food Co-Built by Sichuan and Chongqing, Southwest University, No. 2 Tiansheng Road, Beibei District, Chongqing 400715, China

3 College of Food Science, Southwest University, No. 2 Tiansheng Road, Beibei District, Chongqing 400715, China

4 College of Computer and Information Science, Southwest University, No. 2 Tiansheng Road, Beibei District, Chongqing 400715, China

* Correspondence: zengliangbaby@126.com

† These authors contributed equally to this work.

‡ These authors also contributed equally to this work.

Abstract: Color characteristics are a crucial indicator of green tea quality, particularly in needle-shaped green tea, and are predominantly evaluated through subjective sensory analysis. Thus, the necessity arises for an objective, precise, and efficient assessment methodology. In this study, 885 images from 157 samples, obtained through computer vision technology, were used to predict sensory evaluation results based on the color features of the images. Three machine learning methods, Random Forest (RF), Support Vector Machine (SVM) and Decision Tree-based AdaBoost (DT-AdaBoost), were carried out to construct the color quality evaluation model. Notably, the DT-Adaboost model shows significant potential for application in evaluating tea quality, with a correct discrimination rate (CDR) of 98.50% and a relative percent deviation (RPD) of 14.827 in the 266 samples used to verify the accuracy of the model. This result indicates that the integration of computer vision with machine learning models presents an effective approach for assessing the color quality of needle-shaped green tea.

Keywords: green tea; color characteristic; computer vision system; machine learning

1. Introduction

Green tea is widely consumed around the world due to its pleasant taste and potential health benefits [1,2]. Chongqing, a city located in southwestern China, is recognized as a potential origin of tea [3]. Chongqing lies in the prime region for green tea cultivation and is celebrated for its bountiful yield of needle-shaped green tea. Notably, Yongchuan Xiuya is recognized as a unique Chinese agricultural specialty valued nationwide for its straight appearance, light green color, delicate and aromatic profile, as well as its umami taste. Additionally, it represents one of China's most quintessential needle-shaped green teas. Therefore, developing a swift, objective, and precise assessment approach for evaluating the quality of needle-shaped green tea is significant for the tea industry in Chongqing and nationwide. When evaluating the quality of tea, several attributes are considered, including appearance, aroma, taste, and other sensory factors [4]. Among these attributes, the appearance, particularly the color, plays a crucial role in green tea as it greatly influences purchasing decisions [5]. Traditional methods of assessing tea quality rely on sensory evaluation and quantitative descriptive analysis [6]. However, these techniques require specialized training and have a steep learning curve, resulting in high costs, time-consuming processes and inaccuracy [7]. Consequently, consumers often associate higher prices with higher quality due to a lack of standardized evaluation methods [8]. Therefore, it is essential to establish an accurate, efficient, and objective approach to tea quality evaluation, specifically in terms of assessing appearance. Implementing such standardization would provide valuable guidance to consumers in the market.

1.1. Related Works and Motivation

In recent years, there has been a growing utilization of various sensors, including computer vision [9,10], near-infrared spectroscopy [11], and electronic tongues and noses [12], etc., in the realm of food detection. This trend has fueled the rapid advancement of technologies for fast, accurate, non-destructive detection and assessment, which are presently being studied and applied in the tea industry. These technologies find broad application in monitoring fresh leaves [13], processing processes [14], evaluating tea quality [15]. However, the assessment of the color of tea continues to rely on relatively traditional methods; sensory evaluation [16] and colorimeter assessment methods [17] continue to serve as the primary means of evaluation. Hence, it is essential to develop a fast, objective, and precise assessment method for the color of tea. Computer vision is a field that focuses on the theoretical and algorithmic foundations for deriving meaningful information about objects from images [18]. With its inherent qualities of accuracy and objectivity, it has proven to be a valuable tool for assessing the quality of various products [19], including tea. Ren [20] showcased the application of computer vision technology in the extraction of the morphological parameters of Congou black tea, resulting in the efficient identification of different tea quality grades. In a similar vein, Xu [21] integrated E-nose technology with computer vision to enable rapid and precise determination of the quality grades of Longjing tea. This innovative approach presents a novel idea for achieving high-precision tea quality assessment. High accuracy requires high-performance algorithms, with machine learning playing a pivotal role. In recent studies focused on tea-related topics, numerous machine learning models have demonstrated remarkable success. Widely used machine learning models, such as Support Vector Machine and Random Forest, have exhibited notable advantages in various tasks, including the management of tea plantations [22], processing monitoring [23], tea quality assessment [24], and rapid material detection [25]. Decision Tree (DT) is a simple but powerful classifier that can handle both discrete and continuous features. DT-based Adaboost (DT-Adaboost) can give full play to the classification ability of DT and improve the overall classifier performance [26]. Therefore, DT-Adaboost is well adapted to handle complex data. DT can handle complex situations such as nonlinear relationships, interactions and high-dimensional features, so DT-Adaboost can achieve good classification results when dealing with complex data sets. Meanwhile, DT-Adaboost can reduce the risk of overfitting by iteratively training multiple DT weak classifiers and combining them in a weighted way [27]. For tea images, it is necessary to use DT-Adaboost to improve the performance of tea image recognition in terms of feature complexity and feature similarity. Furthermore, it is important to note that these studies relied on price-based grading of the samples instead of assessments based on sensory evaluations, which may pose challenges in accurately representing the true quality of tea. Additionally, these studies did not evaluate the color characteristics of dried green tea, which is an important aspect of tea quality assessment.

1.2. Contributions and Paper Organization

The objective of this research is to develop a precise, efficient, and unbiased technique for assessing the quality of needle-shaped green tea based on its color characteristics. To achieve this goal, a comprehensive dataset consisting of 885 images was compiled, representing 157 distinct samples of green tea from Chongqing. In order to prepare the images for analysis, preprocessing techniques such as clipping and grayscale transformation were applied. Subsequently, RGB, HSV, and Lab color features were extracted from the processed images. Using sensory evaluation outcomes as ground truth labels, the research focused on constructing an optimal identification model utilizing support vector machine (SVM),

Random Forest (RF), Decision Tree with Adaptive Boosting (DT-Adaboost) integrated learning models. By providing both theoretical foundations and empirical evidence, this study aims to contribute to the digitization and standardization of tea quality assessment practices, regulating the tea market, and enhancing consumer experiences.

The rest of this paper is organized as follows. In Section 2, we outline the materials and methodologies utilized in this study, encompassing tea samples, sensory evaluation techniques, image acquisition and processing approaches, as well as the model training and validation procedures. In Section 3, we conduct a comparison of algorithm performance across different tasks and subsequently discuss the outcomes, and finally, we conclude the paper in Section 4.

2. Materials and Methods

2.1. Sample Preparation

Green tea samples were purchased from various companies located in Chongqing, China. All the samples were processed using the needle-shaped green tea processing method. The standard processing procedures included spreading, fixing, rolling, initial drying, carding, and final drying. Given that color is influenced by fluctuations in moisture content [28], all the green tea samples utilized in this study conformed to the Chinese National Official Standard GB/T 14456.1-2017 [29] (Green tea-Part 1: Basic requirements) for needle-shaped green tea, with a moisture content of less than 7%; the precise moisture content data for each sample can be found in Table S1. A total of 157 samples were obtained over a two-month period (March and April 2022). In order to preserve the color quality of the green tea samples, each sample was sealed in an aluminum foil bag and stored at 4 °C until the analysis process.

2.2. Sensory Evaluation Methods

In the initial stage, color cards (Figure 1) for sensory evaluation were prepared as described, which involved gathering images of all the standard samples and extracting lab parameters from these images. These parameters were then used to determine the starting and ending values of the color scale. These ranges were separated by 10 intervals, respectively. From these defined ranges, the color blocks that most accurately depicted the transition from light green to dark brown were chosen to construct the color card. Thus, objective lab parameters were linked to sensory quality [30]. A team of eight evaluation experts collaborated to perform this step, and the specific lab parameters are displayed in Table S2.

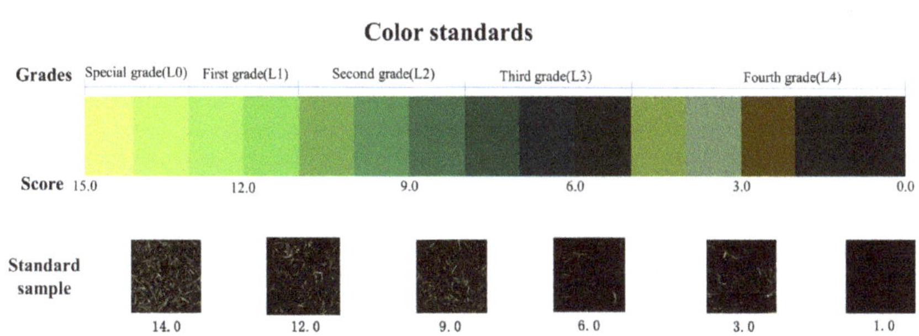

Figure 1. Color standards used for the color sensory evaluation of green tea.

The color qualities of green tea were assessed according to the guidelines specified in the Chinese National Official Standard GB/T 23776-2018 [31], with particular emphasis on the color of dried tea leaves. A reference color card was utilized to aid in color evaluation (Figure 1). The color characteristics of 157 varieties of green tea were assessed by a skilled panel consisting of 8 highly trained panelists from the College of Food Science at Southwest University (Chongqing, China). Each panelist underwent 20 training sessions lasting 1.5 h before each experiment. Before taking part, all participants provided written informed consent [32]. The performance assessment of panelists in the color evaluation group is depicted in Figure S1. Approximately 50 g of each sample was placed in the evaluation tray [24], and a panel of 8 professionally trained panelists assessed the color characteristics. The average score from the panelists was then used as the sample's color score, and subsequent grading was assigned based on these scores. This study was approved by the Ethics Committee of Southwest University (IRB Number: H23151).

2.3. Image Acquisition

The computer vision system was configured with a specialized industrial camera, (ME-EM500C, Shanxi Microvision Intelligent Manufacturing Technology Co., Ltd., Xi'an, China). This camera is designed for high-performance image acquisition in industrial settings. Additionally, a uniform light source (MV-ALP2410, Shanxi Microvision Intelligent Manufacturing Technology Co., Ltd., Xi'an China) was incorporated to ensure consistent lighting during image capture. The system was housed in a darkroom to minimize external interference and ensure accurate image processing. A graphical user interface (GUI) software(Microvision x64 4.0.18.610B, from Shanxi Microvision Intelligent Manufacturing Technology Co., Ltd., Xi'an, China) processing system was utilized to facilitate easy control and analysis of the captured images. The GUI software interface and its associated functions are shown in Figure S2. Together, these components formed a robust and reliable computer vision system for sample image acquisition.

To conduct the image acquisition, 20 ± 0.5 g of tea was extracted from each sample and evenly distributed in the sample pool (Petri dishes) [33]. The sample pool was then placed under the uniform light source to capture images. It should be noted that all images were taken using the camera settings detailed in Figure 2. The specific camera characteristics and lighting conditions are provided in Table S3. To minimize the impact of internal variations among the samples on the results, each sample was randomly sampled five times, resulting in a total of 785 images.

During our sample collection, we observed an imbalance in the number of green tea samples categorized as special grade and fourth grade compared to other categories. This is in line with the proportion of tea product quality distribution in the market, but this data imbalance poses a challenge to the effectiveness of model training. To address this issue, we performed data augmentation by supplementing the green tea images of special grade and fourth grade. Data imbalance is a common occurrence [34], and data augmentation can mitigate this issue. Geometric transformation can be used as a data augmentation technique, is known for its ease of implementation and has been shown to enhance model performance effectively in recent studies in the food-related domain [35,36]. Therefore, we applied various transformations to each green tea image, including rotations of 90, 180 and 270, as well as horizontal flipping [34]. This data enhancement strategy aims to rectify the imbalance in the dataset, ensuring that the model is exposed to a more representative distribution of green tea samples across different grades. By applying these transformations, we generated four additional images for each original image. As a result, the dataset was expanded from its original size of 785 images to a total of 885 images after data augmentation.

2.4. Feature Extraction

As shown in Figure 2, the image underwent preprocessing to extract an 800×800-pixel region of interest (ROI) centered on the raw image's central pixel. Median filtering was applied to diminish image noise [37]. Subsequently, the image required shadow removal. The shadow is caused by underexposure of the lower leaves due to light projecting onto the surface of the above table during the time of shooting. This shadow removal is accomplished by automatically applying a pre-defined grayscale threshold [38]. The OpenCV-Python library is utilized for extracting color features from the region of interest. The mean and standard deviation values were calculated for each channel within the RGB, HSV, and

CIELab color spaces. As a result, a total of 18 color features were acquired, which include the average and standard deviation of the red channel (Ra, Rs), green channel (Ga, Gs), blue channel (Ba, Bs), hue channel (Ha, Hs), saturation channel (Sa, Ss), value channel (Va, Vs), a component channel (aa, as), b component channel (ba, bs), and lightness component channel (La, Ls). To assess the validity of the color variables, a Pearson correlation analysis was performed between these variables and the sensory score using the SPSS software (Statistics 25.0.0, IBM, Armonk, NY, USA).

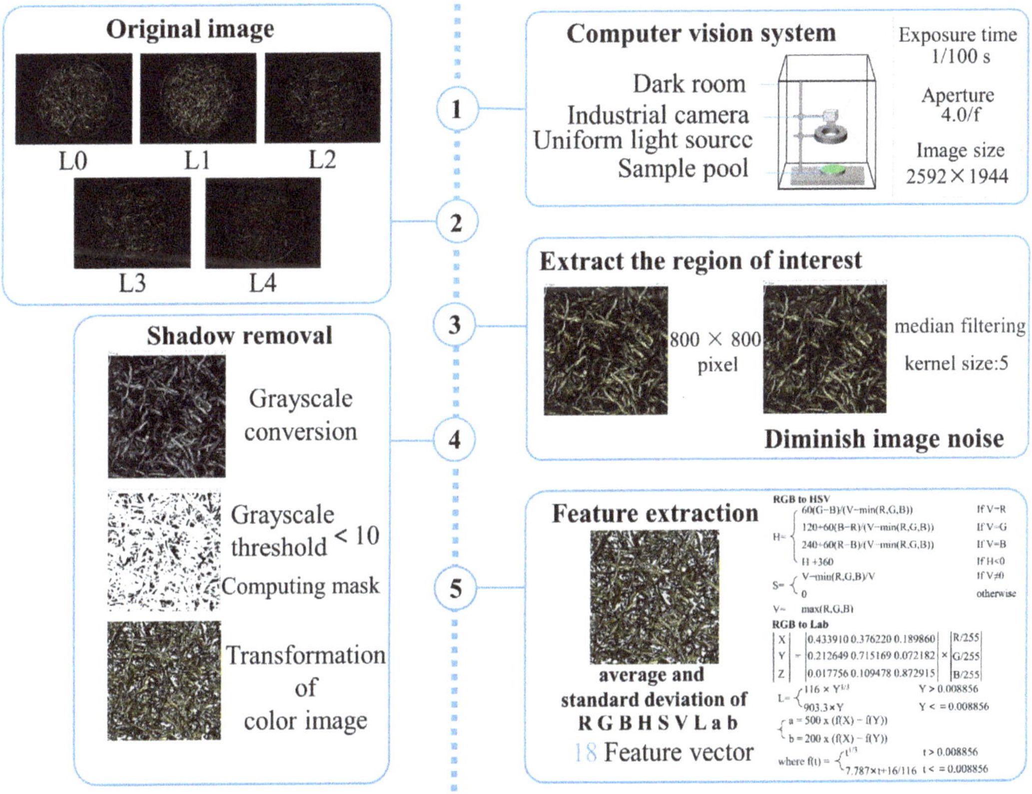

Figure 2. Flow chart of the image acquisition and feature extraction of tea samples.

2.5. Model Establishment and Performance Evaluation

Machine learning was used to build predictive models for sensory quality of the green tea. Three efficient machine learning models were used to predict the tea quality, including SVM, RF, and DT-AdaBoost (Figure 3).

SVM is a widely applied machine learning model in scientific research and industry [20]. The basic concept of the model is to seek a classification hyperplane that can divide the training dataset into a discrete predefined number of classes as the decision surface [20]. In theory and in application, the SVM can obtain accurate classification results by two key parameters, including kernel function and penalty coefficient (c). In this study, we use Radial Basis Function (RBF) as the kernel function of SVM; thus, the key parameters of the model are the optimization penalty coefficients (c) as well as the gamma of the RBF kernel function (g). In order to maximize the accuracy of the model, we used the grid search method to search the parameter pairs (c, g) in the range of 0.01–100.

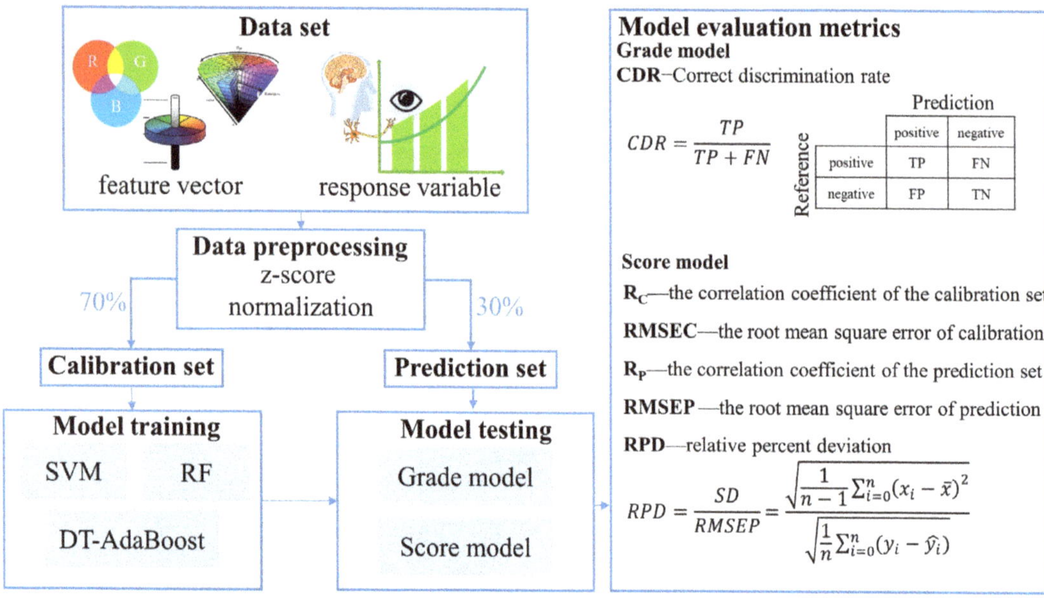

Figure 3. Flow chart of the model training and validation process.

RF [39] is an ensemble learning method which consists of multiple decision trees and uses bootstrap aggregating (bagging). Each decision tree is constructed independently and results in a final classification result or regression value based on majority voting or average prediction [40]. The RF can obtain accurate classification results by two key parameters, including maximum depth of decision tree (depth) and number of estimators (N). The number of estimators denotes the number of decision trees in a random forest. When optimizing the parameters using the grid search method, we set the range of the parameter pairs (depth, N) to 1–12 and 1–1000.

The SVM and RF models have been widely employed in research on evaluating tea quality, demonstrating favorable outcomes in tea classification. In more extensive application contexts and with increased accuracy demands, we endeavored to employ decision tree models for tea quality evaluation to examine their efficacy in both classification and regression undertakings. DT-AdaBoost is another ensemble learning method. It can be used in conjunction with many other types of weak learners to improve performance. Weak learners are a set of learning algorithms or models that individually perform relatively poorly in terms of accuracy or predictive power. In previous studies, the weak learner of AdaBoost is usually a decision tree [41]. Thus, we also chose Decision Tree-based AdaBoost (DT-AdaBoost) as the prediction model. Because the same weak learner as RF is chosen, the key parameters of DT-AdaBoost are the same as those for the RF, including the maximum depth of decision tree (depth) and number of estimators (N), and we also set the search range to 1–12 and 1–1000.

Before establishing the model, the color data set was divided into calibration and prediction sets in the ratio of 7:3 [42]. The objective is to prevent overfitting resulting from additional dataset partitions [43]. Prediction models for color sensory quality grades and scores were then established using SVM, RF, and DT-Adaboost algorithms. Prior to this, the Z-score method [14,44] was applied to preprocess the original data.

The accuracy of the grade model was assessed using the correct discrimination rate (CDR), defined as the ratio of correctly estimated samples to the total number of samples. A higher CDR value indicates a more effective discriminant model. For evaluating the score evaluation model, we employed several indices, including the Correlation Coefficient of the calibration set (Rc), Root Mean Square Error of Cross Validation (RMSECV), Correlation Coefficient of the prediction set (Rp), Root Mean Square Error of Prediction (RMSEP), and the Relative Percent Deviation (RPD). A scoring model with larger Rc, Rp, and RPD values, as well as smaller values of RMSECV and RMSEP, reflects superior prediction performance [44].

3. Results and Discussion

3.1. Sensory Evaluation and Color Feature Extraction Analysis

The sensory evaluation results revealed a wide range of scores, ranging from 3 to 13, indicating the diverse quality grades present in the 157 samples collected for this study. This variability suggests that the data are representative and can be relied upon for thorough analysis and drawing conclusions.

The evaluation of the quality of tea is heavily influenced by the color of dried tea leaves, as it plays a significant role in the initial impression formed by consumers when making purchasing decisions [5]. In order to move towards a more objective assessment method and reduce reliance on subjective sensory evaluation, it is important to explore the relationship between objective physicochemical indicators and sensory quality. The RGB color model, which captures the intensity of light in the red (R), green (G), and blue (B) spectrum, is widely used for color representation [45]. The HSV color space is characterized by H (hue), S (saturation), and V (value), closely mirroring human perception of colors [46]. Furthermore, for color measurements, the Lab* (CIELab) color space is internationally recognized as the standard and is extensively applied in assessing the color of various food products [46]. Therefore, our goal is to provide an unbiased assessment of the color quality of green tea based on these color characteristics.

Pearson's correlation coefficients (r) and two-tailed tests were conducted to analyze the relationship between the color features of green tea and sensory quality, as depicted in Figure 4. The color characteristics of the green tea exhibited significant correlations with sensory qualities. Specifically, the R, G, B, S, V, L, and b* values showed negative correlations, while the a* and H values demonstrated positive correlations with sensory qualities. These findings suggest that greener color and higher saturation levels are generally associated with better sensory qualities, confirming the findings of previous research [47].

To illustrate the overall distribution of data, the comparison of different grades of color variables, such as RGB, HSV, and Lab, is presented in Figure 5. The color features of the first four grades exhibit a general downtrend except for a*, which demonstrates a consistent upward trend. This trend resembles the color changes observed during processing [48]. Therefore, implementing improved techniques can help alleviate darkening and prevent the loss of luster in green tea processing, ultimately leading to higher-quality green tea products. Notably, the fourth-grade tea samples demonstrate a wide range of values in color characteristics other than H, a*, and b*. A low and concentrated H value suggests a color leaning more towards red, orange, or yellow, and less associated with green, cyan, or blue [46]. The concentrated ab values suggest similar hues and potential affiliation with a closely related color group. The substantial variation in brightness and contrast leads to the display of various unappealing color traits in the tea samples, such as gray–green, sere yellow, and dark brown, in line with the outcomes of sensory evaluation.

3.2. Grade Evaluation Model Results

In this study, the evaluation of grades was performed as a classification task in machine learning. SVM, RF, and DT-AdaBoost were used to build classifiers to establish evaluation models for grading tea quality based on color. The purpose was to demonstrate the effectiveness of machine learning classification models in color grade evaluation. These models were implemented using the scikit-learn library in Python.

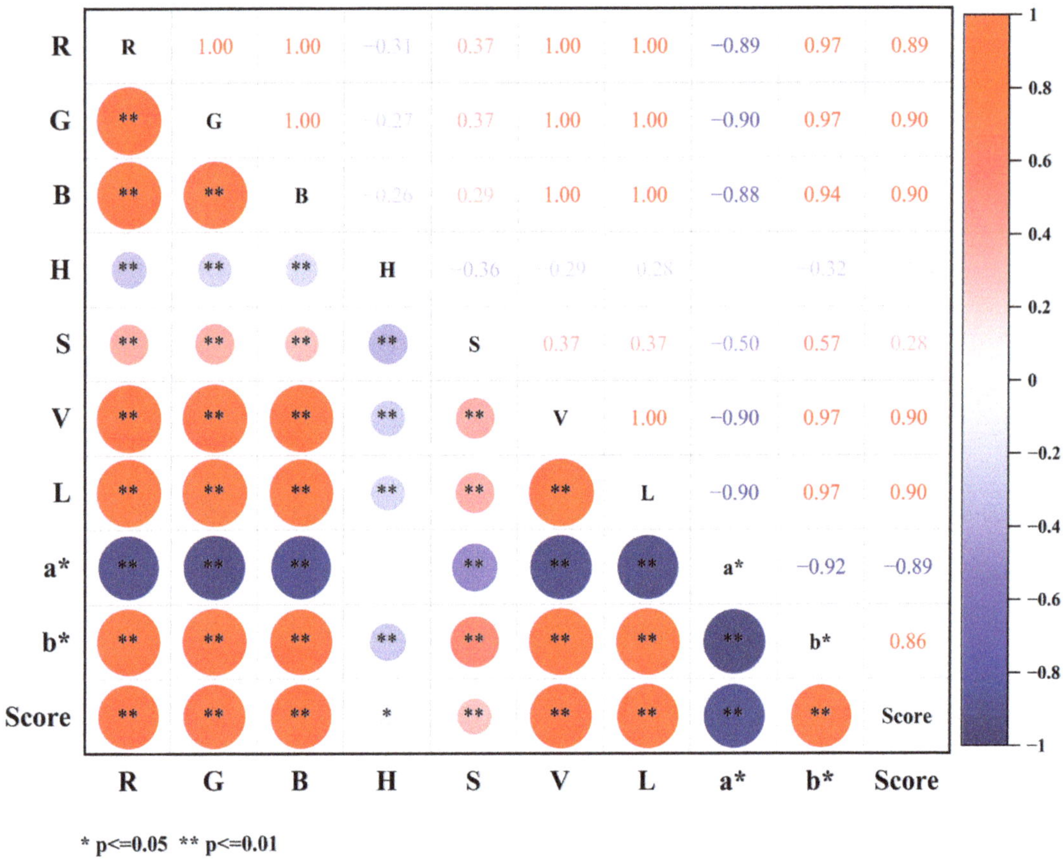

Figure 4. Heatmap for Pearson's correlation coefficients between color characteristics and sensory quality.

Considering that all three models have key parameters that impact the model results, the grid search method was used to optimize these models to obtain the optimal parameters. The results of the hyperparameter search are shown in Figure 6. Table 1 presents the classification results for estimating the color of dried green tea using the various machine learning models. The best result after hyperparameter search was used for each model. As shown in Table 1, all the models demonstrated high accuracy in assessing the color grades of tea. Specifically, the SVM model achieved a CDR of 97.74%, while the RF and DT-Adaboost models achieved CDRs of 98.12% and 98.50%, respectively, in the prediction set. Notably, the DT-Adaboost model achieved the optimal discrimination results. The confusion matrixes of the models are shown in Figure 6. The confusion matrixes indicate that the recall value of L0 and L4 is nearly 100%, which means that there is no misjudgment. The misjudgments mainly occur between L1, L2 and L3. A comparison of the accuracy, precision, and F1 score data reveals that the performance of the three algorithmic models in grade evaluation is closely matched. The identified features are capable of effectively discerning varying grades of tea based on distinct color characteristics.

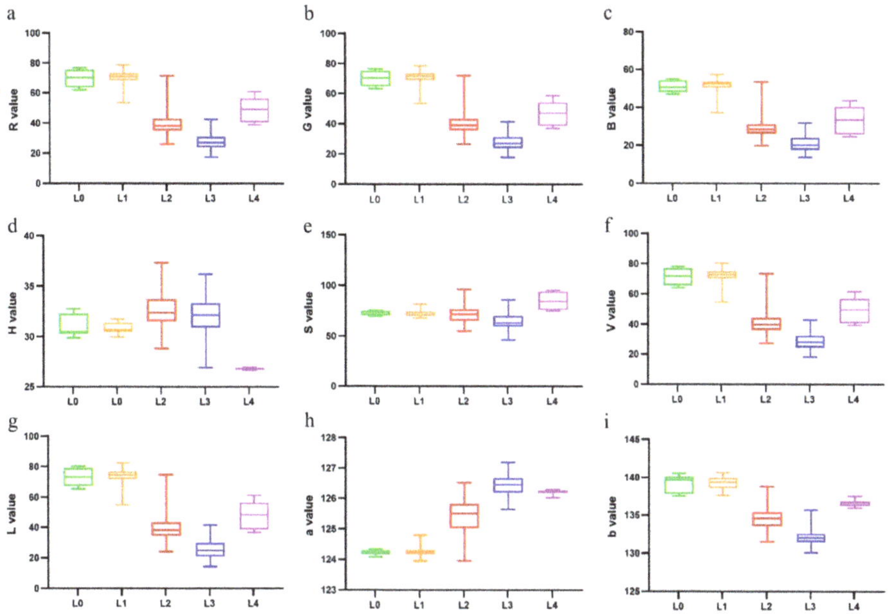

Figure 5. Box plots of image color characteristics: (**a**) R; (**b**) G; (**c**) B; (**d**) H; (**e**) S; (**f**) V; (**g**) L; (**h**) a; and (**i**) b.

Table 1. The classification results obtained using different machine learning models.

Models	Parameters	Accuracy (%)	Precision (%)	F1-Score (%)	CDR (%)
SVM	c = 20, γ = 2	97.74	97.83	97.75	97.74
RF	N = 40, Depth = 10	98.12	98.16	98.12	98.12
DT-Adaboost	N = 70, Depth = 6	98.50	98.56	98.50	98.50

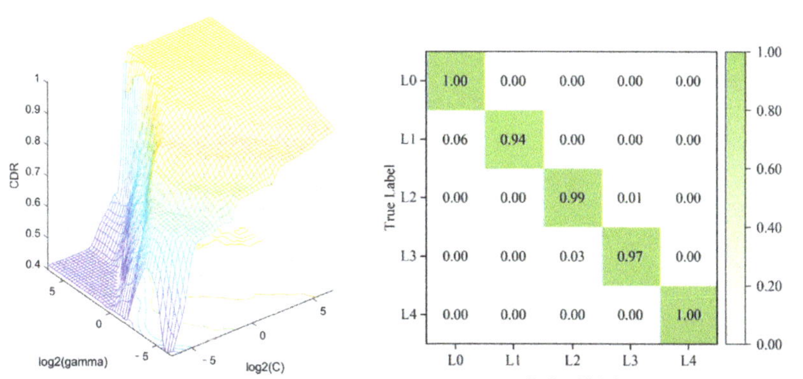

a.

Figure 6. *Cont.*

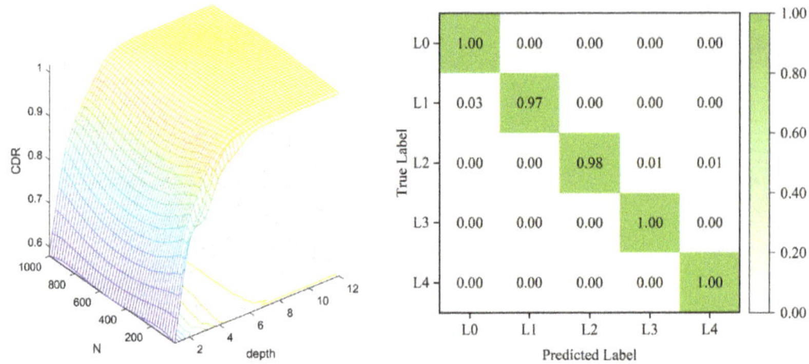

b.

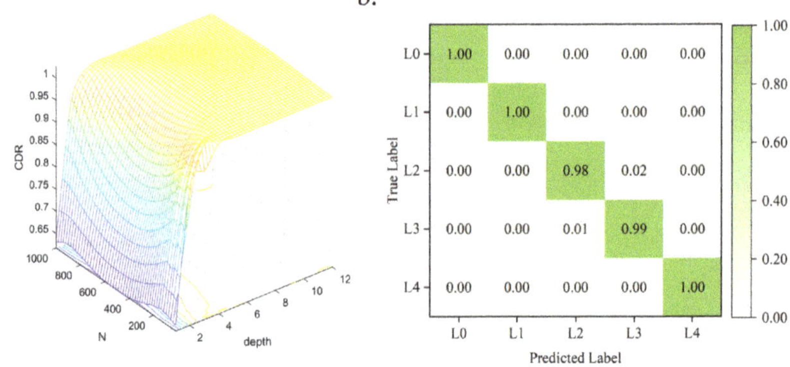

c.

Figure 6. Optimization of parameters for grade prediction models and the confusion matrix of different machine learning methods: (**a**) SVM; (**b**) RF; and (**c**) DT-Adaboost.

3.3. Score Evaluation Model

For score evaluation, our approach is to use a regression task in machine learning. To predict the color sensory scores of green tea, we constructed SVM, RF, and DT-AdaBoost as regressors to establish evaluation models. The methods of model establishment and hyperparameter search are consistent with those in Section 3.2. The results of the hyperparameter search are shown in Figure 7.

The parameters that correspond to the lowest RMSEC value were selected for model training after the hyperparameter search. As shown in Table 2, the SVM, RF, and DT-AdaBoost regression models had respective RMSEC values of 0.313, 0.296, and 0.162, and corresponding RPD values of 7.668, 8.122, and 14.827. These results indicate that all three models exhibit good predictive performance. Notably, the DT-AdaBoost model had the lowest RMSEP value and the highest RPD, with Rp, Rc, and RMSEC values of 0.995, 0.999, and 0.020, respectively.

DT is one of the most widely used machine learning algorithms for solving classification and regression problems [49]. Its versatility and effectiveness make it particularly suitable for the evaluation tasks involved in grading and scoring tea quality. Figure 7 includes the results of score prediction for the samples from different models in the score evaluation, along with the true score results. In Figure 7, the red dots represent the calibration set samples, and the blue dots represent the prediction set samples. The closer the distribution of the dots is to the diagonal line, the closer the predicted value is to the true value. Figure 7 shows that the distribution of red and blue dots of DT-AdaBoost is closest to the diagonal line, followed by RF and SVM.

Table 2. Performance of different models for sensory score using color data.

Models	Parameters	Calibration Set		Prediction Set		RPD
		Rc	RMSEC	Rp	RMSEP	
SVM	C = 7, γ = 0.3	0.991	0.252	0.983	0.313	7.668
RF	N = 90, depth = 12	0.997	0.136	0.985	0.296	8.122
DT-Adaboost	N = 80, depth = 12	0.999	0.020	0.995	0.162	14.827

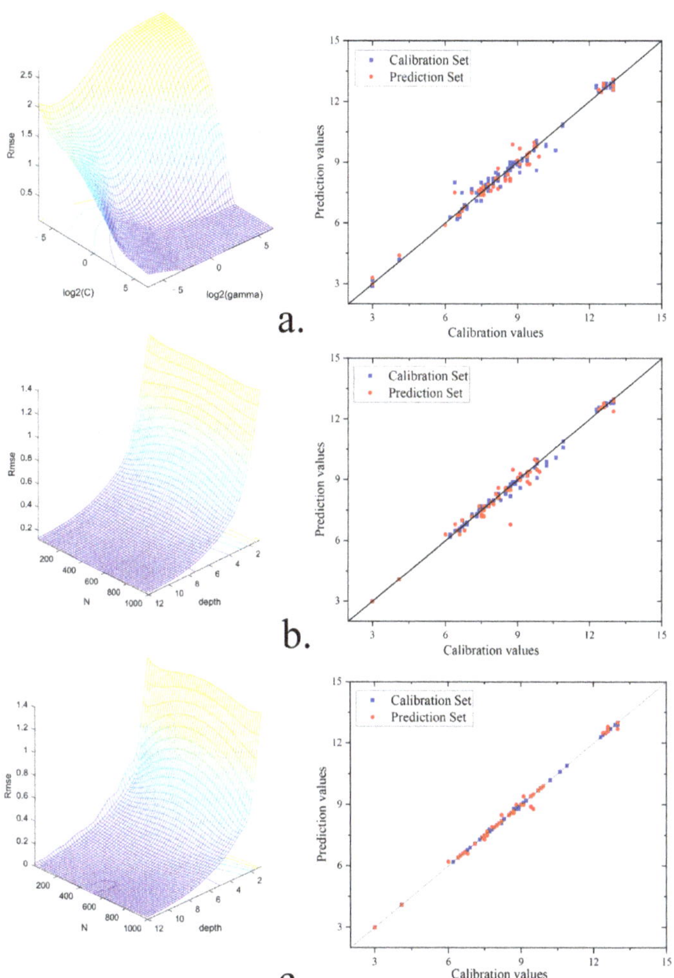

Figure 7. Optimization of parameters for score prediction models and the scatter plot of the predicted and actual values from the sensory model: (**a**) SVM; (**b**) RF; and (**c**) DT-Adaboost.

3.4. Discussion of the Models' Results

To thoroughly assess the quality of tea, it is important to take into account multiple aspects, such as various features, classifications, and regressions. This highlights the necessity of utilizing high-performance models as crucial tools to support the evaluation process. DT-AdaBoost is an ensemble learning model based on Decision Tree (DT), which is one of the most popular ML algorithms and is widely used to solve classification and regression problems. DT-AdaBoost combines DT using AdaBoost algorithms, which is one of the Boosting ensemble learning methods. Models using AdaBoost can have higher accuracy and are less prone to overfitting due to the fact that during Boosting iterations, the weights of the samples are adjusted based on the performance of the previous round of the learner. For samples that are misclassified, their weights are increased, making the next round of training pay more attention to these misclassified samples.

DT-AdaBoost is generally more accurate and has better generalization ability compared to a single classification model like SVM. Usually, integrated learning models outperform single models in terms of both accuracy and generalization [50]. Research has indicated that integrated learning algorithms yield superior outcomes in dealing with classification imbalance challenges, a notion validated by the results of this study [51]. Compared to RF, DT-AdaBoost has the same DT weak classifier but has a different ensemble learning method. The Bagging used by RF focuses more on the generalization ability of the model, and thus, typically makes the model using this method less accurate than the model using boosting. Furthermore, the Boosting model has the capability to improve accuracy significantly by leveraging an ensemble of multiple classifiers [52]. This allows DT-AdaBoost to achieve better results in practical applications of classification and regression tasks compared to RF or SVM. In this study, DT-Adaboost demonstrated superior performance compared to SVM and RF in both classification and regression scenarios, confirming this result and indicating its tremendous potential for application.

Color evaluation in previous studies often involves sensory assessment [16] or the use of a colorimeter [17], leading to differing levels of subjectivity and time-consuming complexities. The classic machine learning models SVM [53] and RF [54] have been extensively utilized in prior research, demonstrating superior accuracy in certain tasks when compared to alternative models. We aimed to investigate the distinctions in tea quality assessment when utilizing the less frequently employed DT-Adaboost method in comparison to SVM and RF. Additionally, to achieve a swift and precise color evaluation approach, computer vision was utilized to capture images of tea leaves, extract color features, and integrate machine learning models for assessing the color quality of needle-shaped green tea. The DT-Adaboost model exhibited optimal accuracy in both classification and regression tasks, bolstering the model's stability. Hence, this method offers a novel solution for evaluating the color quality of needle-shaped green tea.

In this study, computer vision technology was employed to capture images of needle-shaped green tea samples, extract 18 color parameters, and when integrated with machine learning algorithms, enabled precise evaluation of the color quality of green tea. This suggests that a targeted feature selection can enhance the accuracy of the model. Previous studies accurately evaluated the shape quality by using computer vision to identify seven varieties of Congou black tea through extraction of morphological parameters [20], which also demonstrates this point. According to the Chinese National Official Standard GB/T 23776-2018 [31], the appearance quality of tea is determined by its raw materials, shape, color, texture, and other characteristics. This indicates that the use of color and shape features to evaluate the appearance quality is not comprehensive. There is limited research on evaluating the quality of raw materials, texture, and other appearance-related characteristics, which indicates a potential direction for future research. Integrating research findings on appearance, color, texture and other pertinent characteristics would represent a novel approach to accurately and objectively assess the visual quality of tea leaves. This can be achieved by leveraging sensor technology in conjunction with algorithm models, thereby reducing reliance on subjective sensory evaluation methods. Additionally, all samples in the dataset consist of needle-shaped green tea sourced from the Chongqing region, representing a limitation of this study. While the overarching evaluation standards for green tea are largely consistent, variances in quality and assessment criteria arise when considering green tea produced across diverse regions and processed using different methods. These discrepancies necessitate that our model functions merely as a benchmark when appraising additional green tea samples. As such, creating an expanded dataset encompassing various regions and processing techniques would facilitate the development of universally applicable evaluation standards for the wider tea market, by providing a more objective and reliable assessment of tea quality. This could aid tea producers and sellers in making informed decisions regarding pricing and marketing strategies. Ultimately, these results could result in heightened consumer satisfaction and loyalty, alongside improved economic performance for the entire tea industry.

4. Conclusions

This study utilized computer vision technology and machine learning to develop a model to assess the color of green tea. By creating a dataset comprising 885 images derived from 157 samples of tea, dividing the image into 800 × 800-pixel regions surrounding the central pixel, and then extracting 18 color features, the color data were split into calibration and prediction sets in a 7:3 ratio. Machine learning models such as SVM, RF, and DT-Adaboost were applied to correlate color information with sensory evaluation and construct color-based evaluation models, achieving accurate evaluation of tea color quality. When evaluating the tea color grade and score using machine learning techniques such as classification and regression, the DT-Adaboost model exhibited superior performance. Within the prediction set comprising 266 samples, the grading model trained utilizing the DT-Adaboost algorithm demonstrated a CDR of 98.50%, while the score model exhibited an RPD value of 14.827. This study proved that the combination of computer vision and machine learning models facilitates accurate evaluation of tea color quality. Additionally, the Decision Tree model shows promising prospects for practical implementation. Color characteristics, which reflect consumers' initial perception, significantly influence their preferences in purchasing green tea. Providing a more objective and reliable assessment of tea quality can aid tea producers and sellers in making informed decisions regarding pricing and marketing strategies. Ultimately, these results can result in heightened consumer satisfaction and loyalty, alongside improved economic performance for the entire tea industry. In future studies, gathering green tea samples from various regions and using diverse processing methods to enhance the dataset, as well as developing assessment models for attributes beyond color, would facilitate the establishment of a more precise and broadly applicable standardized system for assessing green tea quality.

Supplementary Materials: The following supporting information can be downloaded at: https://www.mdpi.com/article/10.3390/foods13162516/s1, Figure S1: Performance assessment of panelists of color evaluation group; Figure S2: The graphical user interface (GUI) software interface and its associated functions; Table S1: Moisture content of needle-shaped green tea samples in Chongqing area.; Table S2: La*b* values for each color score; Table S3: Camera characteristics and lighting conditions.

Author Contributions: Conceptualization, L.L.; methodology, L.Z. and L.L.; software, Q.L.; formal analysis, J.L., Q.L. and W.L.; investigation, J.L.; data curation, W.L.; writing—original draft preparation, J.L. and Q.L.; writing—review and editing, L.Z. and L.L. All authors have read and agreed to the published version of the manuscript.

Funding: This research was funded by the National Key Research and Development Program of China (No. 2022YFD1601401), Chongqing technology innovation and application development (No. cstc2021jscx-gksbX0066), National Natural Science Foundation of China funded project (No. 32172627) and National Key Research and Development Program of China (No. 2022YFD1600805).

Institutional Review Board Statement: This study was approved by the Ethics Committee of Southwest University (IRB Number: H23151).

Informed Consent Statement: Informed consent was obtained from all subjects involved in the study.

Data Availability Statement: The original contributions presented in the study are included in the article/Supplementary Materials. Further inquiries can be directed to the corresponding author.

Conflicts of Interest: The authors declare that they have no known competing financial interests or personal relationships that could have appeared to influence the work reported in this paper.

References

1. Namal Senanayake, S.P.J. Green Tea Extract: Chemistry, Antioxidant Properties and Food Applications—A Review. *J. Funct. Food* **2013**, *5*, 1529–1541. [CrossRef]
2. Xing, L.; Zhang, H.; Qi, R.; Tsao, R.; Mine, Y. Recent Advances in the Understanding of the Health Benefits and Molecular Mechanisms Associated with Green Tea Polyphenols. *J. Agric. Food Chem.* **2019**, *67*, 1029–1043. [CrossRef] [PubMed]
3. Wambulwa, M.C.; Meegahakumbura, M.K.; Kamunya, S.; Wachira, F.N. From the Wild to the Cup: Tracking Footprints of the Tea Species in Time and Space. *Front. Nutr.* **2021**, *8*, 706770. [CrossRef] [PubMed]
4. Li, M.; Pan, T.; Chen, Q. Estimation of Tea Quality Grade Using Statistical Identification of Key Variables. *Food Control* **2021**, *119*, 107485. [CrossRef]
5. Zhu, H.; Zhang, J.; Liu, F.; Chen, L.; Ye, Y. Using Endogenous Pigments to Recolour Roasted Green Tea. *Food Qual. Saf.* **2021**, *5*, fyab006. [CrossRef]
6. Deng, S.; Zhang, G.; Olayemi Aluko, O.; Mo, Z.; Mao, J.; Zhang, H.; Liu, X.; Ma, M.; Wang, Q.; Liu, H. Bitter and Astringent Substances in Green Tea: Composition, Human Perception Mechanisms, Evaluation Methods and Factors Influencing Their Formation. *Food Res. Int.* **2022**, *157*, 111262. [CrossRef] [PubMed]
7. Zhi, R.; Zhao, L.; Zhang, D. A Framework for the Multi-Level Fusion of Electronic Nose and Electronic Tongue for Tea Quality Assessment. *Sensors* **2017**, *17*, 1007. [CrossRef] [PubMed]
8. Ren, G.; Liu, Y.; Ning, J.; Zhang, Z. Hyperspectral Imaging for Discrimination of Keemun Black Tea Quality Categories: Multivariate Calibration Analysis and Data Fusion. *Int. J. Food Sci. Tech.* **2021**, *56*, 2580–2587. [CrossRef]
9. Kaushal, S.; Tammineni, D.K.; Rana, P.; Sharma, M.; Sridhar, K.; Chen, H.-H. Computer Vision and Deep Learning-Based Approaches for Detection of Food Nutrients/Nutrition: New Insights and Advances. *Trends Food Sci. Technol.* **2024**, *146*, 104408. [CrossRef]
10. Zhu, L.; Spachos, P.; Pensini, E.; Plataniotis, K.N. Deep Learning and Machine Vision for Food Processing: A Survey. *Curr. Res. Food Sci.* **2021**, *4*, 233–249. [CrossRef]
11. Biswas, A.; Chaudhari, S.R. Exploring the Role of NIR Spectroscopy in Quantifying and Verifying Honey Authenticity: A Review. *Food Chem.* **2024**, *445*, 138712. [CrossRef] [PubMed]
12. Tibaduiza, D.; Anaya, M.; Gómez, J.; Sarmiento, J.; Perez, M.; Lara, C.; Ruiz, J.; Osorio, N.; Rodriguez, K.; Hernandez, I.; et al. Electronic Tongues and Noses: A General Overview. *Biosensors* **2024**, *14*, 190. [CrossRef] [PubMed]
13. Wang, Y.; Cui, Q.; Jin, S.; Zhuo, C.; Luo, Y.; Yu, Y.; Ning, J.; Zhang, Z. Tea Analyzer: A Low-Cost and Portable Tool for Quality Quantification of Postharvest Fresh Tea Leaves. *LWT-Food Sci. Technol.* **2022**, *159*, 113248. [CrossRef]
14. Yang, C.; An, T.; Qi, D.; Yuan, C.; Dong, C. Real-Time Discrimination and Quality Evaluation of Black Tea Fermentation Quality Using a Homemade Simple Machine Vision System. *Fermentation* **2023**, *9*, 814. [CrossRef]
15. Wang, L.; Xie, J.; Wang, Q.; Hu, J.; Jiang, Y.; Wang, J.; Tong, H.; Yuan, H.; Yang, Y. Evaluation of the Quality Grade of Congou Black Tea by the Fusion of GC-E-Nose, E-Tongue, and E-Eye. *Food Chem. X* **2024**, *23*, 101519. [CrossRef] [PubMed]
16. Dai, Q.; Liu, S.; Jiang, Y.; Gao, J.; Jin, H.; Zhang, Y.; Zhang, Z.; Xia, T. Recommended Storage Temperature for Green Tea Based on Sensory Quality. *J. Food Sci. Technol.* **2019**, *56*, 4333–4348. [CrossRef] [PubMed]
17. Zou, Y.; Ma, W.; Tang, Q.; Xu, W.; Tan, L.; Han, D.; Tian, Y.; Yuan, Y. A High-precision Method Evaluating Color Quality of Sichuan Dark Tea Based on Colorimeter Combined with Multi-layer Perceptron. *J. Food Process Eng.* **2020**, *43*, e13444. [CrossRef]
18. Wu, D.; Sun, D.-W. Colour Measurements by Computer Vision for Food Quality Control—A Review. *Trends Food Sci. Technol.* **2013**, *29*, 5–20. [CrossRef]
19. Bhargava, A.; Bansal, A. Fruits and Vegetables Quality Evaluation Using Computer Vision: A Review. *J. King Saud. Univ.-Comput. Inf. Sci.* **2021**, *33*, 243–257. [CrossRef]
20. Ren, G.; Gan, N.; Song, Y.; Ning, J.; Zhang, Z. Evaluating Congou Black Tea Quality Using a Lab-Made Computer Vision System Coupled with Morphological Features and Chemometrics. *Microchem. J.* **2021**, *160*, 105600. [CrossRef]
21. Xu, M.; Wang, J.; Gu, S. Rapid Identification of Tea Quality by E-Nose and Computer Vision Combining with a Synergetic Data Fusion Strategy. *J. Food Eng.* **2019**, *241*, 10–17. [CrossRef]
22. Mao, Y.; Li, H.; Wang, Y.; Fan, K.; Shen, J.; Zhang, J.; Han, X.; Song, Y.; Bi, C.; Sun, L.; et al. Low Temperature Response Index for Monitoring Freezing Injury of Tea Plant. *Front. Plant Sci.* **2023**, *14*, 1096490. [CrossRef]
23. Chen, J.; Yang, C.; Yuan, C.; Li, Y.; An, T.; Dong, C. Moisture Content Monitoring in Withering Leaves during Black Tea Processing Based on Electronic Eye and near Infrared Spectroscopy. *Sci. Rep.* **2022**, *12*, 20721. [CrossRef]
24. Jia, J.; Zhou, X.; Li, Y.; Wang, M.; Liu, Z.; Dong, C. Establishment of a Rapid Detection Model for the Sensory Quality and Components of Yuezhou Longjing Tea Using Near-Infrared Spectroscopy. *LWT-Food Sci. Technol.* **2022**, *164*, 113625. [CrossRef]
25. Dai, F.; Shi, J.; Yang, C.; Li, Y.; Zhao, Y.; Liu, Z.; An, T.; Li, X.; Yan, P.; Dong, C. Detection of Anthocyanin Content in Fresh Zijuan Tea Leaves Based on Hyperspectral Imaging. *Food Control* **2023**, *152*, 109839. [CrossRef]

26. Hastie, T.; Rosset, S.; Zhu, J.; Zou, H. Multi-Class AdaBoost. *Stat. Interface* **2009**, *2*, 349–360. [CrossRef]
27. Zhou, Z.-H. *Ensemble Methods: Foundations and Algorithms*; Chapman and Hall/CRC: London, UK, 2012; ISBN 978-0-429-15109-5.
28. Chen, C.; Zhang, W.; Shan, Z.; Zhang, C.; Dong, T.; Feng, Z.; Wang, C. Moisture Contents and Product Quality Prediction of Pu-erh Tea in Sun-drying Process with Image Information and Environmental Parameters. *Food Sci. Nutr.* **2022**, *10*, 1021–1038. [CrossRef] [PubMed]
29. *GB/T 14456.1-2017*; Green tea-Part 1: Basic Requirements. Standards Press of China: Beijing, China, 2017.
30. Quevedo, R.A.; Aguilera, J.M.; Pedreschi, F. Color of Salmon Fillets By Computer Vision and Sensory Panel. *Food Bioprocess. Technol.* **2010**, *3*, 637–643. [CrossRef]
31. *GB/T 23776-2018*; Methodology for Sensory Evaluation of Tea. Standards Press of China: Beijing, China, 2018.
32. Wei, F.; Luo, L.; Zeng, L. Characterization of Key Sweet Taste Compounds in Camellia Nanchuanica Black Tea. *LWT-Food Sci. Technol.* **2023**, *182*, 114858. [CrossRef]
33. Wang, Y.; Li, L.; Liu, Y.; Cui, Q.; Ning, J.; Zhang, Z. Enhanced Quality Monitoring during Black Tea Processing by the Fusion of NIRS and Computer Vision. *J. Food Eng.* **2021**, *304*, 110599. [CrossRef]
34. Shorten, C.; Khoshgoftaar, T.M. A Survey on Image Data Augmentation for Deep Learning. *J. Big Data* **2019**, *6*, 60. [CrossRef]
35. Mojaravscki, D.; Graziano Magalhães, P.S. Comparative Evaluation of Color Correction as Image Preprocessing for Olive Identification under Natural Light Using Cell Phones. *AgriEngineering* **2024**, *6*, 155–170. [CrossRef]
36. Chen, D.; Qi, X.; Zheng, Y.; Lu, Y.; Huang, Y.; Li, Z. Synthetic Data Augmentation by Diffusion Probabilistic Models to Enhance Weed Recognition. *Comput. Electron. Agric.* **2024**, *216*, 108517. [CrossRef]
37. Justusson, B.I. Median Filtering: Statistical Properties. In *Two-Dimensional Digital Signal Prcessing II: Transforms and Median Filters*; Springer: Berlin, Heidelberg, 1981; pp. 161–196. ISBN 978-3-540-38446-5.
38. Dong, C.; Liang, G.; Hu, B.; Yuan, H.; Jiang, Y.; Zhu, H.; Qi, J. Prediction of Congou Black Tea Fermentation Quality Indices from Color Features Using Non-Linear Regression Methods. *Sci. Rep.* **2018**, *8*, 10535. [CrossRef] [PubMed]
39. Breiman, L. Random Forests. *Mach. Learn.* **2001**, *45*, 5–32. [CrossRef]
40. Ho, T.K. Random Decision Forests. In Proceedings of 3rd International Conference on Document Analysis and Recognition, Montreal, QC, Canada, 14–16 August 1995; IEEE Computer Society: Los Alamitos, CA, USA, 1995; Volume 1, pp. 278–282.
41. Wu, Y.; Ke, Y.; Chen, Z.; Liang, S.; Zhao, H.; Hong, H. Application of Alternating Decision Tree with AdaBoost and Bagging Ensembles for Landslide Susceptibility Mapping. *CATENA* **2020**, *187*, 104396. [CrossRef]
42. Zhou, Z.-H. *Machine Learning*; Springer: Singapore, 2021; ISBN 9789811519666.
43. Miraei Ashtiani, S.-H.; Javanmardi, S.; Jahanbanifard, M.; Martynenko, A.; Verbeek, F.J. Detection of Mulberry Ripeness Stages Using Deep Learning Models. *IEEE Access* **2021**, *9*, 100380–100394. [CrossRef]
44. Dong, C.; Ye, Y.; Yang, C.; An, T.; Jiang, Y.; Ye, Y.; Li, Y.; Yang, Y. Rapid Detection of Catechins during Black Tea Fermentation Based on Electrical Properties and Chemometrics. *Food Biosci.* **2021**, *40*, 100855. [CrossRef]
45. León, K.; Mery, D.; Pedreschi, F.; León, J. Color Measurement in L∗a∗b∗ Units from RGB Digital Images. *Food Res. Int.* **2006**, *39*, 1084–1091. [CrossRef]
46. Kang, H.-C.; Han, H.-N.; Bae, H.-C.; Kim, M.-G.; Son, J.-Y.; Kim, Y.-K. HSV Color-Space-Based Automated Object Localization for Robot Grasping without Prior Knowledge. *Appl. Sci.* **2021**, *11*, 7593. [CrossRef]
47. Wu, J.; Ouyang, Q.; Park, B.; Kang, R.; Wang, Z.; Wang, L.; Chen, Q. Physicochemical Indicators Coupled with Multivariate Analysis for Comprehensive Evaluation of Matcha Sensory Quality. *Food Chem.* **2022**, *371*, 131100. [CrossRef] [PubMed]
48. Liu, Z.; Zhang, R.; Yang, C.; Hu, B.; Luo, X.; Li, Y.; Dong, C. Research on Moisture Content Detection Method during Green Tea Processing Based on Machine Vision and Near-Infrared Spectroscopy Technology. *Spectrochim. Acta Part A-Mol. Biomol. Spectrosc.* **2022**, *271*, 120921. [CrossRef] [PubMed]
49. Ji, H.; Pu, D.; Yan, W.; Zhang, Q.; Zuo, M.; Zhang, Y. Recent Advances and Application of Machine Learning in Food Flavor Prediction and Regulation. *Trends Food Sci. Technol.* **2023**, *138*, 738–751. [CrossRef]
50. Jahan, M.S.; Mansourvar, M.; Puthusserypady, S.; Wiil, U.K.; Peimankar, A. Short-Term Atrial Fibrillation Detection Using Electrocardiograms: A Comparison of Machine Learning Approaches. *Int. J. Med. Inform.* **2022**, *163*, 104790. [CrossRef] [PubMed]
51. Galar, M.; Fernandez, A.; Barrenechea, E.; Bustince, H.; Herrera, F. A Review on Ensembles for the Class Imbalance Problem: Bagging-, Boosting-, and Hybrid-Based Approaches. *IEEE Trans. Syst. Man. Cybern. Part C-Appl. Rev.* **2012**, *42*, 463–484. [CrossRef]
52. Banfield, R.E.; Hall, L.O.; Bowyer, K.W.; Kegelmeyer, W.P. A Comparison of Decision Tree Ensemble Creation Techniques. *IEEE Trans. Pattern Anal. Mach. Intell.* **2007**, *29*, 173–180. [CrossRef] [PubMed]
53. Zhang, C.; Wang, J.; Yan, T.; Lu, X.; Lu, G.; Tang, X.; Huang, B. An Instance-Based Deep Transfer Learning Method for Quality Identification of Longjing Tea from Multiple Geographical Origins. *Complex Intell. Syst.* **2023**, *9*, 3409–3428. [CrossRef]
54. Geng, J.; Li, H.; Luan, W.; Shi, Y.; Pang, J.; Zhang, W. Estimation of Daily Actual Evapotranspiration of Tea Plantations Using Ensemble Machine Learning Algorithms and Six Available Scenarios of Meteorological Data. *Appl. Sci.* **2023**, *13*, 12961. [CrossRef]

MDPI AG
Grosspeteranlage 5
4052 Basel
Switzerland
Tel.: +41 61 683 77 34

Foods Editorial Office
E-mail: foods@mdpi.com
www.mdpi.com/journal/foods

www.ingramcontent.com/pod-product-compliance
Lightning Source LLC
Chambersburg PA
CBHW061613220926
51850CB00005B/496
* 9 7 8 3 7 2 5 8 7 3 9 8 2 *